DATE DUE

DEMCO 38-296

Wireless Information Networks

WILEY SERIES IN TELECOMMUNICATIONS AND SIGNAL PROCESSING

John G. Proakis, Editor
Northeastern University

Wireless Information Networks

Kaveh Pahlavan
Worcester Polytechnic Institute

Allen H. Levesque
GTE Laboratories

A Wiley-Interscience Publication
JOHN WILEY & SONS, INC.
New York • Chichester • Brisbane • Toronto • Singapore

This text is printed on acid-free paper.

Copyright © 1995 by John Wiley & Sons, Inc.

All rights reserved. Published simultaneously in Canada.

Library of Congress Cataloging in Publication Data:

Pahlavan, Kaveh, 1951–
      Wireless information networks/Kaveh Pahlavan, Allen H. Levesque.
          p.   cm. — (Wiley series in telecommunications and signal
      processing)
      Includes bibliographical references and index.
      ISBN 0-471-10607-0
      1. Wireless communication systems.   I. Levesque, Allen H.
      II. Title. III. Series.
      TK5103.2.P34 1995
      621.382—dc20   94-22900

Printed in the United States of America

10 9 8 7 6 5 4 3

To those from whom we learned,
To those we taught, and
To those we love.

CONTENTS

Chapter 9   Spread Spectrum for WIN Systems   **359**

PREFACE

This book provides an introduction to the field of wireless information networks. Wireless networking is the enabling communications technology of the 1990s and beyond. Today, mobile and portable telephones and wireless data services are beginning to influence our daily lives, and the 21st century will witness the widespread deployment of wireless networks that will revolutionize the concept of communication and information processing for business, professional, and private applications. The field of wireless communications is experiencing unprecedented market growth, as evidenced by the rapid increase in the size of the cellular and cordless telephone, paging, mobile data, and wireless LAN industries. The initial growth in the market has resulted in important new initiatives in this industry, including the introduction of Digital Cellular, Cellular Digital Packet Data (CDPD), Personal Communication Services (PCS), Mobile Computing, Ad Hoc Networking, and other new systems and services. These developments are all part of a major paradigm shift in the world of telecommunications, a shift away from nearly exclusive reliance on wired networks to an era of "tetherless" communications based largely on wireless technology, and a shift in the computer industry toward integration of high performance distributed computing and portable devices in a mobile computing environment. We adopted the title *Wireless Information Networks* as an encompassing name intended to include all applications related to evolving wireless networks in the telecommunication and the computer industries.

All telecommunication companies that have been involved in traditional wired communications services and product development are now making major investments in wireless technology. Computer companies are investing in wireless communications to add mobile computing and ad hoc networking features to the laptop, palmtop, notepad, and other portable computing devices that are coming into increasingly widespread use. Large corporations, as end-users, are including wireless components in the infrastructure of their networks to extend the computational power of their networks to their field service personnel. Military agencies are developing personal wireless devices for use in tactical environments, as well as portable devices that place a large amount of computational power in the hands of the foot soldier. In addition, many companies that have traditionally been involved in areas other than communications are entering the wireless communications business to be part of this revolution. All of this means that there are a great many engineers, computer science specialists, and managers with a variety of interests who are faced with having to rapidly educate themselves in an area of technology with which they are somewhat or totally unfamiliar. This major new emphasis on wireless communications has also spurred a renewed emphasis on the teaching of principles of wireless communications in colleges and universities. This book is designed to provide students, engineers, scientists, and managers with an introduction to the major technical elements of evolving wireless networks.

The book is written with a "systems engineering" perspective, by which we mean that the various technical topics are presented in the context of ongoing development of specific new systems and services, as well as the key recent developments in national and international spectrum allocations and standards. Our method of presentation is to organize the myriad of emerging wireless technologies into logical categories that reflect the different perspectives that users have toward the various networks and services. The book addresses the major segments of wireless technology: cordless and cellular telephony, Personal Communications Services (PCS), mobile data networks, and wireless local area networks (WLANs). While the book covers technology applicable to a wide range of wireless systems, particular attention is given to indoor wireless communications, as this topic is not covered in any of the books currently in the market.

In writing the book, we have endeavored to bring together treatments of all the major topics to be considered in the design of wireless information networks, but have avoided the presentation of detailed mathematical derivations that are available in other texts. In each instance, we have tried to provide the motivation for various wireless system design choices in the context of overall system considerations. We believe this is an appropriate method for training systems engineers, who should have an overall perspective of the entire system, as well as a working knowledge of how to apply the results of specific research to an engineering problem.

The book covers four categories of topics: characteristics of radio propagation; analysis of various transmission techniques; network access and architecture; and an overview of the major wireless services and products. An overview of systems, standards, and products is provided in Chapters 1, 2, and 12. In these chapters, we provide an overview of major categories of wireless communications and outline the user and market perspectives toward various wireless systems and services. We then briefly review the current state of development of wireless and mobile communications systems, including the important issues of spectrum administration and standards.

In Chapters 3 through 6, we describe the characteristics of radio propagation, as well as measurement and simulation methods used in evaluating existing systems or designing new systems. A detailed description of time and frequency domain statistical channel modeling, a description of the ray tracing algorithm, and a brief overview of direct solution of the radio propagation equations are described in this section of the book.

In Chapters 7 through 10, we discuss transmission techniques. Four major areas of wireless transmission techniques are described: modem technology; signal processing techniques used in wireless networks; spread spectrum technology with emphasis on CDMA; and optical transmission. The fourth category of topics is related to wireless network architectures, which are treated in Chapter 11. A comparative evaluation of various topologies, assigned and random access methods used in voice and data networks, and methods to integrate voice and data are discussed.

The material included in this book has been taught to students and practicing professionals in a number of formats over the last ten years. Various portions of the material have been used in a number of on-site courses in industry, as well as short courses and lecture series sponsored by leading professional societies. The structure and sequence of the chapters are based on a 30-hour course taught and videotaped at the Digital Equipment Corporation during the 1992–1993 academic year. In the Spring semester of the 1993–1994 academic year, the book was used experimentally as the text for a graduate-level course entitled "Wireless Information Networks" taught at the Worcester Polytechnic Institute. Prior to that, portions of the book had been used in other graduate and senior-year undergraduate courses. Therefore, we believe that the book is well-suited for use in classroom teaching or as a self-study text or reference book for professionals in industry. We have included problem sets at the end of each chapter to facilitate teaching in a college or university setting. The problem sets include computer exercises and questionnaires as well as traditional homework problems. The computer exercises can be performed with standard software packages such as MathLAB or Math-CAD, and are designed to encourage the students to use the computer to develop an intuitive understanding of the key concepts. The questionnaire format is used to emphasize the importance of having a general understanding of the overall system at hand and of learning the rationale behind major engineering decisions. The traditional problem sets are exercises for derivation and understanding of the detailed mathematical analysis of various concepts. We have directed these problems toward application-oriented issues. This approach provides students with an understanding of the issues, motivates them to use the computer as a tool in the learning process, and shifts their viewpoint toward real-world engineering problems rather than mathematical drills. We believe this approach is essential for the proper training of engineers for productive careers in the future market-driven telecommunication industry, where simple ideas and added features will often generate greater revenues than will the latest technical inventions.

The book can be used in its entirety for a first- or second-year graduate course in wireless communications networks. As preparation for such a course, students should have an understanding of the elements of probabilistic signal and system analysis and some background in principles of modulation and coding. The first offering of this material at the Worcester Polytechnic Institute was a 14-week course meeting three hours per week. The first two chapters were taught in the first week, Chapters 3 to 6 in the next four weeks, Chapters 7 and 8 in three weeks, and Chapters 9 and 11 in two weeks. The remaining two weeks were spent on student presentations on specific topics, followed by the examinations. The material in Chapters 1 to 5 and Chapters 7 and 9 was covered completely, while Chapters 6, 8, and 11 were covered with more emphasis on the concepts and less emphasis on the details of mathematical derivations. To cover all chapters of the book in full detail, a two-semester course format is advisable, though most of the material might be covered in a fast-paced one-semester course with selective omission of the more specialized topics. With appropriate selection of topics from the book, it can also be used at the undergraduate level. In industry and professional society short courses, the material in the book has been taught in different course formats ranging from 3 to 30 hours of lecture time. When the material was used in a 30-hour training course for engineers and computer specialists, the concepts in all chapters were taught, but the more detailed theoretical topics were omitted. In short courses, material has been taken from selected chapters, depending upon the topics to be emphasized. Chapters 1, 2, and 12 have been used for overview courses, Chapters 3 to 6 for courses on radio propagation modeling, parts of Chapters 7 to 11 with selected material from other chapters for courses related to wireless data systems. The book includes an extensive list of references, that will be especially helpful to the individual using the book for self-study or reference purposes.

Much of the new material in the book is drawn from the published work of the lead author and his students in the Center for Wireless Information Network Studies at WPI. We are pleased to acknowledge the students' contributions to advancing the understanding of wireless channels and networks. In particular, we thank Dr. Steven Howard, for his work reported in Chapters 4, 5, 6, and 8; Dr. Rajamani Ganesh, for his work reported in Chapters 4, 5, and 6; Dr. Ker Zhang, for his work reported in Chapter 11; and Dr. Ganning Yang, for his work reported in Chapters 6 and 8. We thank Dr. Thomas Sexton, Dr. Mitch Chase, Timothy Holt, Aram Falsafi, Glen Bronson, Joseph Meditz, Mudhafar Hassan Ali, and Sheping Li, whose work has been helpful in the preparation of this book. We express our appreciation to the students who helped us by preparing the index and commenting on the manuscript and the problem sets. We thank Duan Wang and Patricia Kush for their assistance in preparing performance data and illustrations for the book. We owe special thanks to the National Science Foundation, NYNEX Science and Technology, WINDATA, Motorola, GTE Laboratories, Raytheon, Digital Equipment Corporation, Apple Computer, COMDISCO Software, and other companies, whose support of the Center for Wireless Information Network Studies at WPI enables graduate students and the staff of CWINS to pursue continuing research in this important field.

The lead author would like to express his deep appreciation to Dr. Phillip Bello, Prof. John Proakis, and Dr. Jerry Holsinger; through them, he has increased the depth of his understanding of the theory and practice of telecommunications. Also, to Prof. James Matthews, for introducing him to the field of radio communications. His co-author would like to express appreciation to Dr. Richard Dean, Mr. David Weissman, Mr. Melvin Landesberg, Dr. Arthur Giordano, and Dr. Osama Mowafi, who have helped in a variety of ways in his work in cellular and mobile communications, and to Mr. Arnold Michelson, who provided constant encouragement throughout his work on this book. Most of all, they are indebted to their families for their patience and support throughout this long and challenging project.

K. P.
A. H. L.

1

OVERVIEW OF WIRELESS INFORMATION NETWORKS

1.1 INTRODUCTION

We are all being exposed to a communications revolution that is taking us from a world where the dominant mode of electronic communications was standard telephone service and voiceband data communications carried over fixed telephone networks, packet-switched data networks, and high-speed local area networks (LANs) to one where a tetherless and mobile communications environment has become a reality. Traditional wireless information networks, which include cordless and cellular telephones, paging systems, mobile data networks, and mobile satellite systems, have experienced enormous growth over the last decade, and the new concepts of personal communication systems, wireless LANs (WLANs), and mobile computing have appeared in the industry.

In conjunction with this revolution, we are witnessing a transition in the infrastructure of our communication networks. After more than a century of reliance on analog-based technology for telecommunications, we now live in a mixed analog and digital world and are rapidly moving toward all-digital networks. The wireless communications industry is one of many that will continue to benefit from the introduction of digital technology. As the demand for communication services continues to increase, manufacturers and service providers are looking towards digital implementations for increased capacity and a wider offering of services to their users. Digital systems have also been advocated for better performance in a wireless environment. Channel coding, interleaving, and other digital techniques can be used to provide additional robustness in radio channels affected by shadowing, fading, and other forms of perturbation.

Quite independent of the use of digital technology in wireless networks is the increasing reliance on data communications in the business, industrial, and governmental sectors as well as for personal and cultural usage. Today's modem-based, circuit-switched data communications connect an ever-growing number of users of personal computers, Fax machines, and a variety of other data terminals. As users become steadily more accustomed to the convenience of wireless voice and message communications services, they will also expect to have the same convenience of access to data services in the mobile and wireless environment.

1

This chapter provides a brief description of wireless information networks and the types of services that are being made available to consumers. We also describe the market trends that are propelling this important and rapidly growing industry. The final section of the chapter summarizes the topics developed in greater detail in subsequent chapters.

1.2   THE INTEGRATION ISSUE

The rapid and accelerating evolution of communications from wired to wireless networks is producing a wide variety of new services, systems, and products. Despite the wide diversity of these developments, the wireless evolution has proceeded along two main paths, which we distinguish as voice-oriented versus data-oriented wireless information networks. This pattern of separation is much like that which existed throughout the development of wired networks and has been caused in large part by the essential differences in the nature of voice, data, and imagery applications, services, and requirements.

In a digital network, voice and data services have different and sometimes contradictory requirements. In order to understand the differences among service requirements, one must first examine the services from the user's point of view. At the outset, it is important to understand that users' expectations are based upon their experience with services provided in the public switched telephone network. Although digitized voice, imagery, and data are all "binary digits," there are different requirements for transmission of each service in a digital network. For example, because of the user's expectation of telephone-quality voice in the public wired network, packetized voice service must be designed with careful attention to minimizing time delays. Inter-packet delays in excess of 100 msec will be noticeable and annoying to the listener. In contrast, delay in a data network, while not desirable, is generally acceptable to the data user. Packetized voice can tolerate packet loss rates of the order of 10^{-2}, or bit error rates of the same order, without a noticeable degradation in service quality. An error rate of 10^{-5} is ordinarily acceptable for uncoded data, but any loss of data packets is totally unacceptable. The lengths of telephone conversations are relatively uniform (approximately 3–20 min), and a few seconds of setup time is therefore acceptable. Each telephone conversation session generates megabytes of digitized information. On the other hand, a communication session for a data service can vary over a wide range from a short electronic mail message carrying only a few bytes of information up to a long file transfer such as the text of a book, which may be as large as a megabyte. On the average, the volume of information involved in a data communication session is much smaller than that of a digitized-voice communication session. The uncertainty in the amount of the information and the low average length of data communication sessions make long setup times undesirable.

While it is true that the major thrust of telecommunications is toward multimedia services, the existing infrastructure of communications networks is still very fragmented. Today we have wired PBXs for local voice communications within office complexes, the public switched telephone network (PSTN) for wide-area voice communications, wired LAN for high-speed local data communications, packet-switched networks and voiceband modems for low-speed, wide-area data communications, and a separate cable network for wide-area video distribution. The process of setting standards for various areas of communications has been similarly fragmented. The standards for voice transmission technology have evolved within the operating companies, the standards for voiceband data modems have been developed by the CCITT, and the standards for LANs have been devised by IEEE 802 and ISO. This separation has come about because each individual network was designed to meet the requirement—be it voice, data, or imagery/video.

The same pattern of separation exists in the wireless information industry as well. The new-generation wireless information networks are evolving around either voice-driven applications such as digital cellular, cordless telephone, and wireless PBX or around data-driven networks such as wireless LANs and mobile data networks. While it is true that all the major standards initiatives are addressing the integration of services, one still sees a separation of the industrial communities that participate in the various standards bodies. That is, we see that GSM, North American Digital Cellular, DECT, and other groups are supported primarily by representatives of the voice-communications industry, whereas IEEE 802.11, WINForum, and HIPERLAN are supported primarily by those with interest in data communications.

Although future personal communications devices may be designed as integrated units for personal computing as well as personal voice and data communications, the wireless access supporting different applications may use different frequency bands or even different transmission technologies. A personal communications service (PCS) may use wideband CDMA in a shared spread-spectrum band, and a digital cellular service may use TDMA or CDMA in another band. Low-speed mobile data may be carried in the gaps between bursts of voice activity, and high-speed, local-area data use another shared wideband channel. At the same time, various services may all be integrated in a metropolitan-area or wide-area network structured with ATM switches. The future direction of this industry depends upon technological developments and a maturity in the spectrum-administration organizations, who must understand the growing massive demand for bandwidth and must in turn develop strategies allowing a fair sharing of increasingly scarce bandwidth. Just as governmental agencies restrict abusive consumption of other limited natural resources such as water, appropriate agencies will have to protect the spectral resources needed for wireless information networks. After all, the electromagnetic spectrum is a modern natural resource supporting the ever-widening array of telecommunications services which are becoming an increasingly important part of the fabric of our personal and professional lives.

1.3 EVOLVING WIRELESS NETWORKS

The mid-1980s saw major new developments in the wireless information industry. The transition to digital cellular technology, led by the Pan-European GSM standard [Hau94], has been followed by the EIA–TIA North American Digital Cellular Standards initiatives and the Japanese Digital Cellular standard [Kin91]. The prime motivation for these initiatives has been to increase the capacity of cellular telephone systems, which have reached the capacity limits of the analog technology in some highly populated metropolitan areas. The extraordinary success of the cordless-telephone market spurred new standardization efforts for digital cordless and CT-2 TelePoint in the United Kingdom [Mot87], wireless PBX, and DECT in Sweden [Buc88, Och89], advanced cordless phone in Japan [Hat88], and the concept of a Universal Digital Portable Communicator in the United States [Cox85, Cox87a, Cox87b, Cox91a]. The success of the paging industry led to development of private wireless packet data networks for commercial applications requiring longer messages [Bro91, Kil92, Par92]. Motivated by the desire to provide portability and to avoid the high costs of installation and relocation of wired office information networks, wireless office information networks were suggested as an alternative [Pah85, Pah85b, Pah85c, Pah88a]. Another major event in this period was the FCC announcement regarding unlicensed ISM bands in May of 1985 [Mar85, Mar87a, Mar91]. This announcement opened the path for development of a wide array of commercial devices ranging from wireless PBX [Kav85] and wireless LANs [IEEE91a, Tuc91] to wireless fire safety devices using spread spectrum technology.

Figure 1.1 distinguishes the various categories of wireless networks we discuss in this book. We first define two broad categories of networks as (1) voice-oriented or isochronous networks and (2) data-oriented or asynchronous networks. Under each main category of networks, we distinguish further between local-area networks and wide-area networks. Each of the resulting four subcategories of networks has a set of characteristics that leads to certain design choices specific to the subcategory. Figure 1.2, which is structured according to the categories defined in Fig. 1.1, depicts various dimensions of today's voice and data communications industries, comparing local cordless voice communication with wide-area cellular voice services, and also comparing wireless LANs with wide-area, low-speed data services

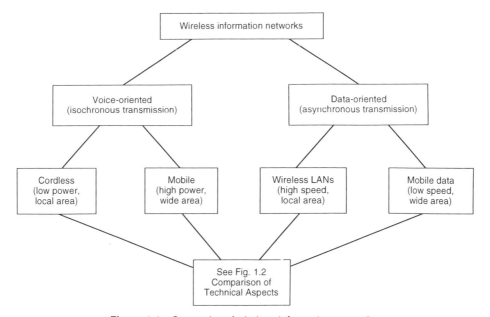

Figure 1.1 Categories of wireless information networks.

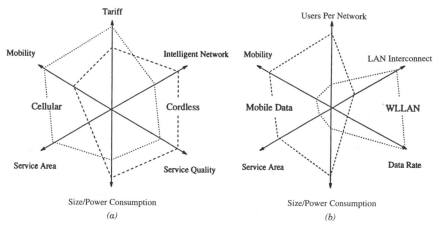

Figure 1.2 Various aspects of the voice (*a*) and data (*b*) oriented services offered by the wireless information network industry. Mobile cellular systems are compared with cordless PCS, and WLLAN data systems are compared with mobile data networks.

[Pah94]. Figure 1.2*a* [Oga94] compares the importance of various issues or parameters related to the wireless voice industry, while Fig. 1.2*b* provides an analogous comparison between wireless data systems. Although from the user's standpoint and the appearance of the handset the digital cellular and PCS systems look very similar, there will in fact be major differences in the operation of the networks supporting the two categories of systems.

A digital cellular system is designed to support mobile users roaming over wide geographic areas, and thus coverage is provided by an arrangement of cells with cell size typically 0.5 to 5 miles in diameter. The radio cell sites for this system are large and expensive, particularly in urban areas where the cost of property is very high. The handset requires an average power of around 1 W, which is reflected in limited battery life and the need for frequent recharging of the batteries or reliance on connection to the car battery. The number of users per cell is large; and to provide as many user channels as possible in the allocated bandwidth, complex speech-coding algorithms are used, which minimize the digitized speech transmission rate but consume a significant amount of electronic power, which in turn places a high demand on battery power.

The PCS systems will be designed for small, low-power devices to be carried and used in and around office buildings, industrial complexes, and city streets. The cell size will be less than a quarter-mile, and the relatively small base stations will be installed on utility poles or attached to city and suburban business buildings. The average radiated power will be 10–20 mW, leading to relatively long battery life. The PCS systems are to replace cordless phones in many market areas, and the quality of voice service is intended to be comparable to wireline phone service. As a result, simple but high-quality speech-coding algorithms such as 32-kbit/sec ADPCM [JTC93, G721] are adopted. While a high-quality voice-coding algorithm of this type does not provide the spectral efficiency of the lower-rate (4–8 kbits/sec) vocoders used in the digital cellular standards, it is far less demanding of digital signal processing complexity, and thus permits the use of very low prime power in the portable units. It is expected that PCS service will distinguish itself from digital cellular service by higher voice quality, longer battery lifetime, and lighter terminals. For more details on the digital cellular and PCS initiatives, the reader can refer to [Goo91a], [Ste90], [Cox90], [Rap91a], [Ram94], [Oga94], [Vit94], and other references cited therein.

Mobile data networks operate at relatively low data rates over well-understood urban radio channels using familiar multiple-access methods. The technical challenge here is the development of a system which makes efficient use of the available bandwidth and the existing infrastructure to support widely separated subscribers. The transmission technology used in mobile data networks is generally rather simple and similar from one network to another. We discuss the major existing and planned mobile data networks and services in Chapter 12.

WLANs and mobile data networks serve somewhat different categories of user applications, and they give rise to different system design and performance considerations. A WLAN typically supports a limited number of users in a well-defined local area, and system aspects such as overall bandwidth efficiency and product standardization are not crucial. The achievable data rate is generally an important consideration in the selection of a WLAN, and therefore the transmission channel characteristics and the application of signal processing techniques are important considerations [Pah88a, Pah88b]. Access methods and network topologies used in WLANs are much the same from one system to another, but the transmission technologies are different. Efficient design of these systems requires evaluation of various transmission techniques and an understanding of the complexities of indoor radio propagation. WLAN manufacturers currently offer a number of nonstandardized products based on conventional radio modem technology, spread-spectrum technology in the ISM bands, and infrared technology.

1.4 MARKET AND USER PERSPECTIVES

1.4.1 Voice-Oriented Services

We can divide voice-oriented services roughly into two subcategories: (1) local-area service using inexpensive low-power subscriber terminals, as typified by cordless telephone, wireless PBX, or PCS, and (2) wide-area services using higher-powered terminals, as typified by cellular mobile telephone service. The PCS systems are commonly viewed as the next generation of cordless telephone systems.

High-Power, Wide-Area Systems (Cellular). One of the most rapidly growing segments of the wireless communications industry is that of cellular mobile telephone service [Bel79a]. Figure 1.3 shows the growth trends for wireline, cordless, and cellular markets in the United States from 1984 through 1994. The rapid growth experienced in the cellular market is very evident. The Cellular Telecommunications Industry Association (CTIA), for instance, has stated that there are currently (late-1994) over 17 million cellular subscribers in the United States, compared with approximately 4.4 million in June 1990 and 90,000 subscribers in 1984 [CTIA]. This trend is apparent across the industry; and with the demand for mobile communications continuing to grow at a rapid pace, problems with high levels of communication traffic are expected to become severe in many service

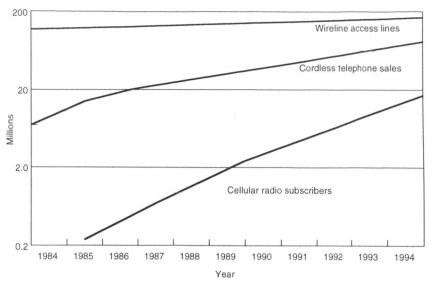

Figure 1.3 Growth trends in the wireline, cordless, and cellular markets in the United States.

areas. As a result, maximizing the utilization of limited radio-frequency resources is now a primary concern to commercial service providers and communication regulatory committees. A discussion of the market growth in the cellular industry can be found in [Cal88]. The rapid growth in the cellular industry has motivated digital cellular standards initiatives in the United States, Europe, and Japan. A chart depicting the evolution of cellular is shown in Fig. 1.4. We discuss the cellular standards initiatives in Chapter 2.

Low-Power, Local-Area Systems (Cordless Telephone and PCS). A cordless telephone uses a low-power radio link as a tetherless extension of fixed wired-network access point. Because a cordless telephone is intended to provide limited portable access to a fixed wireline access point, it is implemented with very simple technology. This simplicity is both advantageous and disadvantageous. Because portable phones evolved from audio technology as mass-market consumer products, they are produced at low cost and are very affordable for everyday home use. At the same time, due to their simplicity, they provide no capability for co-channel interference avoidance or for any control beyond dialing and answering calls. Despite their limitations, cordless phones have become very popular, and in 1993 there were about 60 million cordless phones in use in the United States alone. This growth in popularity, despite the very limited capability of the technology, attests to the great value which consumers place on the convenience of portability.

As the existing forms of wireless communications continue to grow at rapid rates, market studies consistently indicate an enormous future market for widespread new forms of "tetherless" portable radio communications, variably termed *personal communications services* (PCS), *personal communications networks* (PCNs), or *universal portable communications*. In recent years, a great many journal papers and conference presentations have dealt with the market trends,

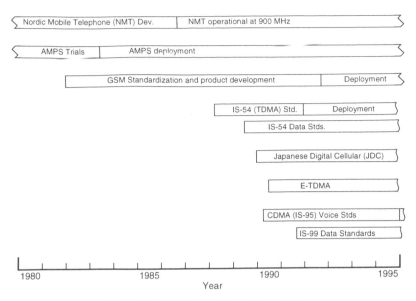

Figure 1.4   Cellular mobile radio evolution.

technological developments, and world-wide standards initiatives related to this coming revolution in communications. (See, for example, [Goo91a], [Goo92], [Cox89], [Cox92a], [Cox92b], [Hei92], [IEEE91a], [IEEE91b], [IEEE92], and other references cited therein.) These new forms of wireless communications are seen as direct evolutions from the highly successful cordless telephone and cellular telephone products and systems, which are depicted in Fig. 1.5. Donald Cox has provided a concise description of this future form of communications: "low-power, exchange-access digital radio, integrated with network intelligence to provide the

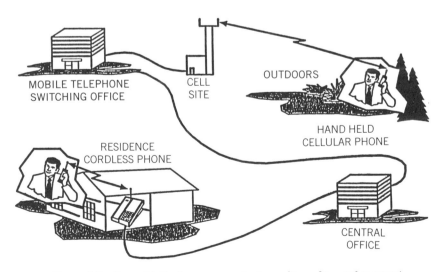

Figure 1.5   Current tetherless communications. (From [Cox 89] © IEEE.)

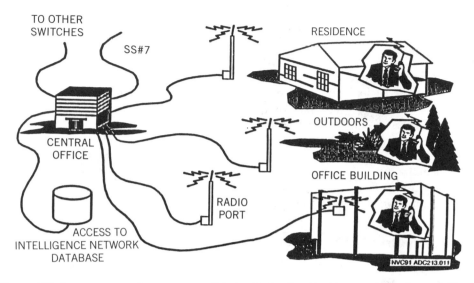

Figure 1.6 Low-power, exchange-access digital radio integrated with network intelligence. (From [Cox 89] © IEEE.)

added feature of widespread portability" [Cox89]. In this concept, depicted in Fig. 1.6, digital radio-access ports would provide access to the local exchange network by way of demand-assigned digital radio links that replace wired customer loops. The radio ports are then connected to a central office (CO), as is conventional practice in local exchanges. The customer's call would be controlled and switched by the CO. Unlike today's wired telephone network, a radio-access customer will not appear on a unique line termination. Instead, when a radio-access call is initiated, the CO switching equipment will recognize the customer and determine the customer's service profile by referring to a database.

The customer's portable communicator will be a small, low-power, lightweight device much like today's cordless telephone, but will provide better voice quality and greater functionality. For radio-access call initiation, or for receiving a call through a radio-access port near the customer's home or place of work, no further network intelligence or action will be required. This is the type of simple digital radio access currently provided by fixed digital radio used for rural telephone service distribution [Cle87]. However, network intelligence implemented in digital switching equipment would provide other functionality to users who move away from their expected locations, provided that a standardized system is in place throughout the area of interest. At a locale away from home or work, the customer would access the network through a radio-access port in that locale. The intelligent network (to be based on Signaling System 7) would route the customer identification to the appropriate database using interoffice common-channel signaling. This would provide the necessary verification and billing for customers on the move. Such portable access to a nonhome exchange could also serve to temporarily update the customer's database to provide call forwarding to the visited locale.

The intelligent network could also provide a beneficial call handoff feature for users moving from room to room in a large building or from street to street in a residential neighborhood. In moving about, the user might experience degraded signal quality on the link to the originally accessed radioport, and better signal quality might be obtained on a connection to another radio port. Such a condition might be detected by either the portable phone or the radio ports, and an automatic request might be made to the CO switching equipment to transfer the call from the original port to the new one. This call transfer is similar to the handoffs done between cell sites in a cellular telephone system.

In this brief discussion we have outlined the chief characteristics of tetherless low-power communications, for which we shall use the name PCS. PCS represents a major paradigm shift in communications, and it is being interpreted in various ways by many segments of the communications industry. Within the United States, and internationally, a number of standards organizations have begun work on PCS in its various forms. We say more about these standards activities in Chapter 2.

1.4.2 Data-Oriented Services

From the data user's perspective, the minimum satisfactory service requirement is low-speed access in wide areas and high-speed access in local areas. The low-speed wide-area access will serve a variety of short-message applications such as notice of electronic or voice mail, while the local-area access will support high-speed local applications such as long file transfers or printing tasks. In the current literature, low-speed, wide-area wireless data communication is referred to as *mobile data*, while local high-speed data communication systems are called *WLANs*. The relationship between WLANs and mobile data services is analogous to the relationship between PCS and digital cellular services. While PCS is intended to provide high-quality local voice communication, the digital cellular services are aimed at wider-area coverage with less emphasis on the quality of the service.

Low-Speed, Wide-Area Systems (Mobile Data). Mobile data radio systems have grown out of the success of the paging-service industry and increasing customer demand for more advanced services. Today 100,000 customers are using mobile data services, and the industry expects 13 million users by the year 2000. This could be equivalent to 10–30% of the revenue of the cellular telephone industry. Today, mobile data services provide length-limited wireless connections with in-building penetration to portable users in metropolitan areas. The future direction is toward wider coverage, higher data rates, and capability for transmitting longer data files.

The data rates of existing mobile data systems are comparable to voiceband modem rates (up to 19.2 kbits/sec). However, the service has a limitation on the size of the file that can be transmitted in each communication session. The coverage of the service is similar to standard land-mobile radio dispatch services, with the difference that the mobile data service must have in-building penetration. Land-mobile radio users typically use the telephone unit inside a vehicle and usually while driving. Mobile data users typically use the portable unit in a building and in a stationary location. Therefore in-building penetration is an essential feature of mobile data services. The major existing mobile data networks, and new networks under development, are discussed in Chapter 12.

Mobile data services are used for transaction processing and interactive, broadcast, and multicast services. Transaction processing has applications such as credit card verification, taxi calls, vehicle theft reporting, paging, and notice of voice or electronic mail. Interactive services include terminal-to-host access and remote LAN access. Broadcast services include general information services, weather and traffic advisory services, and advertising. Multicast services are similar to subscribed information services, law enforcement communications, and private bulletin boards.

There are other low-speed data products using voiceband modems over existing radio channels originally designed for voice communications. Some of these products are used in the VHF land-mobile radio bands for low-speed local data communications in or around buildings. Other products, portable Group 3 facsimile devices with voiceband modems, are used over the analog cellular telephone network to provide wide-area data communications for mobile users without any restrictions on the duration of the call connection.

High-Speed, Local-Area Systems (WLANs). Today, most large offices are equipped with wiring for conventional LANs, and the inclusion of LAN wiring in the planning of a new large office building is done as a standard procedure, along with planning for telephone and electric-power wiring. The WLAN market will very likely develop on the basis of the appropriateness of the wireless solution to specific applications. The targeted markets for the WLAN industry include applications in large indoor areas, offices with wiring difficulties, branch offices and temporary indoor networks. In large indoor areas such as manufacturing floors, warehouses, shopping malls, and stock exchange halls, ceilings are typically not designed to provide a space for distribution of wiring. Also, these areas are not usually configured with walls through which wiring might otherwise be run from the ceiling to outlets. Underground wiring is a solution that suffers from expensive installation, relocation, and maintenance. As a result, the natural solution for networking in most larger indoor areas is wireless communications. Other wide indoor areas without partitioning, such as libraries or open-architecture offices, are also suitable for application of WLANs. In addition, buildings of historical value, concrete buildings, and buildings with marble interiors all pose serious problems for wiring installation, leaving WLANs as the logical solution. WLANs are appropriate for unwired small business offices such as real-estate agencies, where only a few terminals are needed and where there may be frequent relocations of equipment to accommodate reconfiguration or redecoration of the office space. Temporary offices such as political campaign offices, consultants' offices, and conference registration centers provide another set of logical applications of WLANs.

Cost studies in the wired-LAN industry have shown that installation and relocation costs are the largest segments of the LAN-maintenance market. The LAN-maintenance market was over 12 billion dollars in 1990, and installation and relocation costs constituted almost half of that market [PCW91]. Closely related to this cost consideration is the projected growth in numbers of LAN-connected PCs being moved, as contrasted with new PCs on LANs, as shown in Fig. 1.7. Based upon such market studies and projections, the WLAN industry expects to capture 5–15% of the LAN market in the near future (see Fig. 1.8). Several WLAN products currently available in the market are described briefly in Chapter 12.

Although the market for desktop PCs is not growing as it has in past years, the market for portable devices such as laptop, pen-pad, and notebook computers is

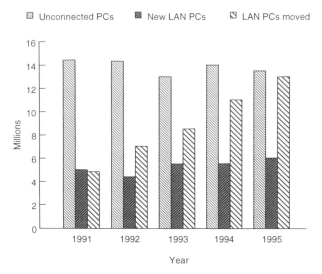

Figure 1.7 PCs in the United States. (Source: IDC.)

growing rapidly. Of greater importance to the wireless data communication industry, the market for networked portables is growing faster than the overall market for portables. Obviously, wireless is the communication method of choice for portable terminals. Mobile data communication services discussed earlier provide a low-speed solution for wide-area coverage. For high-speed and local communications, a portable terminal with wireless access can bring the processing and database capabilities of a large computer directly to specific locations for short periods of time, thus opening a horizon for new applications. For example, one can take portable terminals into classrooms for instructional purposes, or to hospital

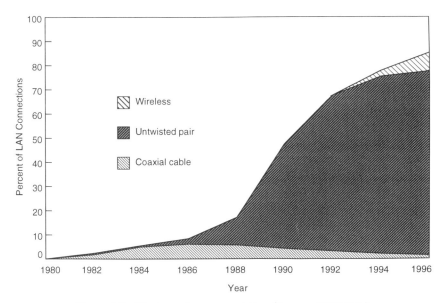

Figure 1.8 Wireless-wire augmentation. (Source: WINDATA.)

beds or accident sites for medical diagnosis. As we look toward the future, a discernible trend seems to be the integration of multiple capabilities into an individual portable communications device. The inclusion of capabilities for multiple data rates, multiple power levels, access to different wireless media, and multimedia applications will provide future users with the means for truly universal portable communications and computing [She92].

1.5 OUTLINE OF THE CHAPTERS

In this section we briefly outline the topics covered in each of the remaining chapters.

The book is focused on four issues:

1. Overview of wireless standards activities and descriptions of the systems and products, which are described primarily in Chapters 2 and 12.
2. Characterization, measurement, and modeling of radio and optical wave propagation. Radio propagation characteristics are covered in Chapter 3, while measurement techniques are covered in Chapters 4 and 5. Modeling techniques are covered in Chapter 6. Optical wave propagation is introduced as a part of Chapter 10.
3. Wireless transmission techniques, covered in Chapters 7 through 10. Wireless multiple access techniques are discussed in Chapter 11.

In Chapter 2 we outline frequency administration issues and wireless standards activities ongoing throughout the world.

We begin Chapter 3 by discussing the key characteristic of radio channels, which is *multipath fading*. We then discuss coverage and *power–distance relationships*, effects of local movements and temporal variations of the channel, and multipath delay spread and data rate limitations. We continue this discussion with a review of the *classical uncorrelated scattering model* of fading multipath channels, developing the key mathematical properties of the model in both the frequency and time domains. The chapter concludes with an introduction to indoor and urban radio channel modeling, to be discussed further in the next two chapters.

Chapters 4 and 5 provide detailed descriptions of channel measurement techniques and present the results of measurements of channel characteristics. Narrowband measurement techniques are discussed in Chapter 4, while wideband measurement techniques are treated in Chapter 5. Results of actual measurements are presented in both chapters.

In Chapter 6 we describe methods used for computer simulation of radio channels. Two general approaches to simulation modeling are discussed: (1) statistical methods based upon previously collected statistical parameters and (2) methods using deterministic analysis of radio propagation for specified configurations. The statistical modeling includes time-domain and frequency-domain approaches to modeling. The emphasis in our discussion of deterministic modeling is on the ray-tracing method, with an introduction to direct solution of Maxwell's equations.

Much of our discussion in these chapters pertains to wideband channel measurement and modeling with emphasis on indoor radio channels, because this topic is not well covered by other current texts. However, the essential characteristics of radio propagation channels are common to both outdoor and indoor channels. Detailed treatments of outdoor radio propagation, with application to cellular and mobile radio systems can be found in books by W. C. Y. Lee [Lee82, Lee86, Lee89], in a collected set of journal papers [Bod84], and in the book by Jakes [Jak74].

In Chapter 7 we discuss various modulator and demodulator (modem) technologies. We begin with a discussion of basic modem techniques—that is, techniques widely employed in standard wireline voice-bandwidth modems. Then we discuss the application of the basic modem techniques to radio channels, where multipath and fading limit performance and give rise to certain design issues and enhancement techniques which are specific to the radio environment.

Chapter 8, which is closely related to Chapter 7, describes signal processing techniques utilized in wireless communications systems. Modern signal processing techniques, together with significant advances in digital technology, have had an important impact on the growth of wireless systems and services, by enabling major simultaneous reductions in the size and cost of mobile and hand-portable communications devices.

Chapter 9 provides an overview of the application of spread-spectrum techniques in both wide-area and local-area wireless networks. The chapter begins with a discussion of the motivation for adoption of spread-spectrum for various applications. Then the basic principles of direct-sequence and frequency-hopped spread spectrum are introduced, and the performance of spread spectrum techniques on fading multipath channels is discussed. The *Code-Division Multiple-Access* (CDMA) technique is introduced, and the issues of power control, fading effects, and system capacity are discussed.

Chapter 10 is devoted to a discussion of wireless optical networks used in WLANs. The chapter begins with a discussion of infrared (IR) implementation issues and then discusses IR channel characteristics and data rate limitations. Modulation techniques for IR channels are then described, and the chapter concludes with a discussion of multiuser-access techniques used in IR WLANs.

In Chapter 11, we turn our attention from links to networks and examine the topologies of wireless networks and the multiuser-access methods employed in wireless networks. The basic network topologies are described, and their relative advantages and disadvantages are discussed. Then channel-access methods are described, including fixed-assignment methods and random-access methods. In the course of the chapter, the application of specific topologies and access methods to existing or developing systems is discussed. The chapter concludes with a discussion of the integration of voice and data services in wireless networks.

In Chapter 12, we discuss existing and planned wireless systems, as well as available WLAN products.

QUESTIONS

(a)   What are the four major categories of evolving wireless information networks, and what is the market drive for the development of each of the systems in each category?

(b)   Why are voice and data networks not integrated?

(c)   Name three major differences between future cordless/PCS and digital cellular.

(d)   Name three major differences between a mobile data system and a wireless LAN.

(e)   Describe the difference between existing analog cordless telephones and second-generation cordless telephones.

(f)   Name three advantages of digital over analog transmission for cellular mobile radio.

(g)   In which country did the idea of cellular radio originate, and in which country was it first deployed?

(h)   Why does a cordless telephone consume less electronic power than a cellular radio?

(i)   Why does a cordless telephone radiate less power than a cellular radio?

(j)   Name three categories of in-building applications in which a wireless LAN is preferred to a wired LAN.

(k)   Name four categories of applications for mobile data services.

(l)   Explain why a cordless PCS phone provides better voice quality than does a present-day cellular phone.

(m)   Which type of terminal has been sold in greater numbers—cellular telephones or cordless telephones?

(n)   Which type of service generates more revenues—cellular telephone or cordless telephone?

(o)   Discuss the radio modem design considerations for a mobile data service, and discuss how they are different for cellular mobile telephone.

(p)   Refer to Figs. 1.7 and 1.8 and discuss the market for wireless LANs.

(q)   What is a typical data rate for a mobile data service?

(r)   From the standpoint of a service provider, what is the difference between a wireless PBX and a PCS system?

(s)   Which industry is expected to generate greater revenues—mobile data or wireless LANs? Why?

2

FREQUENCY ADMINISTRATION AND STANDARDS ACTIVITIES

2.1 INTRODUCTION

The development of large markets for wireless communication services and equipment depends upon the availability of sufficient spectrum in appropriate frequency bands as well as agreements among manufacturers and service providers on standards for wireless user terminals and the supporting network infrastructure. Signal propagation characteristics are different from one frequency band to another, and this leads to differences is signal coverage, which means that different frequency bands are appropriate for different categories of systems. Adequate bandwidth must be allocated to each category of service to allow capacity to be increased in response to growing market demand. Manufacturers generally want to establish standards so that product development and manufacturing resources can be invested in a relatively small number of products, with investments leveraged by economies of scale leading to attractively priced products in a competitive marketplace.

In Chapter 1 we defined four categories of wireless information networks (see Fig. 1.1) and defined various technical and service characteristics (see Fig. 1.2) that distinguish the four categories. In this chapter we address the issues of spectrum administration and standards, with reference to the four categories of networks that we have defined.

For voice-oriented systems, whether they are to support high-power wide-area services or low-power, local-area services, spectrum allocations and interface standards are both important issues, because users will want to have a single portable or personal device usable in any serving system, and manufacturers want to create a consumer market for attractively priced terminals and to have a common set of standards for design and production of infrastructure equipment, such as radio base station transceivers and network switches.

For data-oriented systems, in the developments to date, equipment standards have been a less important issue than spectrum allocation. The major existing mobile data networks have evolved as privately developed and provisioned networks serving relatively narrow markets, such as customer field support. Wireless local area networks (WLANs) typically serve relatively stationary terminals operating in indoor locations, with little need for units in different networks to interoperate with each other. Thus a number of proprietary product designs have evolved in the WLAN market, and attention is only recently being given to developing common interface standards. Spectrum allocation has had somewhat different significance for mobile data networks and WLANs. For mobile data networks, designed to cover wide geographic areas, access to the licensed Specialized Mobile Radio (SMR) bands is necessary, and service providers have been able to acquire the Federal Communication Commission (FCC) licenses in all the market areas they serve. For WLANs, while some products have been developed for licensed microwave bands (18–19 GHz), most WLAN manufacturers have developed products for operation in Industrial Scientific and Medical (ISM) bands, and for infrared (IR) transmission, thereby avoiding the issues of frequency licensing and administration. Industry efforts to define common interface and protocol standards for WLANs have only recently begun, and initial attention in these effort has been given to application using the ISM bands and IR.

In the remainder of this chapter we outline frequency administration and wireless standards activities ongoing throughout the world. In Section 2.2 we discuss frequency administration issues and activities in the United States and worldwide. We then review the evolution of cellular telephone services in the United States, and the frequency administration issues which arose in connection with that evolution. We then briefly describe international spectrum administration activities, with particular attention to activities ongoing under the auspices of the ITU. In Section 2.3 we outline the major wireless standards initiatives in progress throughout the world. We define three generations of wireless systems and briefly describe the major ongoing European and North American activities aimed at defining new digital wireless standards.

2.2   FREQUENCY ADMINISTRATION ISSUES

The major and never-ending problem facing the radio communications industry is the fundamental limitation on availability of frequency spectrum. The task of orchestrating the allocation and licensing of the available spectrum to an ever-growing number of applications and users falls to various frequency administrative bodies throughout the world. The history of frequency administration activities has been characterized by a continual restructuring of frequency band allocations, accompanied by a steady migration toward higher frequency bands, as new systems and services have come into the market. Many of the changes have been surrounded by major controversies among segments of the communications industry. In this section we briefly outline the issues which the frequency administration bodies have been addressing in recent years, and summarize the current status of the spectrum administration rules in the United States and in other regions of the world. We discuss these issues in the context of the ongoing development of wireless information networks.

2.2.1   Spectrum Regulation in the United States

Here we discuss issues of spectrum allocation and regulation in the United States, with reference to four categories of wireless information networks: cellular telephone, personal communications services (PCS), WLANs, and mobile data networks.

Mobile Radio and Cellular Telephone Evolution. In the United States, the FCC establishes regulations for use of the spectrum. From its inception, the FCC has had to deal with crises in spectrum congestion. Mobile radio went into operational use in 1921 with a simple one-way dispatch system operated by the Detroit Police Department. By 1934, there were 194 municipal police radio systems and 58 state police radio stations serving more than 5000 radio-equipped police cars, but only 11 frequency channels allocated for police use. The newly established FCC held hearings in 1936 and in 1937 it granted 29 new channels to law enforcement agencies.

The invention of frequency modulation (FM) by Edwin Armstrong in 1935 transformed the radio industry. The remarkable improvement over the traditional amplitude modulation in received signal quality, resistance to atmospheric noise and signal fading, and ability to operate radios with reduced signal power spurred a rapid growth in the use of mobile radio. By 1940 almost all police radio systems in the country had been converted to FM [Cal88]. During the 1940s, requests for spectrum allocations for mobile radio increased dramatically, and in 1945, Report No. 13 of the Radio Technical Planning Board of the FCC recommended mobile radio spectrum allocations for a wide range of uses, including public safety organizations, public utilities, and transportation services. In 1949 the FCC officially recognized *mobile radio* as a new class of service. The number of mobile users exploded from a few thousand in 1940 to 86,000 by 1948, 695,000 by 1958, and almost 1.4 million by 1963 [Nob62]. It must be noted that the vast majority of these users were not connected to the public telephone network.

True *mobile telephone service*, the interconnection of mobile users with the public telephone network, was introduced in 1946, when the FCC granted AT & T a license to operate such a service in St. Louis. In less than a year, mobile telephone service was being offered in more than 25 cities in the United States. Demand grew rapidly and stayed ahead of capacity in many of the large urban markets. The original FM mobile telephone channels were 120 kHz wide, and by 1950 the FCC decided to split the original channels into 60-kHz channels. The original mobile telephone channels were clustered around 150 MHz, and by the mid-1950s, in response to the steady growth in demand, the FCC authorized new channels in the UHF band—around 450 MHz with channel bandwidth of 50 kHz. By the early 1960s, FM receiver design had been significantly improved, and the VHF channels were split from 60 to 30 kHz, while the UHF channels were halved from 50 to 25 kHz.

In the two decades from the end of World War II through the late 1960s, the FCC was besieged with requests for new spectrum allocations as the demand for mobile telephone services grew. During this period a long struggle was engaged between the mobile telephone industry and the growing television broadcast industry for allocation of frequency bands. A Bell proposal for new mobile channel allocations in the UHF band was rejected and in a major decision in 1949 the FCC established 70 new channels of UHF television. In a second major decision, in 1964 the FCC again rejected proposals to allocate UHF bands to mobile services. Throughout this period, the demand for mobile telephone service grew steadily;

and automatic *trunked radio systems* were introduced in the 1960s, greatly increasing system capacity. A trunked radio system makes a group of radio channels accessible to a much larger group of users by automatically tuning the mobile radio to any available channel as the call connection is established. In spite of such technology advances, it was becoming clear by the late 1960s that the existing spectrum allocations were totally inadequate to support the steadily growing demand for mobile radio services. Finally, in 1975, after 7 years of hearings on Docket No. 18262, the FCC allocated 115 MHz of spectrum in the 806 to 947-MHz band to land mobile communications, 40 MHz for cellular mobile telephone and limited dispatch services, 30 MHz for conventional and trunked dispatch service, and 45 MHz to be held in reserve. This spectrum was cleared by moving prior occupants, educational TV channels, out of those bands.

The 1975 decision allowed only one cellular system per market, and this met with controversy [War83]. After further deliberations, the FCC issued decisions in 1981 and 1982 which in part established two 20-MHz systems per market: One was to be operated by the local telephone company, and the other was to be operated by a non-wireline company. The first commercial cellular system in the U.S. went into operation in 1983, in Chicago. Within a year, some cells in the Chicago system were already saturated, and a petition was filed with the FCC for additional spectrum. An additional cellular allocation of 10 MHz was made in 1986.

The analog cellular system in use throughout the United States, as mandated by the FCC, is based upon the AT & T cellular system design, called the Advanced Mobile Phone Service (AMPS) [Ble80, Bel79a]. The channel structure of the AMPS system is based on 30-kHz channels carrying analog FM transmission [Mac79]. The current state of the cellular industry in many major market areas is that the analog cellular systems have reached their capacity limits. Digital cellular technology provides a way to increase the capacity 3 to 10 times while retaining the existing cellular spectrum allocations. We describe the major digital cellular standards in Chapter 12.

There continues to be widespread use of dispatch-type and trunked radio systems operating in the traditional land-mobile radio (LMR) bands (150 MHz, 350 MHz, and 850 MHz). In effect, these systems operate as private radio networks in designated bands licensed by the FCC. These systems are widely used by police, fire, and other public safety organizations as well as some commercial users such as taxicabs and truck fleets. Some LMR bands are reserved for government use, and they are used by military services, government law enforcement agencies, and the FAA. Private mobile data networks, such as ARDIS and Ram Mobile, operate in the 800- to 900-MHz region. In some LMR bands, capacity limits are being reached, though the market growth there is not as strong as in the cellular market. In the LMR industry, plans are being made to migrate from 25-kHz channels to 12.5-kHz channels with plans for further migration in the future. At the same time, actions are being considered which would force government users to relinquish some of their designated bands to commercial users. All of this means that in these bands efficient spectrum utilization is of the utmost importance, and these systems must be designed to use the available bandwidth to serve the greatest number of users over wide service areas. Thus the emphasis here is on bandwidth-efficient modulation, efficient frequency management schemes, and, in the case of data services, efficient multiuser access protocols.

PCS at 2 GHz. Various concepts for PCS have been under development for the last several years in the U.S. communications industry. In late 1989 the industry alerted the FCC to the potential of PCS by filing a number of applications for experimental licenses. In May 1990 the FCC released a Notice of Inquiry on PCS and received a substantial volume of comments. In late 1991, the FCC released a policy statement finding that an "adequate amount of spectrum to foster the development of innovative and competitive markets for these services" should be allocated from the 1.8- to 2.2-GHz band. Then in early 1992 the FCC released a Notice of Proposed Rule Making proposing a co-primary allocation of the 1.85- to 1.99-, 2.11- to 2.15-, and 2.16- to 2.20-GHz bands to "emerging technologies," including PCS.

On September 23, 1993 the FCC announced its decision on channel allocations for 2-GHz PCS and adopted a Notice of Proposed Rulemaking that will lead to auction procedures for PCS licenses. The FCC decided to provide spectrum for as many as seven wireless communications carriers in every city, and it also decided to let market forces shape the structure of PCS technologies and services. The FCC will award two 30-MHz blocks (several frequency slots from 1.85 to 19.6 GHz) in each of 47 major trading areas (MTAs). In addition, there will be one 20-MHz block (1.88–1.89 and 1.96–1.97 GHz) and four 10-MHz blocks (several slots from 2.13 to 2.2 GHz) in each of 487 basic trading areas (BTAs). The PCS licenses will be awarded by auctions which began in December 1994. License holders will be allowed to aggregate up to 40 MHz of spectrum for PCS service in any area, though a question arises as to whether technology available in the immediate future will allow seamless combination of separated blocks of spectrum.

In addition to the licensed spectrum blocks, the FCC allocated the 1890- to 1930-MHz bands for low-power unlicensed equipment. This includes wireless telephones and PBXs, wireless data networks, and other potential products. Voice equipment will be allowed in the 1890- to 1900- and 1920- to 1930-MHz slots, and data equipment will be allowed in the 1900- to 1920-MHz slot. The WINForum *spectrum etiquette* plan, submitted to the FCC by a group of unlicensed-equipment manufacturers, will govern operation of the unlicensed systems.

While some of the details of the September 1993 channel plan are subject to change, the major outlines of the plan have been made clear, so that manufacturers and service providers can move forward with their plans. It is expected that introduction of PCS services will begin sometime in 1995 or 1996.

WLANs, WLPBXs and Mobile Data Networks. The use of spread-spectrum WLANs in the United States is based on FCC Part-15 regulations, which permit such operation with up to 1 W of transmitted power in the Industrial, Scientific, and Medical (ISM) bands at 900 MHz, 2.4 GHz and 5.7 GHz [Mar85, Mar87a, Mar91]. These bands are also used for some cordless phones. While these bands are attractive for use in WLANs, due to the convenience of unlicensed operation and the ease of providing multiuser access, performance can suffer if careless users cause unnecessary interference to other users. Some progress is being made on this front with the WINforum spectrum-etiquette plan submitted to the FCC [Ste94].

The major technical obstacle still to be overcome in the WLAN industry is the data rate limitation caused by the multipath characteristics of radio propagation. Currently available WLANs provide data rates as high as about 5 Mbits/sec, with some new products promising rates as high as 20 Mbits/sec and more. Achievement of even higher rates may depend upon cooperation from frequency administration organizations in providing wider bandwidth allocations without restriction on the adopted technology, as well as in administering rules of etiquette for compatible use of these bands.

At frequencies around several gigahertz the technology is available for WLAN implementation with a reasonable size, power consumption, and cost. Moving to higher frequencies is the solution for the future. As the frequency increases, the prospect for obtaining a wider bandwidth from spectrum regulatory agencies will improve. However, with today's technology, implementation at a few tens of gigahertz with reasonable product size and power consumption is challenging, particularly when wideband portable communication is considered. At least one WLAN product currently in the market operates in the licensed 18- to 19-GHz band, and ETSI has designated the 17.1–17.3 GHz band for WLAN operation.

At higher frequencies, signal transmission through walls becomes more difficult. For frequencies around a few tens of gigahertz, the signal is mostly confined by the walls of a room. This feature is advantageous in certain applications where confinement of the signal within a room or building is a desirable privacy feature. Also, at higher frequencies the relationship between cell boundaries and the physical layout of the building is more easily determined, facilitating the planning of cell assignments within the building. The technology at higher frequencies is highly specialized and not commonly available within the computer industry. This has encouraged joint ventures among semiconductor, radio, and computer companies to develop new products at these frequencies.

Mobile data is a service shared by a number of users distributed over a large geographic area. Development of a mobile data network requires a major decision on a large investment affordable only by a major industrial organization. When the business need for the network is justified, acquiring the band from the bandwidth allocation agencies is possible. The transmission technology and the access methods are very similar across these systems, and the decisions regarding these factors are overshadowed by the system level decisions that provide a compromise between the cost of implementation and coverage of the network. For example, the concept of overlay of CDPD service onto the existing AMPS system is the single most important technical consideration spurring the development of CDPD. Motivation for this concept is to provide a system that operates compatibly with the channel structure of the existing AMPS systems in the same band. As a consequence, CDPD can be deployed at relatively low cost and can provide wide-area coverage without any need for additional spectrum allocation.

WLANs are designed for a small number of users in a local area. The development of such products does not require large investments or interoperation with other WLANs in assigned bands. As a result, small groups in large companies or small startup companies usually initiate development of these products. Convincing the frequency administration agencies to provide a band for this application has been more difficult, and that has encouraged companies to align the transmission technology with the existing bands rather than petitioning for new bands. The range of coverage is small, which leaves many options open for the transmission technology. The major decision here concerns the choice of the transmission method. An overview of wireless data networks, with emphasis on wireless LANs, is given in [Pah85b] and [Pah94].

2.2.2 Spectrum Administration Worldwide

Two major international bodies have traditionally met the needs of member countries for standards and for coordination of their telecommunications networks. They are the International Telecommunications Union (ITU), an agency of the United Nations, and the Conference of European Posts and Telecommunications Administrations (CEPT). We describe the major wireless-related activities of the ITU in the two following subsections.

The CEPT body brings together the posts and telecommunications administrations of most European countries, both western and eastern. The standardization activities of CEPT have historically supplemented the actions of the CCITT and CCIR. In the last few years, CEPT initiatives have increasingly been taken over by the European Community (EC). The wireless standards activities managed by the EC are discussed in Section 2.3.

ITU Activities Related to Wireless Communications. The ITU, a Geneva-based United Nations organization of about 170 nations, is responsible for communications standards and for treaty-based agreements on spectrum management. Its specific role in spectrum management is to minimize radio interference by establishing international rules standardizing the use of various radio-frequency bands. In discussing the responsibilities of the ITU, it is essential to note that the organization underwent a major structural change on March 1, 1993, which we describe later. Prior to that date, the ITU comprised four permanent groups: the General Secretariat, the International Frequency Registration Board (IFRB), the Consultative Committee on International Radio (CCIR), and the Consultative Committee on International Telegraph and Telephone (CCITT). The responsibilities of these groups included both regulatory and technical functions.

In the regulatory area, the IFRB had two principal responsibilities. First, it administered frequency assignments having international significance. Second, it organizes World Administrative Radio Conferences (WARCs). WARCs, held regularly though infrequently, are organized to update the Radio Regulations and to review frequency registration activities. The two most recent WARCs, held in 1987 and 1992, produced important decisions relative to evolving wireless communications developments.

In the standards area, the CCITT developed *Recommendations* for devices, such as data modems, which operate in "wired" telecommunications networks. The CCITT, through its various study groups, has developed many mobile-related recommendations in such areas as numbering plans, location registration procedures, and signaling protocols.

Wireless standards established by the ITU grew out of recommendations of study groups of the CCIR and, to a lesser extent, the CCITT. The activities of the CCIR developed along two main lines, one concerned with technical aspects of radio spectrum usage, the other concerned with performance criteria and system characteristics for compatible interworking. Study Group 8 of CCIR was responsible for review of recommendations for all mobile communications services including land, aeronautical, satellite, maritime, and amateur radio.

In late 1985, Study Group 8 established a special international group to identify the requirements for globally compatible *Future Public Land Mobile Telecommunications Systems* (FPLMTS) [Cal89, CCI90]. Currently, 27 national administrations and 10 international organizations participate in the work of this group, which is called *Interim Working Party 8/13* (IWP8/13). The term "personal communications" is used by IWP8/13 to describe the next generation (third generation, in our terminology) of mobile communication systems beyond the predominantly vehicular systems in use today. The IWP8/13 defines the *personal station* as a low-power, lightweight unit which can be conveniently carried everywhere. The goal of FPLMTS is *terminal mobility*—that is, the ability of pocket-sized personal stations to access public and private communications networks through interoperable terrestrial and satellite media. For practical purposes, the concept of FPLMTS can be considered to be essentially the same as PCS, a term more commonly used in the United States. Specific study topics in IWP8/13 include network architecture, satellite interworking, network interfaces, radio interfaces, quality of services, network management, and adaptation to the needs of developing countries.

At a meeting of Study Group 8 in April 1988, a new Interim Working Party (IWP8/14) was established to address mobile-satellite requirements for land, sea and air services. It was judged that close coordination with IWP8/13 would be required because these systems are seen as an important part of the vision for future personal communications and FPLMTS. The hope is that the third-generation systems developed under the FPLMTS plan will conform to a mutually compatible set of standards accepted throughout the world instead of the five or more incompatible systems of the second generation. (For example, the existing cellular systems employ different frequency bands in the three ITU regions of the world.) After a thorough study, IWP8/13 concluded that over 227 MHz in the 1 to 3 GHz band would be required for FPLMTS, and based on their studies, recommendations were brought to WARCs held in 1987 and 1992.

The 1987 and 1992 WARCs. After the preliminary work of IWP8/13, it became clear that spectrum management would have to be addressed on a worldwide basis if the goals of FPLMTS were to be realized. For example, the present 800/900-MHz cellular spectrum allocations are different in the three ITU regions, and none corresponds directly with the ITU primary spectrum allocations for mobile communications common to all three regions [Cal89]. This is not a serious problem for today's predominantly vehicular mobile systems, but it is an unsuitable situation for future PCS, which is envisioned to evolve into universal access. With this background, a specialized WARC (WARC-MOB-87) was convened in late 1987 to deal with the specific issue of mobile services. A recommendation of the WARC was that spectrum be allocated for FPLMTS, and that the spectrum needs for land-mobile satellite systems be specifically addressed. WARC-MOB-87 also recommended that another special WARC be convened by 1992 to examine all services in the frequency band 1–3 GHz.

The WARC held in Torremolinos, Spain, in February and March 1992 produced some important decisions relative to future PCS and mobile-satellite communications worldwide. In the proposals brought to WARC-92, there was a marked difference in the emphasis placed by the United States and the European nations on terrestrial-PCS versus land-mobile services. In their presentations to the WARC, the United States virtually ignored PCS and instead made a number of proposals for low earth-orbiting (LEO) satellite services and other mobile-satellite services as well. Some of the U.S. proposals proved to be very controversial, including proposals that the spectrum which Europe proposed for PCS (in the vicinity of 2 GHz) be allocated for mobile satellite services.

One of the most notable aspects of WARC-92 was the unprecedented solidarity of the European nations on the subject of future terrestrial communications services. The Europeans organized themselves within the existing CEPT structure, comprising more than 30 nations, and remained solidly unified in their position. Because of its small land mass and dense population, European spectrum managers believe that additional frequencies for terrestrial, rather than satellite, services are the key to new communications services [Ada92]. Chief among the services the Europeans supported was FLMPTS, for which they requested an allocation of nearly 300 MHz at around 2 GHz. This was in direct conflict with the U.S. proposal that the spectrum in question be allocated to mobile satellite services, and it also conflicted with a further proposal that there be no designation of frequencies for FLMPTS. A compromise was reached in which frequencies were allocated for secondary mobile satellite service within a substantial band allocated to FLMPTS by the following footnote:

> The frequency bands 1885–2025 MHz and 2110–2200 MHz are intended for use on a worldwide basis by administrations [governments] wishing to implement the future public land mobile systems (FLMPTS). Such use does not preclude the use of these bands by other services to which these bands are allocated.

The United States and Europe also compromised on an issue of PCS standards. The Europeans proposed that the ITU develop standards for PCS (having at least two candidates under development), while the United States preferred to leave such standard-setting to commercial entities and the market process. On this point, agreeing that full implementation of PCS was unlikely before the year 2000, the WARC requested that the ITU study and develop standards for roaming characteristics, signaling, and a numbering plan. Detailed discussions of WARC-92 and its conclusions can be found in [Rus92a], [Rus92b], [Wim92], and [Ada92]. The resolution allocating the FLMPTS bands and inviting further studies was considered a victory for the European position in that the new category of service thereby gained official ITU recognition. In summary, the proponents of a new generation of PCS-like services were given strong support by the applicable decisions of WARC-92. However, in the United States and other countries, significant battles remain to be fought over PCS standards and frequencies. U.S. manufacturers interested in worldwide markets will undoubtedly have to participate in European and international standards bodies in order to promote their technologies [Ada92].

2.3 STANDARDS ACTIVITIES

In this section we outline the major standards initiatives aimed at defining and developing new wireless communications systems and services worldwide. The scope of wireless standards activities throughout the world is suggested by Fig. 2.1. An excellent summary of international standards work (as of 1989) in the area of personal communications carried out by CCIR and CCITT can be found in a paper by Callendar [Cal89]. Discussions of standards initiatives in the United States and Europe can be found in several papers by Donald Cox and by David Goodman [Cox92a, Cox92b, Goo91], and Jack Taylor [Tay92]. Discussions of WLAN standards activities can be found in papers by Victor Hayes [Hay91] and Steve Wilkus [Wil91].

2.3.1 Three Generations of Wireless Systems

In discussing the many developments in voice-oriented wireless networks, it has become common usage to refer to different generations of systems. The first generation comprises today's simple cordless telephones and analog cellular telephones. Our present cordless telephones are stand-alone consumer products and do not require any interoperability specifications. Each cordless telephone has its own base station and needs only to be compatible with that base station. However the mobility afforded by this device is very limited. Analog cellular telephones of course provide much greater mobility. In the early 1980s, a *de facto* standard (AMPS) for analog cellular telephones was established in the United States. However, a mobile subscriber cannot be guaranteed the capability to roam from one cellular company's service area to another, unless cellular carriers have established roaming agreements. Matters have been worse in Europe, where there are six different analog cellular telephone standards, and they are not interoperable [Goo91a].

Second-generation cordless telephones, in contrast, will be designed as components of a larger network. These will include Cordless Telecommunications, second-generation networks (CT2), and the Digital European Cordless Telecommunications (DECT) standard, of which we will say more below, and new North American and Pan-European Digital Cellular systems.

As the telecommunications industry begins to implement second-generation networks, researchers and network planners have turned their attention to the more distant future. In a number of organizations, a vision of third-generation wireless information networks is being formed. The central concept here is to create a single network infrastructure that will allow users to exchange economically any kind of information between any desired locations. The new network will merge the separate first- and second-generation cordless and cellular services and also encompass other means of wireless access such as paging, dispatch, public safety, and WLANs. This broad concept of economical, universally interoperable, mobile, and portable telecommunications is coming to be known by the names *personal communications networks* (PCNs) and *personal communications services* (PCS). The term PCN was used first, originating in the United Kingdom. The term PCS was coined later on, in a notice published by the FCC. The term PCS has come to be given a more encompassing meaning.

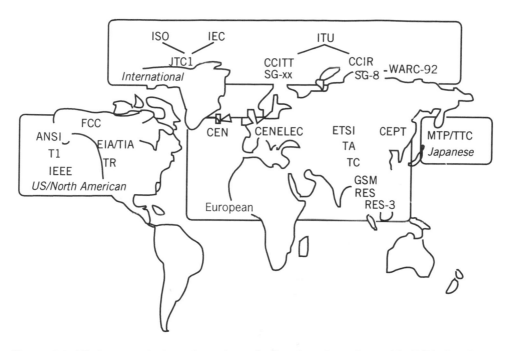

Figure 2.1 Wireless standards and regulatory bodies throughout the world. ANSI, American National Standards Institute; CCIR, Comité Consultatif International des Radiocommunications; CCITT, Comité Consultatif International Télégraphique et Téléphonique; CEN, Comité Européen de Normalisatoin; CENELEC, Comité Européen de Normalisation Electrotechnique; DECT, Digital European Cordless Telecommunications; EIA, Electronics Industry Association; ETSI, European Telecommunications Standards Institute; GSM, Global System Mobile (Groupe Spécial Mobile); IEC, International Electrotechnical Commission; IEEE, Institute of Electrical and Electronic Engineers; ISO, International Organization for Standardization; ITU, International Telecommunication Union; JSA, Japanese Standards Association; JTC1, Joint Technical Committee; NTIA, National Telecommunications & Information Administration, RES3, Radio Equipment and Systems, subgroup 3; T1, Standards Committee T1 — Telecommunications; TA / TC, Technical Assembly / Technical Committee; UDPCS, Universal Digital Portable Communications System; WARC, World Administrative Radio Conference. (From [Wil91].)

As we pointed out earlier, the development of WLANs has not relied upon the prior definition of standards, because WLANs serve small groups of users in relatively fixed locations, and interoperation between separate WLANs is typically facilitated by bridges and routers. However, there are standards initiatives ongoing in the United States (IEEE 802.11) and the EC (HIPERLAN), and the Japanese are monitoring these activities.

2.3.2 European Standards Activities

In Section 2.2 we discussed the role of the ITU in worldwide spectrum administration and commented briefly on the long-standing role of CCITT and CCIR in the establishment of standards. We also noted the traditional role of CEPT in coordinating European telecommunications administration and in supporting the activities of the CCITT and CCIR. We noted further that in the last few years,

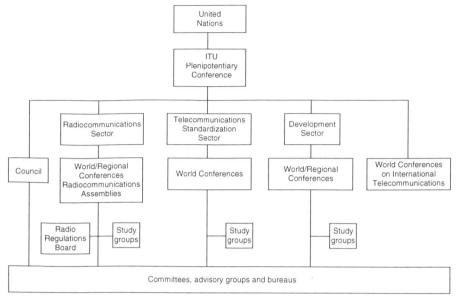

Figure 2.2 ITU organization.

CEPT initiatives have increasingly been taken over by the European Community (EC). In this section we describe some of the major wireless standards activities currently in progress under EC management. However, before discussing the EC's agenda, we describe briefly the new organization of the ITU that was put into place on March 1, 1993.

The ITU organization, shown in Fig. 2.2, now comprises three sectors: (1) the Radio Communications Sector (formerly CCIR and IFRB), (2) the Telecommunications Standardization Sector (formerly CCITT), and (3) a new Telecommunication Development Sector (BDT). While organizational components of the earlier ITU structure are identifiable within the new organization, the changes that went into effect in March 1993 involved more than a rearrangement of the existing components. As part of the reorganization, the mandates of the components were studied and, consequently, their responsibilities were changed considerably. For example, the standardization sector is basically the former CCITT, but certain nonstandardization responsibilities were transferred to the BDT, while the responsibility for standardization of network-related radio communications was taken over from the former CCIR [Irm94]. As a result, after a transition period, essentially all ITU standardization activities, including radio communications, will fall under the mandate of the ITU Telecommunications Standardization Sector, now commonly identified as ITU-T.

We now return to our discussion of the EC's standards activities. The EC's agenda calls for a coordinated approach to the development of future wireless standards. The primary regional standards bodies, ETSI and CEPT, are completing standards for pan-European paging and digital cellular systems. The EC manages its activities through 22 Directorates-General. Specific commission programs affecting person-to-person wireless communications standards are shown in Fig. 2.3 and are discussed in the following subsections.

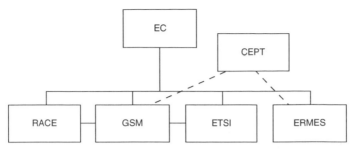

Figure 2.3 Standards bodies of the European Community.

RACE. The R & D in Advanced Communications for Europe (RACE) program, established in 1985, aims to create a technology base in broadband and wireless telecommunications that will enable European companies to capture a significant portion of the worldwide information technology market. The main program now underway supports more than 40 experimental programs in broadband and wireless data applications. The RACE program proposed the merging of three types of systems: a universal mobile telecommunications service (UMTS) with cordless and cellular attributes, a high-bit-rate broadband system that will probably operate at millimeter wave frequencies and be capable of supporting a mobile office, and a public switched telephone network (PSTN), the attributes of which can be enhanced by the portable segments. The main issues are standards, network design and implementation, and terminal development. Within the framework of RACE, 25 organizations are working together to create enabling technologies for third-generation services. A central objective is to provide users of wireless terminals with access to the projected Broadband Integrated Services Digital Network (BISDN). A discussion of the status of RACE activities is given in [Kon93].

GSM. The term GSM originated in the early 1980s as the French acronym for Groupe Special Mobile. This international working group was tasked by most of the European PTT administrations to develop a standard for digital cellular networks allowing international roaming across European borders [Arn93]. The resulting standard is often referred to as the *pan-European standard for digital cellular communications*. The standard is now known by the name *Global System for Mobile Communication (GSM)* [Hau94, Rah93]. The Commission of European Communities (CEC) ensured that common frequency allocations were made in the 900-MHz band in the 12 member states and many other European countries. The GSM system is replacing the six different analog cellular systems now in use in various regions of Europe. The GSM standard is also rapidly gaining adoption outside Europe, notably in the Asia-Pacific region in Singapore, Malaysia, India, Hong Kong, and Australia, where a Pan-Asian Memorandum of Understanding (MoU) is being considered, and in the Middle East and Africa. Moreover, Taiwan, Thailand, and New Zealand are considering the GSM standard, or have already specified it for at least one of their competing cellular networks. As of April 1993, 32 operators in 22 countries were committed to implementing GSM. GSM began operation in 1992, and at the end of 1992, GSM service was being offered in eight countries [Arn93]. Within a year of the introduction of GSM service, most of the major trunk routes in Europe had been reasonably covered. Thus far, only digital-voice service is provided. However, a short-message service is being planned, and a more complete set of data services is being planned for a later phase of implementation. The air-interface specification for GSM is described in Chapter 12.

ETSI. Established by the EC in 1988, the European Telecommunications Standards Institute (ETSI) has taken over and is extending many of the standards responsibilities formerly directed by CEPT. The major goal of ETSI is to break down the borders that still divide Europe's telecommunications systems. Membership in ETSI goes well beyond EC member states to include national administrators, manufacturers, suppliers, and users. ETSI's detailed work is done largely by technical committees, one of which addresses GSM. Its subcommittees cover services and facilities, radio interfaces, network aspects, and data. Other ETSI wireless groups address Universal Personal Communications and the PCN digital cordless system originating in the United Kingdom. An *ad hoc* ETSI body has also been established to work on DECT, which is intended to support wireless office systems. The ETSI has also adopted for standardization the Common Air-Interface (CAI) specification of CT2. ETSI is also developing a set of standards for trunked radio mobile data service, called TETRA. We say more about technical aspects of these standards in Chapter 12.

While most of the wireless standards interest in ETSI is directed toward cellular and cordless telephone, there has also been activity in the area of standards for WLANs. A 20-Mb/sec standard called HIPERLAN (for High-Performance European Radio LAN) is now being defined by the ETSI Sub-Technical Committee RES-10. The standard may be ready by 1995. The initial proposal for spectrum allocation for HIPERLAN is a 150-MHz bandwidth in the 5.2-GHz range and a 200-MHz bandwidth at 17.2 GHz.

ERMES. The European Radio Messaging System (ERMES) will provide a pan-European paging system. As with cellular systems, paging systems have grown up in a piecemeal, uncoordinated fashion, with communications capabilities stopping at national boundaries. ERMES, another initiative of CEPT, introduced common standards for paging (the spectrum used is 169.4–169.8 MHz) in the 12 EC member countries, phased over the period 1991 to 1995.

2.3.3 Standards Activities in North America

Here we describe the major ongoing wireless standards activities in North America. The technical standards being developed under these activities are described in Chapter 12.

TIA Digital Cellular Standards Initiative. Soon after the introduction of cellular telephone service in 1983, the cellular industry saw that cellular service would be a technical and commercial success. However, it was also seen that the widespread acceptance of cellular service would soon pose capacity problems for the service providers. It was becoming clear that in major cellular market areas, the AMPS system, based on analog FM technology, would not provide the capacity needed to meet the rapidly growing demands for service. In 1984, in anticipation of this capacity problem, the cellular industry petitioned the FCC for additional spectrum, and in 1986 the FCC allocated an added 10 MHz to the original 40-MHz cellular spectrum allocation. At the same time, the FCC decided to permit the introduction of spectrally efficient digital radio systems, which offered increased capacity relative to the AMPS system. (The channel plan for AMPS provides 416 30-kHz user channels for each of the two systems operating in each market area. Each cellular call uses two 30-kHz channels, one going in each direction between the mobile phone and the cellular base station.)

The FCC decisions were released in late 1988, and shortly thereafter the Cellular Telephone Industry Association (CTIA) asked the Telecommunications Industry Association (TIA) to begin the process of developing a digital cellular standard to augment and eventually replace the AMPS systems being installed throughout North America. The CTIA is an association of cellular service carriers, whereas the TIA, a branch of the Electronic Industries Association, represents communications equipment manufacturers. (Some carriers, as well as a few government agencies, in the role of service and equipment users, also have membership in the TIA.) As part of its charter, the TIA establishes and organizes committees to develop technical standards for communications equipment, through a process of voluntary participation by members in subcommittees and focused working groups. In the process of developing cellular standards, the TIA standards groups work closely with the CTIA, who define and prioritize market needs which the new standards are designed to meet.

TR45.3. In response to the CTIA request in late 1988, the TIA established the TR45 Committee on Digital Cellular Standards. Major cellular equipment manufacturers from the United States, Canada, Europe and Japan are participating in the work of TR45. The Subcommittee TR45.2, User Requirements, began the task of defining the services and features which would be provided to users of the new digital system. Subcommittee TR45.3 began its work by evaluating spectrally efficient digital modulation techniques, as well as two candidate access methods, FDMA and TDMA. Work was also begun in various working groups to evaluate proposals for signaling, voice digitization, and the provision of data services in the digital system. In January 1992 the EIA and TIA released Interim Standard 54 (IS-54), which defines the air-interface standard for the digital cellular mobile terminals and base stations [EIA92]. Revisions of the standard were subsequently issued, and Revision C was due to be released in late 1994. The standard defines a TDMA system which multiplexes three (in Phase 1) or six (in Phase 2) user channels in each 30-kHz radio channel. The standard specifies that the Phase 1 system will support dual-mode terminals—that is, mobile phones that will operate through the existing AMPS systems as well as the new TDMA systems. This will provide an orderly introduction of digital cellular service by allowing phased replacement of analog channel banks with digital equipment, as well as allowing subscribers to roam freely among AMPS and TDMA service areas. The Phase 2 six-slot TDMA system is intended to be an all-digital system which will not require mobile terminals to continue operation over AMPS.

The IS-54 standard specifies the air-interface transmission structure for digital voice service only. Before the release of IS-54, the TR45.3 subcommittee decided that in the Phase 1 system, data services would be supported with dual-mode terminals by carrying modem line signals over the AMPS analog channels. This is exactly what is done in today's AMPS systems, where many subscribers use modems and Group-3 fax terminals which have been designed specifically for the mobile environment. However, in late 1991 the TR45.3.2.5 Task Group for Digital Data Services was formed and charged with the responsibility for developing standards for data services to be carried in the digital channels of the TDMA system [Sac92, Wei93]. The task group began its work with development of a standard for asynchronous data service and Group 3 facsimile service, and the standard for those services was due to be released in late 1994. In a separate effort, a task group is evaluating methods for providing low-rate packet data service over the digital control channel which is to be incorporated into Revision C of IS-54. (The IS-54 Rev. B system uses an analog control channel compatible with AMPS systems.)

In June 1992, AGT Cellular in Calgary, Alberta, began TDMA cellular operation, becoming the world's first carrier to provide a digital cellular call over a commercially operating cellular system. Shortly thereafter, Rogers Cantel Inc. (Cantel) initiated digital cellular service in Toronto. In the United States, customer trials of TDMA were conducted in 1992, and in mid-1993, TDMA service went into operation in Florida and the Chicago area. The IS-54 TDMA standard is sometimes referred to as the *North American digital cellular (NADC) standard.*

A technique to enhance the capacity of the IS-54 TDMA system has been developed by Hughes Network Systems (HNS). This system, called *Enhanced TDMA* (E-TDMA), would increase the TDMA system capacity by using voice-activity detection and dynamic multiplexing of digitized-voice frames over a set of time slots and frequency channels. At this writing, one cellular carrier, Bell South Mobility, has awarded a contract for acquisition of the E-TDMA equipment.

TR45.5. While the TIA Subcommittee TR45.3 was evaluating digital cellular system concepts based on FDMA and TDMA access methods, Qualcomm, Inc., a San Diego-based company, was developing a digital cellular system based on spread-spectrum code–division multiple-access (CDMA) technology, a development funded in part by several carriers and manufacturers. Qualcomm did not participate in the TR45.3 effort leading to the IS-54 TDMA standard, but instead worked independently to develop their system and brought it to field trials for carrier evaluation in 1990, at about the same time that TDMA field trials were beginning. In 1992 the CTIA established a subcommittee to gather information on spread-spectrum technology, and the country's first customer trials of the TDMA and CDMA cellular systems began in the Chicago area. Subsequently, the TIA formed a new TR45 subcommittee, TR45.5, and began the development of a cellular standard based on the Qualcomm CDMA system design. On July 16, 1993 the CDMA air-interface standard for digital voice service, designated as IS-95, was released [EIA93]. Several U.S. cellular carriers have announced plans to implement the CDMA cellular system, while other carriers continue to conduct field tests and customer trials.

Following release of the IS-95, TR45.5 began work on data services to be provided by the CDMA system. Work began on three data services: asynchronous data, Group 3 facsimile, and packet data. Plans for future work include synchronous data and other services exhibiting fixed time delay [Tie93].

PCS Standards Initiatives. In the United States, there are several ongoing initiatives aimed at developing standards for PCS, including subcommittees T1E1 and T1P1 of committee T1 (Telecommunications) of the American National Standards Institute (ANSI), the TR45.4 Microcellular subcommittee of EIA/TIA TR45, and the U.S. contingent of the ITU's radio standards groups (formerly in CCIR). The ANSI T1 effort is focusing on defining services, signaling structure, network interfaces, and overall systems engineering. The TR45.4 effort is concerned with developing PCS systems based on cellular signaling and technology. Their principal interest is in signaling, radio frequency interfaces, and microcellular architecture. The U.S. ITU responsibility is one of coordinating U.S. positions and contributions to ITU.

In early 1992 the T1P1 subcommittee issued a draft working document on Personal Communications [T1P92]. That document provided definitions of terms applicable to PCS and set forth preliminary descriptions of specific services, from the perspective of the end-user. The document also provided reference models for PCS network architectures and interfaces. The document provided a guide to the work of other technical subcommittees within T1, as well as a basis for discussions with other national and international standards bodies. In 1992 the ANSI T1P1 and TIA TR45.4 and TR45.6 subcommittees formed the Joint Technical Committee (JTC), which was assigned to develop a joint recommendation for a common air interface (CAI) standard for PCS. CAI proposals were submitted to the JTC in November 1993, and at this writing they are under evaluation.

WINForum. The WINForum is an association of companies manufacturing wireless products (in particular wireless LANs and telephones and PBXs) for use in unlicensed frequency bands. The forum started after Apple Computer's petition to the FCC for the so-called data-PCS bands. The initial goal of the WINForum was to establish an alliance to obtain frequency bands for data-PCS. The goals were further expanded to obtain bands for unlicensed voice and data communications. The principal technical focus of this group has been to develop plans for *spectrum etiquette* to be observed in simultaneous operation of different manufacturers' products in common or closely adjacent areas and to provide a fair method of sharing the band. In 1993 WINForum submitted a proposed spectrum etiquette plan to the FCC, to be used by the FCC for the unlicensed PCS bands.

CDPD Consortium. In April 1992, IBM and eight of the largest cellular carriers announced the formation of an alliance to develop the Cellular Digital Packet Data (CDPD) system. These carriers (Ameritech Cellular, Bell Atlantic Mobile Systems, GTE Mobilnet/Contel Cellular, McCaw Cellular, NYNEX Mobile Communications, PacTel Cellular, Southwestern Bell Mobile Systems, and US West) cover 95% of the United States, including all major urban areas. The CDPD system, based on an earlier IBM development called CelluPlan II, is being designed to provide packet data service as a noninterfering overlay to the existing AMPS analog cellular system, using the same 30-kHz channels. CDPD will do this by transmitting data packets in channels which are not being used for voice traffic, and hopping to another channel when the current channel is allocated to a voice call. The compatibility of the CDPD system with the existing AMPS cellular system allows co-existence with any analog cellular system in North America, and it provides for data services that are not dependent upon implementation of a digital cellular standard in the service area. A preliminary field demonstration was conducted in the second half of 1992, and the participating companies subsequently issued a preliminary specification. A final specification was released in July 1993 [CDPD]. The CDPD air-interface specification is an open one, and manufacturers are being encouraged to develop equipment conforming to the specification. Though the CDPD specification has been developed without the governance of any established standards organization, the wide support by major cellular carriers will in effect establish CDPD as a *de facto* standard for cellular packet data services. Several of the carriers supporting the CDPD development began deploying commercial service in 1994. In mid-1994, 69 companies—cellular carriers as well as equipment manufacturers—announced the formation of the CDPD Forum, whose purpose is to foster the widespread implementation of CDPD service throughout the United States. By October 1994, the membership of the forum had grown to 75 companies. While the group is not a formal standards organization, it has begun forming technical working groups that will define various enhancements to the original CDPD specification that, when agreed upon, can be adopted by CDPD service providers. We provide details on the CDPD system design in Chapter 12.

IEEE 802.11. The IEEE working group for WLANs (designated IEEE P802.11) is part of the IEEE Standards Project 802. This project is sponsored by the Technical Committee for Computer Communications, a standards activity of the IEEE Computer Society. Project 802 is recognized as the focal point for the development of LAN standards, and its work is appreciated internationally [Hay91]. The first activity on a wireless medium was started in Working Group 802.4, which worked on the token-passing bus access method starting in 1987. The 802.4L group was formed and works with primary attention to use of radio transmission.

During the studies of modulation techniques, the group's interest focused on the unlicensed radio bands. The intention to use these bands necessitated the use of spread-spectrum modulation, as permitted in the United States by the FCC in the ISM bands (902–928 MHz, 2400–2500 MHz, and 5725–5875 MHz). Other aspects of the studies included gaining an understanding of pertinent radio channel characteristics, such as propagation and noise within buildings. The latter studies included measurements made at a General Motors automobile plant at Ottawa, Ontario, Canada.

In July 1990, the group had come to the conclusion that the token-passing bus protocol was not suitable for controlling a radio medium without incurring inefficiency in use of the radio spectrum. At that point, the Executive Committee of IEEE Project 802 decided to establish Working Group P802.11 with the charter to define an access protocol and a medium-access control specification suited for WLANs. Membership of the group includes representatives from the United States, Canada, Japan, Australia, and Europe. Most major computer manufacturers participate in the group. The plan of the working group is to first establish an IEEE standard, and then submit that standard as a draft to the International Organization for Standardization/International Electrotechnical Commission (OSI/IEC) for consideration as an International Standard.

The term "wireless" includes infrared and visible light, as well as radio transmission. The radio spectrum, however, is a scarce resource, one that is rigidly controlled by the appropriate administrations in each country. It would be most advantageous to obtain sufficient spectrum in various countries, and then harmonize spectrum allocation worldwide. Thus an important part of the 802.11 working group's task is to encourage regulatory agencies to allocate bands for radio LANs. Furthermore, it is desirable to prevent a need for end-user licensing, and thus the means for sharing radio frequencies must be such that no coordination with existing users is required. These considerations led the working group to give initial attention to the ISM bands, and they have in particular concentrated on the 2400- to 2500-MHz band, in large part because that segment of the spectrum is available in the largest number of countries.

The work of 802.11 has concentrated heavily on methods of media access control (MAC), by which multiple users can share a wireless network while minimizing the incidence of collisions. By mid-1993 the MAC proposals offered by various companies were narrowed down to two potential approaches, CSMA/CD (carrier-sense multiple access with collision detection and avoidance) and a version of reservation-TDMA. The CSMA/CD approach was finally selected. The committee has also given attention to the issue of data security (privacy), an important concern in any wireless system where users want to prevent unauthorized access to their network and to safeguard the data they transmit over the network. In addressing this issue, the 802.11 group has worked with a parallel group, IEEE 802.10, that focuses specifically on the data security problem. In mid-1993 the 802.11 working group decided to incorporate into their standard the IEEE 802.10 Secure Data Exchange standard. The IEEE 802.11 working group issued a draft standard in November 1993.

QUESTIONS

(a)   Which agency allocates and regulates communications frequencies in the United States?

(b)   Describe the differences between land-mobile radio and cellular telephone services. Which frequency bands are used by each of these services?

(c)   Which PCS bands are used for licensed and unlicensed operation?

(d)   Explain what is meant by an etiquette plan, and identify the PCS bands in which etiquette is employed.

(e)   Identify the ISM bands and the types of services and products which have been developed for use in these bands.

(f)   Compare the ISM bands with an unlicensed data–PCS (asynchronous) band.

(g)   Identify the frequency bands used by mobile data services.

(h) What is the major international body involved in telecommunication regulations, and which organization has chartered its mission?

(i) Which organization brings together European telecommunication regulations?

(j) Describe the role of the ITU, and describe its structure before, and after, the March 1993 reorganization.

(k) What is the WARC and what is its charter?

(l) Explain the different responsibilities of CCIR and CCIT under the former structure of the ITU.

(m) What is FPLMTS, which frequency bands are expected to be used for this service, and how does it relate PCS?

(n) Explain the three generations of wireless networks, and the differences among them.

(o) Which standards groups are currently involved in wireless LANs, and which countries host these activities?

(p) Which agencies are involved in pan-European paging and digital cellular systems?

(q) What are the four commission programs affecting person-to-person wireless communications in Europe?

(r) What is the UMTS, and how does it compare with FPLMTS?

(s) What is HIPERLAN, and which standards body is involved with it?

(t) What are the data rates and frequencies of operation in HIPERLAN?

(u) Briefly describe the IS-54 air-interface standard. Explain the role of the TIA in development of this standard.

(v) Briefly describe the IS-95 air-interface standard. Explain the role of the TIA in development of this standard.

(w) What is ANSI, and what is its role in the development of North American PCS standards?

(x) What is WINForum?

(y) Explain the principles of operation of CDPD.

(z) What is the mission of the IEEE 802.11 standards group, and to which frequency bands does this standard apply?

3

CHARACTERIZATION OF RADIO PROPAGATION

3.1 INTRODUCTION

The effective design, assessment, and installation of a radio network requires an accurate characterization of the channel. The channel characteristics vary from one environment to another, and the particular characteristics determine the feasibility of using a proposed communication technique in a given operating environment. Having an accurate channel characterization for each frequency band, including key parameters and a detailed mathematical model of the channel, enables the designer or user of a wireless system to predict signal coverage, achievable data rate, and the specific performance attributes of alternative signaling and reception schemes. Channel models are also used to determine the optimum location for installation of antennas and to analyze the interference between different systems.

The wireless networks which we consider in this book operate at frequencies ranging from a few hundred kilohertz (mobile radio dispatch networks) to a few tens of gigahertz [some wireless local area networks (WLANs)]. However, the emerging new systems and services are those operating in a range from around 900 MHz to a few gigahertz. As we described in Chapter 2, cellular systems, second-generation cordless telephones, and some WLANs operate in the 900-MHz region, whereas Handi-Phone, DCS-1800, DECT, and the emerging new personal communications service (PCS) systems are targeted at bands around 2 GHz. In addition, IEEE 802.11-based WLANs operate in the 2.4-GHz ISM band. Given the growing importance of PCS, the convenience of unlicensed operation in the ISM bands, and the fact that several new systems under development in Europe will utilize bands around a few gigahertz, we will give much of our attention to this region of the spectrum.

The frequencies in the region of a few gigahertz have several attractive features for use in the evolving wireless information networks. At these frequencies a transmitter with power less than 1 W can provide coverage for several floors within a building, and if used outdoors it can cover distances of the order of a few miles, as needed for cellular urban radio communications. Furthermore, at these frequencies the size of an efficient antenna can be on the order of an inch, and antenna separations as small as several inches can provide uncorrelated received signals suitable for achieving diversity in the received signal. At lower frequencies, bandwidth is less plentiful, longer antennas and wider antenna separations are required, and there are higher levels of man-made noise interference from ignition systems. Higher frequencies provide more ample bandwidth, but they suffer greater attenuation in transmission through walls. For frequencies in the region of a few tens of gigahertz, signal propagation is largely confined by the walls of a room, and this restricts the applications for some systems. From the standpoint of security, however, confinement in a room can be an attractive feature of these frequencies. Signal coverage can be extended throughout a building by using a leaky cable antenna [Sal87a], and leaky cables are used for communication in tunnels and for paging systems in hospital buildings. Much of our discussion pertains to characterization of indoor radio channels, because this topic is not well covered by other current texts.

Radio propagation in both indoor and outdoor environments is complicated by the fact that the shortest direct path between transmitter and receiver is usually blocked by walls, ceilings, or other objects in an interior space, or by buildings and terrain features outdoors. Thus the signal power is typically carried from the transmitter to the receiver by a multiplicity of paths with various strengths. The arrival times of signals on various paths are proportional to the lengths of the paths, which are in turn affected by the size and the architecture of the environment and locations of objects around the transmitter and receiver. The strengths of such paths depend on the attenuation caused by passage of the signal through, or reflection of the signal by, various objects in the path. The deterministic analysis of propagation mechanisms in such an environment is limited to simpler cases. For more complex cases, statistical analysis is more useful and indeed more typically used. In statistical modeling, the statistics of channel parameters are collected from actual measurements at various locations of the transmitter and the receiver.

The unpredictability of the existing paths between transmitter and receiver in an indoor environment is very similar to the situation with outdoor channels, and in fact the work which has been done in characterization of mobile radio channels provides a useful guideline for modeling indoor channels. In the indoor environment the multipath is caused by reflection from the walls, ceiling, floor, and objects within an office; while in mobile radio, multipath is caused by the ground, as well as the buildings and vehicles in the vicinity of the mobile terminal. Because the distances in an office environment are shorter, the delays between arriving paths are smaller, resulting in a smaller *multipath spread* of the received signal.

We are generally interested in different channel parameters for narrowband and wideband signaling. For narrowband communication applications, such as cordless telephone or low-speed data, we are concerned mainly with the statistics of the received power, while for high data rates or inherently wideband transmission, such as spread-spectrum, the multipath characteristics of the channel are also important.

In this chapter we begin by using simple models to familiarize the reader with the basic radio propagation parameters used in design, analysis, and installation of wireless information networks. The most important issues for the design of a wireless communication system are the achievable signal coverage, the maximum data rate supportable on the channel, and the rate of fluctuations in the channel. For a given transmission power the achievable coverage determines the size of the cells in a cellular system and the range of operation for a system operating with a single base station. The maximum data rate is more important for data communications where one desires high transmission speed for efficient transfer of long messages or data files. The maximum rate of fluctuations in the channel is important in the design of the adaptive parts of the receiver such as timing and phase recovery circuits or power control algorithms. To determine the coverage of a system, the distance–power relationship and the statistics of the power fluctuations at a given distance are needed. The data rate limitations are determined by the multipath structure of the channel. The rapidity of variations in the channel is determined by analyzing the *Doppler spread* of the channel. These concepts are discussed in more detail in Sections 3.2 to 3.4.

In Section 3.5, we describe a more general way of modeling radio channels, one which has been widely accepted and applied in the analysis of a variety of radio systems operating in many different frequency bands. The modeling approach is based on a statistical treatment of time-varying channels, and it yields some useful insights into the key channel characteristics which impact on the way signaling schemes should be designed for such channels. In Section 3.6, we lay the groundwork for a more detailed examination of channel measurement and modeling techniques, which will be described in Chapters 4, 5, and 6.

3.2 MULTIPATH FADING AND THE DISTANCE – POWER RELATIONSHIP

In most radio channels the transmitted signal arrives at the receiver from various directions over a multiplicity of paths. Figure 3.1 provides several examples of multipath fading radio channels. Figure 3.1*a* represents a troposcatter radio communication link used in military applications for communications at long distances. The transmitted signal is directed toward the troposphere layer of the atmosphere, the incident wave is scattered, and some of the scattered signal energy reaches the receiver. The communication between the transmitter and the receiver can be modeled with several paths. Figure 3.1*b* represents a line-of-sight (LOS) microwave radio link, as is widely used in nationwide networks for terrestrial communications. At installation, the antennas are aligned to provide LOS communication. However, for occasional short periods of time, atmospheric conditions can affect radio propagation in such a way that signal components reflected from the ground and the atmosphere become comparable to the LOS component, creating a multipath condition. Figure 3.1*c* represents a mobile radio scenario where the received signal arrives by several paths bounced from large objects such as buildings and local paths scattered from objects close to the receiver, such as ground or trees. Figure 3.1*d* represents a multipath condition for an indoor area.

The phase and amplitude of the signal arriving on each different path are related to the path length and the conditions of the path; this results in considerable amplitude fluctuation of the composite received signal. An exact analysis of the multipath propagation can be done by solving Maxwell's equations with boundary conditions representing the physical properties and architecture of the environment. This method is computationally burdensome; and even with today's most sophisticated computers, only the simplest structures can be treated. A simpler analytical approach is to approximate the radio wave propagation with

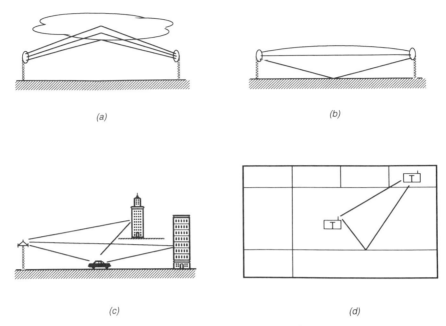

Figure 3.1 Examples of multipath in different radio channels. (*a*) Troposcatter, (*b*) microwave LOS, (*c*) mobile radio, (*d*) indoor radio.

optical wave propagation and to determine the directions of the arriving paths through the rules of geometric optics. This method is commonly referred to as the *ray-tracing method*. The transmitting and receiving antennas are assumed to be radiating points, and each path is modeled as a ray. A ray is the path of an ideal bullet traveling in a straight line and reflecting from the objects according to the rules of geometric optics. Figure 3.2 represents a mobile radio environment where the received signal arrives from two paths: (1) the direct LOS connection between the transmitter and the receiver and (2) the path arriving after reflection from the ground. A more complete ray-tracing algorithm includes (1) the mechanism of transmission through walls and (2) diffraction at the edges of the buildings. Further details of the direct solution to Maxwell's equations and the ray-tracing algorithm will be discussed in Chapter 6. In this chapter we use a simple ray-tracing technique to familiarize the reader with the principles of radio propagation modeling for communication systems applications.

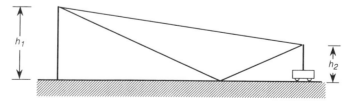

Figure 3.2 Setup for a mobile radio environment.

3.2.1 Narrowband Signals in Free Space

Free space provides the ideal environment for single-path communication. To analyze the multipath condition, we start with a simplified description of radio propagation in a single-path free-space channel. In free space, the relationship between transmitted power P_t and received power P_r is given by

$$\frac{P_r}{P_t} = G_t G_r \left(\frac{\lambda}{4\pi d} \right)^2 \qquad (3.2.1)$$

where G_t and G_r arc the transmitter and receiver antenna gains, respectively, d is the distance between the transmitter and the receiver, $\lambda = c/f$ is the wavelength of the transmitted signal, and c is the velocity of radio wave propagation in free space, which is equal to the speed of light. Defining $P_0 = P_t G_r G_t (\lambda/4\pi)^2$ as the normalized received power at a distance of 1 m, the above equation reduces to

$$P_r = \frac{P_0}{d^2}$$

Over a single path, the received signal power decreases with the square of distance. In logarithmic form (decibel scale) we have $10 \log_{10} P_r = 10 \log_{10} P_0 - 20 \log_{10} d$, which reveals the 20-dB-per-decade (or 6 dB per octave) loss of signal power as a function of distance in free space. The transmission delay is $\tau = d/c \simeq 3d$ nsec or 3-ns delay per meter.

Example 3.1. For a 1-GHz center frequency and dipole antennas with $G_t = G_r = 1.6$, the received power calculated from Eq. (3.2.1) at a distance of 1 m, $d = 1$, from the transmitter is 30 dB below the transmitted power. The received powers at distances of 10 m and 100 m are 50 dB and 70 dB below the transmitted power, respectively. The transmission delays associated with the two distances are 30 nsec and 300 nsec, respectively.

Let us assume that a single cosine with amplitude A_t and frequency f, $\text{Real}(A_t e^{j2\pi ft})$, is transmitted in free space with only the LOS path between the transmitter and the receiver. In practice, achieving LOS transmission usually requires a very narrow transmitter antenna pattern. The received signal is $\text{Real}[A_r e^{j2\pi f(t-\tau)}] = A_r e^{j\phi_r} e^{j2\pi fr}$, where A_r is the amplitude of the received signal and $\phi_r = -2\pi f\tau = -2\pi fd/c$ is the phase of the received signal. Because the power decreases with the square of the distance, the amplitude of the received signal decreases linearly with distance between the transmitter and the receiver. Therefore the received amplitude of the signal at a distance d is $A_r = A_0/d$, where $A_0 = \sqrt{P_0}$ is the amplitude of the received signal at 1-m distance from the transmitter.

3.2.2 Multipath Fading and Narrowband Signals

In a multipath environment, the composite received signal is the sum of the signals arriving along different paths. Except for the LOS path, all paths are going through at least one order of reflection, transmission, or diffraction before arriving at the receiver. At this stage, let us consider only the reflections. Upon each reflection of a path from a surface, a certain fraction of the power is absorbed by the surface and the remainder of the power in that path carries beyond the reflection. If the path has been reflected K_i times before arriving at the receiver, and at each reflection the reflection coefficient is a_{ij}, the overall reflection factor is

$$a_i = \prod_{j=1}^{K_i} a_{ij}$$

where a_{ij} is the reflection coefficient for the jth reflection of the ith path. Therefore, the amplitudes of the signals received from paths other than the LOS path are subject to reflection loss as well as the standard distance-attenuation factor.

If we have L paths and the distance traveled by the ith path is d_i, the amplitude and the phase of the received signal are given by

$$A_r e^{j\phi_r} = A_0 \sum_{i=1}^{L} \frac{a_i}{d_i} e^{j\phi_i}$$

where $\phi_i = -2\pi f d_i / c$. Figure 3.3 shows a phasor diagram representing the

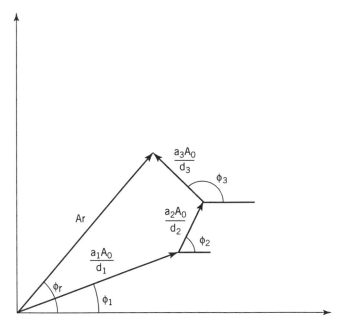

Figure 3.3 Phasor diagram for narrowband signaling on a multipath channel.

signals arriving from different paths as well as the received signal amplitude and phase. The received power is given by

$$P_r = P_0 \left| \sum_{i=1}^{L} \frac{a_i}{d_i} e^{j\phi_i} \right|^2 \tag{3.2.2}$$

The right-hand side of the equation shows the magnitude-square of the vector sum of all paths. If the phase of the first path is used as the reference and the vector sum is taken with the phase of all paths relative to the first path, the result remains the same. Let us consider three examples that use the results of the preceding discussion.

Example 3.2 (Power–Distance Relationship in Mobile Radio Channels). Figure 3.2 depicts a mobile radio environment in which the height of the base station and mobile station antennas are h_1 and h_2, respectively. The distance d between the transmitter and receiver is assumed to be much larger than either antenna height, and it is assumed that there are two signal paths, one the direct LOS path and the other reflected from the ground with a reflection coefficient of $a_1 = -1$, which means that the ground acts as an ideally lossless reflector.

Using Eq. (3.2.2) with the above assumptions, the lengths of the two paths can be assumed approximately the same, d. Then, the received power is given by

$$P_r \simeq \frac{P_0}{d^2} |1 - e^{j\Delta\phi}|^2$$

where $\Delta\phi = 2\pi f \Delta d / c = (2\pi/\lambda) \Delta d$ is the phase difference between the two paths, with Δd being the difference between the two path lengths. The lengths of the two paths are given by

$$d_1 = \sqrt{(h_1 + h_2)^2 + d^2} \simeq d + \frac{(h_1 + h_2)^2}{2d}$$

$$d_2 = \sqrt{(h_1 - h_2)^2 + d^2} \simeq d + \frac{(h_1 - h_2)^2}{2d}$$

Therefore

$$\Delta d = \frac{2h_1 h_2}{d}$$

and

$$\Delta\phi = \frac{2\pi}{\lambda} \times \frac{2h_1 h_2}{d}$$

For small values of $\Delta\phi$ we have

$$|1 - e^{j\phi}| \simeq |1 - (1 - j\Delta\phi)| \simeq |\Delta\phi|$$

Then the received power is given by

$$P_r = \frac{P_0}{d^2}|\Delta\phi|^2 = \frac{P_0}{d^2}\left(\frac{2\pi}{\lambda}\right)^2 \times \frac{4h_1^2 h_2^2}{d^2} = P_t G_t G_r \times \frac{h_1^2 h_2^2}{d^4}$$

Note that the gradient of the distance–power relationship is increased to four. Thus the power will decrease 40 dB per decade of distance, in contrast with the 20 dB per decade found for the case of LOS transmission in free space.

The first conclusion to be drawn from this example is that the multipath changes the distance–power relationship. The second conclusion is that for mobile radio communications, when the cause of multipath is reflection from the ground, 40 dB per decade is a reasonable model for the path loss characteristic.

Example 3.3 (Fading Caused by Multipath). In this example we consider a hypothetical indoor environment, shown in Fig. 3.4. We assume that we have a very large space (such as an auto assembly line area) and that two mobile units [e.g., mobile robots or automatic guided vehicles (AGVs)] are communicating over a wireless link. Furthermore, we assume that the antennas are horizontally polarized so that the electromagnetic fields radiated toward the ceiling and the floor are the same as those along the LOS path, and the power of signals reflected from the walls is negligible with respect to the signal arriving along the LOS path or reflected from the ceiling or floor. In this example, the ceiling height is assumed to be 5 m and the antennas are 1.5 m above the floor. The received power in this case is given by

$$P_r = P_0 \left| \sum_{i=1}^{3} \frac{a_i e^{j\phi_i}}{d_i} \right|^2$$

in which the reflection coefficients are assumed to be $a_1 = +1$ (the LOS path) and $a_2 = a_3 = -0.7$ and where the path distances are related by

$$d_2 = 2 \times \sqrt{\frac{d_1^2}{4} + (1.5)^2}$$

and

$$d_3 = 2 \times \sqrt{\frac{d_1^2}{4} + (3.5)^2}$$

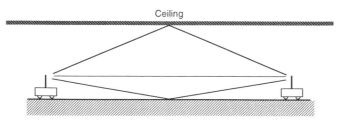

Figure 3.4 A hypothetical large indoor environment.

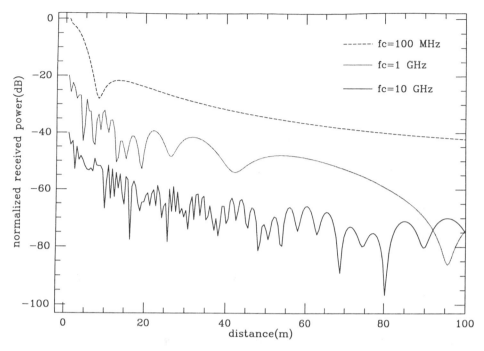

Figure 3.5 The normalized received power versus distance for distances between 1 and 100 m.

Figure 3.5 shows the normalized received power versus distance calculated for distances ranging from 1 to 100 m. The plot shows power in decibels, and distance on a linear scale.

It can be seen from Fig. 3.5 that while the average power decreases with distance, the power also fluctuates as much as 20–30 dB. The reason for this power fluctuation is that the relative phases of the arriving paths are changing as we move from one location to another. Therefore, there is a randomness in the summation of these paths. At certain locations all the paths are essentially in phase alignment, producing relatively large received power; and in some other locations the paths are nearly canceling each other, producing a drastic reduction of the received power. These fluctuations constitute distance-dependent fading observed by mobile users. The previous example did not exhibit these power fluctuations because the assumption was made there that the distance between the terminals is much greater than the height of the antenna. As a result, the phase difference between the two paths always remained small but nonzero, preventing complete cancellation of the two paths and a resulting deep fade in the received signal.

From Example 3.3 we can conclude that the multipath causes extensive power fluctuation in the received signal, producing deep fades at particular locations.

Example 3.4 (Two-Dimensional Ray Tracing Inside a Room). In this example we consider a more complicated situation in order to study the effects of multiorder reflected paths. Here the transmitter and receiver both have vertically polarized

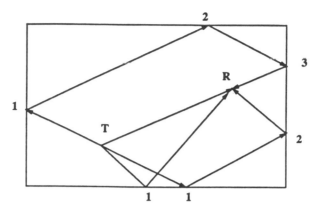

Figure 3.6 Reflections for ray tracing in a rectangular room.

omnidirectional antennas, and the antenna pattern prevents strong components from being reflected from the ceiling or the floor. As a result, only the paths reflected from the walls will contribute significantly to the received signal. To find all the paths under these circumstances, we need only trace the paths in two dimensions. Figure 3.6 depicts a two-dimensional map of the inside walls of a room with examples of LOS and first-, second-, and third-order reflected paths. The propagation paths between the transmitter and the receiver are determined by simple rules of geometric optics. The walls are assumed to be dark mirrors reflecting a portion of the signal energy and absorbing the remainder.

Figure 3.7 shows the received power in decibels versus distance, where the receiver is located at the center of a 50-m × 50-m room and the transmitter is moved along a straight line from 2 m to 20 m from the receiver. Figure 3.7a gives results obtained for the LOS path, and Figs. 3.7b, 3.7c, and 3.7d include all the first-, second-, and third-order reflections. The second- and third-order reflections contribute very little to the received power. In this example, tracing the first-order paths is adequate to show the distance-dependent fading caused by the multipath. The gradient of the power–distance relationship remains very nearly the same as for free space (1.90 vs. 2).

Figure 3.8 shows the received power in one-quarter of a 30-m × 30-m room for different locations of a receiver when the transmitter is fixed at the center of the room. The results shown in this figure assume a reflection coefficient of −0.7, with reflections of up to third order being considered.

With four walls there are 4 first-order, 12 second-order, and 48 third-order reflections. These components arrive with different amplitudes and phases; and invoking the central limit theorem, the summation of these signal components should approximate a zero-mean complex Gaussian random variable. The LOS path always exists and adds a nonzero mean to the complex Gaussian variable. The amplitude of the complex Gaussian variable in general obeys a Rician distribution, which reduces to a Rayleigh distribution when the mean is zero. Based on these considerations, it is typically assumed in the literature that the received amplitude

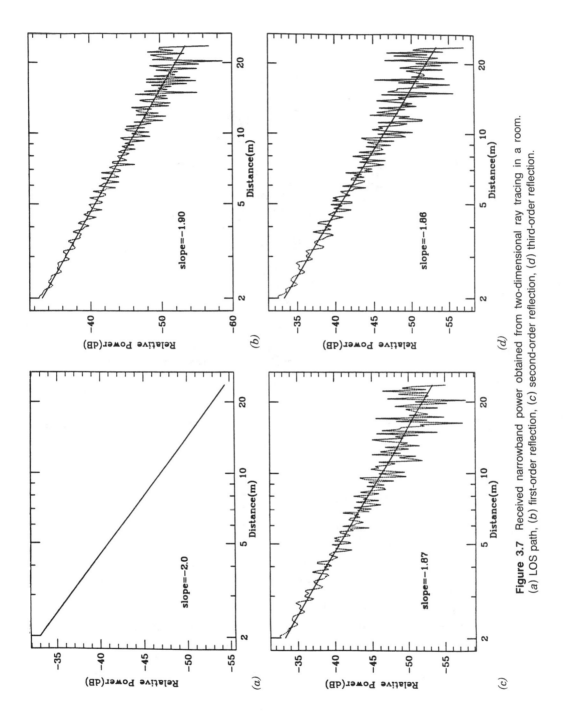

Figure 3.7 Received narrowband power obtained from two-dimensional ray tracing in a room. (a) LOS path, (b) first-order reflection, (c) second-order reflection, (d) third-order reflection.

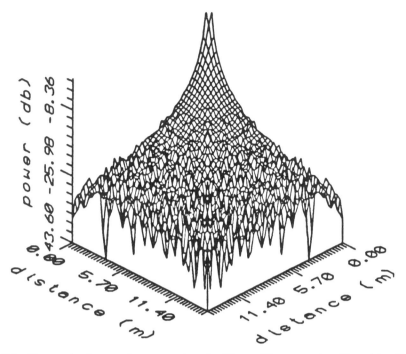

Figure 3.8   The received power in one-quarter of a 30-m × 30-m room for different locations of a receiver when the transmitter is fixed in the center of the room. Optical reflections of up to third order are considered with a reflection coefficient of 0.7.

in the absence of an LOS signal component is Rayleigh, while in LOS environments the received signal is assumed to be Rician.

3.3   LOCAL MOVEMENTS AND DOPPLER SHIFT

In the previous section we developed a simple description of radio wave propagation by analyzing the reception of a narrowband signal (a sine wave) transmitted over a multipath channel. In this section we examine the behavior of the signal in the frequency domain to show the effects of movements on the characteristics of the received signal.

It is well known from the fundamentals of physics that whenever a transmitter and a receiver are in relative motion, the received carrier frequency is shifted relative to the transmitted carrier frequency. This shifting of frequency is the Doppler effect of wave propagation between nonstationary points. We shall now show how the Doppler effect constitutes a source of signal fading in a multipath environment.

Figure 3.9 shows a typical example in which a fixed and a portable terminal are communicating over a radio link. The distance between the transmitter and the receiver is d_0 and the portable terminal is moving with speed v_m toward the fixed terminal. Let us assume that the portable terminal is transmitting a tone at

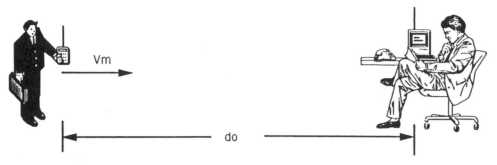

Figure 3.9 A typical example in which a fixed and a portable terminal are communicating over a radio link. The distance between the transmitter and the receiver is d_0, and the portable terminal is moving with speed v_m toward the fixed terminal 2.

frequency f_c and the amplitude of the received signal is A_r. If the transmitter is stationary, the received signal is represented by $r(t) = \text{Real}[A_r e^{j2\pi f_c(t-\tau_0)}]$, where $\tau_0 = d_0/c$ is the time required for the radio wave to propagate from the transmitter to the receiver with velocity c.

As the transmitter moves toward the receiver, the propagation time will change with time as

$$\tau(t) = \frac{d(t)}{c} = \frac{d_0 - v_m t}{c} = \tau_0 - \frac{v_m}{c}t$$

The received signal is then given by

$$r(t) = A_r e^{j2\pi f_c[t-\tau(t)]} = A_r e^{j[2\pi(f_c+f_d)t-\phi]}$$

where $\phi = 2\pi f_c \tau_0$ is a constant phase shift and

$$f_d = \frac{v_m}{c} f_c \tag{3.3.1}$$

is a shift in the frequency observed at the receiver, commonly referred to as the *Doppler frequency shift*. The Doppler frequency shift is either positive or negative depending on whether the transmitter is moving toward or away from the receiver.

Example 3.5. If Eq. (3.3.1) is applied to a typical indoor environment, a person walking at 3 miles/hr (1.34 m/sec) will cause a maximum Doppler shift of ± 4 Hz for a carrier frequency of 910 MHz. For a mobile user with a speed of 60 miles/hr the associated Doppler shift for the same frequency is ± 80 Hz.

In a realistic indoor environment, the received signal arrives from several reflected paths with different path distances, and the velocity of movement in the direction of each arriving path is generally different from that of another path. Thus a transmitted sinusoid, instead of being subjected to a simple Doppler shift, is received as a spectrum, which is referred to as the *Doppler spectrum*. This effect, which can be viewed as a spreading of the transmitted signal frequency, is referred to in a general way as the *Doppler spread* of the channel. Doppler spread also occurs with a fixed transmitter and receiver when a person or an object moves within the propagation path, producing time-variant multipath characteristics. In indoor and outdoor communication applications, as the terminals move about, or other objects move around the terminals, the received signal level fluctuates. The width of the Doppler spread in the frequency domain is closely related to the rate of fluctuations in the observed signal. The adaptation time of algorithms used in receivers (e.g., for automatic gain control or adaptive equalization) must be faster than the Doppler spread of the channel in order to accurately track the fluctuations in the received signal. Classical modeling of the Doppler spread will be explained in the next section, and the results of Doppler spread measurements in the indoor and outdoor radio channels will be presented in Chapter 4.

3.4 MULTIPATH FOR WIDEBAND SIGNALS

In the previous two sections we developed a simple description of radio wave propagation by analyzing the reception of a narrowband signal (a sine wave) transmitted over a fading multipath indoor radio channel. We also showed that the relative movement of the terminals while transmitting a sine wave causes Doppler shifts of the various multipath signal components, and this results in Doppler spread of the received signal. In this section we extend our analysis to the case of a wideband signal. If we regard a sinusoid as an ideal narrowband signal, then the analogous ideal signal for the wideband case is an impulse function, which has infinite bandwidth. We analyze some simple cases of transmission of an impulse on an indoor radio channel to provide some insight into the effect of multipath on wideband communications.

Given the same multipath situation that we examined earlier, a transmitted impulse $\delta(t)$ will arrive at the receiver as the sum of several impulses with different magnitudes and phases. The composite impulse response for given locations of the transmitter and receiver is then represented by

$$h(\tau, t) = A_0 \sum_{i=1}^{L} \frac{a_i}{d_i} e^{j\phi_i} \delta(t - \tau_i) \qquad (3.4.1)$$

where the τ_i and ϕ_i are determined in the same way as they were for narrowband signaling, and $A_0 = \sqrt{P_0}$. If we define $\beta_i = A_0 a_i / d_i$, we have

$$h(\tau, t) = \sum_{i=1}^{L} \beta_i e^{j\phi_i} \delta(t - \tau_i) \qquad (3.4.2)$$

where β_i and ϕ_i represent the amplitude and phase of the ith path arriving at delay τ_i. Equation (3.4.2) is widely used for statistical modeling of both indoor and outdoor radio propagation. Figure 3.10 shows a block diagram that is helpful for computer simulation of the wideband characteristics of the channel. For ideal

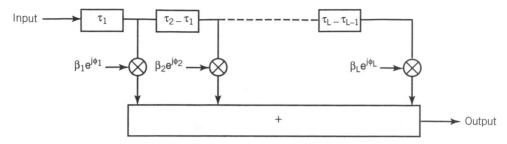

Figure 3.10 Block diagram for the discrete delay channel model.

wideband communication, the paths are isolated and independent of one another, and therefore the phase differences between arriving paths do not change the amplitude characteristics of the channel. In other words, impulses arriving at different times do not interact with each other. The received power in this case is given by

$$P_r = P_0 \sum_{i=1}^{L} \left| \frac{a_i}{d_i} \right|^2 = \sum_{i=1}^{L} |\beta_i|^2 \qquad (3.4.3)$$

Here, the received signal power is the sum of squares of all path amplitudes. In the case of narrowband signaling, Eq. (3.2.2), the amplitudes were added vectorially and the overall power was the square of the resulting vector magnitude. As a result, the normalized received power of a narrowband signal is less than or equal to that of a wideband signal. In simple terms, for the wideband transmitted signals, the received paths are in effect isolated by the correlation properties of the signal, and the powers from different paths add algebraically. With narrowband signaling the paths are added together vectorially in accordance with their individual phases and this interaction among the paths reduces the normalized received power relative to the wideband case.

In practice, the bandwidth of the channel is finite and realistic impulsive signals are represented by pulses of very short but nonzero duration. Figure 3.11*a* shows a sample of a ray-traced impulse response in a typical square room discussed in our examples, with 2-nsec pulses replacing the ideal impulses. We see that the multipath channel has spread the transmitted signal in the time domain, just as multipath with motion had spread the transmitted sinusoid in the frequency domain in the narrowband case examined earlier. Figures 3.11*b* and 3.11*c* represent the response when the transmitted impulse function is replaced by a narrow pulse of width of 5 nsec and 10 nsec, respectively. Figure 3.12 represents a sample wideband indoor radio channel measured in both time and frequency domains. The resolution in the time domain is 5 nsec, which accounts for a transmission bandwidth of 200 MHz. Note that the frequency response varies by as much as 40 dB from one frequency to another.

Example 3.6. The environment of this example is the same as that of Example 3.4 for narrowband signaling. Figure 3.13 shows received wideband power versus distance from 1 to 25 m for a 50-m × 50-m room for different numbers of

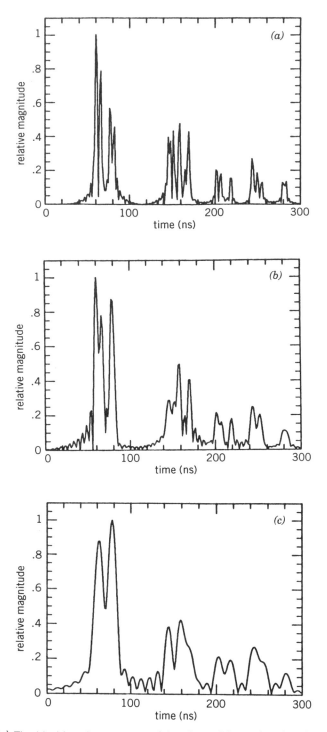

Figure 3.11 (*a*) The ideal impulse response of the channel for a given location of the transmitter and the receiver. (*b, c*) The response if a narrow pulse with width of 5 or 10 nsec, respectively, is used.

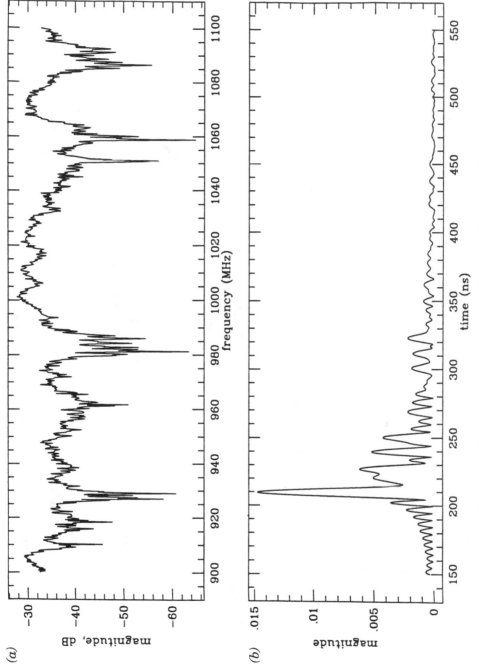

Figure 3.12 Sampled measured time (*a*) and frequency (*b*) response in an indoor area.

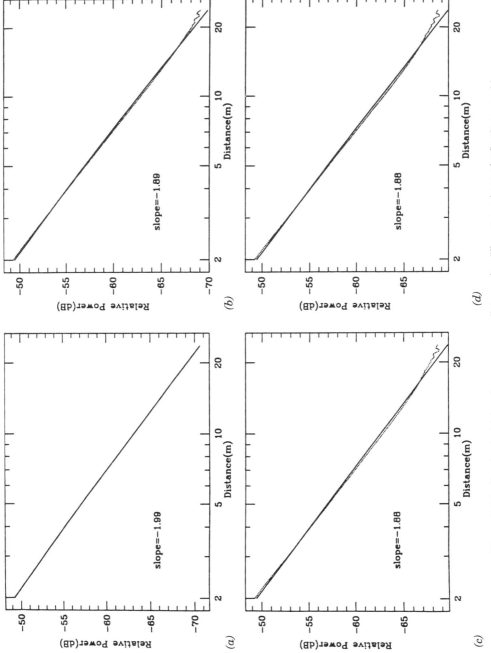

Figure 3.13 Received wideband power versus distance for different numbers of reflections used in the two-dimensional ray-tracing algorithm. (a) LOS path, (b) first-order reflection, (c) second-order reflection, (d) third-order reflection.

54

reflections, together with the best-fit line to the calculated signal power. This figure should be compared with Fig. 3.7 for narrowband signaling. In general, the best-fit line and the gradient of the distance–power relationship are nearly the same for wideband as for narrowband signals. If none of the paths with reflections are considered, the LOS path provides the same power for both cases as well. If we include the reflected paths, fluctuations in power for the narrowband signal are significantly more than those for the wideband signal. This characteristic is due to the fact that in wideband signaling the phase of the received signal does not play a role in the calculation of the power, while the received power in narrowband signaling is the result of phasor summation of several vectors, which is very sensitive to the phases of the arriving paths.

3.4.1 Multipath Delay Spread

In order to be able to assess the performance capabilities of various wireless systems, we want to have a convenient numerical measure of the time dispersion, or *multipath delay spread* of the channel. The simplest measure of multipath delay spread is the overall span of path delays (i.e., earliest arrival to latest arrival) which is sometimes referred to as the *excess delay spread*. However, this is not necessarily the best indicator of how any given system would perform on the channel. This is because different channels with the same excess delay spread can exhibit very different profiles of signal intensity over the delay span, and different intensity-delay profiles will have greater or lesser impact on the performance of any given system.

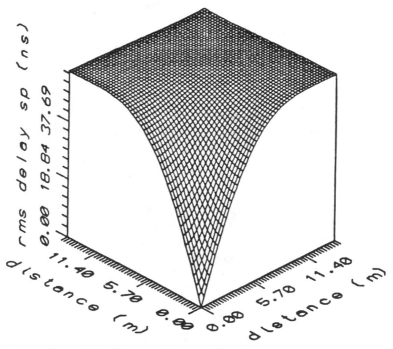

Figure 3.14 The rms delay spread in a 30-m × 30-m room.

Thus a better measure of delay spread is the root mean square (*rms*) *delay spread*, τ_{rms}, which is the second central moment of the channel impulse response. It is given mathematically by

$$\tau_{\mathrm{rms}} = \sqrt{\overline{\tau^2} - (\bar{\tau})^2} \tag{3.4.4}$$

where, given L propagation paths,

$$\overline{\tau^n} \equiv \frac{\displaystyle\sum_{i=1}^{L} \tau_i^n |\beta_i|^2}{\displaystyle\sum_{i=1}^{L} |\beta_i|^2}, \qquad n = 1, 2$$

Figure 3.14 shows the rms multipath spread as a function of location in one-quarter of a 30-m × 30-m room. One can see from the figure that a maximum rms multipath delay spread of less than 60 nsec is observed in this case.

3.5   THE CLASSICAL UNCORRELATED SCATTERING MODEL

In the preceding discussion we showed how changes in the relative phases among multiple reflected signal paths cause fluctuations in the power of the received composite signal. On indoor and urban radio channels these changes are caused either by movements of the transmitter or the receiver or by the movement of people or vehicles near the transmitter or the receiver. Without such movements, given fixed locations of the transmitter and the receiver, the channel impulse response remains constant. As the location of the transmitter or the receiver is changed or some object moves close to the transmitter or the receiver, the impulse response of the channel will change. The rate of change depends on the speed of the movements. The classical model, which we describe next, provides us with a certain mathematical structure that relates to one another the key parameters that we described earlier in this chapter.

The classical method of channel modeling was developed to describe signal-transmission over a variety of radio channels having randomly time-varying impulse responses. In communication over such channels, even when transmitter and receiver are stationary, signals are subject to time dispersion and random fluctuations, caused by the constantly changing characteristics of the transmission media. Common examples are (a) long-distance ionospheric communications in the 3- to 30-MHz high-frequency (HF) band and (b) beyond-the-horizon tropospheric scatter communications in the 300- to 3000-MHz ultra-high-frequency (UHF) and 3000- to 30,000-MHz super-high-frequency (SHF) bands.

In the case of HF communications, long-distance propagation is achieved by refraction of the transmitted signal at various layers of the ionosphere. The heights, thicknesses, and ion densities of the ionospheric layers, together with the constant random motion of the ions within each layer, cause time dispersion and random amplitude and phase fluctuations in the signal as it is bent back to earth. In the case of the tropospheric scatter (troposcatter) channel, it is more accurate to describe the received signal as consisting of a continuum of multipath components created by the physical characteristics of the troposphere, such as (a) meteorological effects and (b) the constantly changing interaction among the multipath components producing random fading in the received signal.

In other frequency bands, the details of the propagation mechanisms might be different, but in each case the overall effect is some combination of time dispersion and apparently random amplitude and phase fluctuations in the received signal, a set of characteristics commonly termed *multipath fading*. In order to assess the effectiveness of some signal design and the corresponding performance of a receiving system operating on a given multipath fading channel, it is important to be able to mathematically characterize the behavior of the channel. Because, to the observer, the variations in the received signal are not predictable, but apparently random, the variations are best described in statistical terms. In particular we want to characterize a multipath fading channel in terms of *correlation functions* and *power spectral density functions*.

3.5.1 Correlation Properties in the Delay Variable

We begin by assuming that the effects of the transmission medium are sufficiently random, and the number of multipath signal components sufficiently large, that we can invoke the central limit theorem. We can then assume that the overall impulse response of the channel is accurately represented by a complex Gaussian process $h(\tau, t)$, where inclusion of the variable t in the argument indicates that in general the channel impulse response is time-varying. The channel impulse response for the indoor and outdoor applications defined in Eq. (3.4.2) was a discrete function of the delay variable τ, while here the impulse response is a continuous function of τ. For a transmitted waveform with complex envelope $p(t)$, the complex envelope of the received signal in the case of the continuous delay function is given by

$$r(t) = \int_{-\infty}^{\infty} h(\tau, t) p(t - \tau) \, d\tau$$

If we were to use the discrete channel model of Eq. (3.4.2), the received signal would be

$$r(t) = \sum_{i=1}^{L} \beta_i e^{j\phi_i} p(t - \tau_i) \tag{3.5.1}$$

The block diagram of Fig. 3.10, adapted for the continuous delay channel, is shown in Fig. 3.15. The transmitted signal is passed through a tapped delay line with delay values of $d\tau$ and with tap gains of $h(\tau, t) \, d\tau$.

As a way of modeling such a channel, Bello [Bel63a] suggested the assumption of *wide-sense stationary uncorrelated scattering* (WSSUS). This assumption leads to several interesting and useful conclusions. The physical meaning of the assumption, which is valid for most radio transmission channels, is that the signal variations on paths arriving at different delays are uncorrelated and the correlation properties of the channel are stationary; that is, they do not change with time. In mathematical terms, the assumption results in the following simplification. The

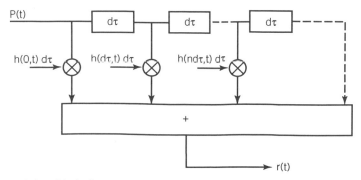

Figure 3.15   Block diagram for the continuous delay channel impulse response.

autocorrelation of the observed impulse response at two different delays and two different times is given by

$$R_{hh}(\tau_1, \tau_2; t_1, t_2) = E\{h(\tau_1; t_1)h(\tau_2; t_2)\} = R_{hh}(\tau_1; \Delta t)\delta(\tau_1 - \tau_2)$$

Given the assumption of *uncorrelated scattering*, the only nonzero value of the correlation is observed when the delays are the same; given stationarity, the correlation values depend only on the difference in time of occurrence of the two impulse responses, not the time of occurrence of each event. For $\Delta t = 0$, this function is represented by $Q(\tau)$ and is referred to as the *delay power spectrum* of the channel:

$$Q(\tau) = R_{hh}(\tau; 0) \tag{3.5.2}$$

The delay power spectrum represents the received power as a function of time delay, given that an impulse function is transmitted. It represents the received power at different delays averaged over time. The overall range of values of τ for which $R_{hh}(\tau; 0)$ has significant nonzero value is referred to as the *excess delay spread* or simply the *delay spread* of the channel. The second central moment of this function is referred to as the *rms delay spread* and is defined as

$$\tau_{\text{rms}}^2 = \frac{\int_{-\infty}^{\infty} (\tau - \bar{\tau})^2 R_{hh}(\tau)\, d\tau}{\int_{-\infty}^{\infty} R_{hh}(\tau)\, d\tau} \tag{3.5.3}$$

where

$$\bar{\tau} = \int_{-\infty}^{\infty} \tau R_{hh}(\tau)\, d\tau$$

The rms delay spread represents the effective value of the time dispersion of a transmitted signal, as caused by the multipath in the channel. For reliable digital communication over the channel, the time duration of each transmitted symbol should be much longer than this value in order to minimize the distortion of the symbol shape observed at the receiver. Because the duration of a transmitted symbol is inversely proportional to the data rate, the inverse of the rms delay spread can be taken as a measure of the data rate limitations of a fading multipath channel. If we refer to the inverse of the rms delay spread as the *coherence bandwidth* of the channel, we can state that the rate of transmitted symbols should be much smaller than the coherence bandwidth of the channel in order to minimize the distortion of the transmitted pulse shapes. (*Note:* Some authors choose to define the coherence bandwidth as the inverse of the overall delay spread of the channel [Pro89].) There are systems that operate reliably over radio channels with symbol durations near the rms delay spread (signal bandwidth close to the coherence bandwidth), but these systems require the use of *adaptive equalization* or other anti-multipath techniques to compensate for the distortions introduced by multipath and fading. We say more about adaptive equalization in Chapter 8.

Figure 3.16 shows an example of delay power spectra derived analytically for a troposcatter channel [Bel69]. Figure 3.17 shows the experimentally measured values of this function for a real troposcatter link [She75]. Accurate measurement of the delay power spectrum is possible only when the channel impulse response varies slowly with time. For such a slowly varying channel the correlation properties of the channel remain the same during the measurement time Δt and we have

$$R_{hh}(\tau; \Delta t) \simeq R_{hh}(\tau; 0) = Q(\tau) \qquad (3.5.4)$$

When modeling indoor and outdoor radio channels with the discrete channel-impulse response of Eq. (3.4.2), it is often assumed that the channel does not

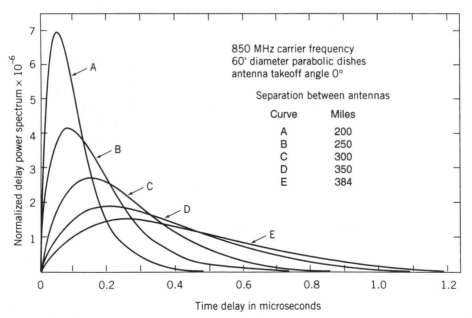

Figure 3.16 Analytically predicted delay power spectra for a troposcatter link. (From [Bel69] © IEEE.)

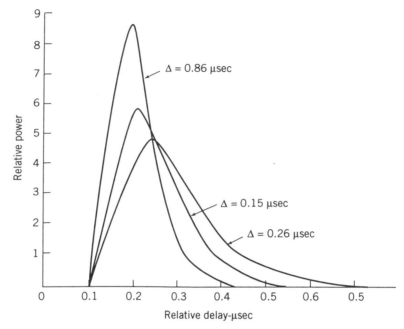

Figure 3.17   Experimentally measured delay power spectra for a troposcatter link. (From [She75] © IEEE.)

change with time, in which case the average of the channel impulse response is the same as the impulse response itself. The delay power spectrum in this case is simply the square of the magnitude of the channel impulse response. The rms multipath delay spread is then given by Eq. (3.4.4).

3.5.2   Multipath Delay Characteristics in the Frequency Domain

Using the WSSUS assumption, we can derive several useful and mathematically interesting properties of the channel correlation function in the frequency domain. Given a channel with impulse response $h(\tau; t)$, the frequency response is defined as the Fourier transform of this function on the argument τ, which is written as

$$H(f;t) = \int_{-\infty}^{\infty} h(\tau;t)e^{-j\omega\tau}\,d\tau$$

For ideal measurement of the impulse response, an impulse function is transmitted through the channel, whereas for ideal measurement of the frequency response, sinusoids at different frequencies should be transmitted. Now, given the assumptions of the WSSUS model, the channel impulse response $h(\tau; t)$ is a wide-sense stationary zero-mean Gaussian process in the time variable t. Therefore the frequency response $H(f; t)$, being obtained as a linear operation on $h(\tau; t)$, is also a wide-sense stationary zero-mean Gaussian process in t. Figure 3.12 shows channel time- and frequency-domain responses measured on a channel in a typical indoor area. The time-domain response shows the arrival of the multiple paths, while the frequency response exhibits amplitude variations from one frequency to another. The cause of these variations is the multipath structure of the channel, which causes constructive interference and signal enhancement at certain frequencies but causes destructive interference and deep fades at other frequencies. This channel characteristic is referred to as *frequency-selective multipath fading*.

If we were to show additional frequency responses measured at various points in time, we would see the positions of the highs and lows in the frequency response vary randomly from one measurement to another. To characterize these variations statistically, we can compute the correlation between values of the frequency response taken at various frequency spacings. The correlation in the frequency domain is defined as

$$R_{Hh}(f_1, f_2; \Delta t) = E\{H^*(f_1; t)H(f_2; t + \Delta t)\}$$

$$= \int_{-\infty}^{\infty}\int_{-\infty}^{\infty} E\{h^*(\tau_1; t)h(\tau_2; t + \Delta t)\}e^{j2\pi(f_1\tau_1 - f_2\tau_2)} \, d\tau_1 \, d\tau_2$$

$$= \int_{-\infty}^{\infty} R_{hh}(\tau_1; \Delta t)e^{j2\pi \Delta f \tau_1} \, d\tau_1 = R_{Hh}(\Delta f; \Delta t) \qquad (3.5.5)$$

where $\Delta f = f_1 - f_2$ and the channel is assumed to be WSSUS. The new function $R_{Hh}(\Delta f; \Delta t)$ is referred to as the *spaced-time, spaced-frequency correlation function* of the channel. As shown above, this function is the Fourier transform of the spaced-time correlation function $R_{hh}(\tau; \Delta t)$ on the delay variable. Equation (3.5.5) shows that this process is wide-sense stationary over both time and frequency variables.

For a slowly time-varying channel, the value of $R_{Hh}(\Delta f; \Delta t)$ calculated with observation times separated by Δt is the same as that found with no time separation, and thus we have

$$R_{Hh}(\Delta f; \Delta t) \simeq R_{Hh}(\Delta f; 0) = R_{Hh}(\Delta f)$$

which can be measured by transmitting two frequencies Δf apart and determining the correlation between the received signals. The inverse Fourier transform of this function is the delay power spectrum $Q(\tau)$.

3.5.3 Correlation Properties in the Time Variable

In the preceding paragraphs we discussed the correlation properties in the delay variable of the channel impulse response. We introduced the delay power spectrum and its Fourier transform, which is the spaced-frequency spaced-time autocorrelation function of the channel, and we showed how these functions are related to channel measurements in the time and frequency domains. We also discussed the special case of a slowly time-varying channel, where the time variable has no effect on the derivation of the correlation functions and power spectral density functions or, as a practical matter, on the measurements of these channel characteristics. In the following discussion we further analyze the WSSUS channel model with attention to fluctuations in time.

We first take the Fourier transform of the spaced-time spaced-frequency correlation function on the time variable, which yields

$$R_{HH}(\Delta f; \lambda) = \int_{-\infty}^{\infty} R_{Hh}(\Delta f; \Delta t) e^{-j2\pi\lambda \Delta t} d(\Delta t)$$

Now, for $\Delta f = 0$ we have $R_{Hh}(0; \Delta t)$ under the integral and the transform gives

$$D(\lambda) = R_{HH}(0; \lambda) \tag{3.5.6}$$

which is called the *Doppler power spectrum* of the channel. The Doppler power spectrum represents the strength of the Doppler shift at different frequencies caused by movements of the terminals or the objects close to them. To measure $R_{Hh}(0; \Delta t)$ we can transmit a single sinusoid ($\Delta f = 0$) and determine the autocorrelation function of the received signal. The Doppler power spectrum is the Fourier transform of this autocorrelation function. On the other hand, we know that the Fourier transform of the autocorrelation function of a time series is the magnitude-squared of the Fourier transform of the original time series. Therefore, we may simply transmit a sinusoid and use Fourier analysis to generate the power spectrum of the received signal amplitude; this power spectrum is the Doppler power spectrum of the channel. The width of the Doppler power spectrum is referred to as the *Doppler spread* of the channel and provides a measure of the *fading rate* of the channel. We might regard the Doppler power spectrum as the frequency-domain dual of the delay power spectrum, which we discussed near the beginning of Section 3.5.1. In a manner similar to the treatment of delay spread, the second central moment of the Doppler spread function, the *rms Doppler spread*, is sometimes used as a measure of the fading rate in the channel. However, in the design of communication receivers the maximum rate of variations of the channel is important, and therefore the more commonly used parameter is the overall Doppler spread rather than rms Doppler spread.

The reciprocal of the Doppler spread is called the *coherence time* of the channel, which is a measure of the time interval over which a transmitted symbol will be relatively undisturbed by channel fluctuations. For slowly time-varying

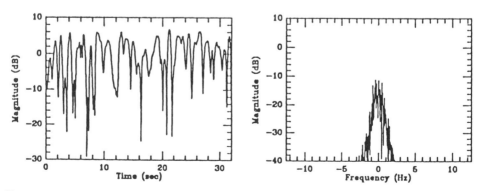

Figure 3.18 A sample of measured amplitude fluctuation on an indoor radio channel and its Fourier transform. The maximum Doppler spread in this sample is around 4 Hz.

channels, the long coherence time is beneficial to accurate measurement of the channel characteristics, as one can use long observation times during which the measured channel response does not change significantly. Figure 3.18 [How90a] shows an example of amplitude fluctuations measured on an indoor radio channel, together with the calculated Fourier transform. The maximum Doppler spread in this example is around 4 Hz.

3.5.4 The Scattering Function

The inverse Fourier transform of $R_{HH}(\Delta f, \lambda)$ on the Δf variable, which is the Fourier transform of $R_{hh}(\tau, \Delta t)$, taken over Δt, is called the *scattering function*:

$$S(\tau, \lambda) = R_{hH}(\tau; \lambda) \tag{3.5.7}$$

It represents the rate of variations of the channel at different delays. To measure the scattering function the received signal in individual taps of a tapped delay line is analyzed in the frequency domain. In practice it is usually assumed that the time and frequency components of the scattering function are independent. With this assumption the scattering function is decomposed into the delay and Doppler power spectra:

$$S(\tau, \lambda) = Q(\tau) \times D(\lambda) \tag{3.5.8}$$

Figure 3.19 [Par89] shows a three-dimensional description of this function measured in an urban radio environment. Figure 3.20 summarizes all the correlation functions we have discussed here and shows the relationships among them.

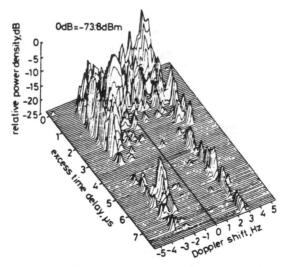

Figure 3.19 The measured scattering function on a troposcatter channel. (From [Par 89], © Blackie, with permission.)

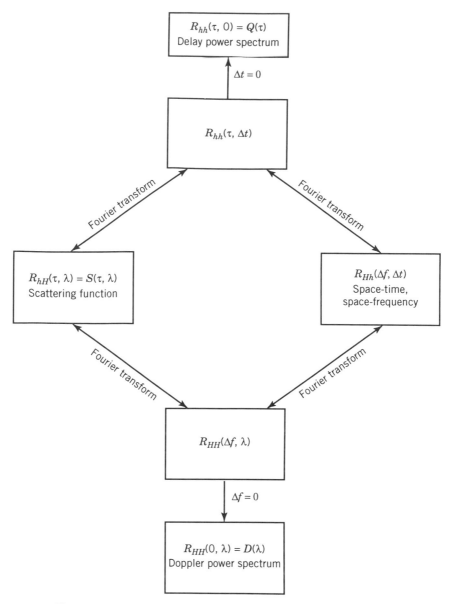

Figure 3.20   Summary of the correlation functions in classical modeling.

3.6   INDOOR AND URBAN RADIO PROPAGATION MODELING

The first step in constructing a channel model is to classify the physical characteristics of the channel. Then, the models for narrowband and wideband signaling are developed for different environments. In narrowband modeling we are interested only in the measurement and modeling of the received power. In each physical environment we relate the path loss to the distance between the transmitter and the receiver. In wideband applications, we are interested in modeling the multipath structure or frequency-selective behavior of the channel in different physical environments. The modeling is based either on the statistics of the measured channel profiles or on the direct solution of radio propagation equations. The approaches to modeling indoor and outdoor radio propagation environments are quite different, as we discuss in the following subsections.

3.6.1 Physical Operating Environments

Wireless information networks operate in a variety of different environments requiring attention to various aspects of radio propagation. WLANs and wireless PBX systems are designed for use in offices and commercial buildings. PCS systems are to operate in and around buildings in residential as well as office areas. Cellular telephone systems are designed for use in all outdoor areas and particularly in moving vehicles. Mobile data services are intended to cover metropolitan areas, and most applications require in-building penetration. Table 3.1 shows a classification of the physical operating environments for various wireless information networks. This table provides a guideline for development of channel models needed for different applications. Generally, environments are either indoor or outdoor areas. The indoor areas include residential, office, and commercial buildings. The outdoor areas are categorized as urban high-rise, urban/suburban low-rise, and residential areas. As we discussed earlier in this chapter, the characteristics of the radio channel change when terminals move or when other objects are moving in the vicinity the transmitter or the receiver. The rapidity of change is proportional to the speed of the movements. Basically, there are two ranges of speed of primary interest for wireless networks: pedestrian speeds of around 3 miles/hr (1.34 m/sec) and vehicular speeds up to about 55 miles/hr (24.6 m/sec). For networks in indoor areas we are concerned only with pedestrian movements, whereas in outdoor environments we must deal with either pedestrian or vehicular movements.

Indoor residential areas are typified by wooden-frame single-family houses with one or two stories. The interior walls are typically covered with a thin layer of

TABLE 3.1 Physical Operating Environments for Wireless Information Networks

General Environment	Specific Environment	Doppler Shift Range (Hz)	
		Minimum	Maximum
Indoor	Residential	2	10
	Office	2	10
	Commercial	2	10
Outdoor			
Pedestrian	Urban high-rise	2	10
	Urban / suburban low-rise	2	10
	Residential	2	10
Vehicular	Urban high-rise	5	150
	Urban / suburban low-rise	5	200
	Residential	5	100

plaster inside cardboard (gypsum board). The exterior frame is filled with insulation and covered by plywood and then wooden siding or brick. With a typical residence, having many windows, there is significant radio penetration from outside the structure. Indoor office areas typically consist of large spaces partitioned into cubicles. In each cubicle there are several metallic objects such as bookshelves and desks. The frame of the building is usually constructed with metallic studs and sometimes concrete frames, while the insulation and the exterior walls can be similar to residential construction. The ceilings and floors are usually heavier than in residential construction and they include significant amounts of metal and concrete, presenting a stronger barrier to radio wave penetration from one floor to another. The indoor commercial areas include large open spaces such as manufacturing floors, shopping malls, storage areas, and transportation stations. These areas usually have high ceilings, thick layers of concrete, and heavy metallic framing.

The outdoor urban high-rise area is typified by the downtown area in any large city, often referred to as the "urban canyon." Thick layers of concrete and heavy metallic frames in the exteriors of buildings restrict radio wave propagation into and through the buildings. The rooftops are high, and therefore signal propagation is aided very little by diffraction. The radio waves are guided through the streets by the mechanism of reflection, with a significant power loss. The large number of moving vehicles causes continual changes in the channel characteristics. The urban/suburban low-rise areas typically include wide streets bordered by low-rise buildings. Here, propagation is aided by diffraction from the roofs of buildings. Vehicle speeds are typically much higher than in the downtown high-rise areas. The outdoor residential areas are the streets of the indoor residential areas. The roads are usually two lanes wide, cars are parked alongside the streets, and the volume of vehicular traffic is usually low. In this kind of environment, the trees along the street can also influence the radio propagation characteristics.

3.6.2 Methods for Modeling

There now exists an extensive body of literature on radio propagation prediction and modeling, beginning with papers published as early as the mid-1930s. Many researchers have developed a variety of experimentally or theoretically based models to predict radio propagation in various frequency bands, and for various physical characteristics of the transmission path. A number of prediction models have been developed that take into account antenna height, path length, earth curvature, terrain irregularity, foliage, urban streets and buildings, tunnels, and so on. Widely used propagation models include those of Bullington [Bul47, Bul77], Longley and Rice [Lon68], Okumura [Oku68], and Lee [Lee82]. Some of the key papers in this field are reprinted in [Bod84], where many other pertinent references can be found as well. Also, in 1988 the IEEE Vehicular Technology Society published a special journal issue devoted to radio propagation prediction and modeling for the 806- to 947-MHz region of the radio spectrum [IEE88a]. That special issue includes papers reviewing and comparing several well-known propagation models. In addition, a book by Jakes [Jak74] provides an extensive treatment of radio propagation for mobile radio systems, with emphasis on the frequency range from 450 MHz up to 10 or 20 GHz. The Jakes text includes the description of a simulation model for signal fading in the mobile environment, a model which is used extensively in the mobile communications industry. Narrowband models for mobile radio propagation are also treatcd in [Lee82, 89]. Broadly speaking, the modeling work we have cited in this paragraph essentially addresses the narrowband communication case. That is, the associated models provide predictions or simulations of received signal strength but do not provide detailed information regarding the time dispersion imposed on the signal by multipath effects. Work by Turin [Tur72], which we shall next refer to, does address time-domain characteristics. We shall also describe recent work on channel modeling for PCS applications in Chapters 4 and 6.

The most commonly used statistical models for indoor radio propagation are the time-domain statistical models. These models, originally suggested by Turin [Tur72] for modeling urban radio channels, assume that the channel impulse response is in the form of Eq. (3.4.2); and based on measured data they provide statistics for the amplitudes, delays, and phases of the arriving paths. Details of the analysis for urban radio channels are available in [Suz77], [Has79], and [Par89]. Various methods of regenerating the time-domain response of indoor radio measurements are described in [Sal87b], [Gan89], [Gan91a], [Gan91b], [Gan92], [Gan93], [Rap91b], [Yeg91], [Has93a], and [Has93b]. Another approach to reproducing the measured channel responses is to use the frequency response of the channel for statistical modeling. The frequency response of the channel shown in Fig. 3.12 is assumed to be an autoregressive process. The poles of the process at different locations are calculated from the sample measurement of the channel frequency response in different locations. The statistics of the locations of the poles over a set of measurements represent the model. The poles are then used in a filter driven by complex Gaussian noise. The output of the filter is used as the frequency response, and its inverse Fourier transform is used as the impulse response of the channel [Pah90b, How90b, How92, How91, Mor92]. The relationship between the arriving paths and locations of the poles is more complex than in the time-domain approach. However, evaluation of the parameters for the autoregressive model is simpler and it requires fewer statistical parameters to represent the channel.

Statistical models cannot relate radio propagation characteristics to the exact locations of the transmitter and the receiver; rather, they provide only a collection of possible channel profiles. Deterministic radio propagation modeling relates the radio propagation to the physical layout of the building by solving the radio propagation equations. The statistical models are based on actual measurements in specific buildings. The deterministic models are based on a simplified layout of a building, omitting the details of furniture and the exact properties of the structural materials. The deterministic models are much more demanding of computational power, as compared with the statistical models. A relatively simple approximate solution to indoor radio propagation is obtained by the ray-tracing algorithm [Des72, Gla89]. In this method, walls, ceilings, and floors are assumed to be dark mirrors. The paths between the transmitter and the receiver are determined through transmission, reflection, and diffraction mechanisms. Computational time with the ray-tracing algorithm grows exponentially with the complexity of the building. For applications in which directions of the arriving paths are important, such as analysis of systems using sectored antennas, ray-tracing provides a more reasonable model for the channel. Several groups of investigators are developing ray-tracing techniques for indoor radio propagation, as reported in [Law92], [Rus91], [McK91], [Hol92a], [Hol92b], [Hol92c], [Hon92], [Rap92], [Ho94], [Yan93a, b], and [Bro93]. Using numerical analysis methods, one can also carry out direct solutions of Maxwell's equations. In particular, the *finite-difference time-domain* (FDTD) method can be used to solve the equations. The advantage of the FDTD method is that it simultaneously provides a complete solution for all points in the map. This is very important when signal coverage throughout a given area is to be determined. The FDTD method solves the equations over the area with a grid on the order of magnitude of the wavelength. As a result, memory requirements increase with the increase in the frequency of operation and the size of the area. Some results for the indoor radio propagation using the FDTD method are available in [Yan93b].

In the next three chapters we delve further into radio propagation for wireless networks. In Chapter 4 we describe results of measurement and modeling for narrowband signals. In Chapter 5 we analyze wideband measurement systems and the results of measurements in indoor and outdoor areas. In Chapter 6 we provide the details of statistical and building-specific methods of modeling and simulating radio propagation.

QUESTIONS

(a)   Name two major classes of wireless applications in which a channel model is needed either for design or for performance evaluation of the system.

(b)   Describe the power loss in decibels per octave of increase in distance, as a function of the distance–power gradient α.

(c)   What causes signal fading?

(d)   Why do signal arrivals from different paths causes a narrowband signal to fade?

(e)   Why is the multipath spread greater in outdoor areas than in indoor areas?

(f)   Explain why the Doppler spread is greater on a mobile radio channel than it is in indoor areas.

(g)   Why are the power fluctuations for wideband signals smaller than for narrowband signals?

(h)   Is the distance–power gradient for narrowband signals the same as that for wideband signals? Explain.

(i)   List the basic assumptions underlying the WSSUS channel model.

(j)   What are the principal methods used for indoor and outdoor radio propagation modeling?

(k)   What parameter is commonly used to represent the multipath delay spread on a radio channel?

(l)   What parameter is commonly used to represent the Doppler phenomenon?

PROBLEMS

Problem 1. Give an equation for exact calculation of P_r in Example 3.2.

(a) Assuming $h_1 = 100$ m, $h_2 = 3$ m, $P_0 = 0$ dBm, and $f_c = 800$ MHz, sketch and label the exact value of P_r in decibels versus d in logarithmic form (similar to Fig. 3.7) for $10 < d < 100$ m. Use 100 points for your plot, and use MathCAD or MatLAB for calculation and drawing.

(b) Compare the results of part (a) with the approximated results of Example 3.2.

(c) Repeat (a) and (b) for $1000 < d < 10,000$ m.

Problem 2. Give an equation for calculation of $\Delta\tau$, the delay between the arrival of the two paths in the Example 3.2.

(a) Assuming $h_1 = 100$ m, $h_2 = 3$ m, $P_0 = 0$ dBm, and $f_c = 800$ MHz, sketch and label the $\Delta\tau$ versus d in logarithmic scale for $10 < d < 100$ m. Use 100 points for your plot, and use MathCAD or MatLAB for calculations and drawings.

(b) Repeat (a) for $1000 < d < 10,000$ m.

(c) Repeat (a) and (b) with τ_{rms} replacing $\Delta\tau$.

(d) If the maximum data rate of a modem, R, is related to the rms multipath spread of the channel by $R \simeq 0.1/\tau_{rms}$, sketch the maximum data rate versus distance for $10 < d < 100$ m and $1000 < d < 10,000$ m.

Problem 3. The circular scattering model for a mobile radio channel assumes that the paths from a mobile transmitter are scattered from a uniform circle around the transmitter before they arrive at the base station antenna. Figure P3.1 shows a typical situation and relevant parameters used in the model. Associated with each path in this model we have several parameters: the distance between the transmitter and the receiver d, the radius of the scattering circle $R \ll d$, the angle of arrival of a path θ, and the rate of variation of the distance (velocity) in the direction of arrival of a path v_θ.

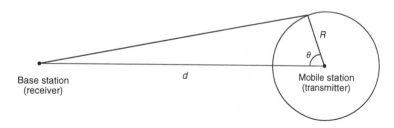

Figure P3.1

(a) Show that if the transmitter moves toward the receiver with a velocity of v m/sec the velocity in the direction of the path with angle θ is given by

$$v_\theta = v \cos \theta \quad \text{m/sec}$$

(b) Give $f(\theta)$, the Doppler shift of the path with the angle of arrival of θ, in terms of the velocity of the vehicle, the frequency of operation, and the angle of arrival. Sketch $f(\theta)$ as a function of θ for $0 < \theta < 2\pi$. What are the angles that provide the minimum and maximum Doppler shifts?

(c) Because $d \ll R$, all the path lengths are approximately R and the path loss associated with all paths is the same. As a result, the amplitudes of all arriving paths are the same. However, because the rate of variation of the distance for different paths is not the same, the Doppler shift associated with each path is different. The difference in the Doppler shift for different paths causes a phase difference among the arriving paths, and the phasor representing the complex envelope of each received path is given by

$$x_\theta(t) = A e^{j2\pi f(\theta)t}$$

where A is the fixed path amplitude. The complex envelope of the received signal from all paths is then given by

$$r(t) = A \int_0^{2\pi} x_\theta(t)\, dt$$

Plot the envelope of the received signal (magnitude of the received phasor) in decibels as a function of time, and use 100 samples at sampling intervals of 0.1 msec. Assume $v = 80$ km/hr, $f = 800$ MHz, and $A = 1$.

(d) Sketch the probability density function of the linear magnitude of the 100 amplitude samples of the received signal. Name a distribution function that fits the observed samples and determine its mean and variance.

Hint: Use MathCAD or MatLAB to do the computations.

Problem 4. Use the circular scattering model of Problem 3 to generate 200 samples of the magnitude and phase of the channel. Using MatLAB or MathCAD, calculate and plot in decibels the magnitude of the Fourier transform of the generated samples at sampling intervals of 0.1 msec. Assume $v = 80$ km/hr, $f = 900$ MHz, and $A = 1$.

Problem 5. Starting with Eq. (3.5.3), show that if the channel impulse response is represented by Eq. (3.4.2), the rms delay spread is given by Eq. (3.4.4).

Problem 6. Using MathCAD or MatLAB, repeat Example 3.3 (the three-path indoor model) and plot the wideband receive power in decibels versus distance $(1 < d < 100$ m) in log scale for the center frequency of 1 GHz. Determine the distance–power gradient by finding the slope of the best-fit line to the received power plot.

Problem 7. Repeat Problem 6 for a center frequency of 10 GHz. What is the difference in the distance–power gradient obtained in Problem 6 and this problem.

Problem 8

(a) For Example 3.3, sketch the rms multipath delay spread in nanoseconds versus the distance on a log scale for $1 < d < 100$ m. Assume $f_c = 1$ GHz.

Do you see any relationship between the distance and the rms multipath delay spread? Can you generalize your conclusion?

(b) Repeat part (a) for a center frequency of 10 GHz. Is there any difference in the rms multipath due to this change in the center frequency?

Problem 9. Consider a fading channel with scattering function

$$S(\tau, \lambda) = Q(\tau) \times D(\lambda)$$

where $Q(\tau)$ and $D(\lambda)$ are uniformly distributed functions within the ranges $0 < \tau < 100$ msec and $0 < \lambda < 10$ Hz.

(a) What are the multipath spread and the rms multipath spread of the channel?

(b) What are the maximum Doppler spread and the rms Doppler spread of the channel?

(c) What is the coherence bandwidth of the channel?

(d) What is the coherence time of the channel?

Problem 10. Repeat Problem 9 for

$$Q(\tau) = e^{-\tau/T}$$

and

$$D(\lambda) = \frac{1}{f_m} \times \left[1 - (\lambda/f_m)^2\right]^{-1/2}, \qquad |\lambda| < f_m$$

Assume $T = 10$ nsec and $f_m = 10$ Hz.

Problem 11. Repeat Problem 10 for

$$Q(\tau) = 0.7\delta(\tau) + 0.3\delta(\tau - 20 \times 10^{-9})$$

and

$$D(\lambda) = \frac{0.8}{f_m} \times \left[1 - (\lambda/f_m)^2\right]^{-1/2} + 0.2\delta(\lambda), \qquad |\lambda| < f_m$$

Problem 12. Rederive Eq. (3.5.5) and explain the details of the derivation.

4

CHANNEL MEASUREMENT
AND MODELING
FOR NARROWBAND SIGNALING

4.1 INTRODUCTION

In the preceding chapter we showed that due to the constructive and destructive interference of multipath components received at the different locations, multipath propagation causes substantial variations in the amplitude of a received radio signal. We also showed that the Doppler shifts imparted to the various multipath signals, due to movement of the terminals or movement of people or objects around the transmitter and the receiver, cause a spectral spreading of the received signal. Then we discussed how the multipath and Doppler effects place limitations on the rate of signaling achievable over the channel, and we showed these effects to be related to three parameters:

1. The distance–power gradient (α)
2. The root mean square (rms) delay spread (τ_{rms}) of the channel
3. The Doppler spread of the channel (f_d)

The distance–power gradient is used for the determination of power decrease as a function of distance from the transmitter. As a simple rule, 10α is the average attenuation per decade of increase in the distance. The Doppler spread is related to the aggregate of Doppler shifts of multipath components; each shift is approximated by v_m/λ, where v_m is the effective closing velocity of the path and λ is the wavelength of the carrier frequency. The rms multipath delay spread limits the symbol transmission rate R of a simple modulation technique to an approximate value $R \simeq 0.1/\tau_{rms}$. In general, measurements are performed using either narrowband or wideband techniques and equipment, and the results are used to develop narrowband or wideband models, respectively. Narrowband measurements can provide parameters α and f_d, while τ_{rms} can be determined from the results of wideband measurements.

In this chapter we describe measurement and modeling techniques used to determine the narrowband characteristics of radio propagation and present some results obtained in such measurements. Narrowband measurements are made when the transmission rate of the intended application is well below the coherence bandwidth of the channel. As an example, as will see later, the coherence bandwidth of the indoor radio channel for the distances less than a 100 m between the transmitter and the receiver is around a few megahertz, which means that transmission rates on the order of several hundred kilobits per second are considered to be narrowband. For digital cordless applications the transmission rates are always below these values. As a result, cordless telephone applications have provided the main motivation for narrowband measurements and modeling in indoor areas [Ale82, Ale83a, Ale83b].

The received power always varies with small local changes, on the order of the wavelength of the carrier frequency, in the location of the transmitter and receiver or the movement of the objects around them. However, the average received power over a small area is related to the distance from the transmitter to the center of the receiving area. The channel characteristics extracted from narrowband channel measurements are: (1) the relationship between distance and the average received power, (2) the statistics of the fluctuations in received signal power in local and extended areas, and (3) the Doppler spread, which provides a measure of the rate of fading in the channel.

Modems are designed to operate with certain tolerance to fluctuations in the power of the received signal. The range of operation of the receiver and, consequently, the size of the cells in a cellular architecture, depend on the distance–power relationship. This relationship in indoor areas is related to the layout of the building and the materials used in its construction. The statistics of the amplitude fluctuations provide information for the calculation of probability of error and probability of outage for different modulation techniques. The Doppler spread is helpful in the specification and design of adaptive algorithms such as automatic gain control and timing- or phase-recovery circuits.

4.2   DISTANCE – POWER RELATIONSHIP AND SHADOW FADING

The simplest method of relating the received signal power to the distance is to state that the received signal power P_r is proportional to the distance between transmitter and receiver d, raised to a certain exponent which is referred to as the *distance–power gradient*; that is,

$$P_r = \frac{P_0}{d^\alpha}$$

where P_0 is the received power at distance 1 m from the transmitter. For a free-space path, $\alpha = 2$; and for the simplified two-path model of an urban radio channel given as Example 3.2 in the preceding chapter, $\alpha = 4$. For indoor and urban radio channels the distance–power relationship will change with the building and street layouts, as well as with construction materials and density and height of the buildings in the area. Generally, variations in the value of the distance–power gradient in different outdoor areas are smaller than variations abserved in indoor areas. The results of indoor radio propagation studies show values of α smaller than 2 in corridors or large open indoor areas and values as high as 6 in metal buildings.

The distance–power relationship (in decibels) is given by

$$10 \log_{10} P_r = 10 \log_{10} P_0 - 10\alpha \log_{10} d$$

where $10 \log_{10} P_r$ and $10 \log_{10} P_0$ are transmitted and received power at 1 m in decibels, respectively. The last term in the right-hand side of the equation represents the power loss in decibels with respect to the received power at 1 m, and it indicates that for a one-decade increase in distance the power loss is 10α dB, and for a one-octave increase in distance it is 3α dB. For a free-space path the power loss is 20 dB per decade or 6 dB per octave of distance. In urban areas, given the two-ray approximation discussed in Chapter 3, attenuation is 40 dB per decade or 12 dB per octave. If we define the path loss in decibels at a distance of 1 m, as $L_0 = 10 \log_{10} P_t - 10 \log_{10} P_0$, the total path loss L_p in decibels is given by

$$L_p = L_0 + 10\alpha \log_{10} d \qquad (4.2.1)$$

which represents the total path loss as the path loss in the first meter plus the power loss relative to the power received at 1 m. The received power in decibels is the transmitted power in decibels minus the total path loss L_p. This normalized equation is occasionally used in the literature to represent the distance–power relationship.

To measure the gradient of the distance–power relationship in a given area, the receiver is fixed at one location and the transmitter is placed at a number of locations with different distances between the transmitter and the receiver. Either the received power or the path loss is plotted in decibels against the distance on a logarithmic scale. The slope of the best-fit line through the measurements is taken as the gradient of the distance–power relationship. As we saw in Examples 3.4 and 3.6 in Chapter 3, the distance–power gradient obtained from narrowband and wideband measurements have the same values. Calculation of the power from the result of wideband measurements provides an average received power in a local area, resulting in smaller deviations from the best-fit line. Figure 4.1 shows a set of wideband measurements of averaged received power taken in an indoor area at distances from 1 to 20 m, together with the best-fit line through the measurements.

The earliest statistical measurements of signal amplitude fluctuations in an office environment were reported by Alexander of British Telecom for a cordless telephone application [Ale82]. The measurements were made by fixing the transmitter while moving the receiver to various locations in a multiple-room office. The measurements were made using a small hand-held 30-mW transmitter operating at 941 MHz, with a vertically polarized quarter-wave dipole antenna. The receiver used a half-wave vertically polarized dipole antenna and had a dynamic range of 60 dB.

The first experiment of a series was performed in a building with steel partitioning [Ale82]. The receiver was fixed in one room, and the transmitter was moved among 13 other rooms. The distance–power gradient measured in this experiment was 5.7. That experiment was followed by a set of measurements made in buildings constructed with various materials [Ale83a]. Table 4.1 shows the results of the

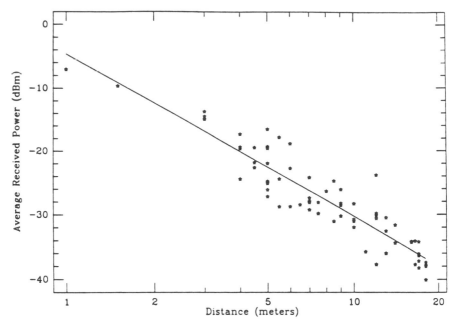

Figure 4.1 A scatter plot of the power (dBm) versus distance on a logarithmic scale for a wideband indoor radio measurement experiment.

measurements made in the different buildings. For each building or area within a building, the table shows the coverage for 1 mW of transmitted power, the distance–power gradient, the correlation coefficient, and the decibel spread from the best-fit line. The maximum values of the distance–power gradient are around 6, which corresponds to buildings with concrete and metal structures, with communication among several floors. Values around 2 or smaller are shown for open areas and plasterboard partitions, where the gradient is close to or even better than that of free space. The gradients lower than free space are observed in areas such as hallways, where the structure acts as a waveguide between the transmitter and the receiver. The measurements in brick buildings give intermediate values around 4. The maximum spread of power is ± 16 dB from the best-fit line. The correlation has its highest value where the building architecture provides a linear arrangement of rooms, whereas the minimum correlation is observed in houses with brick or breeze-block construction.

Since those earliest measurements, many researchers have performed narrowband measurements within buildings, primarily to determine the distance–power relationship and the distribution of the received signal envelope. Arnold et al. [Arn89] reported copolarized attenuation measurements made at 815 MHz within two office buildings. The measurements were made using a modified hand-held 815-MHz transceiver, with an integral vertical half-wavelength coaxial sleeve dipole, as the signal source. The received signal was captured with a similar antenna attached to a 9.5-ft mast, connected to a modified 815-MHz FM communications receiver and a digital storage oscilloscope. Narrowband radiowave propagation measurements into and within a building were also reported by Barry and

TABLE 4.1 Coverage Areas and Distance – Power Relationships in Several Buildings [Ale83a]

Building	Construction	1 mW Distance[a]	Distance – Power Relationship (Gradient)	Power Correlation[b]	Spread ± dB[c]
1 Offices	Brick	17 m	3.9	0.97	8
2 Offices first floor[d]	Brick	12 m	3.9	0.86	10
Offices ground floor	Brick	12 m	3.9	0.96	6
3 Offices	Brick / block / plasterboard, reinforced concrete shell	25 m	6.1	0.89	16
4 Offices	Brick / plasterboard	> Floor	5.3	0.99	1
Ground floor[d]		16 m	4.3	0.94	12
First		12 m	4.8	0.95	8
Second					
Through floors 1, 2, 3, 4, 5	Reinforced concrete floors	10 m	5.1	0.98	3
5 Offices 1st floor[d]	Plasterboard with	27 m	6.2	0.95	9
ground floor	metal support studding	8 m	3.1	0.93	6
6 Laboratory	Block plus some metal faced partitioning	20 m	6.5	0.96	8
7 Offices	Plasterboard	30 m within 60 m outside	2.8	0.75	16
8 Offices	Plasterboard	32 m	3.7	0.96	7
9 Offices	Steel	10 m	5.7	0.92	10
10 House	Brick / breeze / plasterboard	> Building	1.4	0.54	7
11 House	Brick / breeze block	> Building	4.0	0.76	7
12 House	Brick / breeze block	> Building	2.2	0.70	12
13 Workshop	Open plan	60 m	2.5	0.97	4
14 Hangar	Open plan	> Building	1.2	0.99	1

[a]Radial extent of coverage area with 1-mW source power.
[b]Correlation is the degree of fit of the best-fitting straight line computed by linear regression.
[c]Spread is the maximum scatter of the points about the line.
[d]Base receiver remains on this floor when measuring other floors in this building.

Williamson at Auckland University [Bar86]. In this work, measurements were made using a 17-W, 927-MHz transmitter feeding a half-wavelength dipole antenna. The mobile receiver had a similar antenna mounted 1.6 m above floor level and connected to field-strength measuring equipment.

To extend this model for application to multistory buildings, signal attenuation by the floors in the building is included as a constant independent of the distance [Mot88b]. The path loss in this case is given by

$$L_p = L_0 + nF + 10\alpha \log_{10} d \qquad (4.2.2)$$

where F represents the signal attenuation provided by each floor and n is the number of floors through which the signal passes. The received power is plotted versus distance, and the best-fit line is determined for each of different values of F. The value of F which provides the minimum mean-square error between the line and the data is taken as the value of F for the experiment. For indoor radio measurements at 900 MHz and 1.7 GHz, Motley and Keenan have reported values of $F = 10$ dB and $F = 16$ dB, respectively [Mot88a].

The preceding equation suggests an exact relationship between path loss and distance. But, in general, buildings are not symmetric and the furnishing are not the same in all directions, and therefore we expect to find somewhat different path losses in different directions. A deterministic model for this variation is not feasible, and therefore we usually resort to statistical models. The cause of this power loss is the obstruction by other objects around the receiver, and it is usually referred to as *shadow fading* or *large-scale fading*. To determine the statistics of the shadow fading, the results of Eq. (4.2.2) are compared with the measured average path loss in a large area. The distribution of the error between the results of measurement of the average path loss and the prediction given by Eq. (4.2.2) provides the model for shadow fading. Result of measurements on indoor [Mot88a, Mot88b, Gan91, How91] and urban [Jak74, Lee89] radio channels show that a lognormal distribution best fits the large-scale variations of the signal amplitude. In [Mot88b], the variations of the mean value of the signal were found to be lognormal with a variance of 4 dB.

In Eq. (4.2.2) the relationship between the path loss and the number of floors is linear. However, results of measurements in [Rap92], [Sei92], and [Ake88] do not agree with this assumption. Therefore we may improve upon Eq. (4.2.2) with

$$L_p = L_0 + L_f(n) + 10\alpha \log_{10} d + l \qquad (4.2.3)$$

where $L_f(n)$ represents the function relating the power loss to the number of floors n, and where l is a lognormal-distributed random variable representing the shadow fading. Table 4.2 gives a set of suggested parameters in decibels for the path loss calculation using Eq. (4.2.3). The columns of the table provide the path loss in the first meter, the gradient of the distance–power relationship, the equation for calculation of multifloor path loss, and the variance of the lognormal shadow fading parameter. It is assumed that the base and portable stations are inside the same building. The parameters are provided for three classes of indoor areas, residential, offices, and commercial buildings. This table is taken from an NTIA/ITS recommendation for radio-frequency (RF) channel modeling for personal communications service (PCS) applications [JTC94].

4.2.1 Partitioned Modeling for Indoor and Microcellular Environments

The area between the transmitter and the receiver is often not homogeneous with a single distance–power gradient. In these cases the power loss should be described with multiple distance–power gradients, each associated with a segment of

TABLE 4.2 Recommended Parameters for Path Loss Calculations in PCS Indoor Radio Environments [JTC94]

Environment	Residential	Office	Commercial
A (dB)	38	38	38 dB
B	28	30	22
$Lg(n)$ (dB)	$4n$	$15 + 4(n - 1)$ dB	$6 + 3(n - 1)$ dB
Lognormal fading (Standard deviation dB)	8	10	10

the path between the transmitter and the receiver. Results of wideband measurements in a partitioned indoor area show significant differences among the values of the distance–power gradient in different parts of a building [Gan91a]. Figure 4.2 depicts the middle part of the third floor of the Atwater Kent Laboratories at Worcester Polytechnic Institute. The receiver was located at the center of Room 317, and the transmitter was moved to different locations in various rooms for measurements. The area was divided into three segments: the interior of a small laboratory (Room 317), corridors around the laboratory, and offices on the opposite side of the corridor. Three different gradients of 1.76, 2.05, and 4.21 were calculated from the results of the measurements made in the three subareas. Inside the small laboratory all the locations provide a strong line-of-sight (LOS) connection and the gradient is 1.76, which is less than the free-space gradient. This is consistent with the results of two-dimensional ray tracing in a similar environment presented in Examples 3.4 and 3.6 of the previous chapter. In the corridors there is at least one plaster wall with metal studs between the transmitter and receiver, and the gradient is close to that of free-space propagation. The third subarea, with gradient 4.21, includes at least two walls, one of which contains a number of metal doors. Also, inside the rooms are several metal shelves, cabinets, and desks.

In [Ake88], based on measurements in a multistory building, the path loss was modeled with three different gradients. In these measurements the transmitter was fixed in the middle of a corridor and the receiver was moved away from the transmitter to other corridors and rooms. The model developed from the measurements suggests a gradient $\alpha = 2$ for the distances $1 < d < 10$ m, a value of 3 for $10 < d < 20$ m, a value of 6 for $20 < d < 40$ m, and a value of 12 for $d > 40$ m. This leads in the following equations for the path loss:

$$L_p = L_0 + \begin{cases} 20\log_{10} d, & 1 < d < 10 \text{ m} \\ 20 + 30\log_{10} \dfrac{d}{10}, & 10 < d < 20 \text{ m} \\ 29 + 60\log_{10} \dfrac{d}{20}, & 20 < d < 40 \text{ m} \\ 47 + 120\log_{10} \dfrac{d}{20}, & d > 40 \text{ m} \end{cases}$$

4.2.2 JTC Path Loss Model for Microcell – Macrocell

Here we briefly describe a path loss model recommended by a technical working group of the TIA/ANSI Joint Technical Committee (JTC) on the Air-Interface Specification for PCS [JTC94]. It is assumed here that the distance between the base and mobile stations is less than 1 km and that the base station antenna height is below the rooftop level. The working group recommends the following path loss model for microcells with known physical geometry. The model divides the distances into two LOS and one obstructed line-of-sight (OLOS) regions. The first LOS region is the Fresnel zone defined by the break-point distance [Jen65]:

$$d_{\text{bp}} = \left(\frac{4h_b h_m}{\lambda} \right)$$

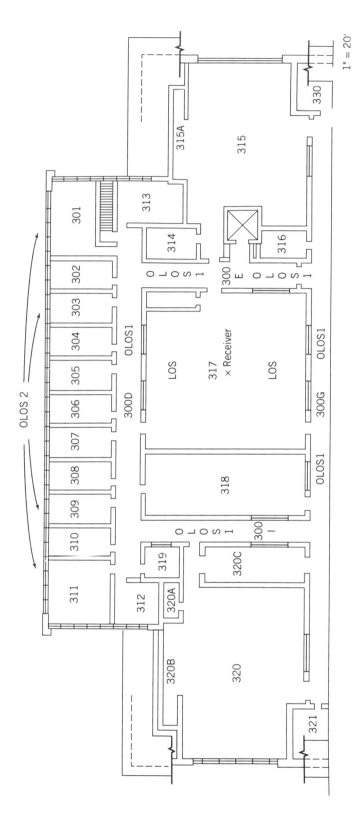

Figure 4.2 Layout of the third floor of the Atwater Kent Laboratories at the Worcester Polytechnic Institute, used for partitioned measurements.

where h_b and h_m are the heights of the base and mobile station antennas, respectively, and λ is the wavelength of the carrier frequency. In this region the power received from the LOS path dominates the total power of the other paths and the propagation loss is the same as for free-space propagation. The second LOS region starts at d_{bp} and continues to d_{cor} where the mobile unit turns a corner and loses the LOS path. In this region the gradient is assumed to be 4, to include the direct LOS path as well as the path reflected from the ground. The third region starts from d_{cor} where the mobile loses the LOS path. The gradient in this region is assumed to be 5, and an additional path loss of L_{cor} is added to compensate for the immediate power drop after turning the corner. The formula for calculation of the path loss in then given by

$$
L_p = 38.1 + \begin{cases} 20 \log_{10} d, & d < d_{bp} \\[2em] 20 \log_{10} d_{bp} + 40 \log_{10} \dfrac{d}{d_{bp}}, & d_{bp} < d < d_{cor} \\[2em] L_{cor} + 20 \log_{10} d_{bp} + 40 \log_{10} \dfrac{d_{cor}}{d_{bp}} + 50 \log_{10} \dfrac{d}{d_{cor}}, & d > d_{cor} \end{cases}
$$

where the path loss in the first meter of distance from the base station is assumed to be $L_0 = 38.1$ dB.

For cases where a detailed description of the microcellular environment is not available, the JTC document recommends the following general path loss model:

$$
L_p = 38.1 + \begin{cases} 25 \log_{10} d, & d < d_{bp} \\[1.5em] 25 \log_{10} d_{bp} + 45 \log_{10} \dfrac{d}{d_{bp}}, & d > d_{bp} \end{cases}
$$

The JTC recommendation for the macrocell environment, where the antenna height is above the rooftop level, is given by

$$
L_p = \max[A + B \log_{10} d, 38.1 + 20 \log_{10} d]
$$

where

$$
A = 88 - 13.82 \log_{10} h_b + C
$$
$$
B = 49 - 6.55 \log_{10} h_b
$$

Table 4.3 provides the clutter correction factor C, building penetration loss, and variance of the lognormal shadowing parameters, as suggested by the JTC working group.

4.2.3 Path Loss for Mobile Radio Application

One of the most commonly used path loss models for urban radio propagation is the model originally developed by Okumura et al. [Oku68] based on extensive

**TABLE 4.3   Recommended Parameters for Path Loss Calculations in PCS
Outdoor Radio Environments [JTC94]**

Environment	Urban High-Rise	Urban / Suburban Low-Rise	Residential	Rural
Clutter correction factor C(dB)	0	-6	-12	-18
Building penetration loss (dB)	15	15	10	10
Lognormal shadowing	10	10	10	10

radio propagation studies made in Tokyo. This model was further adapted for computer simulation by Hata [Hat80]. The path loss in this model is given by the expression

$$L_p = 69.55 + 26.16 \log_{10} f - 13.82 \log_{10} h_b - A(h_m)$$

$$+ (44.9 - 6.55 \log_{10} h_b) \log_{10} d$$

where the range of frequency is 150 MHz $< f <$ 1500 MHz, the range of the height of the base station antenna is $30 < h_b < 300$ m, and the distance range is given by $1 < d < 10$ km. The function $A(h_m)$ in decibels for a small or medium size city is

$$A(h_m) = (1.1 \log_{10} f - 0.7)h_m - (1.56 \log_{10} f - 0.8)$$

with $1 < h_m < 10$ m. For a large city we have

$$A(h_m) = 3.2[\log_{10}(11.75h_m)]^2 - 4.97, \qquad f \geq 400 \text{ MHz}$$

Over the restricted range of parameters, Hata's equations provide a simple but very accurate approximation to Okumura's method. These equations have evolved out of experimental results by taking into account various parameters causing attenuation. More detailed treatments of models for path loss in urban radio channels are available in [Lee89], [Jak74], and [Par89].

The path loss modeling methods described in this section are based on generalizations of results obtained in certain specific measurement programs. However, there is no universally accepted model for path loss. One important limitation of these modeling methods is that they do not include the specification of building characteristics. As a consequence, much attention is being given to building-specific radio propagation models such as ray tracing, and these techniques are emerging as the leading techniques for the future. However, there are drawbacks in the use of building-specific radio propagation models: the complexity of computation, the need for large amounts of computer memory, and the enormous cost for creating a detailed electronic map. With the growing availability of electronic maps and the continual increase in computational power and memory capacity of computers, it is expected that increasingly accurate building-specific radio propagation models will evolve. This subject is discussed further in Chapter 6.

4.3 LOCAL AND TEMPORAL ENVELOPE FADING

In the previous section we modeled the average received power over a large area with a deterministic and a random component. The deterministic component was a function of distance, a function that changed from one physical environment to another. The random component, shadow fading, was modeled using a lognormal-distributed random variable with a variance that can be slightly different in different environments. This slow fading component represents the difference in the overall characteristics of the environment. It remains stationary in areas with dimensions on the order of a wavelength of the carrier frequency, and it changes as the receiver moves from an area to another. We use the models developed in the last chapter to determine the average received power at and around a location for the receiver. However, in and around each location as the receiver moves on the order of a wavelength or other objects move close to the transmitter or the receiver, the received narrowband signal power fluctuates significantly due to *multipath fading*. As we showed in Chapter 3, multipath fading causes power fluctuations on the order of 30–40 dB. The statistical fluctuation of the amplitude of the received power is the superposition of fast local multipath fading over slow shadow fading. The slow shadow fading component causes changes in the mean value of the received power as the terminal moves from one area to another. The fast fading component changes rapidly as the transmitter or the receiver moves slightly or other objects are moved in the vicinity of the transmitter or the receiver. To model the multipath fading characteristics of the channel, we analyze the statistics of the temporal and local amplitude variations as well as the spectrum of the variations, termed the *Doppler spectrum*.

As we discussed in Chapter 3, the complex envelope of the received narrowband signal on indoor and urban radio channels is represented by the complex addition of individual phasors representing the magnitudes and the phases of the individual paths. Small movements of the transmitter and receiver or the movement of objects around them will cause random changes in the magnitude and the phase of the individual paths; and according to the Central Limit Theorem, the sum of all paths will form a complex Gaussian random variable. In the absence of a dominant LOS path, the Gaussian process has zero mean and in the presence of a dominant path it will have a nonzero mean value. The magnitude of a complex Gaussian random process obeys a Rayleigh distribution if the mean of the process is zero and obeys a Rician distribution otherwise. The phase of a complex Gaussian process always has a uniform distribution. As a result, for both indoor and urban radio channels we may assume that the multipath fading is generally Rayleigh unless there exists a strong LOS component in which case multipath fading is Rician. To examine the accuracy of this model we will examine the results of a few indoor radio propagation experiments in the following sections.

A simple and useful model for mobile radio channels has been proposed by Clarke [Cla68]. This model assumes a dense array of randomly oriented scattering objects located around the mobile unit. In the definition of this model, Clarke makes the simplifying assumption that all the scatter components arrive with the same amplitude (termed isotropic scattering) but the components are distinguished from one another by the angles of arrival and the phases of the components. The angles of arrival and the phases of the received signals are both assumed to be

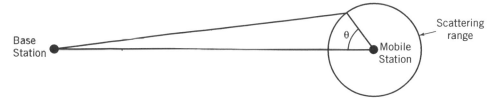

Figure 4.3   The ring scattering model for mobile radio communications.

uniformly distributed, and the arrival angle and phase of each component are assumed to be statistically independent of each other. With uniform fixed amplitude of the signal components, the addition of phasors with uniformly distributed phase angles will result in a Rayleigh distribution for the magnitude of the complex sum of all the paths. This will change to a Rician distribution in the presence of a strong LOS path. In [Cla68] Clarke shows that the scatter-only model provides an accurate representation of mobile radio signals in heavily built-up areas such as New York City, when the signal energy propagates from transmitter to receiver largely by way of scattering, either by reflection from the sides of buildings or by diffraction around buildings or other man-made or natural obstacles. In suburban areas, the received signals are often a combination of a scattered signal and a direct plane-wave signal, a condition represented by the Rician model.

Figure 4.3 provides a simplified description of the isotropic scattering model for mobile radio and is very useful for analyzing the Doppler spectrum of the channel. Assuming that a mobile terminal is moving toward the base station at a constant velocity v_m, the Doppler shifts associated with different paths are not the same. The direct path from the transmitter to the receiver has the maximum positive Doppler shift of $f_m = v_m/\lambda$, whereas the path arriving from a reflection behind of the mobile terminal has the maximum negative Doppler shift of $-f_m$. Other paths have Doppler shift values between these two limits, given by $f(\theta) = f_m \cos \theta$, where θ represents the angle of arrival of the path. Therefore, the path with angle of arrival θ is represented by a phasor of the form $A_\theta \exp[\omega_m \cos \theta]$, where A_θ is the magnitude of the path associated with the arriving angle θ. With the assumption of equal magnitudes $A_\theta = A$ for the arriving paths, and taking the model to the case of a continuum of arriving signal components, we can define the *azimuthal distribution of signal power* as

$$Z(\theta) = \frac{1}{2\pi}, \qquad -\pi \leq \theta \leq \pi$$

where the total arriving signal power is normalized as 1.0. As was described in Problem 3 of Chapter 3, if the distance between the two stations is much larger than the radius of the scattering circle, the Doppler frequency is related to the azimuth angle by the relationship $f_d = f_m \cos \theta$, or $\theta = \cos^{-1}(f_d/f_m)$, we can derive the Doppler spectrum as

$$D(f) = R_{HH}(0, f) = Z(\theta) \times \left| \frac{d\theta}{df} \right|$$

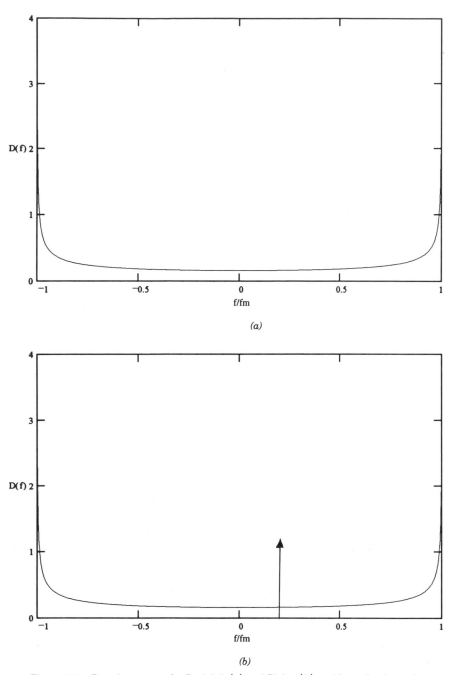

Figure 4.4 Doppler spectra for Rayleigh (*a*) and Rician (*b*) mobile radio channels.

where we use f in place of f_d. Now, because

$$\frac{d}{dx} \cos^{-1}(y) = -1/\sqrt{1 - y^2} \; \frac{dy}{dx}$$

we can write the Doppler spectrum as

$$D(f) = \frac{1}{2\pi f_m} \times \left[1 - (f/f_m)^2\right]^{-1/2}, \qquad |f| \le f_m \qquad (4.3.1)$$

Figure 4.4a shows the typical spectrum, in which for values of f close to $\pm f_m$ the height of the Doppler component rises to two high peaks at the edges of the spectrum. In the presence of a strong component with Rician distributed envelope fading, as shown in Fig. 4.4b, the spectrum has an additional impulse representing the shift associated with the strong component. The spectra described above have been shown to match experimental data gathered for mobile radio channels [Jak74]. For the indoor radio channel, the assumptions of equal component amplitudes and uniform distribution of the angles of arrival do not hold; and as we will see in the following subsection, the Doppler spectra have shapes different from the mobile radio case. The JTC channel model for indoor areas assumes $D(f)$ to be a flat spectrum [JTC94].

4.4   MEASUREMENT OF DOPPLER SPREAD

We now present some results of measurements of Doppler spreading on indoor radio channels, as caused by traffic and local movements of the communication terminals. These were controlled experiments in which the only movements were those for which we were trying to determine the resulting Doppler spread [How90a]. As we saw earlier, the characteristics of the channel are influenced by the existence of a strong LOS path, and so we consider both LOS and OLOS experiments. The measurements reported here were made on the third floor of the three-story Atwater Kent Laboratories at Worcester Polytechnic Institute.

In order to determine the Doppler spread of a radio channel, we require a system capable of measuring the short-term variations of the channel. The system should be capable of sampling the amplitude of the received signal at the Nyquist rate associated with the highest Doppler shift caused by the movements of the equipment or neighboring objects.

A simple and accurate method of measuring the short-term variations of the narrowband characteristics of the indoor radio channel is to use a network analyzer in an experimental configuration shown in Fig. 4.5. Here a 910-MHz signal generated by the network analyzer is power-split and used as both the reference input to the network analyzer and, after passing through 100 ft of coaxial cable, the input to a transmitter RF amplifier having 45-dB gain. The output of the RF power amplifier is radiated by the transmit dipole antenna. The signal from the receive dipole antenna is passed through an attenuator and a series of amplifiers with an overall gain of 60 dB. The output of the amplifier chain is returned to the

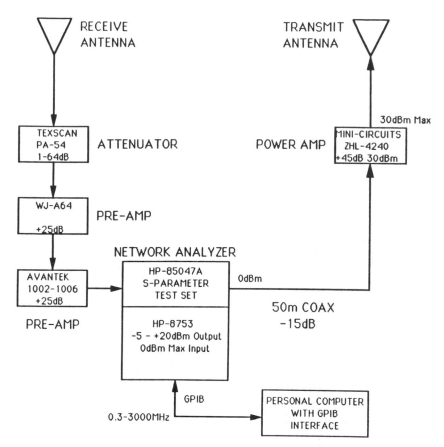

Figure 4.5 Measurement system used for narrowband indoor radio propagation measurements.

network analyzer, where the time variations of the channel relative to the fixed reference input are measured. The measurement data files are then read and stored in a PC controller for subsequent analysis. In a 32-sec interval, the network analyzer samples the received amplitude and phase at the rate of 25 samples/sec. Therefore, the maximum Doppler shift measurable is 12.5 Hz and the resolution is 0.03125 Hz. As we will show later, the same system, with a different network analyzer configuration, is used for wideband measurements.

While for the case of mobile radio the Doppler spectrum has the relatively regular shape shown in Fig. 4.4, the Doppler spectra for other wireless applications have a variety of shapes. The indoor radio wireless local area network (WLAN) user in a small room may observe a stationary channel with no Doppler spread. However, the same user may observe a Doppler spectrum associated with the movements of people around the transmitter and the receiver if the system operates in a more populated larger indoor area such as a manufacturing floor or an office building. A cordless telephone user observes a Doppler spectrum associated with the random motions of the device as the person speaks on the phone. Figure 4.6 shows four time-domain plots of received signal amplitude variations, along with the four corresponding Fourier transforms $|H(f_c; t)|$, measured using

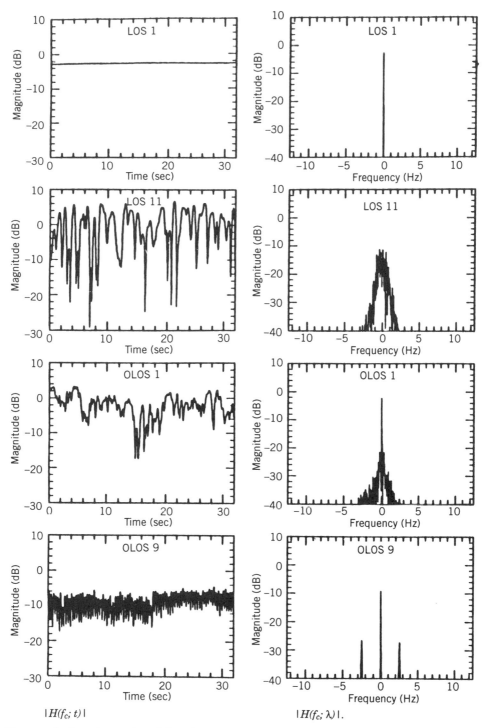

$|H(f_c; t)|$

$|H(f_c; \lambda)|.$

Figure 4.6 Sample results of Doppler spectrum measurements on an indoor radio channel. (a) LOS 1: $B_D = 0$, no movement. (b) LOS 11: $P_{\text{dev}} = 35$ dB, $B_D = 4.9$ Hz, random motion of Tx. (c) OLOS 1: $P_{\text{dev}} = 20$ dB, $B_D = 5.7$ Hz, traffic close to Tx. (d) OLOS 9: $P_{\text{dev}} = 10$ dB, $B_D = 5.2$ Hz, cyclic motion of Tx [How90a].

the system shown in Fig. 4.5. Plots are shown for two LOS channels and two OLOS channels. Figure 4.6*a* shows data from an LOS experiment in which the environment was kept constant and the distance between the transmitter and the receiver was 1 m. There was no time-domain variation of the received signal, and consequently the computed Fourier transform shows an impulse at zero frequency with no Doppler spread. In Fig. 4.6*b* the transmitter was moved randomly around a fixed point 12 m away from the receiver. The time-domain measurements show maximum power fluctuations of 35 dB, and the corresponding Fourier transform has a bell shape with a width of $B_D = 4.9$ Hz. Figure 4.6*c* shows data from an OLOS experiment in which the transmitter and the receiver were at fixed locations 4 m apart, and there was random pedestrian traffic close to the transmitter. The maximum power fluctuation is seen to be 20 dB, and the Doppler spread has a two-sided exponential shape with $B_D = 5.1$ Hz. Figure 4.6*d* shows the results of cyclic motion of the transmitter to change the orientation and consequently the polarization of the antenna. The results show a maximum power fluctuation of 10 dB with a $B_D = 5.2$ Hz. The spectrum shows two strong components representing the rate of the cyclic motion. Generally, the shape of the Doppler spectrum is related to the nature of the movement, but the maximum Doppler shift is related to the fastest motion of the human hand and body, which remained almost the same throughout all experiments.

Table 4.4 summarizes the parameter values found in the 11 LOS measurements, which were made in an electronics laboratory. The results are ordered by increasing distance between the transmitter and receiver. To provide a reference, measurement 1 was taken at 1 m with no movements. Three measurements were taken at each of the three distances 3 m, 6 m, and 12 m. Measurements 2, 5, and 8 were taken with people moving in the path between the transmitter and receiver. Measurements 3, 6, and 9 were taken with pedestrian traffic close to the transmitter and receiver. Measurements 4, 7, and 10 were taken with small cyclic movement of the transmitter. Measurement 11 was taken at 12-m separation with

TABLE 4.4 Results of Measurement of the Doppler Spread and Its rms Values in Several LOS and OLOS Indoor Areas [How90a]

	LOS				OLOS			
Number	Distance (m)	$B_{D\text{-rms}}$ (Hz)	B_D (Hz)	Power (dB)	Distance (m)	$B_{D\text{-rms}}$ (Hz)	B_D (Hz)	Power (dB)
1	1	0.016	0.0	0	4	0.373	5.7	20
2	3	0.610	6.1	30	4	0.190	5.1	10
3	3	0.424	4.8	35	4	0.199	4.7	4
4	3	0.092	0.4	4	8	0.873	4.9	30
5	6	0.665	1.9	25	8	0.559	3.6	35
6	6	0.424	3.3	30	8	0.761	4.8	8
7	6	0.236	0.3	3	13	0.461	4.4	25
8	12	0.217	2.0	15	13	0.257	3.0	10
9	12	0.247	3.9	20	13	0.649	5.2	10
10	12	0.130	4.9	4	13	0.288	1.0	8
11	12	0.531	4.9	35				

random motion of the transmitter, to simulate the typical movements of a cordless-phone user.

For the OLOS measurements, the receiver was placed in the Communication Research Laboratory (CRL). A total of 10 measurements, summarized in Table 4.4, were taken in three different locations. Location 1 was in the corridor next to the CRL, with one wall separating the transmitter and receiver. Location 2 was in the microwave laboratory at the other side of the corridor, with two walls separating the transmitter and receiver. Location 3 was in an electronics laboratory adjacent to the microwave laboratory, with three walls separating the transmitter and receiver. Measurements 1, 4, and 7 were taken with pedestrian traffic close to the transmitter. Measurements 2, 5, and 8 were taken with traffic close to the receiver. Measurements 3, 6, and 9 were taken with small cyclic movement of the transmitter. Measurement 10 was taken at the third location with traffic between the transmitter and receiver, but not in the same room as either the transmitter or the receiver.

Other measurements in similar configurations are Bultitude's measurements [Bul87], which include periods of no movement as well as periods of local movements. His analysis of the movement data, which he extracts from the overall sequence of measurements, shows the channel to be wide-sense stationary for time periods of at least 3.4 sec. Rappaport [Rap89] measured temporal fading of the received signal envelope over a 100-sec period during the normal working hours in a factory. His analysis of the temporal fading data showed the dynamic range to be about 10 dB. Both researchers compared the temporal fading data with the Rayleigh and Rician distributions and showed that the Rician distribution fits the data well.

4.5 DOPPLER SPREAD AND RMS DOPPLER BANDWIDTH

As we discussed earlier, the maximum Doppler frequency shift imparted to an unmodulated carrier is related to the velocity of movement v_m and the wavelength of the carrier λ by $f_d = v_m/\lambda$. If we use this equation as an approximation to the Doppler spread B_D of the indoor radio channel, a person walking at 3 miles/hr (1.34 m/sec) will produce a maximum Doppler shift of ± 4 Hz at a carrier frequency of 910 Hz. These values are consistent with the measurements of Doppler spread B_D shown in Table 4.4.

To relate the results of these measurements to the classical wide-sense stationary uncorrelated scattering (WSSUS) model described in Chapter 3, we note that our time-domain measurements represent $H(f_c; t)$. The complex autocorrelation function of $H(f_c; t)$ is $R_{Hh}(0; \Delta t)$, and the Doppler power spectrum defined in Eq. (3.5.6) is given by

$$D(\lambda) = R_{HH}(0; \lambda) = \int_{-\infty}^{\infty} R_{Hh}(0; \Delta t) e^{-j2\pi\lambda \Delta t} d(\Delta t) = |H(f_c; \lambda)|^2$$

where

$$H(f_c; \lambda) = \int_{-\infty}^{\infty} H(f_c; t) e^{-j2\pi\lambda t} dt$$

In summary, to determine the Doppler power spectrum, $D(\lambda)$, we transmit a single tone at frequency f_c and we measure the amplitude fluctuations of the received signal in time, $H(f_c; t)$. Magnitude square of the Fourier transform of the measured $H(f_c; t)$ provides us with the $D(\lambda)$. The Doppler spread B_D is the range of frequency λ over which the Doppler power spectrum $D(\lambda) = |H(f_c; \lambda)|^2$ is nonzero. In practice, $|H(f_c; \lambda)|$ is never zero and a threshold is applied to $|H(f_c; \lambda)|$ to determine B_D. The threshold applied for the experiments described in this section is -40 dB [How90a].

A more specific measure of the Doppler spread is the rms Doppler spread given by

$$B_{D\text{-rms}} = \left[\frac{\int_{-\infty}^{\infty} \lambda^2 D(\lambda)\, d\lambda}{\int_{-\infty}^{\infty} D(\lambda)\, d\lambda} \right]^{1/2} \tag{4.5.1}$$

which is a weighted measure of the spectral distribution of signal power rather than simply the overall width of the spectrum. The values of rms Doppler spread measured in the LOS and OLOS experiments are included in Table 4.4.

4.5.1 Fading Rate and Fade Duration

For a Rayleigh fading envelope distribution, the average number of downward crossings of a level A per second, N, is given by

$$N(\rho) = \sqrt{2\pi}\, B_{D\text{-rms}} \rho e^{-\rho^2} \tag{4.5.2}$$

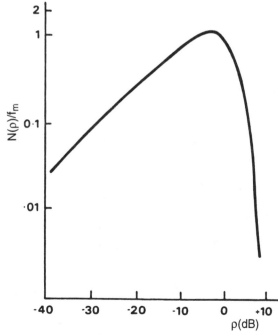

Figure 4.7 Normalized level-crossing rate versus the normalized threshold for Rayleigh envelope fading.

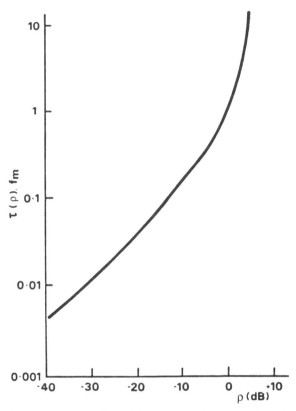

Figure 4.8 Normalized average duration of fade versus the normalized threshold for Rayleigh envelope fading.

where $\rho = A/A_{\text{rms}}$ is the ratio of the threshold level to the rms amplitude of the fading envelope, and $B_{D\text{-rms}}$ is the rms Doppler shift of the signal. We define *fade rate* $= 1.475B_{D\text{-rms}}$, which is the average rate of downward crossings of the median $(A = \sqrt{2\ln 2}\,A_{\text{rms}})$ signal level [Ric48].

The average fade duration for a given threshold ρ is given by

$$\tau(\rho) = \frac{\text{Prob}[\alpha < \rho]}{N(\rho)} = \frac{e^{\rho^2} - 1}{\rho B_{D\text{-rms}}\sqrt{2\pi}} \qquad (4.5.3)$$

Figures 4.7 and 4.8 represent plots of the average number of fades and average fade duration, respectively, normalized to the Doppler shift, versus ρ in decibels.

4.6   EXPERIMENTAL EVALUATIONS OF ENVELOPE FADING

A systematic approach to determining the distribution function of the received amplitudes is to compare the results of amplitude measurements with a few candidate distributions. Each candidate distribution is represented by a function with a few parameters. The functions are selected to have relevance to the physical environment, and the parameters are determined from the measured data. The cumulative distribution function (CDF) of the resulting curves with the parameters obtained from the measurements is compared with the CDF of the empirical data. The average mean-squared error (AMSE) between the two curves is used as a measure of goodness of fit of the curve to the empirical data.

The probability density functions are defined for linear values of the amplitudes, while the measurements are usually documented on a decibel scale. Transferring the logarithmically measured data to a linear scale and performing the curve fitting with linear data will reduce the accuracy of the curve-fitting operation. Therefore, for greater accuracy, we should find the CDF in decibels, which involves an algebraically tedious change of variable, an operation not analytically tractable for all distribution functions. The CDFs of the Rayleigh, Weibull, Nakagami, lognormal, and Suzuki distribution functions in decibel scale and the relationship between the parameters of the function and the mean and variance of the empirical data are available in [Lor79], [Gan91a]. An analytical solution for the Rician distribution is not available. Therefore, the curve fitting with Rician distribution involves transferring the empirical data to a linear scale, which reduces the reliability of the conclusions drawn from the curve-fitting operation.

4.6.1 Distribution Functions for Amplitude Modeling

In this section we introduce the linear forms of distribution functions most commonly used for fitting to envelope fading data [Has93a]. We provide a brief description of the Rayleigh, Rician, lognormal, Suzuki, Weibull, and Nakagami distributions. The transformation of these distributions to decibels scale and their parameters is discussed in Appendix 4A of this chapter.

The Rayleigh distribution is the most popular distribution function used for statistical modeling of envelope fading of radio signals. The Rayleigh probability density function for the amplitude a is given by

$$f_A(a) = \frac{a}{\Gamma}e^{-a^2/2\Gamma}$$

which is described by a single parameter Γ. The mean and variance of the Rayleigh distributed random variable are given by $\sqrt{\Gamma\pi/2}$ and $(2 - \pi/2)\Gamma$, respectively.

The square of the magnitude of a Rayleigh distributed random variable, representing the signal power, has an exponential distribution, the chi-square distribution with two degrees of freedom. If we let $\gamma = a^2$, the probability density function of γ is

$$f_\Gamma(\gamma) = \frac{1}{\bar{\gamma}}e^{-\gamma/\bar{\gamma}}$$

where $\bar{\gamma}$ is the average received power. We will use this distribution in several instances in this book for calculation of error rates of modulation techniques and throughput of contention-based protocols operating over Rayleigh fading channels.

The Rayleigh distribution is also used for calculation of fading characteristics such as the fading rate and average fading duration, as given by Eqs. (4.5.2) and (4.5.3).

The Rician distribution, commonly used to model the amplitude variation in presence of a strong dominant LOS path, is described by

$$f_A(a) = \frac{a}{\sigma^2} e^{-(a^2-v^2)/2\sigma^2} I_0\left(\frac{av}{\sigma^2}\right), \qquad a \geq 0$$

where σ^2 represents the variance of the random component, v is the amplitude of the fixed component, and $I_0(\cdot)$ is the modified Bessel function of the first kind. The parameter $k = v^2/\sigma^2$ is the ratio of the deterministic to the random component of the process. Usually k and σ are used as the parameters identifying the Rician distribution function. Values of k around 6 dB are typical in modeling indoor radio channel amplitude fluctuations [Bul87].

The lognormal probability distribution function is used for modeling the large-scale variations of the received power in indoor and urban radio channels. The model suggests that the decibel value of the average received power over a large area forms a normal (Gaussian) distribution function. The probability density function of a lognormal distributed random variable is given by

$$f_A(a) = \frac{1}{\sqrt{2\pi}\,\sigma a} e^{-(\ln a - \mu)^2/2\sigma^2}$$

where μ and σ are the mean and standard deviation of the random variable, respectively. In indoor and urban radio applications the mean of the lognormal random variable is assumed to be zero and the variance is the only parameter needed to describe the distribution function.

In portable and mobile radio channels, the local distribution of the signal amplitude in areas with dimensions on the order of the wavelength is Rayleigh and the wide area coverage is represented by a lognormal distribution. The overall distribution of the received signal amplitude is then represented by the integral of the Rayleigh distribution over all possible values of σ represented by the lognormal distribution. This new distribution was first suggested by Suzuki and is named for him [Suz77]. The *Suzuki random variable* is defined by the following probability density function:

$$f_A(a) = \int_0^\infty \frac{a}{\sigma^2} e^{-a^2/2\sigma^2} \frac{1}{\sqrt{2\pi}\,\sigma\lambda} e^{-(\ln\sigma-\mu)^2/2\lambda^2}\, d\sigma$$

where λ^2 is the variance of the lognormal distribution. This distribution has a very clear physical interpretation, but the complicated mathematical form has limited its practical applications.

Two other distributions used for modeling of the envelope fading in indoor and urban radio channels are those of Weibull and Nakagami [Lor79, Nak60]. These distributions form a superset of other distribution functions. The Weibull probability density function is given by

$$f_A(a) = \frac{sr}{\sigma}\left(\frac{ra}{\sigma}\right)^{s-1} e^{-(ra/\sigma)^s}, \qquad a \geq 0$$

where s is the shaping parameter, σ is the rms value of the random variable, and $r = [(2/s)\Gamma(2/s)]^{1/2}$ is a normalization factor [Has93a] based on the gamma function. For $s = 2$ the Weibull distribution function reduces to the Rayleigh, and for $s = 1$ it reduces to the exponential distribution.

The Nakagami distribution is defined as

$$f_A(a) = \frac{2m^m a^{2m-1}}{\Gamma(m)\Omega^m} e^{-(ma^2/\Omega)}, \qquad a \geq 0$$

where Ω is the mean-square value of the random variable and $m = \Omega^2/\mathrm{Var}[a^2]$, which is constrained to be equal to or larger than $1/2$. For $m = 1$ the Nakagami distribution reduces to Rayleigh, and for $m = 1/2$ it is a one-sided Gaussian distribution. With proper adjustment of the parameters it can also fit Rician and lognormal distributions very tightly.

4.6.2 Measurement Results

In this section we describe the results of experimental curve fitting for the LOS and OLOS narrowband indoor radio measurement experiments described in Section 3.5. We will compare the CDF of the 801 samples of the received signal envelope $|H(f_c;t)|$ with the lognormal, Weibull and Rayleigh distributions [How90c]. We will use the AMSE between the measured CDF and the analytical CDF as a measure of the closeness of fit to the analytical model. To quantify our comparisons, we will say that a chosen distribution is a close fit to the measured data if the AMSE is on the order of 10^{-4}, a marginal fit to the measured data if the AMSE is on the order of 10^{-3}, and a poor fit to the measured data if the AMSE is on the order of 10^{-2}. Figure 4.9 shows the CDF of OLOS measurement 10 and also shows the theoretical CDFs for lognormal, Weibull, and Rayleigh distributions. The Rayleigh distribution with an AMSE of 3.5×10^{-2} is a poor fit, but the Weibull distribution with an AMSE of 1.94×10^{-4} gives a very close fit to the measured data. The lognormal distribution with an AMSE of 1.45×10^{-3} is a marginal fit. The AMSEs for all the measurements are given in Table 4.5. For envelope fading variations of less than 10 dB, the Rayleigh distribution provides a poor fit to the data with AMSEs on the order of 10^{-2}. The Rayleigh distribution provides a close fit for the measurements where the power variations are 30 dB or greater. Out of the 21 measurements, the Rayleigh distribution is a close fit for 6 measurements, a marginal fit for 5 measurements, and a poor fit for 10 measurements. The lognormal distribution is a close fit for 9 measurements and a marginal fit for 12 measurements. The Weibull distribution is a close fit for 15 measurements and a marginal fit for 6 measurements.

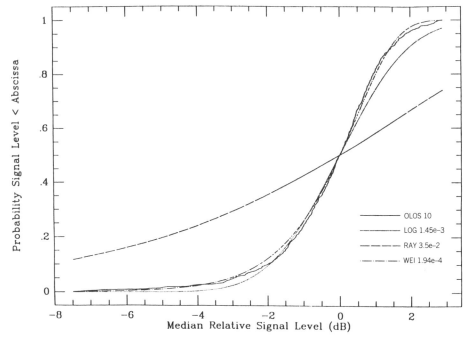

Figure 4.9 Sample result of curve fitting for evaluation of the envelope fading statistics.

**TABLE 4.5   The Error Associated with Curve Fitting to Rayleigh,
Weibull, and Lognormal Distributions in Several LOS and OLOS Experiments**

Number	LOS			OLOS		
	Lognormal	Weibull	Rayleigh	Lognormal	Weibull	Rayleigh
1	2.63×10^{-4}	5.68×10^{-4}	8.17×10^{-2}	1.68×10^{-3}	2.51×10^{-4}	7.44×10^{-3}
2	6.00×10^{-4}	8.83×10^{-4}	3.47×10^{-4}	1.43×10^{-3}	8.40×10^{-4}	2.36×10^{-2}
3	5.50×10^{-4}	2.16×10^{-3}	3.60×10^{-3}	3.27×10^{-3}	5.93×10^{-4}	6.51×10^{-2}
4	1.36×10^{-3}	4.86×10^{-4}	6.15×10^{-2}	3.29×10^{-4}	5.50×10^{-4}	3.28×10^{-4}
5	9.40×10^{-4}	1.46×10^{-4}	8.01×10^{-4}	4.59×10^{-4}	5.35×10^{-4}	9.51×10^{-4}
6	2.36×10^{-3}	2.04×10^{-4}	2.16×10^{-4}	2.02×10^{-4}	1.50×10^{-3}	2.42×10^{-2}
7	4.01×10^{-4}	4.17×10^{-4}	5.76×10^{-2}	1.73×10^{-3}	2.63×10^{-4}	4.93×10^{-3}
8	4.11×10^{-4}	1.41×10^{-3}	1.89×10^{-2}	3.07×10^{-3}	5.63×10^{-4}	2.49×10^{-2}
9	5.21×10^{-3}	1.83×10^{-3}	6.92×10^{-3}	1.52×10^{-3}	1.44×10^{-3}	9.84×10^{-3}
10	1.50×10^{-3}	4.04×10^{-3}	5.23×10^{-2}	1.45×10^{-3}	1.94×10^{-4}	3.50×10^{-2}
11	3.98×10^{-3}	9.71×10^{-4}	9.60×10^{-4}			

In calculating the average probability of error for digital signaling over a fading channel, the probability distribution of the fading envelope is used to average the conditional probability of error at each envelope level [Ale82, Pro89]. It is the near-zero values of the signal envelope, where the conditional probability of error approaches 0.5, that most heavily influence the average probability of error and therefore is the region where the assumed theoretical distribution should best fit

TABLE 4.6 The Error Associated with Curve Fitting to Rayleigh, Weibull, and Lognormal Distributions in Several LOS and OLOS Experiments

Number	LOS			OLOS		
	Lognormal	Weibull	Rayleigh	Lognormal	Weibull	Rayleigh
1	1.30×10^{-4}	1.38×10^{-4}	8.14×10^{-2}	1.46	3.45×10^{-5}	5.17×10^{-3}
2	2.41×10^{-4}	7.46×10^{-5}	2.62×10^{-4}	1.41	7.87×10^{-4}	1.84×10^{-2}
3	1.03×10^{-3}	1.31×10^{-3}	6.97×10^{-3}	6.20	1.05×10^{-4}	5.93×10^{-2}
4	1.07×10^{-3}	6.24×10^{-4}	5.72×10^{-2}	9.39	6.90×10^{-5}	4.27×10^{-4}
5	9.12×10^{-5}	7.29×10^{-5}	7.75×10^{-4}	3.46	1.25×10^{-4}	9.83×10^{-5}
6	6.38×10^{-4}	1.06×10^{-4}	2.29×10^{-4}	1.89	1.12×10^{-4}	2.47×10^{-2}
7	1.68×10^{-4}	8.97×10^{-5}	5.50×10^{-2}	2.44	4.09×10^{-4}	3.93×10^{-3}
8	3.78×10^{-4}	6.74×10^{-4}	1.98×10^{-2}	1.19	4.88×10^{-4}	1.54×10^{-2}
9	4.40×10^{-4}	1.96×10^{-4}	2.15×10^{-3}	2.00	1.11×10^{-3}	7.26×10^{-3}
10	2.32×10^{-3}	3.07×10^{-3}	5.98×10^{-2}	2.02	2.13×10^{-4}	3.06×10^{-2}
11	1.38×10^{-3}	5.79×10^{-4}	1.24×10^{-3}			

the measurements. In Table 4.6 we show the AMSE calculated for only envelope levels below the median for each measurement. In general, the AMSE computed below the median is less than the corresponding AMSE computed over the full range of the distribution. For the 21 measurements, the Rayleigh distribution is a close fit for 5 measurements, a marginal fit for 6 measurements, and a poor fit for 10 measurements. The lognormal distribution is a close fit for 14 measurements and a marginal fit for 7 measurements. The Weibull distribution is a close fit for 18 measurements and a marginal fit for 3 measurements.

4.7 METHODS TO SIMULATE NARROWBAND CHANNELS

To simulate a narrowband channel, we need to generate a random process with specific envelope fading density function and a specific Doppler spectrum. All the computer programming languages used for the development of computer simulations for telecommunication applications, such as C, Pascal, FORTRAN, and even BASIC, have uniform and Gaussian random number generators. More modern block-oriented simulation software, such as SPW™, provide these random variables as building blocks. The channel simulation software in these cases should generate other random variables from these distributions and shape the Doppler spectrum of the signal. Generating a new random variable from an old random variable involves a mapping with specific rules. Methods for computer simulation of Rayleigh, Rician, lognormal, Suzuki, Weibull, and Nakagami random variables are described in [Med93], [Pre91], and [Jer92].

Spectral shaping can be done by passing the random variable through a filter with the specific spectral shape. If it is inconvenient to develop the needed spectrum by filtering, one may instead generate a series of oscillators with different frequencies and add the outputs to form the specific spectrum. The first approach has been used extensively in simulation of HF and troposcatter fading channels. The second approach is often used in simulation of mobile radio channels, based on the Clarke assumption of isotropic scattering [Cla68].

4.7.1 Filtered Gaussian Noise Fading Models

A widely used approach to simulation of fading radio channels is to construct a fading signal from in-phase and quadrature Gaussian noise sources. Because the envelope of a complex Gaussian noise process has a Rayleigh probability density function (PDF), the output of such a simulator will accurately simulate Rayleigh fading. In this approach, the Doppler spectrum of the channel of interest is provided by applying the appropriate filtering to the Gaussian noise sources. Figure 4.10 shows a block diagram of the basic technique for simulating Rayleigh fading using two filtered Gaussian noise processes. In some applications of this technique, a detailed specification of the channel Doppler spectrum is not available, and any of a number of convenient filter transfer functions may be implemented, with an appropriate choice of the filter bandwidth. When the *fade rate* of the channel is specified, one can simply set the rms filter bandwidth $B_{D\text{-rms}}$ in accordance with the relationship *fade rate* = 1.475 $B_{D\text{-rms}}$ for the Rayleigh fading process [Ric48]. Because the simulator of Fig 4.10 simply multiplies the signal by a single complex Gaussian variable, this constitutes a simulation of *frequency-non-selective fading* or *flat fading*. This technique is appropriate for simulation of channels where the signal bandwidth is considerably smaller than the coherence bandwidth of the channel. This technique is often used for simulation of ionospheric and tropospheric scatter channels, and has been used for simulation of mobile radio channels, as will be shown in the following example.

Example 4.1. Reference [Arr73] describes a device designed to simulate Rayleigh fading characteristics of mobile radio channels. The hardware simulator used two Gaussian noise sources with identical shaping filters to generate the quadrature components of a Rayleigh fading signal. The shaping filters were designed to approximate the theoretical spectrum of the mobile radio channel. The theoretical spectrum, given earlier in Eq. (4.3.1), is shown here in Fig. 4.11*a*. The frequency response of each shaping filter is shown in Fig. 4.11*b*. The shaping filter consisted of two active filters in cascade: a low-pass filter and a peaking amplifier. The filter was designed to the desired fading rate by equating the second moments of the theoretical and simulated spectra. The degree to which the level-crossing rate compared with theory is shown in Fig. 4.12. In the figure, the measured level-crossing rate $N(\rho)$ in $\sec^{-1}$ (normalized to $B_{D\text{-rms}} = 1$ Hz) is plotted against the crossing level ρ in decibels relative to the rms envelope level. The theoretical curve in the figure is given by Eq. (4.5.2). It can be seen from Fig. 4.12 that the measured level-crossing rate agrees with theory to within 3 dB over a range extending down to 30 dB below rms.

For some applications, the radio channel is characterized by a combination of a fading signal component and one or more nonfading or *specular* components. Figure 4.13 shows a simulation model incorporating a fading component and a single specular component. Such a model is appropriate for simulation of cases such as LOS microwave channels, where there is a nonfading signal arriving on a direct path, as well as a fading signal produced, for example, by atmospheric

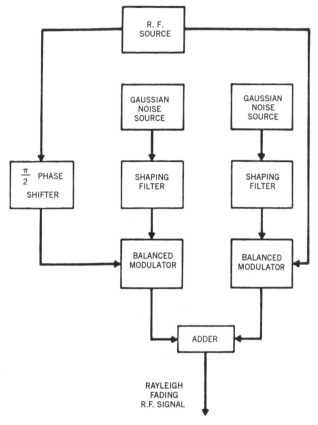

Figure 4.10   Block diagram of a filtered Gaussian noise fading simulator.

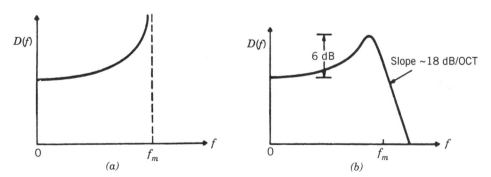

Figure 4.11   Theoretical mobile radio spectrum and a simulator shaping filter. (a) Theoretical spectral density. (b) Shaping filter frequency response. [From [Arr73] © IEEE.]

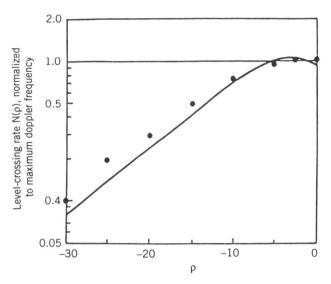

Figure 4.12 Simulated and theoretical level-crossing rates. The level-crossing rate is normalized to $f_m = 1$ Hz. (From [Arr73] © IEEE.)

effects. The inclusion of a specular component in this simulation model is equivalent to adding a nonzero mean value to each of the quadrature Gaussian noise sources, and therefore the simulator produces Rician rather than Rayleigh fading.

Sometimes it is necessary to simulate a frequency-selective fading channel, a channel in which the multipath spread is greater than the symbol duration. This is readily accomplished by replicating the basic complex Gaussian fading model as shown in Fig. 4.14. In this model the signal is fed into a tapped delay line, with taps spaced $1/W$ sec apart, where W is the signal bandwidth [Ste65]. Each tap is

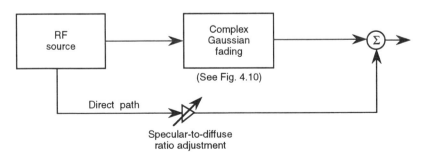

Figure 4.13 Simulation of a combined specular and fading channel.

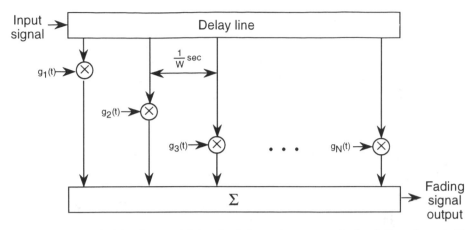

Figure 4.14 Tapped delay line model for simulating a frequency-selective fading channel. The $\{g_i(t)\}$ are independent complex Gaussian fading processes.

multiplied (complex multiplication) by a filtered complex Gaussian fading process, and the faded multiplier output is adjusted by a fixed attenuation factor before summation of the faded components. In setting up this model, one chooses the number of taps so that the length of the delay line approximates the delay spread of the channel. The attenuations applied to the multiplier outputs can be set to provide any specified "shaping" of the multipath delay-power profile of the channel. For example, modeling of some high-frequency (HF) propagation channels would require inclusion of multiple fading signal components arriving with specific delays related to reflections of the signal from well-defined layers of the ionosphere. As with the simple flat fading model of Fig. 4.10, the tapped delay line model can be made to include a specular component simply by adding a nonfading signal component to the summation in Fig. 4.14, thus producing a Rician frequency-selective fading model.

Several other variations of these basic filtered-noise simulation models are discussed in [Jer92]. We shall discuss tapped-delay-line models more extensively in Chapter 6, where we consider simulation of wideband channels.

4.7.2 The Jakes Model

As an alternative to modeling with filtered complex Gaussian noise, one may instead approximate the Rayleigh fading process by summing a set of complex sinusoids. The number of sinusoids in the set must be sufficiently large that the PDF of the resulting envelope provides an acceptably accurate approximation to the Rayleigh PDF. With this modeling method, the sinusoids are weighted so as to produce an accurate approximation of the desired channel Doppler spectrum. One technique of this type is that proposed by William Jakes of Bell Laboratories for the simulation of fading mobile radio channels [Jak74]. This simulation technique, based on the isotropic scattering model studied earlier by Clarke [Cla68], has come to be known as the *Jakes model* and is widely used in the mobile communications industry. The technique was originally developed for a hardware simulator implementation, but it is often implemented in software as well. Software realizations of the Jakes model have been adopted by standards groups for use in testing candidate speech coding schemes as well as radio-link error-control protocols [Ses91, Lev93].

Jakes [Jak74] shows that the theoretical Doppler spectrum for the isotropic scattering mobile radio channel, given earlier in Eq. (4.3.1), can be well approximated by a summation of a relatively small number of sinusoids, with the frequencies and relative phases of the sinusoids set according to a specific formulation. Following our notation of Section 3.3, the maximum Doppler shift frequency is $f_m = v_m/\lambda$, where v_m is the velocity of the mobile and λ is the wavelength of the carrier frequency. In the model described by Jakes, the ideal isotropic continuum of arriving scatter components is approximated by N plane waves arriving at uniformly spaced azimuthal angles. Jakes restricts $N/2$ to be an odd integer and defines another integer $N_0 = (1/2)(N/2 - 1)$. This leads to a simulation model having one complex frequency oscillator with frequency $\omega_m = 2\pi f_m$ plus a summation of N_0 complex lower-frequency oscillators with frequencies equal to the Doppler shifts $\omega_m \cos \theta_n$, where θ_n is the arrival angle for the nth plane wave (see Fig. 4.3) and where $n = 1, 2, \ldots, N_0$. Each oscillator has an initial phase, and these phases are to be chosen as part of initializing the simulation. We can express the complex envelope $T(t)$ of the fading signal in the form

$$T(t) = \frac{E_0}{\sqrt{2N_0 + 1}}[x_c + jx_s]$$

where

$$x_c(t) = 2\sum_{n=1}^{N_0} \cos \phi_n \cos \omega_n t + \sqrt{2} \cos \phi_N \cos \omega_m t$$

$$x_s(t) = 2\sum_{n=1}^{N_0} \sin \phi_n \cos \omega_n t + \sqrt{2} \sin \phi_N \cos \omega_m t$$

and where $\omega_n = \omega_m \cos(2\pi n/N)$, $n = 1, 2, \ldots, N_0$. In the equations above, ϕ_N is the initial phase of the maximum Doppler frequency sinusoid, while ϕ_n is the initial phase of the nth Doppler-shifted sinusoid. The quantities x_c and x_s are the in-phase and quadrature components, respectively, of the model output. Note that the amplitudes of all the components are made equal to unity except for the one at maximum Doppler frequency ω_m, which is set to $1/\sqrt{2}$.

In a hardware implementation of the simulator intended to operate with RF equipment, the outputs of the individual oscillators, with the appropriate gain factors, are first summed to produce x_c and x_s, which are then multiplied by in-phase and quadrature signal carrier components, respectively, and then summed to produce the final output signal, as acted upon by fading. In a software realization of the model, one would apply x_c and x_s to the in-phase and quadrature components of a baseband signal representation. In a software simulation, one might generate the trigonometric functions using look-up tables [Cas90].

In using this simulation method, one must choose the initial phases (ϕ_n and ϕ_N) of the Doppler-shifted components in such a way that the phase of the resulting fading process will exhibit a distribution as close as possible to uniform. This is discussed in some detail in [Jak74], where rules are given for initializing the phases of the sinusoids. The number of Doppler-shifted sinusoids is chosen large enough that $T(t)$ provides a good approximation to a complex Gaussian process (via Central Limit Theorem), and therefore the envelope $|T(t)|$ is approximately Rayleigh. Jakes suggests that $N_0 = 8$ provides an acceptably accurate approximation to the ideal case of Rayleigh fading.

Example 4.2. Consider communication at a carrier frequency of 900 MHz (cellular band) and a vehicle closing velocity of 100 km/hr (27.8 m/sec). For these conditions the maximum Doppler frequency is $f_m = v_m/\lambda = 83.3$ Hz. Therefore in simulation with the Jakes model, the highest frequency sinusoid has frequency $f_m = 83.3$ Hz, and the frequencies of the N_0 remaining sinusoids are $83.3 \cos(2\pi n/N)$, $n = 1, 2, \ldots, N_0$. In [Jak74], Jakes suggested two methods for setting the initial phases of the f_m sinusoid and the N_0 lower-frequency sinusoids. Here we use Case 2, $\phi_N = 0$ and $\phi_n = \pi n/(N_0 + 1)$, where $n = 1, 2, \ldots, N_0$. Figure 4.15 shows samples of the output of the fading process produced with a Jakes model simulation using $N_0 = 8$. In that simulation the carrier frequency was 900 MHz, and a vehicle speed of 100 km/hr was assumed. Figure 4.15a shows the measured spectrum. Figure 4.15(b) shows a histogram of samples of the output envelope. Figure 4.15(c) shows the measured histogram of phases of the fading signal output. Using this simulation, the bit-error performance for $\pi/4$-QDPSK modulation was evaluated and was found to agree with theoretical performance in flat Rayleigh fading to within about 0.3 dB at BER $= 10^{-3}$ [Lev93].

4.7.3 Flat-Spectrum Fading Model

Another model often used to simulate fading channels is a very simple model called the *flat spectrum fading model*. The model is based on an assumption of scatterers having a uniform distribution in three dimensions. As the name implies, the Doppler spectrum defining this model is flat over a range of Doppler shifts symmetric about the carrier frequency:

$$D(f) = \frac{1}{2\pi f_m}, \qquad |f| \leq f_m$$

Where f_m is the maximum Doppler frequency. This model is often used in

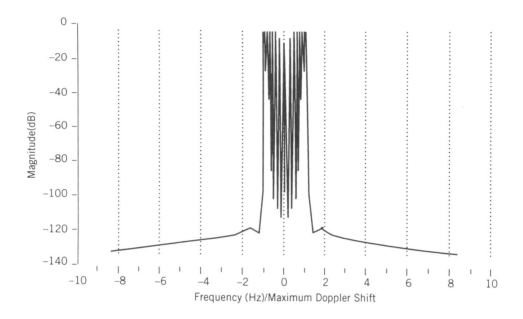

(a)

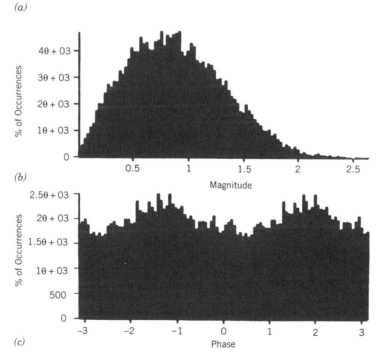

(b)

(c)

Figure 4.15 Example of cellular mobile radio channel simulation using the Jakes model. $V = 100$ km / hr, $f_c = 900$ MHz, $N_0 = 8$. (a) Measured power spectral density. (b) Envelope histogram. (c) Phase histogram.

applications where the multipath fading results from random movements of scattering elements in the area of the communication path, from random movements of the transmitter or receiver, or from both causes. A good example of this type of application would be a WLAN operating in an office environment, or on a factory floor, where there is a good deal of random pedestrian traffic. In such an application the maximum Doppler shift f_m would be set in accordance with an estimate of maximum pedestrian walking speed. For example, for the RF channel model being proposed for use in design of the air-interface specification for PCS (services to operate at 2 GHz), it is recommended that most of the indoor pedestrian communication environments be modeled with the flat-spectrum model, using a maximum Doppler frequency of 9.6 Hz [JTC94]. Channel modeling for PCS comes under the category of wideband channel characterization, which we discuss in Chapter 6.

As with the mobile radio channel model, this model can be implemented using either the filtered complex Gaussian noise method or the sum-of-sinusoids method. With the Gaussian noise method, the low-pass shaping filter would be chosen to have a relatively flat amplitude function and an rms bandwidth approximately equal to f_m. With the sum-of-sinusoids method, the complex envelope of the fading signal is simulated as a summation of uniformly spaced sinusoids, with the maximum frequency set equal to f_m. The number of sinusoids is chosen to provide an acceptably accurate approximation to Rayleigh fading, and the initial phases of the sinusoids are chosen to provide an approximately uniform distribution of the fading signal phase over $(0, 2\pi)$.

APPENDIX 4A: DISTRIBUTION FUNCTIONS AND THEIR PARAMETERS

In Section 4.6, Rayleigh, Weibull, Nakagami-m, lognormal, and Suzuki distributions were introduced as potential models for path amplitudes, and their distribution functions over a linear amplitude variable were given. On fading indoor and urban radio channels, signal amplitude fluctuations typically exhibit a wide dynamic range, and consequently the signal amplitude level measurements are recorded on logarithmic scales. The empirical moments are determined from the measured amplitudes of the signal, and the parameters of the theoretical distribution functions are then determined by equating theoretical and empirical moments on the log scale. A selected formula for the probability density function $f_A(a)$ of the linear signal amplitude A should be transformed by

$$f_B(b) = f_A(a)\left|\frac{da}{db}\right| \qquad (4A.1)$$

to determine the probability density function $f_B(b)$ of the equivalent signal amplitude a measured in decibels. A reference signal level should be defined identically for all distribution functions. The rms value of the signal $\overline{a^2}$ is usually chosen for this purpose. With the rms value as the reference, the transformation equations can be written as

$$b - b_0 = 10 \log_{10}\left(\frac{a^2}{\overline{a^2}}\right) \tag{4A.2}$$

where b_0 is the rms value of the signal b. Taking the derivative, we have

$$\frac{da}{db} = \frac{\sqrt{\overline{a^2}}}{M} \exp\left\{\frac{b - b_0}{M}\right\} \tag{4A.3}$$

where $M = 20 \log_{10} e \simeq 8.686$.

Using Eqs. (4A.1) and (4A.3), the probability density function $F_B(b)$ can be determined from the known density function $F_A(a)$. The remainder of this Appendix gives a summary of the procedure and the equations used to determine the required distribution parameters.

The error criterion of the Cramer–Von Mises test [Suz77] is used to compare the goodness of fit among the distribution functions. This criterion is defined as

$$\omega^2 = \int_{-\infty}^{\infty} \left[F_X(x) - F_X^*(x) \right]^2 dF_X(x) \tag{4A.4}$$

where $F_X(x)$ is a theoretical CDF and $F_X^*(x)$ is an experimental CDF.

Using Eqs. (4A.1) and (4A.3), the probability density function $F_B(b)$ of the signal amplitude b measured in decibels, was calculated from the known probability density function $F_A(a)$ of the equivalent linear signal amplitude a. The rms value of the signal B_0 was chosen as the reference signal level for all five selected distributions. The parameters of the five theoretical distribution functions were obtained by equating theoretical and empirical moments on the log scale.

The parameters of each distribution can be calculated from the mean (first moment) of the measured values

$$\hat{b} = \frac{1}{N} \sum_{i=1}^{N} x_b \tag{4A.5}$$

and the second central moment of the measured values

$$\mu_{\hat{2}} = \overline{(b - \hat{b})^2} = \frac{1}{N} \sum_{i=1}^{N} \left(x_b - \hat{b}\right)^2 \tag{4A.6}$$

where N is the number of measured events, x_b is a single measured amplitude, and $\hat{\,}$ denotes the estimate determined from the empirical data.

All five distribution functions are described unambiguously by one or two parameters. The following section summarizes the equations for the probability distribution functions and the parameters of each distribution (the constant $M = 20 \log_{10} e \simeq 8.686$) [Lor79], [Gan91a].

Rayleigh Distribution

$$F_B(b) = 1 - \exp\left\{-\exp\left[\frac{2(b - b_{OR})}{M}\right]\right\}$$ (4A.7)

where

$$b_{OR} \simeq \hat{b} + 2.506 \text{ dB}$$ (4A.8)

Lognormal Distribution

$$F_B(b) = 1 - \frac{1}{2}\operatorname{erfc}\left\{\frac{b - b_{OL} + \dfrac{\sigma_l^2}{M}}{2\sigma_l^2}\right\}$$ (4A.9)

where $\sigma_L = \sqrt{\hat{\mu}_2}$ and

$$b_{OL} \simeq \hat{b} + \frac{\sigma^2}{M}$$ (4A.10)

Weibull Distribution

$$F_B(b) = 1 - \exp\left\{-g(w)\exp\left[\frac{w(b - b_{OW})}{M}\right]\right\}$$ (4A.11)

where

$$w \simeq \frac{11.14}{\sqrt{\hat{\mu}_2}}$$ (4A.12)

$$b_{OW} \simeq \hat{b} + \frac{5.013}{w} + \frac{4.343}{w}\log_e\left[\Gamma\left(1 + \frac{2}{w}\right)\right]$$ (4A.13)

and

$$g(w) = \left[\Gamma\left(1 + \frac{2}{w}\right)\right]^{w/2}$$ (4A.14)

where $\Gamma(x)$ is the gamma function [Abr68].

Nakagami-*m* Distribution

$$F_B(b) = \int_{-\infty}^{b} f_B(\beta) \, d\beta \tag{4A.15}$$

where

$$f_B(b) = \frac{2m^m}{M\Gamma(m)} \exp\left\{ m\left[\frac{2(b - b_{0N})}{M} - \exp[2(b - b_{0N})] \right] \right\} \tag{4A.16}$$

$$b_{0N} \simeq \hat{b} - \frac{M}{2}\{\Psi(m) - \log_e(m)\} \tag{4A.17}$$

and where $\Psi(x)$ is the digamma function [Abr68], $\Gamma(x)$ is the gamma function, and

$$m = \frac{4.4}{\sqrt{\hat{\mu}_2}} + \frac{17.4}{\hat{\mu}_2^{1.29}} \tag{4A.18}$$

Suzuki Distribution

$$F_B(b) = 1 - \frac{1}{\sqrt{2\pi}\,s} \int_{-\infty}^{\infty} \exp\left\{ -\exp\left[\frac{2(b - b_{0R})}{M} \right] \right\} \exp\left\{ -\frac{b_{0R} - b_{0S} + \dfrac{s^2}{M}}{2s^2} \right\} db_{0R} \tag{4A.19}$$

where s is the Suzuki parameter, b_{0R} is the rms value of the Rayleigh distribution, and b_{0S} is the rms value of the Suzuki distribution. The parameters S and b_{0S} are given by

$$s = \sqrt{\hat{\mu}_2 - 31.025} \tag{4A.20}$$

$$b_{0S} \simeq \hat{b} + 2.51 + 0.115s^2 \tag{4A.21}$$

QUESTIONS

(a)   What does the lognormal element of a path loss model represent?

(b)   Explain the difference between shadow fading and multipath fading. What is the typical envelope distribution associated with each of the two forms of fading?

(c)   What is the break-point distance for the Fresnel zone, and how do radio waves propagate in this zone?

(d)   What is a typical value of power loss observed when a mobile turns a corner and loses the LOS connection with the base station?

(e)   What are the typical values of the measured distance–power gradient in indoor areas as reported in this chapter, and how do they compare with the JTC recommendations?

(f)   Discuss the difference between the radio propagation characteristics for mobile and PCS applications.

(g)   What are the typical values for Doppler spread in indoor areas?

(h)   What are the shapes of the measured Doppler spectra in indoor areas as reported in this chapter, and how do they compare with the JTC recommendation?

(i)   What are typical values of Doppler spread for mobile radio applications?

(j)   Describe the difference between the shapes of the Doppler spectra for indoor and outdoor areas?

(k)   What methods are used for simulation of the narrowband signal fluctuations in mobile radio channels?

(l)   Describe the difference between the Rayleigh and Rician Doppler spectra in mobile radio channels.

(m)   What distribution functions are typically used for modeling the amplitude fluctuations in portable and mobile radio applications?

(n)   Explain how to obtain the Rayleigh distribution as a special case of the Weibul distribution.

(o)   How do we reduce the Nakagami distribution to the Rayleigh distribution?

(p)   Which other distribution functions can also be represented by the Nakagami distribution?

(q)   Which distributions are represented by the Susuki distribution?

(r)   Describe the difference between the Rician and Rayleigh distributions.

PROBLEMS

Problem 1. Consider a mobile data communication network in which the minimum required received SNR for proper operation is 10 dB, the background noise level in the band is -120 dBm, and the in-building penetration loss is 15 dB. If the transmitter and the receiver antenna gains are 2, the frequency of operation is 800 MHz, the height of the base station and mobile station antennas are 100 m and 1.4 m, respectively, and the maximum transmitted power is 10 W, determine the coverage of base station using the following:

(a)   The free-space propagation equation given by Eq. (4.2.1).

(b)   The simple two-path model analyzed in Example 3.2.

(c)   Okumora's model for a medium size city (Section 4.2.3).

(d) Okumora's model for a large city.

(e) How much difference exists among various approaches, and how can it be explained?

Problem 2. Consider the floor plan of Fig. 4.2. We want to predict the path loss using different path loss models. Assume that the transmitter and the receiver are located in the center of Rooms 311 and 317, respectively the transmitter and receiver antenna gains are 1.6, and the frequency of operation is 2 GHz. Calculate the path loss using the following:

(a) The experimental partitioned model based on this floor plan (with three partition gradients of 1.76, 2.05, and 4.21), which was described at the beginning of Section 4.2.1.

(b) Akerberg's generalized partitioned model for the indoor propagation loss described at the end of Section 4.2.1.

(c) The JTC model with the building classified as an office area.

Problem 3. A 10-mW transmitter operates with a receiver having sensitivity of 90 dBm. We want to determine the coverage in various environments using the JTC model.

(a) What is the coverage in indoor residential, office, and commercial environments?

(b) What is the coverage in an outdoor microcell with a distance of 10 m between the transmitter and the corner where the LOS connection is lost? Assume that the height of the transmitter and the receiver antennas are 12 m and 2 m, respectively.

(c) What is the coverage in a microcellular environment without a detailed description of the environment?

Problem 4. A 10-W transmitter operates with a receiver having sensitivity of 90 dBm. We want to determine the coverage in a macrocellular environment.

(a) What is the coverage if we use the macrocellular JTC transmission loss model with transmitter and receiver antenna heights of 100 m and 2 m, respectively? Repeat (a) for all four macrocellular environments shown in Table 4.3.

(b) Compare the results of part (a) with the predicted value from Okumora's model within a medium size city and a center frequency of 1.5 GHz.

Problem 5. The Doppler power spectrum $D(\lambda)$ of the indoor radio channel is often assumed to be uniformly distributed with a maximum Doppler shift of 10 Hz.

(a) Determine the rms Doppler shift of the channel.

(b) Determine the average number of fades per second and the average fade duration if the threshold for fading is chosen to be 10 dB below the average rms value of the signal.

(c) Repeat (b) for a threshold of 20 dB below average rms.

Problem 6. Implement the simulator of Example 4.1 and regenerate Figs. 4.11 and 4.12.

Problem 7. Repeat Example 4.2 using MatLAB.

(a) Sketch 512 consecutive samples of the amplitude of the simulated channel fluctuations in decibels. Take the samples for the sketch at eight times the maximum Doppler shift f_m of the channel.

(b)   Using the results of the simulation, sketch the experimental probability density function of the amplitude fluctuations on a linear scale.

(c)   Sketch the Fourier transform of the linear amplitudes to observe the shape of the Doppler spectrum. Does it follow the spectrum predicted by Clarke's circular scattering model?

Problem 8. Simulate a flat spectrum narrowband fading channel with $f_m = 10$ Hz.

(a)   Sketch 512 consecutive samples of the amplitude of the simulated channel.

(b)   Take the Fourier transform of the samples, and show that the results of your simulation comply with the model specifications.

Problem 9. Sketch the CDFs of the Rayleigh distribution for variance of 1, and the Rician, and lognormal distributions for mean and variance of 1.

Problem 10

(a)   Show that a Rayleigh distributed random variable, z, can be generated from two independent Gaussian distributed random variables x, y from the following relation:

$$z = \sqrt{x^2 + y^2}$$

(b)   Simulate 100 samples of a Rayleigh distributed random variable with variance 1 using MathCAD or MatLAB. Create the probability density function (PDF) and the CDF of the 100 simulated samples, and compare the results with the theoretical PDF and CDF of the Rayleigh distribution.

Problem 11

(a)   Give a transformation that generates an exponentially distributed random variable from a uniformly distributed random variable.

(b)   Simulate 100 samples of an exponentially distributed random variable with variance 1 using MathCAD or MatLAB. Create the probability density function (PDF) and the CDF of the 100 simulated samples, and compare the results with the theoretical PDF and CDF of the exponential distribution.

Problem 12

(a)   Show that the distribution function of the square of a Rayleigh distributed random variable is exponential.

(b)   Simulate 100 samples of a Rayleigh distributed random variable with variance 1 using MathCAD or MatLAB. Create the probability density function (PDF) and the CDF of the 100 simulated samples, and compare the results with the theoretical PDF and CDF of the Rayleigh distribution.

Problem 13. Generate a Rician distributed random variable, and check its CDF against the theoretical CDF of the Rician distribution. Assume that mean and variance of the random variable are both normalized to 1.

Problem 14. The lognormal distributed random variable, y, is generated from a Gaussian random variable, x, by the following transformation:

$$y = e^x$$

(a) Show that the probability distribution function of the lognormal random variable is given by

$$f_Y(y) = \frac{1}{\sqrt{2\pi}\,\sigma_x y}\, e^{-(\ln y - \mu_x)^2/2\sigma_x^2}$$

where μ_x and σ_x are the mean and standard deviation of the Gaussian random variable, respectively.

(b) Show that the mean and variance of the lognormal distribution are given by

$$\mu_y = e^{\mu_x + \sigma_x^2/2}$$

and

$$\sigma_y^2 = e^{2\mu_x + \sigma_x^2}\left(e^{\sigma^2 - 1}\right)$$

(c) Show that the mean and variance of the Gaussian random variable whose logarithmic form generates the lognormal distribution are given by

$$\mu_x = \ln\left(\frac{\mu_y}{\sqrt{\dfrac{\sigma_y^2}{\mu_y^2} + 1}}\right)$$

and

$$\sigma_x^2 = \ln\left(\frac{\sigma_y^2}{\mu_y^2} + 1\right)$$

Problem 15

(a) Using the results of Problem 14, describe a step-by-step procedure to simulate a lognormal random variable with specific mean and variance from a Gaussian random variable generator.

(b) Use MathCAD or MatLAB and the results of part (a) to generate 100 samples of a lognormal distributed random variable with mean and variance of 1.

(c) Compare the results of simulations from part (a) with the theoretical CDF of the lognormal distribution.

Problem 16. Start from the definition of the Rayleigh distribution and derive Eqs. (A.4.11) and (A.4.12).

5

MEASUREMENT OF WIDEBAND CHANNEL CHARACTERISTICS

5.1 INTRODUCTION

In narrowband measurements, we analyze the response of the channel at or around a single frequency, and from these measurements we are able to extract the power fluctuations caused by the signal arriving from a number of different paths. Narrowband measurements do not provide any information on the magnitude or the time delay of any individual path. Rather, they reflect the vector addition of the complex amplitudes of the arriving paths as observed in the power fluctuations in the received narrowband signal. Wideband measurements, in contrast, provide information on the multipath delay spread and structure of individual paths as well as the frequency selectivity of the channel. Stated in simple terms, if we assume that the channel is fixed during a measurement interval, narrowband measurements resemble measurements of the channel response to a single frequency whereas wideband measurements resemble measurements of the impulse response or overall frequency response of the channel.

Wideband measurements can be performed either in the time domain by direct measurement of the impulse response of the channel, or in the frequency domain by direct measurement of the frequency response of the channel. In theory, using Fourier transform techniques the measured time and frequency responses should provide identical results. However, as we will see later, there are some shortcomings in using the Fourier transform of the results of measurements, particularly if the measurement system does not provide both magnitude and phase of the measured characteristics.

In this chapter we describe measurement techniques used to determine the wideband characteristics of radio propagation and present some results obtained in such measurements. Systematic measurements for portable or mobile applications are done in several ways, which we list as follows:

1. *Spatial or large-scale measurements* in which one of the terminals is held fixed and the other terminal is moved to different locations, spaced at least several wavelengths apart.

2. *Local or small-scale measurements* in which the transmitter or the receiver is moved about in an area surrounding a specific location, to collect a number of measurements.

3. *Traffic-effect* or *temporal measurements* in which the transmitter and the receiver are held fixed and measurements are made with traffic moving between or around the terminals.

4. *Partitioned measurements*, in which the effects of dividing walls on the characteristics of the channel are studied. The overall measurement area is partitioned, and characteristics and parameters of the channel in the smaller areas are measured and compared.

5. *Frequency-dependence measurements* in which characteristics measured at different frequencies are compared.

5.2 TIME-DOMAIN MEASUREMENT TECHNIQUES

The impulse response of the channel is measured either by transmitting a wideband spread-spectrum signal and correlating the received signal with the transmitted sequence, or by direct transmission of a short radio-frequency (RF) pulse and observing the received signal arriving from different paths. In both cases the time resolution of the measurements is inversely proportional to the bandwidth of the measurement system. The spread-spectrum method sends a steady stream of bits, and the ratio of peak to average transmitted power is unity. With the pulse transmission method, an RF pulse is transmitted periodically with a low duty cycle and the ratio of peak to average power is very high. As a result, given amplifiers designed for identical peak power operation, we can achieve greater coverage with the spread-spectrum approach. In practical implementations of the two systems described in this chapter, we will achieve (a) better coverage with the spread-spectrum technique and (b) better resolution and acquisition time with the direct pulse sounding technique. Consequently, for areas less than 100 m in radius the pulse sounding technique is more popular, and for larger areas the spread-spectrum technique is more typically used. For most indoor applications such as wireless local area networks (WLANs) or wireless PBX systems, path distances of interest are typically no greater than 100 m, and thus the pulse sounding technique has been applied extensively. For mobile radio and personal communications service (PCS) applications used in outdoor areas, path distances are longer, and the spread-spectrum technique is more typically used.

5.2.1 Measurements Using Spread-Spectrum Signals

The traditional method for wideband measurement of the multipath spread in radio channels is the use of principles of spread-spectrum technology. This method has been used for wideband measurement of the mobile radio channel [Cox72, Par89] as well as other radio channels such as troposcatter [She75]. The earliest wideband measurements of multipath spread in building environments [Dev84] were made using a spread-spectrum receiver adapted from a measurement system used for the mobile radio channel [Cox72]. The same approach was used in [Bul89] to study and compare indoor radio propagation characteristics at 910 MHz and 1.75 GHz. In this section we outline the basic principles of spread spectrum communications as applied to wideband channel measurement. Further details of spread-spectrum technology and its applications to wireless information networks are provided in Chapter 9.

To understand the principles of this measurement technique, assume we have a symbol shape $f(t)$ of duration T_s consisting of a sequence of N narrower pulses (chips) $p(t)$ with binary amplitudes $\pm b_i$ and duration $T_c = T_s/N$:

$$f(t) = \sum_{i=1}^{N} b_i p(t - iT_c)$$

Ideally, the pattern of the sequence $\{b_i\}$ of length N is selected so that it is orthogonal to any circularly shifted version of itself. Because all elements of the sequence are ± 1, the sum of squares of the sequence is N. Sequences with the orthogonality property are referred to as *pseudonoise (PN) sequences*, and they are treated extensively in the spread-spectrum communication literature [Sim85].

Furthermore, assume that $x(t)$ is the periodic form of $f(t)$ repeated every T_s seconds:

$$x(t) = \sum_{n} f(t - nT_s)$$

The function $x(t)$ is a periodic function, and therefore its autocorrelation function (ACF) is also periodic with the same period T_s. With the orthogonality condition on the sequence $\{b_i\}$, one period of the ACF of $x(t)$ is given by

$$R_{xx}(\tau) = \frac{1}{T_s} \int_0^{T_s} x(t)x(t - \tau)\, dt = \frac{N}{T_s} \int_0^{T_c} p(t)p(t - \tau)\, d\tau = \frac{N}{T_s} R_{pp}(\tau)$$

where $R_{pp}(\tau)$ is the ACF of the pulse $p(t)$. Figure 5.1 shows an example of $x(t)$ and its correlation function for the case of rectangular $p(t)$ pulses.

Let $x(t)$ be the complex envelope of a transmitted signal and assume that it passes through a multipath channel with equivalent baseband channel impulse response given by

$$h(\tau, t) = \sum_{i=1}^{L} \beta_i \delta(t - \tau_i) e^{\phi_i}$$

where β_i and ϕ_i are the magnitude and phase, respectively, of the path arriving at delay τ_i. If the complex envelope of the received signal

$$r(t) = \sum_{i=1}^{L} \beta_i x(t - \tau_i) e^{\phi_i}$$

is cross-correlated with the $x(t)$, the resulting cross-correlation function is a periodic function with period T_s. One period of this cross-correlation function is

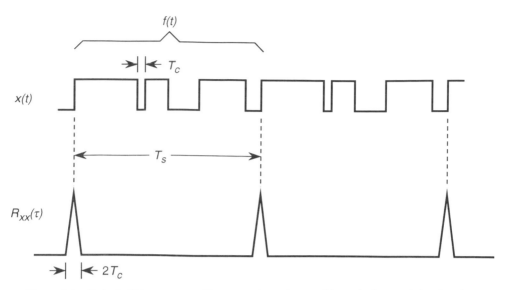

Figure 5.1   Periodic PN sequence with rectangular pulses and its periodic correlation function.

given by

$$R_{xr}(\tau) = \sum_{i=1}^{L} \beta_i R_{xx}(\tau - \tau_i)e^{j\phi_i} = \frac{N}{T_s} \sum_{i=1}^{L} \beta_i R_{pp}(\tau - \tau_i)e^{\phi_i}$$

If we assume that $2T_c$, the width of the correlation function $R_{pp}(\tau)$, is narrow enough to resolve all paths, and the multipath delay spread $\tau_L - \tau_0$ is less than T_s, then one period of the received signal is identical to the channel impulse response with impulses replaced by $R_{pp}(\tau)$ and a normalization factor N/T_s included in the result.

This mathematical concept provides a basis for the design of a system for measurement of the channel impulse response, as depicted by the block diagram in Fig. 5.2. The PN sequence of length N with chip rate $R_c = 1/T_c$ is modulated onto a carrier at frequency f_c, and the modulated signal after power amplification is fed to the transmit antenna. The receiver consists of a *sliding correlator* and demodulator. The received signal is multiplied by a replica of the transmitted sequence at a slower rate $R_c - \Delta R$ and integrated over T_s to generate samples of the ACF. The integration interval t_s should cover a number of chips adequate to provide a reliable measure of the ACF. In practice, integration is usually done over a set of intervals from fractions of T_s up to T_s. The inherent dynamic range of the system is limited by the residual value of the ACF of the PN sequence; therefore, longer codes provide a wider dynamic range.

To understand how the sliding correlator works, assume that transmitter and receiver codes start at the same reference time and $T_s \simeq t_s$. At the end of the transmission of the first symbol the receiver chips are delayed $\Delta R \times t_s$ chips relative to the transmitter chips. If $\Delta R \times t_s \ll 1$, the first sampled output of the

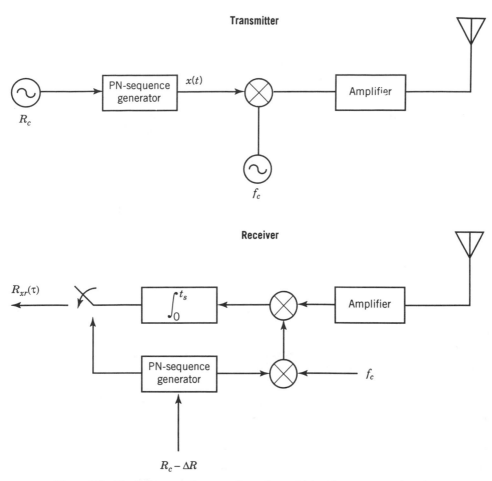

Figure 5.2 Block diagram of a spread-spectrum wideband measurement system.

receiver represents $R_{xr}(\Delta R \times t_s/R_c)$, which is a good approximation to $R_{xr}(\tau)$, at $\tau \approx 0$. The next sampled output of the signal represents the next sample of $R_{xr}(\tau)$ evaluated $\Delta R \times t_s/R_c$ seconds later. Following the same pattern, the sampled outputs of the receiver taken every t_s seconds in real time will represent samples of $R_{xr}(\tau)$ taken at effective sampling delays of $\tau_s = \Delta R \times t_s/R_c$. In other words, the samples of the output taken every t_s seconds represent samples of $R_{xr}(\tau)$ taken every τ_s seconds in the delay variable. Because $R_{xr}(\tau)$ is a periodic function with period T_s in the delay variable, every

$$k = \frac{R_c}{\Delta R} \times \frac{T_s}{t_s}$$

samples, the delay between the receiver and the transmitter is increased by T_s seconds in the delay variable, and the codes return to their initial time alignment. In real time, it requires t_s seconds to generate a sample of $R_{xr}(\tau)$ and we need k

samples to construct one period of $R_{xr}(\tau)$. Therefore it requires

$$T_m = k \times t_s = \frac{R_c}{\Delta R} \times T_s$$

seconds to take one set of samples of the ACF needed to construct a channel profile. During the measurement time T_m, the channel must be stationary, and therefore the value of T_m should be less than twice the inverse of the maximum Doppler spread of the channel in order that the measured ACF not be significantly affected by the changes in the channel due to the movements of the measurement terminal or of nearby objects or people.

The earliest use of this method for indoor radio measurements at 850 MHz was described by Devasirvatham [Dev84]. The PN sequence was applied to the carrier using biphase modulation, resulting in triangular autocorrelation pulses. The parameters were

$$R_c = 40 \text{ MHz}, \qquad \Delta R = 4 \text{ kHz}, \qquad N = 1023$$

Using these parameters, resolution (the width of the ACF) was $2/R_c = 50$ nsec, the symbol duration $T_s = N/R_c$ was 25.6 μsec, the measurement time T_m was 256 msec, and $k = 10,000$ samples were taken for each channel impulse response measurement. The transmit and receive antennas were sleeve dipoles, and the power into the transmit antenna was 26 dBm. The highest ratio of the output signal to the correlation noise level of the PN sequence was 40 dB.

The same parameters were used by Bultitude, Mahmoud, and Sullivan [Bul89] for extensive indoor radio measurements at 910 MHz and 1.75 GHz. Figure 5.3 shows the details of equipment used in the 910 MHz measurements. At the transmitter, the HP 5065A provided the reference signal for the transmitter and the receiver. The 40-MHz clock from the Rockland 5600 was used with the HP 3760A data generator to generate the PN sequence which in turn modulated the 70-MHz IF signal provided by the Fluke 6160B. The output of the Fluke was also passed through a 12-times frequency multiplier followed by an amplifier to generate an 840-MHz carrier. The carrier was mixed with the 70-MHz modulated IF signal and passed through an amplifier and a filter with 80-MHz bandwidth centered at 910 MHz. The modulated signal at 910 MHz was then amplified and fed through a monopole antenna. Using the reference 5-MHz signal from the transmitter and also using circuits similar to those in the transmitter, the reference 840-MHz carrier, a 70-MHz IF reference, and a PN sequence at the rate of 40 MHz − 4 kHz = 39.996 MHz were generated. The 840-MHz carrier was mixed with the amplified arriving signal, and the resulting signal was passed through a low-pass filter (LPF) followed by a chain of amplifiers with voltage-controlled gain to generate a 70-MHz IF modulated signal. At the two-channel sliding correlator, the PN sequence was modulated onto inphase (I) and quadrature (Q) 70-MHz carriers to provide the correlation reference. The received IF signal was mixed with the I and Q references and passed through the LPF integrators. These filters were single-pole RC filters having 3-dB cutoff frequencies of 4 kHz. The output provided the I and Q components and the squared envelope of the received demodulated signal.

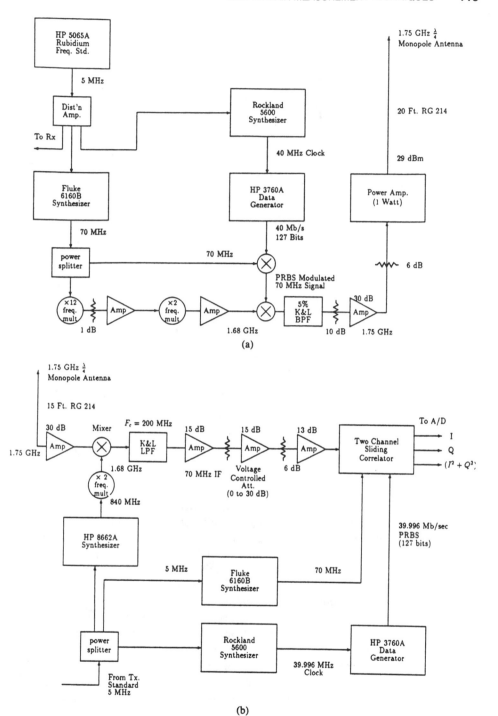

Figure 5.3 Details of a spread-spectrum measurement system used for indoor radio propagation studies. (*a*) Transmitter, (*b*) receiver. (From [Bul89] ©IEEE.)

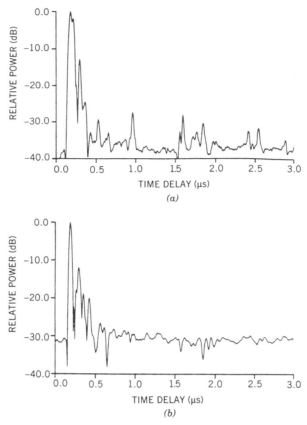

Figure 5.4 Sample of measured channel delay profile using a spread-spectrum system. (*a*) The 900-Hz band. (*b*) The 1.7-GHz band. (From [Bul89] ©IEEE.)

The I and Q signals were sampled at 16 kHz with a 12-bit A/D converter, and the sampled signal was stored in a digital computer. This sampling rate provided an effective sampling interval of 6.25 nsec for the recorded channel impulse responses. Considering the 50-nsec base width of the basic autocorrelation pulses, eight samples of each pulse are represented in the measurement results. Figure 5.4 shows samples of measurements taken with this system.

For mobile radio channel measurements the differences between the path lengths are larger, and therefore longer excess delay has to be measured, but coarser resolution is acceptable. Typical values of $R_c = 10$ MHz resulting in a resolution of around 0.2 μsec are used to measure the excess delay of up to $T_s = 102.3$ μsec for a PN-sequence length of $N = 1023$ [Par89]. The ΔR of 5 kHz provides $k = 2000$ samples per profile taken in $T_m = 204.6$ msec. With this rate we can capture accurate channel impulse responses for Doppler shifts up to about 2.5 Hz.

5.2.2 Measurements Using the Direct Pulse Transmission Method

An obvious way to measure the impulse response of the channel is to transmit a very short RF pulse and observe the multiple pulses received. This method was originally used for urban radio channel measurements [Tur72]. Recently, it has attracted a lot of attention for indoor radio propagation studies [Sal87b, Pah89, Rap89]. The pulse measurement systems discussed in this section are very similar in design. All incorporate noncoherent receivers, so that a power-delay profile is measured rather than an impulse response. Because the phase is not available, it is not convenient to find the exact frequency response of the channel by using the Fourier transform. The resolution and the measurement time for these systems are better than for the systems described in the previous section, but the dynamic range of measurements is more restricted.

In [Sal87b], a 1.5-GHz CW signal was modulated by a train of 10-nsec pulses with a 600-nsec repetition period. A vertically polarized omnidirectional discone antenna was used to transmit this signal. At the receiver, a similar antenna was followed by an amplifier chain and a square-law detector. A computer-controlled oscilloscope was then used to collect the received power delay profile. A coaxial cable was used to trigger the oscilloscope from the transmitter's pulse generator to guarantee a stable timing reference. Using measurements made with this system in one office building, rms delay spreads and power–distance relationships were calculated, and a statistical model for indoor multipath propagation was proposed. For the measurements and the model, the phases of the multipath components are assumed *a priori* to be statistically independent uniform random variables over $(0, 2\pi)$.

In [Rap89] a similar wideband measurement system was used to collect propagation data in factory settings. A 1.3-GHz carrier was modulated by a train of 10-nsec pulses with a 500-nsec repetition period. Discone antennas were used at both the transmitter and receiver. The receiving oscilloscope was triggered internally by the first received pulse in each power-delay profile measurement. Using measurements made with this system in five factory environments, rms delay spreads and power–distance relationships were calculated, and a statistical model for indoor multipath propagation was proposed. Comparing these measurements and some narrowband measurements made using the same measurement system [Rap89], an argument is made for the phases of the multipath components to be statistically independent uniform random variables over $(0, 2\pi)$.

We now describe the simple measurement system used in [Pah89] for wideband indoor radio propagation studies. Figure 5.5 shows a schematic diagram of the measurement setup. The setup operates at a 910-MHz carrier frequency in accordance with the FCC band allocations in the United States for secondary spread spectrum communication. The modulated carrier is fed to a 45-dB amplifier, and the output is transmitted with an omnidirectional quarter-wave dipole antenna placed about 1.5 m above the floor level. The stationary receiver uses the same type of antenna at the same height, which is approximately the height of an antenna mounted on top of a desktop PC. The antenna is followed by a step attenuator and a low-noise, high-gain ($\simeq$ 60 dB) amplifier chain. The signal is then detected using a square-law envelope detector whose output is displayed on a digital storage oscilloscope coupled to a PC with a GPIB instrument bus. The components used in the measurement setup have a flat frequency response in the 900 MHz range. The dynamic range of the receiver display is limited to 25 dB owing to the linear scale on the digital storage oscilloscope. However, the actual dynamic range of the measurement setup is more than 100 dB, which is achieved by manually adjusting the step attenuators at the receiver. A coaxial cable was

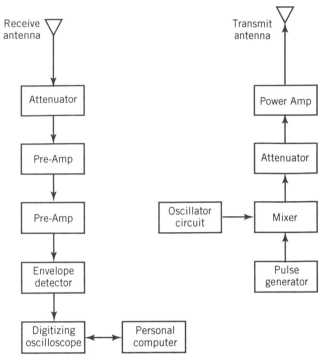

Figure 5.5 A simple pulse transmission measurement sytem used for wideband time-domain measurements of indoor radio propagation.

used to trigger the oscilloscope from the pulse generator of the transmitter, to guarantee a stable timing reference. Typically, 64 repetition periods are averaged by the oscilloscope to form a power-delay profile.

As in [Sal87b] and [Tur72], this measurement system is noncoherent and thus does not show the phases associated with the arriving paths. These phases are reasonably assumed *a priori* to be statistically independent uniform random variables over $(0, 2\pi)$ [Tur72, Sal87b]. Also the base width of the pulses in the received profiles in this system is 5 nsec and has better resolution than the apparatus used in [Sal87] and [Rap87]. Figure 5.6 shows three samples of measurements taken with this system in different locations. Figure 5.7 shows a three-dimensional plot of 20 multipath profiles in one line-of-sight (LOS) location. The system is capable of measuring and storing up to 20 complete profiles in 1 sec, which is adequate for observation of the effects of Doppler spread in the wideband signal.

If we assume the use of coherent detection rather than envelope detection and also assume that the transmitted pulses $p(t)$ are narrow enough to resolve all paths and that the multipath delay spread $\tau_L - \tau_0$ is less than repetition period of the transmitted pulses, one period of the received signal $r(t)$ is identical to the channel impulse response with impulses replaced by $p(t)$:

$$r(t) = \sum_{i=1}^{L} \beta_i p(t - \tau_i) e^{\phi_i} \simeq \sum_{i=1}^{L} \beta_i \delta(t - \tau_i) e^{\phi_i} = h(t)$$

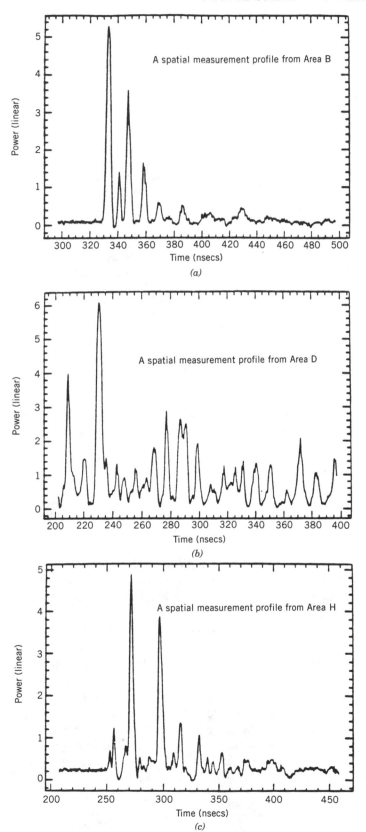

Figure 5.6 Samples of indoor multipath delay profiles measured using the time pulse transmission technique. (*a*) Area B, (*b*) Area D, (*c*) Area H.

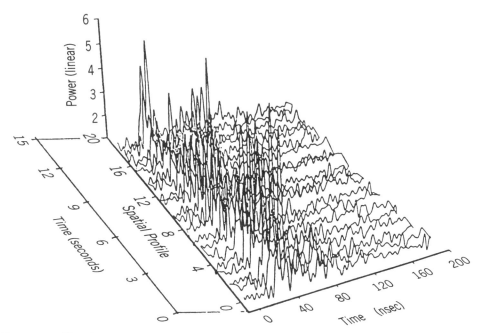

Figure 5.7 Plot of 20 multipath delay profiles measured to represent the short-time variations of the channel in one location [Gan93].

where β_i, τ_i, and θ_i represent the magnitude, arrival time, and phase associated with the ith path. Assuming that the envelope detector used in the measurement system resembles a square-law detector, the measured profile stored in the scope is represented by

$$|r(t)|^2 = \sum_{i=1}^{L} \beta_i^2 p^2(t - \tau_i)$$

in which the information related to the phase is eliminated by envelope detection. If we detect the peak of each individual arriving path in the stored profiles, the square root of its magnitude represents the amplitude β_i and its occurrence time, the arrival time τ_i.

5.3 RESULTS OF TIME-DOMAIN WIDEBAND MEASUREMENTS

The three parameters of interest extracted from the results of time-domain wideband measurements are the received power, mean excess delay, and the rms multipath delay spread. For the calculation of these parameters we only need the magnitude and arrival time of the paths, which can be found from the results of coherent or noncoherent measurements. To calculate these parameters from the magnitude and the delay of the arriving paths we use Eqs. (3.4.3) and (3.4.4).

In this section we examine the results of wideband measurements in order to compare the wideband characteristics of various building structures and to analyze the effects of movement and partitioning. In the last few years, several researchers have performed wideband measurements of indoor radio propagation at different frequencies. Here we provide an overview of these measurements with particular emphasis on measurements performed at Worcester Polytechnic Institute. For similar results in outdoor areas, for mobile radio applications, the reader can refer to [Par89] and [Tur72], and for a survey of other research in indoor radio propagation the reader can refer to [Has93a].

The behavior of radio propagation in outdoor and indoor areas follows the same general pattern. The mobile radio channel is characterized by higher multipath spread because of the longer distances involved in mobile communications. The mobile channel also exhibits higher values of Doppler spread because the terminals are intended to operate at vehicular rather than pedestrian speeds. Furthermore, path losses in different buildings exhibit wider variations than those observed in typical outdoor areas.

The measurements performed in various environments are broadly categorized as spatial, local, and mixed measurements. In a spatial experiment, the measurements are taken at points distributed throughout the test area, such as the floor of a building. This is done by fixing the receiver in a central location and moving the transmitter to various locations. The locations are selected based on the placement or likely placement of communication equipment for planned wireless networks. The surrounding environment is kept stationary during the measurement acquisition time by preventing movements close to the transmitter and receiver. The objective of spatial experiments is to determine the effect of location on the propagation parameters. The objective of local experiments is to determine the effect of local movements on the propagation parameters. Local movements are either (a) traffic in the vicinity of the measurement equipment or (b) movements (on the order of a wavelength) of the transmitter or receiver between measurements. A mixed experiment is a combination of a spatial and local experiment; that is, multiple measurements are taken around each location in a spatial distribution of locations.

5.3.1 Spatial Measurements

In this section we compare results of spatial measurements taken in five areas in three manufacturing floors, as well as in an office environment partitioned into three areas. We compare the measurements in terms of the distance–power gradient and the characteristics of the rms multipath delay spread. In a spatial measurement, the receiver was fixed in a central location. The transmitter was moved to various locations in each site, such as the likely positions of planned wireless terminals throughout the floor of a building. The received multipath profiles were measured and stored in a computer using the system described in Fig. 5.5. Each stored profile is a time average of 64 profiles collected during 15–20 sec at one location. During the measurement time, care was taken to prohibit any movements in the vicinity of the transmitter and the receiver. The distance between the transmitter and the receiver varied between 1 and 100 ft. A total of 526 profiles were collected from these measurements.

A manufacturing floor environment is typically characterized by large open areas, containing various items of machinery and equipment of different sizes. There are usually no walls between the transmitter and receiver, and a "direct" path is available for most of the locations. To indicate representative statistical characteristics of such an environment, we show the results of measurements made in five different manufacturing areas [Pah89]. Area A (Infinet Inc., North Andover, MA) is a typical electronics shop floor having a wide open area containing circuit board design equipment, as well as soldering and chip mounting stations. Area B (also at Infinet Inc.) includes test equipment and storage areas for common electronic equipment, partitioned by metallic screens. Area C (Norton Company, Worcester, MA) is a large open area containing grinding machines, huge ovens, transformers, and other heavy localized machinery. Area D (General Motors, Framingham, MA) is a car assembly line "jungle" floor having a dense array of welding and body shop equipment of all kinds. The environment, as presented to radio waves, results in many obstructions by, and signal reflections from, the various objects. Area E (also in the General Motors plant) is a vast open area used for final inspection of new cars coming off the assembly line. This area has many LOS paths between the transmitter and the receiver. The numbers of averaged profiles collected from the five areas were 54, 48, 75, 45 and 66, respectively, resulting in a total of 288 profiles representing the manufacturing environments. Table 5.1 provides short summary descriptions of the five manufacturing areas.

The typical office environment has less open space, and for most sites the "direct" path is obstructed by the presence of one or more walls. The environment, as presented to radio waves, has many reflections from the walls and ceilings. To represent the statistical characteristics of such an environment, three different office areas are considered. The office areas (Areas F–H) discussed in this set of spatial measurements are located on the third floor of the Atwater Kent Laboratories at the Worcester Polytechnic Institute. The floor plan was shown in Fig. 4.2. For these measurements, the receiver is located in the center of room 317, an electronics laboratory comprising typical equipment such as oscilloscopes, voltmeters, and power supplies on wooden benches. Area F is inside this laboratory, and hence all test locations in this area have a direct LOS to the receiver. Area G is the corridor (300I, D, G, and E) around this electronics laboratory,

TABLE 5.1   Short Description of the Results of Wideband Time-Domain Measurements in Five Manufacturing Areas and Three Offices

Measure-ment Area	Number of Measurement Locations	Distance – Power Gradient α	Max rms Delay Spread (nsec)	Median rms Delay Spread (nsec)	Mean rms Delay Spread (nsec)	Range of Power Fluctuations (dB)
A	54	2.348	40	15.29	16.64	30.34
B	48	3.329	60	31.62	29.03	39.85
C	75	2.185	152	48.90	52.38	35.50
D	45	2.196	150	52.57	73.13	28.02
E	66	1.398	146	19.37	33.13	24.97
F	54	1.76	48	12.40	15.75	18.0
G	96	2.05	55	44.19	39.53	24.50
H	88	4.21	146	50.3	55.19	28.53

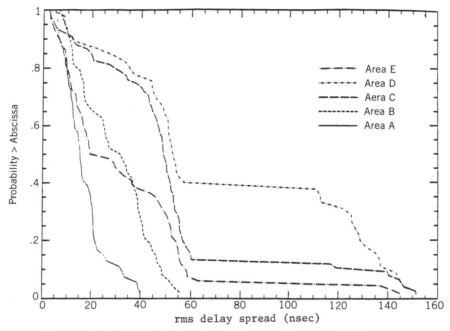

Figure 5.8 The cumulative distribution function of the rms delay spread of the measurements in five manufacturing areas.

separated in most parts by a sheetrock wall with metal studs and some glass windows. Area H consists of all the office rooms, 301–311, on the other side of the corridor, each having typical modern office equipment including a personal computer and a printer. This area is separated from the receiver by at least two walls of sheetrock and some glass windows in each wall. All the rooms in this area are very similar in structure and size. A total of 234 profiles collected from these areas included 84 in the offices, 96 in the corridor, and 54 inside the electronics laboratory.

Figures 5.8 and 5.9 [Gan93] show the cumulative distribution functions of the rms delay spread for the manufacturing floor areas (A–E) and the college building areas (F–H), respectively. Table 5.1 lists the maximum, median, and mean values of the rms delay spread measured in all the areas. Area A, which is a very open space, has the lowest mean and maximum rms delay spread among all the manufacturing floor areas. Area B exhibits higher values of the mean and the maximum rms delay spread. This is because, in most instances, the direct LOS path is obstructed by metallic objects, and therefore the received signal is composed of several reflections. Areas C and D have numerous pieces of machinery in a small localized area, and thus higher values for the rms delay spread. Areas E and F have very few metallic structures in the large open area and thus lower median values for the rms delay spread. Areas G and H are for the most part obstructed by one or two walls and metal doors, which results in higher values of the rms delay spread. Because of dense local reflections from the body shop and the welding equipment, area D has the highest average rms delay spread.

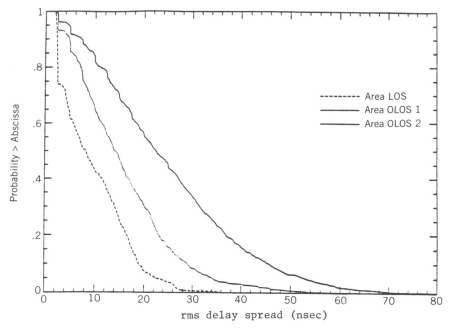

Figure 5.9 The cumulative distribution function of the rms delay spread of the measurements in three office areas.

The average value of the rms delay spread is thus dependent on (a) the availability of a "direct" LOS path between the transmitter and the receiver, (b) the size of the site, (c) materials used for the walls and the ceiling of the building, and (d) the objects in the area surrounding the transmitter and the receiver, locally and globally. For example, in Figs. 5.8 and 5.9, for the points to the left of the dashed "horizontal" line segments, an unobstructed "direct" path between the transmitter and the receiver was available. On the other hand, the LOS path between the transmitter and receiver was blocked by metallic objects for the points to the right of the dashed "horizontal" line segments, leading to higher values of the rms delay spread. Another interesting observation made in regard to areas F and G is the increase in the average delay spread value, due to signal propagation through one wall. When the signal had to propagate through two walls, the average rms delay spread increased further. This increase in the rms delay spread, observed when signal propagation occurs through one or more walls, is useful in assigning data rates to each cell, when a cellular indoor radio system is designed.

Table 5.1 also lists the values of the distance–power gradient α obtained for all the areas. The line-fitting method described earlier for narrowband measurements is used again here for the wideband power data. As described in Chapter 3 the distance–power gradients obtained from narrowband and wideband measurements are expected to be the same. Areas A, C, D, and G exhibited values of α between 2 and 2.5. This is due to the open areas available for unobstructed signal propagation and relatively few local surrounding objects taller than the antennas. Areas E and F had vast open areas, and they exhibited an α less than the theoretical α for free space. These open areas had very few objects in the local vicinity of the transmitter and receiver and generally afforded LOS paths between the transmitter and receiver. Area B included partitioning by metallic screens, and the transmitted signal was scattered locally by numerous pieces of equipment in the vicinity of the transmitter. This contributed to a higher value of α. Area H was separated from the receiver by two walls, and there were many local reflections by the walls and the ceiling inside the rooms. The highest value of α was observed in Area H. In manufacturing floor environments, though signal propagation through the walls is very unlikely, presence of a significantly large number of local reflecting objects may cause the value of α to rise. On the other hand, in office environments, presence of a large open space is less likely, but signal propagation through the walls greatly influences the value of α.

Most of the wideband indoor radio propagation studies in various buildings report maximum rms multipath delay spreads of around 100 nsec [Sal87b, Has93b, Dev91]; higher values are also reported in [Rap89a] and [Dev87]. The rms delay spreads reported for mobile radio channels are on the order of microseconds without distant reflectors such as hills [Cox72, Cox73] and are on the order of several tens of microseconds if there are distant reflectors [Rap90]. The excess delay spread for an indoor radio channel is usually on the order of several hundred nanoseconds, typically on the order of several microseconds [Par89, Tur72] without distant reflectors, and around 100 μsec with distant reflectors [Rap90].

5.3.2 Temporal Variations of the Wideband Characteristics

Local measurements are performed to determine the channel variations observed over a short time at a fixed location of the terminal. Such variations are experimentally induced by having people moving about in the vicinity of the fixed transmitter/receiver antenna or by manually shaking the antenna on its stand. The objective of these experiments is to compute and compare the statistics of rms delay spread and received wideband power observed in the multipath profiles for these variations. We now describe two sets of experiments performed to induce such variations, in one LOS and one obstructed LOS (OLOS) environment [Gan91a, Gan91b, Gan93]. The first set involved two persons walking briskly around the transmitter and the receiver, labeled as experiments A (LOS) and C (OLOS). The second set involved a person "shaking" and "wiggling" the transmitter antenna on its base, labeled as experiments B (LOS) and D (OLOS). These experiments were made with both transmitter and receiver stationary on the third floor of the Atwater Kent Laboratories. For the LOS experiments, both the transmitter and receiver were located in the central electronics laboratory. For the OLOS experiments, the receiver was located in the communications research lab, comprising typical office furniture and computers as well as radio communication equipment; the transmitter was located in a computer laboratory separated from the receiver by two walls having glass windows. The walls were made of plasterboard with metal studs.

For all four sets of data, the distance between the transmitter and the receiver was fixed at 10 m. A total of 400 profiles were collected from the four experiments. An adequate sampling rate of 20 samples per second was used to properly sample

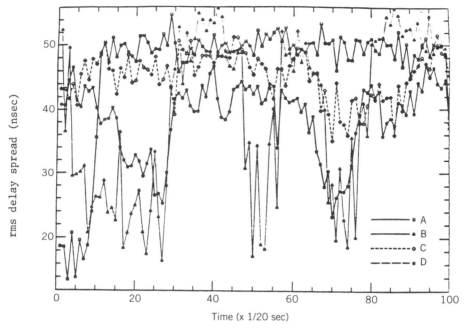

Figure 5.10   Temporal variations of the rms delay spread measured during four experiments.

the short-time variations. While the profiles were being stored during the experiment, care was taken to prohibit any other kind of activity or movement in the vicinity of the experiment.

The rms delay spread and the received power versus time were computed for each of the 100 profiles obtained within each set of experiments (A, B, C, and D). Figure 5.10 shows the variations of the rms delay spread for the four experiments. The variations in the rms delay spread for the LOS experiments A and B are about 40 nsec, while for the OLOS experiments C and D they are about 20 nsec. The standard deviations of the rms delay spread for the LOS sets A and B are 9.2 nsec and 12.8 nsec, while for the OLOS sets C and D they are 3.7 nsec and 5.7 nsec. Thus local variations of the rms delay spread in LOS channels, caused by pedestrian traffic, are greater than those observed in OLOS channels. Also, on the average, variations in the delay spread caused by local pedestrian traffic near the antennas were smaller than those observed for movements of the antenna.

Figure 5.11 shows the temporal fluctuations of the received multipath power for the four experiments. The range of short-time fluctuations in the multipath power was 7–9 dB in LOS and 5–6 dB in OLOS experiments. The standard deviations of fluctuations in multipath power for all the data sets were around 1 dB. These variations are far below the variations observed for similar experiments for narrowband communications discussed in the last chapter. This conclusion is consistent with the observation made from the results of two-dimensional ray tracing in Examples 3.4 and 3.6. Generally, local and temporal variations of the power are affected by the bandwidth of the communication system. As the bandwidth increases, the local and temporal variations decrease. The CDF of

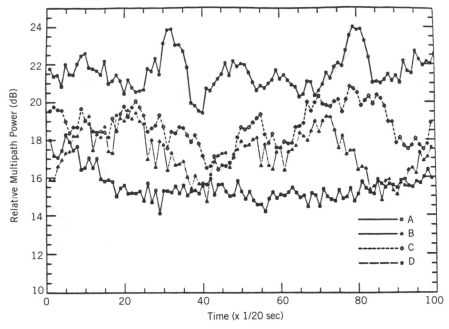

Figure 5.11 Temporal variations of the wideband received power measured during four experiments.

the multipath power from each of the above experiments was compared with the lognormal and Rayleigh distributions, and the results were shown to fit the lognormal distribution [Gan93]. The results of temporal and local variations of wideband signals did not exhibit Rayleigh characteristics, however. The multipath power at each location is the sum of squared magnitudes of the path amplitudes, which is independent of the phases of the paths. In narrowband signaling, the phase differences among the arriving paths produce Rayleigh-distributed multipath fading. The power in the wideband signals is averaged over faded and unfaded frequencies, and thus the frequency-selective fading is averaged over the entire band.

5.4 FREQUENCY-DOMAIN MEASUREMENTS

In frequency-domain measurements of radio propagation characteristics, the frequency response of the channel is measured directly. In indoor areas these measurements are performed conveniently with the aid of a network analyzer. The measurement system shown in Fig. 4.5, used for the narrowband Doppler-spread measurements discussed in Section 4.4, is used with a different network analyzer setup for wideband frequency-domain measurements. The main component of the measurement system, the network analyzer with a Fourier analysis option, measures the frequency response of the channel and takes its inverse Fourier transform to determine the channel impulse response.

A network analyzer system consists of a sine-wave source, signal separation devices, a receiver, and a display. In our measurement system, the network analyzer generates a linear stepped-sweep signal. The signals are fed to a receiver for measurement, processing, and display. The HP 8753B vector network analyzer integrates a high-resolution synthesized RF source and a dual-channel, three-input receiver to measure and display magnitude, phase, and group delay of the transmitted and reflected power. The HP 8753B option 010 has the additional capability of transforming measured data from the frequency domain to the time domain, using the Fourier analysis option.

The transmit portion of the frequency-domain measurement system consists of the network-analyzer-synthesized HP 8753B source. The built-in synthesized source of the HP 8753B produces a -15- to 20-dBm swept RF signal in the range of 900 MHz to 1.1 GHz. The output signal is passed through a cable to be used as the input of a Mini-Circuits ZHL-4240 power amplifier that amplifies the signal to 30 dBm (1 W). The output of the power amplifier is transmitted using a dipole antenna. The receiving portion of the frequency-domain measurement system consists of the receiving antenna, an attenuator pair, a low-noise amplifier, a high-gain amplifier, and the receiver portion of the network analyzer. The receiving antenna is of the same design as was used for the transmitter. The first attenuators are HP 8494A and HP 8496A passive step attenuators, and the cascaded attenuators provide 0–121 dB of attenuation in 1-dB steps. The second attenuator output is fed to an Avantek AWT-2071 low-noise, 2.2-dB noise figure and 12-dB gain amplifier followed by the Avantek AWT-2054 amplifier with 3.2-dB noise figure and 49.2-dB gain at 1.0 GHz. The output of the AWT-2054 is fed to the receiver portion of the network analyzer.

The network analyzer is controlled by a PC with a general-purpose instrumentation bus (GPIB) board. The PC initializes the network analyzer preceding each measurement, and it collects the data at the completion of the measurement. The magnitude and phase of the measured frequency response are the results typically stored for each measurement.

The measurements were made at a set of evenly spaced frequencies,

$$f_i = f_0 + if_s$$

where $f_0 = 900$ MHz is the lowest frequency in the band of interest and $f_s = 0.25$ MHz is the frequency-sample spacing. Each frequency sample is measured with the network analyzer dwelling for 0.5 msec at the selected sample frequency. The IF filter bandwidth of the network analyzer is set at 3 kHz. The frequency response consists of 801 complex samples, which require a collection time of 400 msec. From this frequency response a time response of 4000-nsec duration is derived using the Fourier transform option. The time response is truncated to show only the portion with significant energy. The 200-MHz bandwidth gives an equivalent resolution of 5 nsec in the time domain. This time resolution is comparable to the best resolution obtainable with time-domain measurement systems [Dev87, Bul89, Sal87b, Pah89, Rap89].

Figure 5.12 shows a plot of the magnitude and phase of a typical frequency response and the corresponding magnitudes of the time-domain response obtained by the inverse Fourier transform. The magnitude of the frequency response is

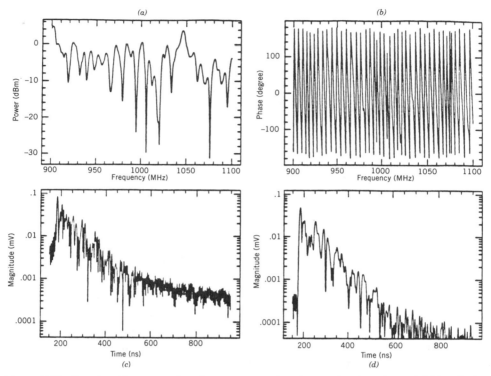

Figure 5.12 Results of frequency-domain measurements using a network analyzer. (*a*) Magnitude of the frequency response. (*b*) Phase of the frequency response. (*c*) Multipath delay profile obtained from rectangular windowed frequency response. (*d*) Multipath delay profile obtained from Hamming windowed frequency response.

shown in decibels, the phase of the frequency response is shown in degrees, and the magnitude of the time response is plotted on a logarithmic scale. In obtaining the time response from the results of frequency-domain measurements, the noise floor (height of the sidelobes) and the resolution of the pulses can be controlled by the type of window applied in the frequency domain before taking the inverse Fourier transform. Figures 5.12*c* and 5.12*d* show the time responses obtained from a frequency response which has been multiplied by a rectangular window and a Hamming window, respectively. The rectangular window provides the best resolution in the time response (5 nsec), but the detection of paths is more difficult because of the − 13-dB sidelobes. The smoother Hamming window provides a time response with sidelobes that are 40 dB down, but it provides a poorer resolution (around 8 nsec).

5.5 SOME FREQUENCY-DOMAIN MEASUREMENT EXPERIMENTS

Here we provide the results of some limited spatial, local, and mixed frequency-domain measurements using the system described in the previous section [How90c]. The spatial measurements were taken in 128 locations in two buildings. The area covered in each building is around 50 m $\times$ 50 m. The first set of spatial measurements (G1) was made in an IBM office, located on the sixteenth floor of the 32-story Shawmut Bank building in downtown Worcester, MA. The office consists of a central open area surrounded by adjoining smaller offices. The receiver was placed in a central location, and the transmitter was moved to 70 different locations for the various frequency response measurements. The second spatial measurement experiment (G2) was performed on the second floor of the three-story Atwater Kent Laboratories at Worcester Polytechnic Institute. The receiver was placed in the central computer terminal room. Measurements were taken from 58 locations near computer terminals in the same room, the adjacent laboratories, a laboratory across a hallway, and offices across another hallway.

For the local-measurement database, two experiments comprising a total of 60 frequency responses were performed. The first local experiment (L1) consisted of 28 measurements taken at one location. Measurements were made on the third floor of the Atwater Kent Laboratories with the receiver residing in the Communication Research Laboratory and the transmitter placed in the Electronics Laboratory. The shortest path between the transmitter and receiver passed through three walls. The measurements were taken with people moving in the vicinity of the transmitter or receiver so as to create the maximum amount of variation of the received time response. The second local experiment (L2) consisted of 32 measurements taken with the transmitter in the Electronics Laboratory and the receiver in the Communication Research Laboratory. For the first 16 local measurements the transmitter was fixed, and the receiver was moved over the 16 vertices of a 67.5-cm $\times$ 67.5-cm square with 22.5-cm grid spacing. For the second 16 measurements, the transmitter was moved on a similar grid, while the receiver was fixed.

A mixed experiment (M1) consisting of 621 measurements, nine measurements at 69 spatially distributed locations, was performed on the same floor as was experiment G2. Floor plan and locations for measurements are shown in Fig. 6.36. The receiver was placed in room 1 and the transmitter was moved to 69 different locations, some in the same room as the transmitter but most in the surrounding rooms. At each location, nine measurements were taken at the positions defined by the nine vertices of a 2-ft $\times$ 2-ft square with 1-ft grid spacing.

5.6 RESULTS OF WIDEBAND CHARACTERIZATION

The average power for each measurement was calculated by taking the received power at each frequency and averaging over the 801 measurement frequencies. The received wideband power in decibels versus the logarithmic distance was then fitted to a line to determine the distance–power gradient. The areas covered by G1, G2, and M1 give distance–power gradients of 2.6, 2.5, and 2.2, respectively. These results are comparable to results obtained from wideband time-domain measurements in similar areas.

To determine the cumulative distribution function (CDF) of the received power in different experiments, the results of all measurements in an experiment were compared with the theoretical CDFs for lognormal, Weibull, and Rayleigh probability distributions. The lognormal distribution provided the best fit for all the

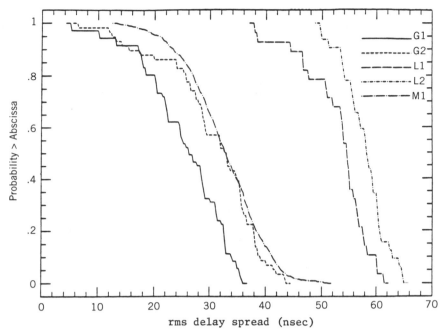

Figure 5.13 CDF of the rms delay spread obtained from the results of five experiments using the frequency-domain measurement system.

experiments. This observation is consistent with the wideband time-domain measurements reported in [Gan91a, b]. The spatial power fluctuations due to shadow fading were on the order of 30 dB, which corresponds well with the distances covered. As expected, for the local wideband measurements, the power fluctuations due to multipath fading were small.

Figure 5.13 shows the CDF of the rms delay spread for the five experiments. The means and variances of the rms delay spreads in different experiments are shown in Table 5.2. Using a rectangular window, the mcan rms delay spreads obtained for different areas are between 19 and 47 nsec. These results agree closely with those of time-domain measurements reported for similar buildings [Bul89, Sal87b, Pah89, Gan91a, b]. With the Hamming window, the mean rms delay spreads increase by 7–10 nsec for each area; however, choice of windowing shows no effect on the standard deviations or the received power. Sidelobes of the Hamming window are considerably lower than those of the rectangular window, allowing more paths to be detected. These paths have small amplitude and large delays. Therefore, they can affect the rms delay spread significantly while having a neglible effect on power.

The agreement between the CDFs for experiments G2 and M1 is consistent with the overlap of the measurement areas for those experiments. The average of the rms delay spreads for the experiments L1 and L2 is higher than for experiments G1, G2, or M1. In all measurements in the L1 and L2 experiments there were three walls between the transmitter and receiver, and the measurements were purposely taken in a location with a large τ_{rms}.

TABLE 5.2 Summary of the Results of Two Global, Two Local, and One Mixed Measurement Experiment Using the Wideband Frequency-Domain Measurement System

Experiment	No. of Meas- ure- ments	3-dB Width Median (MHz)	3-dB Width Mean (MHz)	3-dB Width S.D. (MHz)	rms Delay Spread Mean (nsec)	rms Delay Spread S.D. (nsec)
G1	70	5.75	7.74	6.51	19.11	7.56
G2	58	5.25	7.41	7.47	21.03	8.06
L1	28	2.75	3.11	1.13	39.86	9.1
L2	32	2.5	2.56	0.22	46.92	4.74
M1	621	4.75	5.57	2.84	23.1	7.46

5.6.1 Frequency Correlation Function

The measured samples of the frequency response $H(f_n, t)$ can be interpreted as a random process. The autocorrelation function of this process,

$$R_{Hh}(k, 0) = \frac{1}{N} \sum_{i=1}^{N-k} H^*(f_i, t) H(f_{i-k}, t), \qquad k \geq 0$$

is an important function. This function provides the average received power

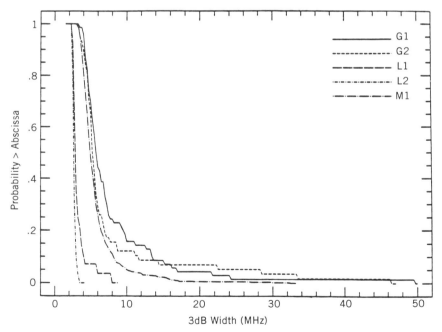

Figure 5.14 CDF of the 3-dB width of the frequency correlation function obtained from the results of five experiments using the frequency-domain measurement system.

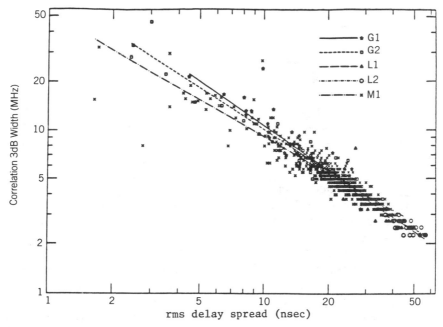

Figure 5.15 The 3-dB width of the frequency correlation function versus the associated rms delay spread and the best-fit line.

$[P_r = R_{Hh}(0,0)]$, and it is used for the *autoregressive modeling* discussed in the next chapter. It is also related to the rms delay spread of the channel. The 3-dB width of $|R_{Hh}(k,t)|$ is a measure of the similarity or coherence of the channel in the frequency domain, which is inversely proportional to the delay spread of the channel.

Figure 5.14 shows the CDF of the 3-dB width B_c for each of the five experiments. Table 5.2 gives the median, mean, and standard deviation of the 3-dB width for each experiment. Because it is only one location which is repeatedly measured for each local experiment, variations of 3-dB widths are small compared with those found in spatial experiments. We also note that the local movements of the transmitter in experiment L2 yield the smallest variation.

Using the inverse relationship between width in the frequency domain and duration in the time domain, a relationship of the form $B_c = C\tau_{rms}^{-\beta}$ between the 3-dB width (MHz) of the frequency correlation function and the rms delay spread (nsec) of the channel is determined from a linear regression (on logarithmic scales). Figure 5.15 shows the scatter plot, on a log–log scale, of the 3-dB width of the frequency correlation function versus the corresponding rms delay spread for all experiments. It can be seen that for the spatial experiments, the β values are around 1. Generally the inverse relationship between the multipath delay spread and the width of the frequency correlation function indicates the direct relationship between the coherent bandwidth of the channel and the 3-dB width of the frequency correlation function.

5.7 EFFECTS OF THE FREQUENCY ON THE RESULTS OF MEASUREMENTS

Most of the measurements and modeling methods discussed in this book are obtained for frequencies around 1 GHz. The evolving wireless information networks use frequencies from around 1 GHz up to 60 GHz. Operating frequencies of most of the mobile services are near 1 GHz, and the evolving PCS systems mostly operate around 2 GHz. The ISM bands at 910 MHz, 2.4 GHz, and 5.7 GHz and the frequency bands around 17–19 GHz are used by commercially available WLANs. Frequency bands at 60 GHz are also being considered for high-speed indoor radio applications. In this section we compare the results of measurements at around 1 GHz with measurements made at other frequencies.

In free space, as shown by Eq. (3.2.1), an increase in the frequency of operation affects only the power loss in the first meter but does not change the distance–power gradient giving the relative power loss with increase in distance. This conclusion remains the same for real LOS environments. In an OLOS environment all the paths arrive through mechanisms of reflection, transmission, or diffraction. All these mechanisms are frequency-dependent, and therefore the magnitudes and the phases of the arriving paths will change with frequency. In principle, changes in the magnitudes and phases of the arriving paths should affect the distance–power gradient, delay spread, and Doppler spectrum of the channel.

The results of measurements made at frequencies from 0.5 to 3.0 GHz on mobile radio channels show no significant variations in the statistical parameters as observed over that frequency range [Tur72, Suz77]. This conclusion holds as well for indoor radio propagation studies reported for 1–2 GHz [Bul89, Dev91]. The depth of penetration of the signal reduces as the frequency increases, therefore the power loss due to the walls or the floors increases with an increase in the frequency [Mot88a, Mot88b]. As the frequency increases by an order of magnitude to several tens of gigahertz, the reflected, transmitted, and diffracted paths become weaker. This results in confinement of the signal in the room and reduction in the multipath spread of the channel [Dav91].

Doppler shift increases linearly with increasing frequency of operation. Therefore, the maximum Doppler shift frequency, representing the maximum rate of variations in the channel, increases linearly with increasing frequency. The results of two-dimensional ray tracing in a single room shown in Fig. 3.5 demonstrate the increase in the rate of fluctuations in the channel as the frequency increases. The range and the statistics of the amplitude fluctuations, however, remain essentially constant with frequency.

5.8 COMPARISON BETWEEN MEASUREMENT SYSTEMS

Frequency-domain measurement using a network analyzer is the simplest method of channel measurement for use in small areas. For short distances on the order of several meters, the network analyzer and two antennas can be used, without the need for any additional amplifiers, to measure wideband indoor radio propagation. As the distances increase, we need to add transmitter and receiver amplifiers to the measurement system. The noncoherent pulse transmission technique with envelope detection is also easy to implement. In addition to the amplifier, we need a pulse generator, a mixer, and an oscillator at the transmitter, and an envelope detector at the receiver. The most expensive component in this setup is a digital high-frequency scope used to display and store the channel impulse response. If the phase information is needed, the in-phase and quadrature-phase circuitry are also needed, which adds to the complexity of the system. Without phase information the Fourier transform does not provide the correct frequency response of the channel. The spread-spectrum system requires all the circuitry used for the pulse transmission setup, with additional circuits to spread the signal at the transmitter and perform correlation at the receiver.

Among the three specific systems that we described in this chapter, the pulse measurement system provides the finest time resolution, around 3 nsec; while the resolution with the frequency-domain measurement system is around 5 nsec, and the resolution with the spread-spectrum system is about 40 nsec. The network analyzer in the frequency-domain measurement system had the Fourier transform option, providing the time- and frequency-domain measurements simultaneously. However, the time-domain measurements made with the network analyzer show a higher number of detected paths per measurement as compared with the results of pulse measurement techniques.

The ratio of the peak to average transmission power observed with the pulse measurement technique is much higher than with the spread-spectrum and frequency-domain measurement systems. As a result, with the same transmitter and receiver amplifiers, the spread-spectrum and the frequency-domain measurement systems have better coverage than the pulse transmission technique. The acquisition time of the pulse transmission technique in the systems mentioned in this chapter was the lowest among the three, and it was the only method with which we could measure the temporal variations of the wideband indoor radio channel. The frequency-domain measurement system could only measure the temporal variations of the narrowband signals.

QUESTIONS

(a) How do we measure the distance–power gradient on a radio link?

(b) How do we measure the Doppler spread?

(c) How do we measure multipath delay profiles?

(d) If the transmitter and the receiver are fixed, can the channel have any Doppler spread? Explain.

(e) What is the simplest system for measurement of indoor radio propagation?

(f) Which measurement technique is most traditional?

(g) Considering the results of indoor radio propagation measurements reported in this chapter, what is a reasonable number for the maximum rms delay spread in indoor areas? How does it compare with the delay spreads adopted for the JTC model?

(h) What is the range of distance–power gradients for indoor radio propagation measurements as reported in this chapter? How does it compare with those given by the JTC model?

(i) What is the maximum Doppler shift due to measurement of the local short-term variations, as reported in this chapter? What type of movement has caused this maximum fluctuation?

(j) What is the maximum variation in rms delay spread due to measurement of local short-term variations, as reported in this chapter? What type of movement has caused this maximum fluctuation?

(k) Explain how we use the results of time-domain measurements, made with the pulse transmission method depicted in Fig. 5.5, together with the FFT, to produce an accurate representation of the channel frequency response. Consider the question again using the time-domain measurement approach depicted in Fig. 5.3.

(l) Discuss the resolution capability of the three measurement systems reported in this chapter.

(m) Which of the three measurement techniques is most difficult to implement, and which is the simplest?

(n) Given equal transmission powers, which measurement system provides the largest coverage area?

(o) In the spread-spectrum measurement system, if we keep the bandwidth fixed, how can we increase the resolution?

(p) How does the frequency correlation function relate to the data rate limitations of the channel?

(q) What are the advantages and disadvantages of the spread-spectrum, time-domain measurement method as compared with the pulse transmission method?

PROBLEMS

Problem 1. We want to use the spread-spectrum technique with the sliding correlator for measurement of the multipath characteristics of the urban radio channel. Assume that the maximum delay spread to be measured is 10 μsec, the resolution of the measurement system (the base of the autocorrelation function of the spreading code) is 100 nsec, and the maximum Doppler shift of the channel is 100 Hz. If the system is transmitting at 800 MHz, the pulses are ideal rectangular, and the chip values are independent from one another, give a block diagram for the measurement system. Specify the bandwidth expansion factor N, the chip rate R_c, the difference between the chip rates at the transmitter and the receiver, and the required transmission bandwidth of the system. If the transmission power is 10 W, what is the coverage of the system if it is used in a macrocellular environment (use JTC model) with transmitter height of 20 m and receiver height of 4 m?

Problem 2. How can we modify the block diagram shown in Fig. 5.3 so that the system can be used for direct pulse transmission measurement? In this case, only one pulse with duration of one chip is transmitted every N chip durations. How does the average transmitted power and the transmitted bandwidth differ from that of the original spread-spectrum measurement system?

Problem 3. Give a block diagram and detailed specification of equipment specifications for a frequency-domain measurement system that duplicates the measurements taken by the spread-spectrum system described by Fig. 5.3.

Problem 4. The frequency-domain measurements for shorter distances can be done with only a network analyzer and two antennas. Simple antennas used in portable phones or cellular phones can be used for this purpose. These antennas are usually omnidirectional monopoles with antenna gains of around 2 (3 dB). The typical maximum transmitted power from the network analyzer is -20 dBm, and the noise level at the receiver end of the analyzer is -90 dBm. If we use this simple setup for measurement, what are the typical distances that will enable us to measure the channel characteristics at 1 GHz? (If a network analyzer is available, set up this simple experiment and confirm the accuracy of your calculations.)

COMPUTER SIMULATION
OF WIDEBAND RADIO CHANNELS

6.1 INTRODUCTION

In the previous three chapters we defined the parameters that characterize
multipath fading, described the systems used for measuring these parameters, and
presented the results of measurements made on various radio channels. The
measurement results were divided into two categories, narrowband and wideband
measurements. The results of narrowband measurements addressed signal coverage through the power–distance relationship, and they related the Doppler spread
to the movements of objects in the coverage area and the movements of portable
or mobile terminals. We then discussed the statistics of the amplitude fluctuations
of a received narrowband signal and its Doppler spectrum in indoor and outdoor
areas and also described methods for computer simulation of narrowband amplitude fluctuations. For the case of wideband signaling, we introduced various
methods for measuring the multipath characteristics of the radio channel for
portable and mobile users and presented results of some time- and frequency-
domain measurements. We also discussed results of measurements of the root
mean square (rms) multipath spread and the 3-dB width of the frequency correlation function in different buildings, as well as the effects of partitioning and
short-time variations. Finally we provided a comparison among different measurement techniques.

In this chapter we describe methods for computer simulation of wideband radio
propagation, which includes the multipath characteristics of the channel. In the
past, performance evaluations of communication systems were typically based
upon simple statistical models of the channel and closed form solutions providing
bit-error rates (BERs) for different modulation techniques. With the rapid increase
in computational power of computers and drastic reduction in their cost, computer
simulation is becoming an increasingly popular approach to performance evaluation. Many versatile software products are also becoming available for use in
developing communication system simulations, including "block-oriented" simulation packages which offer many conveniences to the user.

Ideally, a simulation should provide "snapshots" of the wideband channel
response, in either the time or frequency domain, at a rate twice the Doppler

spread of the channel. A complete simulator of this form provides both static and dynamic behavior of the channel. Computer simulation of the channel is used for performance evaluation of modems, analysis of multiple access methods, placement of base stations in a cellular system, and analysis of interference in various networks. For most of these applications the channel response as a function of location is of primary importance, and a description of the static behavior of the channel is adequate. The channel model provides static snapshots of the channel impulse response at different locations, to be used for evaluations under various performance criteria, such as probability of outage or average probability of error of a specific modulation technique over a prescribed area. The dynamic behavior of the channel is needed primarily for detailed analysis of the behavior of the adaptive functions of modems, such as automatic gain control (AGC), equalization, and timing recovery.

There are two basic approaches to simulating wideband radio propagation characteristics: (1) measurement-based statistical modeling and (2) direct analytical solution of the radio propagation equations. Measurement-based statistical models are based on a mathematical description using several parameters. The parameter values are evaluated for each individual measurement of the wideband channel characteristics, and the statistics of the parameters over a large database are used to complete the model for a given coverage area. Statistics gathered from measurements in typical areas are extended to develop a more generalized model for all coverage areas. Statistical models generally do not incorporate details of the siting of buildings in an outdoor coverage area or the layout of rooms within a building. Instead they classify all areas into a limited number of broadly designated environments and all buildings into a few classes of buildings.

In modem performance evaluations, the system designer is usually concerned with the overall performance over typical areas or typical buildings, and statistical models usually serve the purpose reasonably well. In some other application such as microcellular or indoor installations, where proper siting of antennas is an important issue, building-specific radio propagation models offer a more precise tool for determining optimum antenna locations. Building-specific radio propagation models are based on direct solution of the radio propagation equations with boundaries defined by a map of a coverage area or the layout plan of a building. The technique known as *ray tracing* provides a simple approximation for analysis of radio wave propagation. Another approach is numerical solution of the Maxwell equations using the *finite-difference time-domain* (FDTD) technique.

To compare the results of various computer simulation techniques, several approaches might be taken. The most obvious is to compare the measured and simulated channel responses in typical locations. This method is not well suited to evaluation of statistical models, because statistical models do not relate the channel response to a specific location. However, for assessing building-specific radio propagation models, this method is very useful. Another approach to evaluating the results of a simulation method is to compare empirical data with the cumulative distribution functions (CDFs) of the rms delay spread and multipath power produced by the simulation. Yet another approach to comparing radio propagation models is to evaluate the performance of a particular modem over the measured and modeled channels. Standard modulation techniques such as BPSK and wideband techniques such as direct-sequence spread-spectrum or non-spread signaling with adaptive equalization can be used as benchmarks in these evaluation approaches.

In the following sections we describe various methods for simulation of the radio channel and we compare simulation results with the results of wideband measurements.

6.2 WIDEBAND TIME-DOMAIN STATISTICAL MODELING

Time-domain techniques using wideband statistical models are the most popular methods for computer simulation of indoor and outdoor radio systems, for both mobile and personal communications service (PCS) applications. Standards-setting bodies usually recommend a generalized and simple time-domain wideband statistical model for simulation of the radio channels.

The mathematical model used to describe the time-domain characteristics of the channel assumes the propagation medium to act as a linear filter and it defines the impulse response, $h(\tau, t)$, as a function of the delays $\{\tau_i\}$, amplitudes $\{\beta_i\}$, and phases $\{\phi_i\}$ of the signals arriving along different paths, as follows:

$$h(\tau, t) = \sum_{i=1}^{L} \beta_i \delta(t - \tau_i) e^{j\phi_i} \tag{6.2.1}$$

This mathematical formulation was first suggested for statistical modeling of the urban radio channel by Turin [Tur72] and later used for statistical modeling of the indoor radio channel [Sal87b, Gan91a, Gan91b, Rap91b, Yeg91, Has93a]. The simple and generalized models recommended by the GSM standards body for mobile radio channel modeling and by the Joint Technical Committee (JTC) for PCS channels are based on the same mathematical formulation [GSM91, JTC94]. To develop a statistical model for computer simulation using this formulation, we need the statistics of the arrival delays, amplitudes, and phases of the signals received along different paths. In the Chapter 5 we showed how to evaluate a single set of path arrival delays and amplitudes. In this section we discuss the statistics of these parameters.

6.2.1 Path Arrival Times

A simple statistical model for the path arrivals is a *Poisson process*, the model typically used for characterizing random arrivals in queuing theory analysis. On indoor and outdoor radio links, if the objects causing the multipath are located randomly throughout the space surrounding the link, the Poisson distribution should provide a good model for the path arrivals. However, the results of several studies of urban [Tur72, Suz77] and indoor [Gan89, Yeg91, Has93b] radio environments have shown that the Poisson distribution does not closely match the results of empirical measurements. This observation suggests that on indoor and urban radio channels the spatial distribution of the objects causing multipath cannot be accurately described as totally random. In this section, closely following the experimental results of Ganesh and Pahlavan [Gan89], we provide an explanation of this phenomenon.

To evaluate the accuracy of the Poisson model for path arrivals, we examine the results, described in Chapter 5, of wideband indoor radio propagation measurements made in manufacturing and office areas. The path arrival distribution given by the Poisson model is compared with the empirical data to determine the degree of closeness. The time axis of each measured time domain channel profile is divided into bins of width 5 nsec, which is the pulse width used by the measurement system. The existence of a path in a bin is determined by comparing the peak value of the signal in each bin with a certain threshold set according to the level of the background noise. If the peak value is higher than the threshold, we declare that a path exists in the bin. The number L of paths in the first N bins of each measured profile is determined. Then the probability of having L paths over all the measured profiles is calculated (the first path, which always exists, and serves as the reference for the delay times, is not included in the calculation). To determine the empirical path index distribution, the probability of receiving l paths in the first N bins $P_N(l)$ is plotted against l. This procedure is repeated for $N = 5, 10, 15,$ and 20 bins.

The Poisson process is a one-parameter model of "totally random" events occurring at a fixed average rate λ. The probability $P_N(l)$ for the theoretical Poisson path index distribution is given by

$$P_N(l) = \frac{\lambda^l}{l!} e^{-\lambda} \qquad (6.2.2)$$

where l is the path index, and λ is the mean path arrival rate, given by

$$\lambda = \sum_{i=1}^{N} r_i \qquad (6.2.3)$$

In this equation, r_i is the path occurrence probability for bin i, defined as the ratio of the number of times we have detected a path in bin i to the total number of profiles used for statistical modeling.

Figures 6.1 and 6.2 provide a comparison between the empirical path index distributions and the theoretical Poisson path index distributions [Gan89] for $N = 5, 10, 15,$ and 20 bins. The figures correspond to the manufacturing floor areas and the office areas, respectively. For clarity the results are plotted as continuous curves, though they have values only for integer path numbers. We observe considerable discrepancy between the empirical and Poisson distributions for all values of N, irrespective of the environment. This discrepancy reflects a tendency of the paths to arrive in groups, rather than in a random manner.

To explain these discrepancies, a modified Poisson model was proposed by Suzuki [Suz77] for characterizing urban radio channels. This model was subsequently extended to indoor radio propagation [Gan89]. Figure 6.3 summarizes the description of the modified Poisson process. For the modified Poisson process, the probability of having a path in bin i is given by λ_i if there was no path in the $(i - 1)$st bin, or by $K_N \lambda_i$ if there was a path in the $(i - 1)$st bin. The "underlying" probabilities of path occurrences λ_i are related to the empirical path occurrence

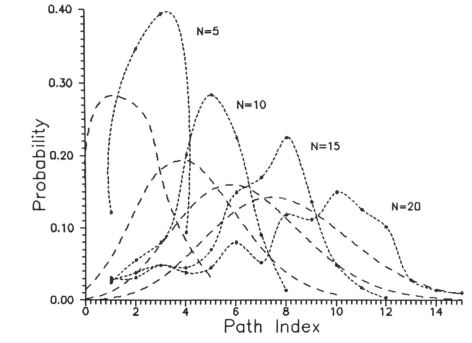

Figure 6.1 Empirical ($\cdots$) and theoretical (---) Poisson path index for manufacturing floors.

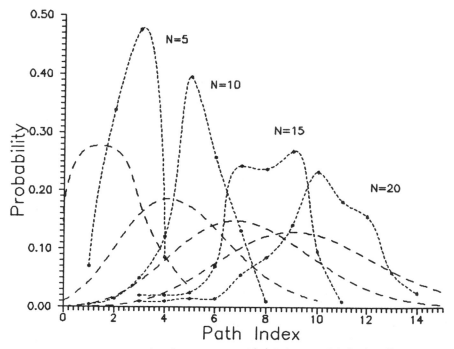

Figure 6.2 Empirical ($\cdots$) and theoretical (---) Poisson path index for offices.

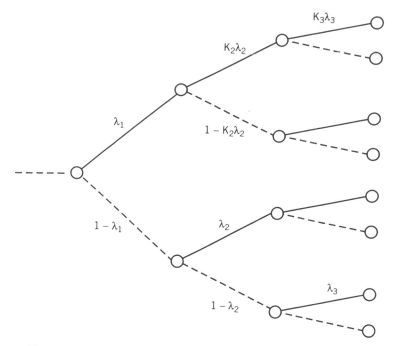

Figure 6.3 Tree structure representing the modified Poisson process.

probabilities r_i by

$$\lambda_i = \frac{r_i}{(K_N - 1)r_{i-1} + 1}, \qquad i \neq 1 \tag{6.2.4}$$

where $\lambda_1 = r_1$.

The modified Poisson path index distribution is related to the $\{\lambda_i\}$ by the following recursive equations [Suz77]:

$$P_i(l) = P_{1,i}(l) + P_{2,i}(l)$$

$$P_{2,i+1}(l) = P_{2,i}(l-1)K_N\lambda_{i+1} + P_{1,i}(l-1)\lambda_{i+1}$$

$$P_{1,i+1}(l) = P_{2,i}(l)(1 - K_N\lambda_{i+1}) + P_{1,i}(l)(1 - \lambda_{i+1})$$

where $P_i(l)$ is the probability of having l paths in the first i bins, $P_{1,i}(l)$ is the probability of having l paths in the first i bins conditioned on having no path in the ith bin, and $P_{2,i}(l)$ is the probability of having l paths in the first i bins conditioned on having one path in the ith bin. The process begins in bin 1, where $P_{1,1}(0) = 1 - \lambda_1$, $P_{2,1}(1) = \lambda_1$, $P_{1,1}(l) = 0$ for $l \geq 1$, and $P_{2,1}(l) = 0$ for $l \geq 2$ or $l \leq 0$. Starting with a small value of K_N and minimizing the mean-square error between the empirical distribution and theoretical modified Poisson path index distribution found using the equations above, optimum values of K_N are found from the data for $N = 5$, 10, 15, and 20 bins. To aid in simulation, the optimum values of K_N, which are functions of the "number of bins N," are replaced by new

TABLE 6.1 Parameters K and λ for the Modified Poisson Process in Manufacturing Floors and Office Areas

Bin Number	Manufacturing Floors		College Floors	
	K_i	λ_i	K_i	λ_i
1	0.964	0.4513889	0.697	0.3636364
2	0.932	0.5418182	0.666	0.6408464
3	0.900	0.5217417	0.634	0.6288999
4	0.868	0.4942650	0.602	0.7164743
5	0.836	0.5146308	0.570	0.7313251
6	0.804	0.5442811	0.539	0.7230882
7	0.772	0.5484897	0.507	0.7350161
8	0.740	0.5127087	0.475	0.8171411
9	0.708	0.5711924	0.443	0.7717720
10	0.676	0.3649115	0.411	0.6645216
11	1.340	0.3444129	0.589	0.6733878
12	1.440	0.2928466	0.595	0.6594670
13	1.540	0.3592564	0.602	0.6470692
14	1.636	0.2782416	0.608	0.6743552
15	1.732	0.1192904	0.614	0.4608082
16	5.796	0.1300228	1.056	0.4402552
17	6.388	0.1044048	1.117	0.4432566
18	6.980	0.1019357	1.177	0.3898038
19	7.572	0.0901229	1.238	0.3576177
20	8.164	0.0810876	1.299	0.3297661

parameters $\{K_i\}$ ($i = 1, 2, \ldots 20$), which are functions of the bin numbers $\{i\}$ [Has79]. The $\{K_i\}$ are determined by linear interpolation. For the final calculation of the modified Poisson path index distribution, the equations above are used again with interpolated $\{K_i\}$ ($i = 1, 2, \ldots, 20$) replacing K_N ($N = 5, 10, 15, 20$). Table 6.1 shows he optimum values of K_i and λ_i calculated for the measurements made in the manufacturing floors and college office areas.

Figures 6.4 and 6.5 provide a comparison between the empirical path index distributions and the modified Poisson distributions, for the manufacturing floors and the college office areas, respectively. The curve fittings show considerable improvement over those shown in Figs. 6.1 and 6.2 for the Poisson model. This suggests that the paths do not arrive randomly but in groups, and the presence of a path at a given delay is greatly influenced by the presence or absence of a path in earlier bins. In mathematical terms, the modified Poisson model utilizes the empirical probability of occurrence for each bin, while the Poisson model simply uses the sum of the probabilities of occurrence for all bins.

Two other approaches are used to modify the Poisson arrival model. The first approach, suggested in [Sal87b] for indoor radio propagation, assumes that the paths arrive in clusters. The path arrivals in each cluster, and the arrivals of the clusters, are both assumed to be Poisson processes. The problem encountered with this approach is that there is no reliable way of directly identifying the clusters from the results of measurements. This prevents us from developing a logical and systematic means of determining the model parameters from measurement data. In only a small fraction of measurements can one recognize a pattern of having

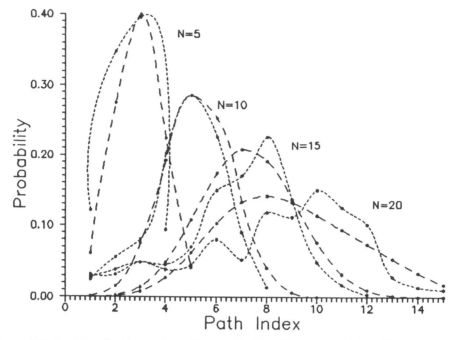

Figure 6.4 Empirical ($\cdots$) and theoretical modified (---) Poisson path index for manufacturing floors. Asterisk denotes that clustering property is exhibited.

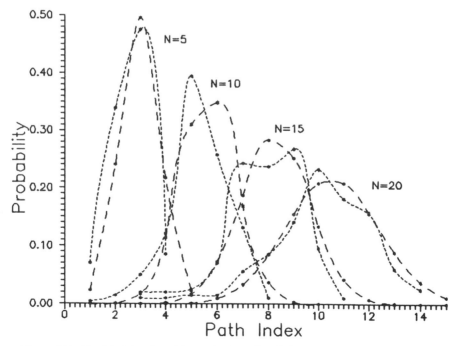

Figure 6.5 Empirical ($\cdots$) and theoretical modified (---) Poisson path index for offices.

more than one cluster. We should also caution that the model developed in [Sal87b] is based on a rather limited set of measurement data.

The second approach to modifying the Poisson process is to analyze the interarrival delays. The interarrival delay for the Poisson distribution is exponentially distributed, but one may consider other distributions for the interarrival times. It is shown in [Yeg91], based on extensive measurements in manufacturing floors, that the Weibull distribution best fits the interarrival delays of the time-domain model. The Weibull distribution has multiple parameters and therefore can provide a better fit to any empirical distribution than the one-parameter Poisson model. However, unlike the Poisson model and its variations, the Weibull-based model lacks an obvious physical interpretation.

6.2.2 Path Amplitudes

The simplest method of modeling the path amplitudes is to assume that each measured path is the phasor sum of several paths arriving so close to one another that they are not distinguishable by the measurement system. With this assumption, the amplitude fluctuation of each path follows a statistical pattern similar to that of the amplitude of a narrowband signal. The small-scale variations form Rayleigh and Rician distributions for the obstructed line-of-sight (OLOS) and LOS cases, respectively. The large-scale variation of the mean of the amplitude fluctuations is modeled by a lognormal distribution. The Doppler spectrum of each path would follow the Jakes spectrum for mobile radio applications and would follow uniform spectra for indoor wireless applications.

To justify the validity of these assumptions we examine the distribution of path amplitudes measured in the two wideband indoor radio propagation experiments discussed in the previous section. The discussion closely follows the experimental results reported by Ganesh and Pahlavan [Gan91b]. We divide the time scale into 5-nsec bins and record the path amplitudes in the bins found to contain paths. The statistics of the amplitude fluctuations are then analyzed for each individual bin. The curve-fitting approach introduced in Chapter 4 is used to find the distribution of the path amplitudes in each bin. Figures 6.6 and 6.7 give the comparisons between the theoretical and the experimental distributions for bin 1 in the manufacturing floor areas and bin 5 in the college office areas, respectively. The horizontal axis is normalized to the measured median signal value in decibels. Evaluation of the distribution over individual bins reveals that the lognormal and Suzuki distribution functions provide the closest fit to the measured data. The lognormal assumption is consistent with the previous models, and computer simulation with this approach is simpler than with the other models. Therefore we judge this method to be the preferred approach to modeling of received amplitudes.

The inhomogeneities of the radio channel result in variations in the mean and variance of the amplitudes from one delay to another. To simulate these changing parameters, we need to know the distribution of their variations. The scatter plots of mean and standard deviation of the lognormal distribution versus delay were fitted to decaying exponentials of the form $Ae^{-\tau/T} + B$, where T is the decay rate, τ is the delay, and A and B are constants [Gan91a]. Note that the decay rate is defined with respect to bin number, which models the arrival delay relative to the

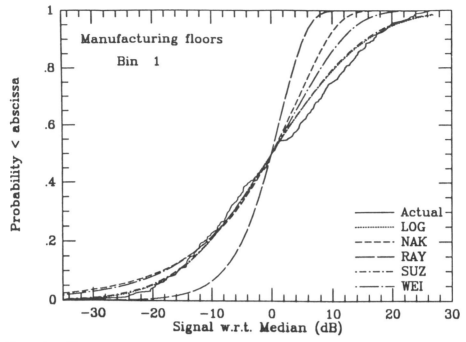

Figure 6.6 Theoretical amplitude CDFs and the empirical CDF for the measured data in bin 1.

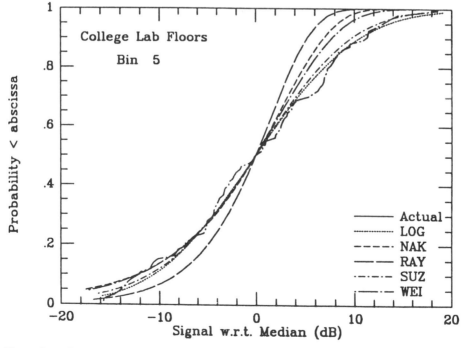

Figure 6.7 Theoretical amplitude CDFs and the empirical CDF for the measured data in bin 5.

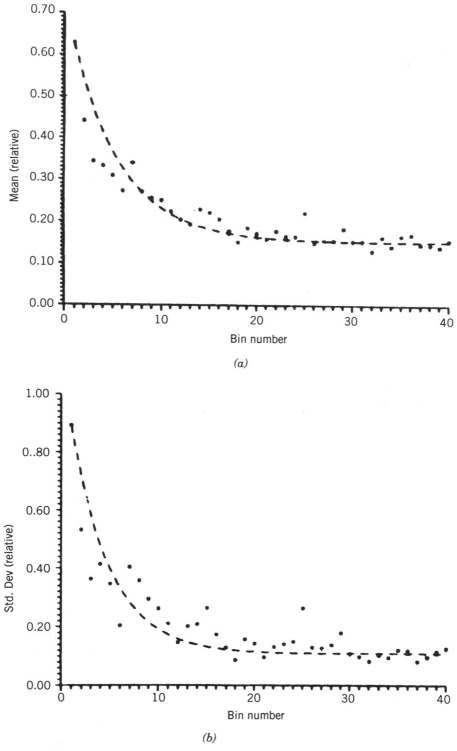

Figure 6.8 Mean (*a*) and standard deviation (*b*) of the arriving path amplitudes as a function of the path arrival delay, and best exponential fit, for manufacturing floors.

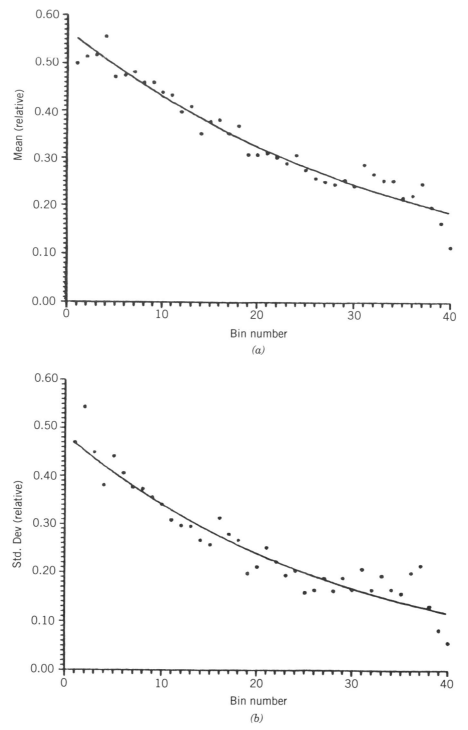

Figure 6.9 Mean (*a*) and standard deviation (*b*) of the arriving path amplitudes as a function of the path arrival delay, and best exponential fit, for office areas.

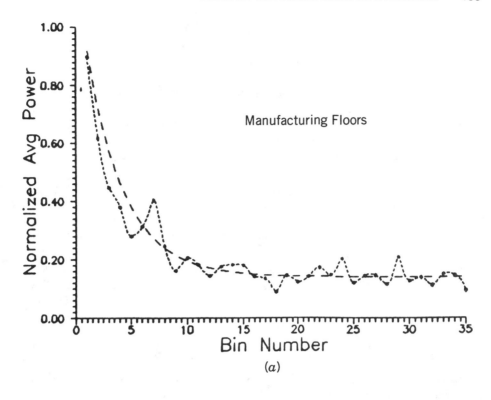

(a)

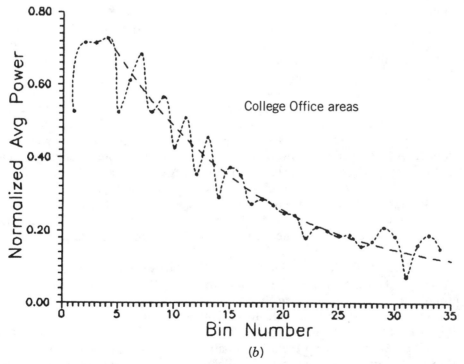

(b)

Figure 6.10 Average received power versus delay path arrival, and best exponential fit. (a) Manufacturing floors. (b) College office areas.

delay of the first arriving path. Figures 6.8 and 6.9 show the mean and standard deviation and their exponential fits for the manufacturing floor and the college office areas, respectively. For the manufacturing areas, the decay rate T was 5.0 for the mean and 4.0 for the standard deviation. In these areas most of the received power is concentrated in the earliest-arriving paths, resulting in a faster decay of the received power with delay. On the other hand, the college office areas exhibited a wider spread of power in delay and thus a slower decay for the received power with delay. In the college offices, the decay rate T was 35.5 for the mean and 28.3 for the standard deviation. Note that the scatter plots give the values of the mean and standard deviation for the path amplitude, given the existence of a path at that delay. The modified Poisson process determines whether or not a path exists at a given delay. Figure 6.10 shows the average received power versus the delay of the path arrival and the best-fit exponential function found for the measurements made in the manufacturing and college office areas. This function is the equivalent spatial delay power spectrum in which the power at different delays is determined by averaging over the measurements in an area rather than by classical averaging over time. As we will see in the following section, the equivalent delay power spectrum is useful in calculation of modem BERs.

6.2.3   Simulation of the Channel Impulse Response

In this section we first examine the sensitivity of computer simulation results to the measurement-based statistical models used to represent the path delays and amplitudes in the simulation. The database used here is the set of measurements made in the manufacturing and office areas discussed earlier in this chapter. We then introduce the models recommended by some standards organizations for the simulation of indoor and outdoor radio channels for PCS and mobile radio applications.

To observe the influence of a statistical model on computer simulation results, we consider two statistical models and compare the CDFs of the rms delay spreads derived from simulations with the results from empirical data. The first statistical model assumes Poisson and Rayleigh distributions for the arrival times and amplitudes of the paths, respectively. This model is similar to the model suggested in [Sal87b] except that the statistics of the magnitudes and arrival times of the paths are extracted from the empirical data. Because we were unable to directly identify clusters from the individual measurements, we have assumed only one cluster. The second model, suggested in [Gan91b], assumes a modified Poisson and a lognormal distribution for the arrival times and amplitudes of the paths, respectively.

For the Poisson/Rayleigh model, the mean path arrival rates obtained from the measurement data gathered in the manufacturing and office areas were used to determine the presence or absence of a path in any bin. The measured signal powers in each bin were then used to determine the Rayleigh amplitude of an existing path. For the modified Poisson/lognormal model, Fig. 6.3 and Table 6.1 were used to determine the presence of a path in a bin, and the amplitudes of the paths were determined by exponential fits for the mean and variance of the lognormal distribution discussed in last section.

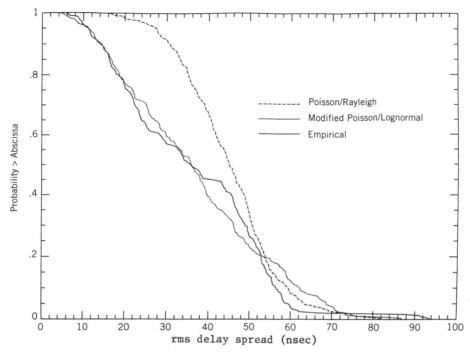

Figure 6.11 The CDFs of the rms multipath delay spread for Poisson / Rayleigh and modified Poisson / lognormal models and the results of empirical measurements in manufacturing areas.

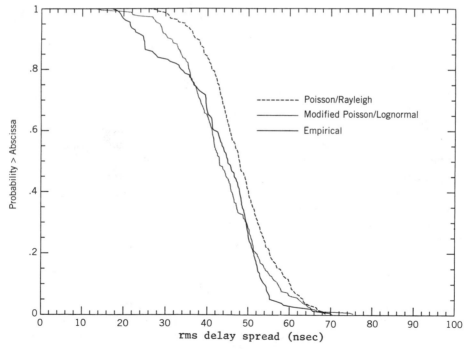

Figure 6.12 The CDFs of the rms multipath delay spread for Poisson / Rayleigh and modified Poisson / lognormal models and the results of empirical measurements in office areas.

The channel profiles were simulated for the Poisson/Rayleigh and modified Poisson/lognormal distribution models. The rms delay spread of the results of the two simulations and the results from the empirical data for manufacturing and office areas are shown in Figs. 6.11 and 6.12, respectively. The dashed lines in these figures are the cumulative distributions of the rms delay spreads computed from the two simulations, while the solid lines represent the results of actual measurements. The match between the empirical and the simulated distributions for the Poisson/Rayleigh distribution is seen to be very good. The match between the modified Poisson/lognormal simulation and the empirical results is even better than that provided by the Poisson/Rayleigh simulation.

Models Recommended by Standards Bodies. A standards committee usually recommends a set of propagation conditions or channel characteristics to be used for hardware or software simulation of the channel. Such a recommendation provides a common basis for comparative evaluation of alternative modulation schemes, adaptive equalization techniques, link-layer protocols, and access methods under consideration for adoption into the particular standard. These recommendations generally comprise two parts: a path loss model and a wideband propagation model. Path loss models were discussed in Chapter 5, and they provide equations to relate the average received power to the distance between transmitter and receiver in different environments. The wideband propagation models provide a procedure for modeling the wideband characteristics in different areas. The wideband models generally assume that the channel is subject to wide-sense stationary uncorrelated scattering (WSSUS), defined by its scattering function:

$$S(\tau, f) = R_{hH}(\tau, f) = Q(\tau)D(f) \qquad (6.2.5)$$

where $Q(\tau)$ is the discrete delay power spectrum and $D(f)$ is the continuous Doppler power spectrum of the channel. The discrete delay power spectrum is defined by a set of taps with specified arrival delays and average relative powers. The Doppler power spectrum is defined by a continuous frequency function that specifies the distribution function and the spectrum of the local shadow fading. The application environments are separated into different classes, and for each class a numerical table specifies the characteristics of individual taps. Each tap is implemented using the techniques described in Chapter 5 for simulation of narrowband signal characteristics. As we discussed earlier, for both indoor and urban radio channels the path arrivals are random and correlated, which contradicts the WSSUS fixed-tap model. However, for all practical purposes these simplified models are adequate to represent the channel for evaluation of the various techniques incorporated into wireless standards.

The GSM-Recommended Model. Table 6.2 gives model parameters for "typical rural areas," as recommended in the GSM standard [GSM91]. This model defines the discrete delay power spectrum with six taps, each with two alternative tap settings. The values for the delay and average relative power for the two choices are shown in the columns labeled (1) and (2). The Doppler spectrum choices are either Rician or the classical Rayleigh. In a manner similar to the simulation of narrowband signals, the Doppler power spectrum for the classical Rayleigh model

TABLE 6.2   Typical Values of the Arrival Delay and Average Power for Rural Areas, Recommended by the GSM [GSM91]

Tap Number	Relative Time (μsec)		Average Relative Power (dB)		Doppler Spectrum
	(1)	(2)	(1)	(2)	
1	0.0	0.0	0.0	0.0	Rice
2	0.1	0.2	−4.0	−2.0	Class
3	0.2	0.4	−8.0	−10.0	Class
4	0.3	0.6	−12.0	−20.0	Class
5	0.4	—	−16.0	—	Class
6	0.5	—	−20.0	—	Class

is

$$D(f) = \frac{1}{2\pi f_m}\left[1 - (f/f_m)^2\right]^{-1/2}, \qquad -f_m < f < f_m \qquad (6.2.6)$$

where $f_m = v_m/\lambda$ is the Doppler spread, v_m is the mobile vehicle velocity, and λ is the wavelength at the carrier frequency. The Rician spectrum is the sum of the classical Doppler spectrum and one direct path, weighted so that the total multipath power is equal to that of a direct path alone:

$$D(f) = \frac{0.41}{2\pi f_m}\left[1 - (f/f_m)^2\right]^{-1/2} + 0.91\delta(f - 0.7f_m), \qquad -f_m < f < f_m,$$

$$(6.2.7)$$

To simulate the channel, the absolute power at each location is determined from the path loss model and each tap is implemented using the methods described for simulation of the narrowband signals. Appendix 6A provides GSM-recommended tables for typical rural, hilly, and urban areas, along with the tap settings to be used for testing receivers employing equalization. Using these tables

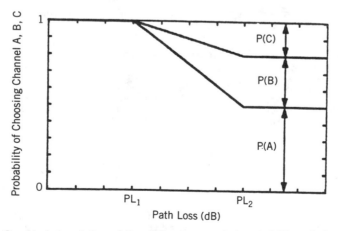

Figure 6.13 Graphical description of the JTC-recommended probabilities of choosing multipath channels A, B, and C as a function of path loss [JTC94].

TABLE 6.3 Delay Spread Parameters and the Probability of Selection of a Class of Impulse Delay Profile in All Areas as Recommended by the JTC [JTC94]

Environment	τ_{rms-A} (nsec)	$P(A)$ (%)	τ_{rms-B} (nsec)	$P(B)$ (%)	τ_{rms-C} (nsec)	$P(C)$ (%)	PL_1 (dB)	PL_2 (dB)
Indoor residential	18	60	70	35	150	5	50	75
Indoor office	35	50	100	45	450	5	60	100
Indoor commercial	55	50	150	45	500	5	60	100
Outdoor urban high-rise; low antenna	100	40	750	55	1,800	5	60	120
Outdoor urban / sub-low-rise; low antenna	100	40	1,000	55	2,000	5	60	120
Outdoor residential; low antenna	68	40	500	55	1,000	5	60	120
Outdoor urban high-rise; high antenna	500	40	4,000	55	8,000	5	70	120
Outdoor urban / sub-low-rise; high antennas	400	40	5,000	55	12,000	5	70	120
Outdoor residential high antenna	350	40	2,500	55	8,000	5	70	120

one can implement hardware or software simulation of the mobile radio channel as recommended by the GSM standard.

The JTC Recommendation for PCS. A more elaborate and comprehensive model is recommended by the PCS Joint Technical Committee (JTC) for simulation of radio propagation in different areas for PCS and mobile users [JTC94]. This recommendation includes parameters for both indoor and outdoor channels. The path loss model for this recommendation was discussed in Chapter 4. Here we discuss the multipath profile structures. The general structure of the JTC model is the same as the GSM model, but the JTC model is more comprehensive. The JTC model divides the environments into one indoor and two outdoor classes. The indoor areas are in turn divided into residential, office, and commercial areas. The outdoor areas include urban high-rise, urban/suburban low rise, and outdoor residential areas. Each class of outdoor areas is divided into other classes specified

TABLE 6.4 Typical Arrival Delay and Average Power for the Taps in the Three Channel Models Suggested for the Residential Indoor Areas by the JTC [JTC94]

	Channel A		Channel B		Channel C		
Tap	Relative Delay (nsec)	Average Power (dB)	Relative Delay (nsec)	Average Power (dB)	Relative Delay (nsec)	Average Power (dB)	Doppler Spectrum
1	0	0	0	0	0	−4.6	Flat
2	50	−9.4	50	−2.9	50	0	Flat
3	100	−18.9	100	−5.8	150	−4.3	Flat
4			150	−8.7	225	−6.5	Flat
5			200	−11.6	400	−3.0	Flat
6			250	−14.5	525	−15.2	Flat
7			300	−17.4	750	−21.7	Flat
8			350	−20.3			Flat

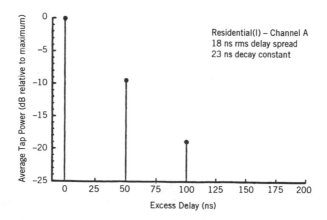

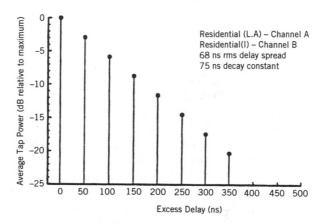

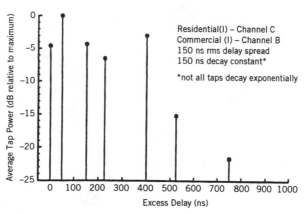

Figure 6.14 Samples of different classes of the JTC-recommended channel profiles for some indoor areas [JTC94].

by the transmitter-antenna height with respect to the tops of buildings. Each tap is simulated in the same way as we described for narrowband signals. The model defines three types of Doppler spectra—classical Jakes, Rician, and flat—for each tap of the discrete-time model. The classical Jakes and the Rician spectra are similar to the spectra used in GSM model. The flat spectrum is used for simulation of the Doppler spectrum in indoor areas and is defined by

$$D(f) = \frac{1}{2\pi f_m}, \qquad -f_m < f < f_m \tag{6.2.8}$$

Because in the same area the multipath characteristics can be quite different from one radio link to another, this model suggests three different types of channel profiles for each environment, providing a wide variety of rms multipath delay spreads for each class of area. The model assumes that the rms multipath spread and the path loss are correlated and provides a statistical procedure for selecting one of the three channel profiles based on the path loss of the channel. Figure 6.13 shows the piecewise linear curves that relate the probability of choosing one of the three channel profiles to the path loss. The curves are identified with five parameters PL_1, PL_2, $P(A)$, $P(B)$, and $P(C)$. PL_1 and PL_2 represent the path losses associated with the two elbows of the curves. $P(A)$, $P(B)$, and $P(C)$ represent the probability of selecting profile A, B, or C, respectively. Table 6.3 shows these parameters as well as the expected rms delay spreads of individual classes of profiles for all nine environments. Table 6.4 shows the relative delay, average power, and Doppler spectrum of the recommended taps for the indoor residential areas. The tap gains are selected to generate the recommended rms delay spreads given in Table 6.3. Figure 6.14 shows the discrete delay power profiles related to the same areas. Appendix 6B provides the details of the tap gains and delay power profiles for all of the other areas. Generally, the average power relative to the first path is assumed to form an exponential function:

$$\tilde{Q}(\tau) = e^{-\tau/\tau_d}$$

TABLE 6.5 Decay Constants for Power Delay Profile in Different Environments, as Recommended by the JTC [JTC94]

Environment	τ_{d1}	τ_{d2}	τ_{d3}
Indoor residential	23	75	150
Indoor office	60	140	475
Indoor commercial	75	150	525
Outdoor urban high-rise low antenna	140	750	2,000
Outdoor urban / suburban low-rise low antenna	140	1,500	2,000
Outdoor residential low antenna	75	525	1,500
Outdoor urban high-rise high antenna	525	4,000	10,000
Outdoor urban / suburban low-rise high antenna	350	5,000	15,000
Outdoor residential high antenna	325	3,000	10,000

where values of τ_d for different areas are given in Table 6.5. The average power of each tap is determined by sampling this function at the appropriate delays. Some variations of the rules are defined by including additional paths to achieve the desired delay spread for a particular area.

6.3 WIDEBAND FREQUENCY-DOMAIN CHANNEL MODELING

In frequency-domain modeling, channel frequency response measurements are used to develop a statistical model for computer simulation of the channel. In this section we describe a particular approach to frequency-domain modeling, based on autoregressive modeling. With this approach, a statistical autoregressive model is developed from measurement data and is used for computer simulation of the channel frequency response. Here we describe this method in the context of modeling an indoor radio channel and the discussion follows closely the experimental work of Howard and Pahlavan [Pah90b, How92].

Autoregressive modeling of time-domain signals is a standard technique in the field of digital signal processing [Mar87b]. Here we apply the technique to samples of the frequency response of a channel. In mathematical terms, an autoregressive (AR) model maps a large set of data points representing a sample of a stochastic process onto a limited number of filter poles representing an *AR process*. To develop the model, we first determine the locations of the poles of the AR process for each measured frequency response. Then we determine the statistics of the poles over the measurement set. With this approach the entire data base obtained from a measurement experiment is mapped onto a few parameters representing statistics of the locations of the poles of the frequency-domain AR process. To produce a sample profile in a simulation, the poles are regenerated from the statistics, and the AR model defined by these poles is driven with white noise to produce a sample frequency response. As we will show in the following discussion, the pole locations are related to the amplitude and arrival time of a cluster of paths in the time domain.

6.3.1 Autoregressive (AR) Modeling

With the AR model, the frequency response at each location is a realization of an AR process of order p given by the equation

$$H(f_n; t) - \sum_{i=1}^{p} a_i H(f_{n-i}, t) = V(f_n) \tag{6.3.1}$$

where $H(f_n; t)$ is the nth sample of the complex frequency-domain measurement at a given location and $V(f_n)$ is a complex white noise process representing the error between the actual frequency response value at frequency f_n and its estimate based on the last p samples of the frequency response. The parameters of the AR model are the complex constants $\{a_i\}$. Taking the z-transform of Eq. (6.3.1), we can view the AR process $H(f_n; t)$ as the output of a linear filter with transfer function

$$G(z) = \frac{1}{1 - \sum\limits_{i=1}^{p} a_i z^{-i}} = \prod_{i=1}^{p} \frac{1}{(1 - p_i z^{-1})} \qquad (6.3.2)$$

driven by a zero-mean white noise process $V(f_n)$. Given the mathematical form of $G(z)$, the AR model is often referred to as an *all-pole model*. Using the AR or all-pole model, the channel frequency response represented by the N measurement samples is described by the p parameters of the AR model or the locations of the p poles of $G(z)$ where typically $N \gg p$.

The AR parameters $\{a_i\}$ are the solutions of the Yule–Walker equations [Mar87b]:

$$R(-l) - \sum_{i=1}^{p} a_i R(i - l) = 0, \qquad 1 < l < p \qquad (6.3.3)$$

in which $R(k) = R_{Hh}(k, 0)$ is the frequency correlation function defined as

$$R(k) = \frac{1}{N} \sum_{i=1}^{N-k} H^*(f_i, t) H(f_{i-k}, t), \qquad k \geq 0 \qquad (6.3.4)$$

The variance of the zero-mean white noise process $V(f_n)$ is the same as the minimum mean-square error of the predictor output, which is given by

$$\sigma_v^2 = R(0) - \sum_{i=1}^{p} a_i R(i) \qquad (6.3.5)$$

Mapping a Time-Domain Profile to an AR Model. In time-domain modeling, the channel is characterized by the power-delay profile such as those shown in Fig. 6.14. To find the appropriate poles related to a power-delay profile, we need to extract the correlation in frequency $R(k)$ from the time-delay profile $|h(\tau_i; t)|^2$. Since the frequency correlation function is the Fourier transform of the magnitude square of the power-delay profile, these two functions are related by

$$R(k) = \sum_{i=1}^{L} |h(\tau_i, t)|^2 e^{-j2\pi/Nk\tau_i}$$

where N is the total number of points used in the frequency domain measurement. The $R(k)$s obtained in this manner can be used in Eq. (6.3.3) to determine the AR parameters. Then using Eq. (6.3.2), the locations of the poles are identified. In [Ali94] this method is used to map the JTC model to an equivalent frequency-domain model identified with the locations of the poles of an AR process.

The Order of the AR Process. To examine the accuracy of the model, results of frequency-domain measurements introduced in the previous chapter are examined. In general, the order of the AR process will be different for measurements made at different locations. Therefore the first step in setting up a model is to determine the minimum order of the process that accommodates all the measurements. Using standard AR *order-selection criteria*, such as the Akaike information criteria [Mar87b], the fifth-order process has been shown to provide an upper bound on the order of the frequency-domain measurements of an indoor radio channel [How91]. This bound is very conservative, and to determine a more realistic order we examine the results of measurements directly. Figure 6.15 shows the five pole locations of the fifth-order AR process for four frequency-domain measurement experiments described in Chapter 5. For a fifth-order model, typically, the magnitude of the largest pole is 0.97 or greater, the magnitude of the second pole is between 0.57 and 0.98, and the magnitudes of the remaining three poles are around 0.5, making them insignificant.

In conventional parametric spectral estimation, a pole close to the unit circle represents significant power at the frequency related to the angle of the pole. The frequency is calculated as $f = |p_i|/2\pi T_s$ and the spectrum is periodic with period $1/T_s$, where T_s is the sampling time. In our application, the model is applied in the frequency domain. The time-domain response obtained from the inverse Fourier transform is a periodic function and a pole close to the unit circle represents significant power at the path delay related to the angle of the pole. The time-domain response is a periodic function with period $1/f_s$ and the corresponding delay is calculated as $\tau = -|p_i|/2\pi f_s$, where f_s is the sampling rate in the frequency domain—that is, the distance between two consecutive frequencies at which the frequency response was measured. For the measurements and modeling discussed in this chapter, $f_s = 0.25$ MHz, which covers a span of $1/(0.25$ MHz$) = 4000$ nsec for the time-domain measurements. In all the measurements the significant paths were found to lie in the interval of delays from 150 nsec to 550 nsec, which corresponds to the angular range $-3\pi/40$ to $-11\pi/40$. As shown in the scatter plots in Fig. 6.15, at most two significant poles exist in this range. Based on these observations we conclude that a two-pole model is adequate to represent the indoor radio channel for these experiments.

Accuracy of the Second-Order Process. To demonstrate that a two-pole model is adequate, the first two poles of the fifth-order model are used to regenerate the frequency responses of the channel. Figures 6.16 and 6.17 represent a sample frequency response and a sample time response, respectively, obtained from a second-order AR model. The time response also represents the inverse transform of the two-pole transfer function of the AR process. The presence of two "humps" corresponds to two major poles, with each hump representing a cluster of arriving paths. Thus two major poles can be interpreted as two significant clusters of multipath arrivals. The interpretation of a pole as defining a cluster of paths, and the distance from the unit circle as defining the power in the cluster, provides a useful physical interpretation for this AR model. The observation of more than one cluster in time-domain measurements of indoor radio channels was first reported in [Sal87b]. However, as we discussed earlier, it is difficult to distinguish adjacent clusters in direct examination of the data, because the clusters may overlap, or the trailing cluster may be too small to detect [Sal87a]. Careful examination of the time-domain measurements in [Sal87] had shown that the existence of two clusters can be observed in only a fraction of the measured

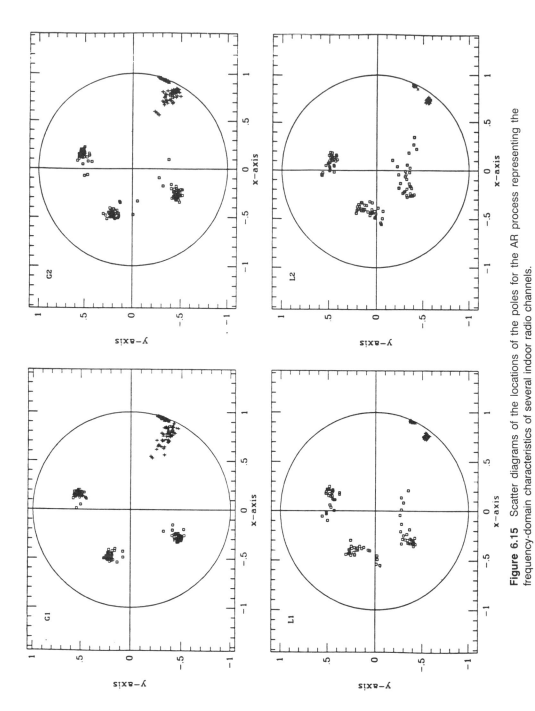

Figure 6.15 Scatter diagrams of the locations of the poles for the AR process representing the frequency-domain characteristics of several indoor radio channels.

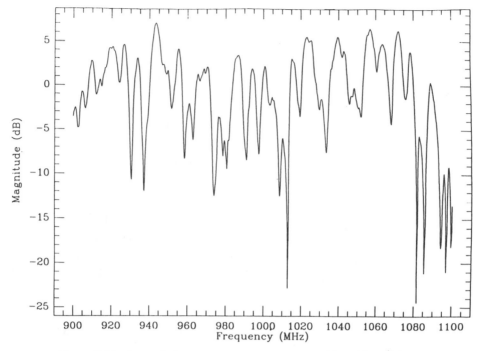

Figure 6.16 A sample frequency response regenerated from AR modeling.

time-domain channel responses. However, by using frequency-domain AR modeling, two clusters are identified for all the measurements.

Figure 6.18 shows the CDF of the 3-dB width of the frequency correlation function for both the original measurements (solid line) and the measurements regenerated from the two-pole model (short-dashed line) of an experiment done in a small office. Figure 6.19 shows the CDF of the rms delay spread for both the original measurements (solid line) and the regenerated measurements (short-dashed line). Because the parameters of the model are determined from the frequency-domain autocorrelation function, the frequency-domain statistics agree more closely with the original measurements than do the time-domain statistics. This analysis is extended to the remainder of the measured data, and the results support the above conclusions.

The third CDF shown in Figs. 6.18 and 6.19 (long-dashed lines) corresponds to the normalized AR model. In the normalized AR model, the angle of the first pole for all the measurements is fixed at $-18°$, which is equivalent to normalizing the arrival of the first cluster at 200 nsec. The angle of the second pole is adjusted to maintain the original angular separation in the AR model, while the magnitudes of the poles are left unchanged. This normalization does not affect the rms delay spread or the 3-dB width of the frequency correlation function, and it reduces the complexity of the model. After normalization, the channel that was represented by two complex parameters (magnitude and angle for both pole 1 and pole 2) will be represented with one real and one complex parameter (magnitudes of poles 1 and 2, and the angular difference).

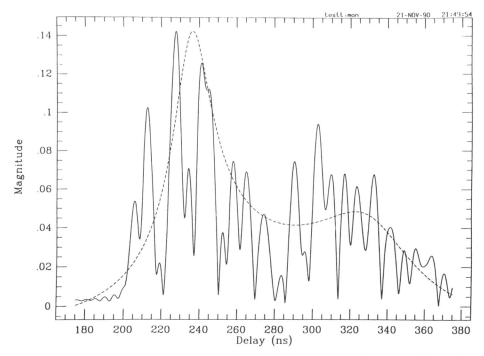

Figure 6.17 A sample time response obtained by taking the inverse Fourier transform of the frequency response shown in Figure 6.16. The dashed line represents the inverse transform of the transfer function of the AR process. The two peaks correspond to the two poles.

Table 6.6 summarizes the statistics of the locations of the first two poles for each of the five experiments described in Chapter 5. Experiments G1 and G2 were spatial measurement made in two different offices, L1 and L2 were two local experiments in which the distance between the transmitter and receiver was relatively fixed, and M1 was a mixed experiment in which nine measurements were taken around each location and the locations were chosen in several rooms of an office area. As shown in Table 6.6, for all the experiments, the magnitude of pole 1 has a larger mean value and smaller variations than the magnitude of pole 2, indicating that the first cluster is always present whereas the second cluster is not always significant. The poles for local experiments have smaller variations than the poles for spatial experiments, indicating smaller changes in the channel multipath characteristics in the local environment.

To relate the rms delay spread to the pole locations for our measurements, we calculate the correlation between these parameters. The correlation of the magnitude of pole 1 with the rms delay spread reveals high negative values; thus as pole 1 approaches the unit circle the rms delay spread decreases. In contrast, the magnitude or angle of pole 2 shows weak correlation with the rms delay spread. Given two clusters of arrivals, the minimum rms delay spread occurs when the paths of cluster 1 are much stronger than the paths of cluster 2, while the maximum rms delay spread occurs when the amplitudes of the paths are more nearly equal. Because the second cluster represents energy arriving at a later time

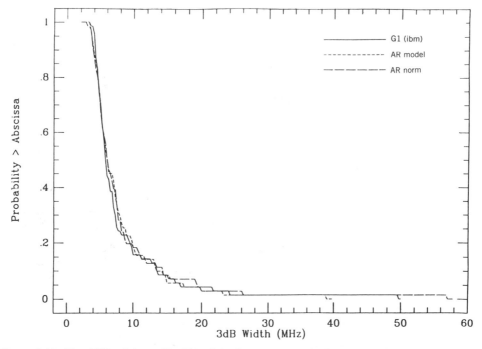

Figure 6.18 The CDFs of the 3-dB width of the frequency correlation function for the results of AR modeling, normalized AR modeling, and the empirical data collected in area G1.

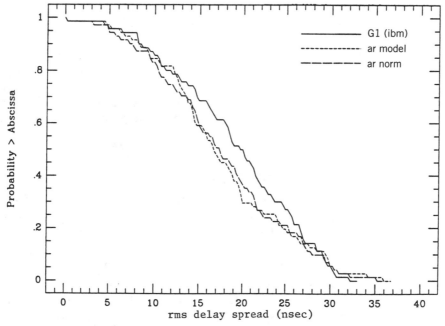

Figure 6.19 The CDFs of the rms multipath delay spread for the results of AR modeling, normalized AR modeling, and the empirical data collected in area G1.

**TABLE 6.6   Statistics of the Location of the First Two Poles
for Each of the Five Frequency-Domain Measurement Experiments**

Experiment	Parameter	Mean	Standard Deviation	Minimum	Maximum
G1	$\|p_1\|$	0.9871	0.006832	0.9648	0.9982
	$\text{ANG}(p_1)$	-18.82	1.989	-22.75	-14.73
	$\|p_2\|$	0.8380	0.08995	0.5701	0.9780
	$\text{ANG}(p_2) - \text{ANG}(p_1)$	-6.355	1.355	-9.376	-2.459
G2	$\|p_1\|$	0.9878	0.005093	0.9750	0.9986
	$\text{ANG}(p_1)$	-19.04	1.839	-22.44	-14.94
	$\|p_2\|$	0.8624	0.07776	0.6224	0.9599
	$\text{ANG}(p_2) - \text{ANG}(p_1)$	-8.239	1.477	-9.213	-3.549
L1	$\|p_1\|$	0.9850	0.004760	0.9758	0.9942
	$\text{ANG}(p_1)$	-22.80	0.9938	-25.36	-21.42
	$\|p_2\|$	0.9332	0.01783	0.8979	0.9599
	$\text{ANG}(p_2) - \text{ANG}(p_1)$	-12.95	0.7876	-14.70	-11.48
L2	$\|p_1\|$	0.9797	0.004941	0.9632	0.9871
	$\text{ANG}(p_1)$	-24.59	0.7449	-27.34	-23.57
	$\|p_2\|$	0.9268	0.01551	0.8880	0.9533
	$\text{ANG}(p_2) - \text{ANG}(p_1)$	-12.52	0.7932	-13.84	-10.96
M1	$\|p_1\|$	0.9860	0.005735	0.9660	0.9964
	$\text{ANG}(p_1)$	-18.50	1.41	-22.9	-15.17
	$\|p_2\|$	0.8614	0.07405	0.5997	0.9739
	$\text{ANG}(p_2) - \text{ANG}(p_1)$	-8.483	1.667	-12.98	-0.176

(from greater distance), the magnitude of pole 1 is almost always larger than the magnitude of pole 2. Therefore, in the two-pole model, if the magnitude of pole 1 decreases toward the magnitude of pole 2, this increases the rms delay spread. Whereas the magnitude of pole 2 may in some cases increase to create the same effect, there are also cases where pole 2 is insignificant relative to pole 1 and its movement is uncorrelated with the rms delay spread value. These observations will be exploited in the statistical modeling that we describe in the next section. Given the inverse relationship between rms delay spread and the 3-dB width of the frequency correlation function, similar results regarding the 3-dB widths are observable.

6.3.2   Statistical Modeling in the Frequency Domain

Assuming that the frequency response of the channel forms a second-order AR process, modeling of the channel can be reduced to statistical characterization of the two poles of the model and the variance of the white noise process driving the two-pole filter. Five real values are required to represent the two complex poles and the variance of the driving noise. In this section we introduce four models, and in each model it is assumed that the four values used to represent the two complex poles are statistically independent Gaussian random variables defined in a pre-scribed range. The mean, variance, and range of each value are determined from the empirical data for a given experiment. The angle of the first pole, representing the arrival time of the first cluster, is normalized to the arbitrary value of $-18°$, which corresponds to the first cluster being centered at 200 nsec. Note that in time-domain modeling the modeled arrival delays are the delays relative to arrival of the first path. In the four models described here, we reference the delays to the arrival of the first cluster.

The four second-order models are defined as follows:

1. Model 1 assumes that the magnitude of the first pole and the angle and magnitude of the second pole are all Gaussian random variables with means and variances determined from the values measured in an experiment.

2. Model 2 assumes that the angle of pole 2 is fixed at its average as determined from an experiment, and the magnitudes of the two poles are treated as random variables.

3. Model 3 assumes that the second pole is fixed at the average angle and magnitude of the second pole determined from an experiment, and only the magnitude of the first pole is considered as a random variable.

4. Model 4 assumes that both poles are fixed at the average magnitudes and angles of the poles determined from an experiment.

These models reflect the relative importance of the locations of poles 1 and 2 in their influence on the rms delay spread and 3-dB width of the frequency correlation function. Model 4 with two fixed poles is the simplest of the four models, whereas model 1, with random magnitude of the two poles and angle of the second

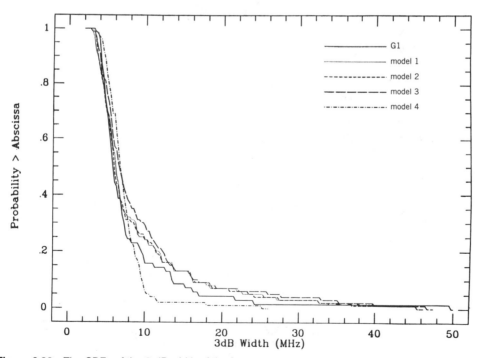

Figure 6.20 The CDFs of the 3-dB width of the frequency correlation function for the four statistical AR models and the empirical data collected in area G1.

pole, is the most complex. Table 6.6, discussed earlier, shows the mean, standard deviation, and range of all parameters obtained from the five experiments.

For computer simulation of the channel, any one of the four models can be used, with the means and standard deviations provided in Table 6.6, to generate the two poles (the poles lying outside the measurement range are discarded). The poles are then used to form the AR process. The variance of the noise driving the process is determined from

$$\sigma_v^2 = \frac{R(0)}{R_g(0)} \tag{6.3.6}$$

where the received power, $R(0)$, is determined from a lognormally distributed random variable with mean and standard deviation given in Table 6.6. The $R_g(n)$ is the inverse z-transform of $G(z)G^*(z^{-1})$ with the $G(z)$ defined in Eq. (6.3.2) and is the autocorrelation of $g(n) = Z^{-1}[G(a)]$. $R_g(0)$ is related to pole 1 (p_1) and pole 2 (p_2) by

$$R_g(0) = \frac{1 - |p_1|^2|p_2|^2}{\left(1 - |p_1|^2\right)\left(1 - |p_2|^2\right)|1 - p_1 p_2^*|^2} \tag{6.3.7}$$

The frequency response measurement is generated by using Eq. (6.3.1) with $p = 2$, $a_1 = p_1 + p_2$, and $a_2 = -p_1 p_2$.

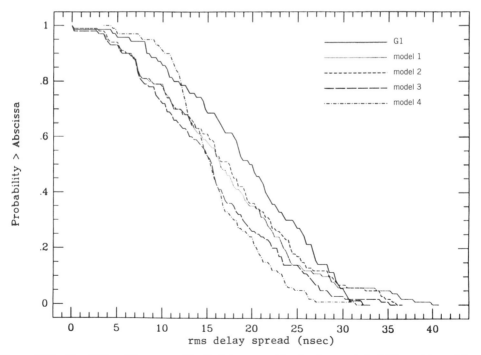

Figure 6.21 The CDFs of the rms multipath delay spread for the four statistical AR models and the empirical data collected in area G1.

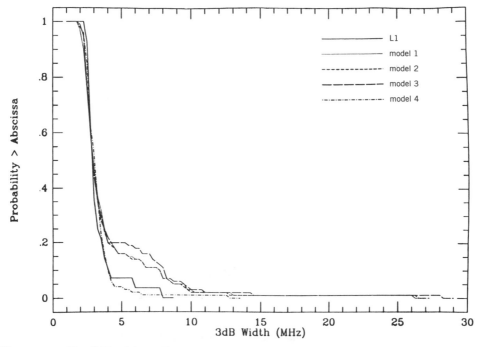

Figure 6.22 The CDFs of the 3-dB width of the frequency correlation function for the four statistical AR models and the empirical data collected in area L1.

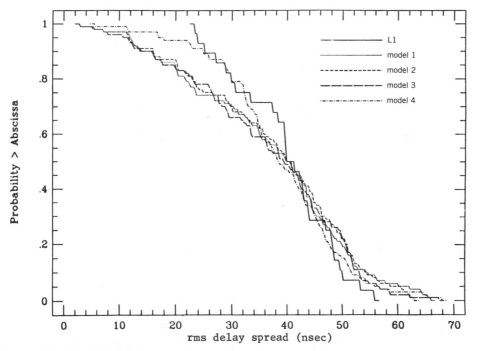

Figure 6.23 The CDFs of the rms multipath delay spread for the four statistical AR models and the empirical datacollected in area L1.

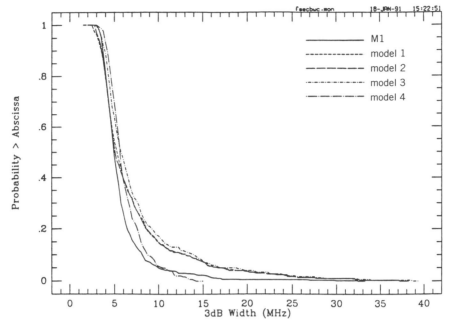

Figure 6.24 The CDFs of the 3-dB width of the frequency correlation function for the four statistical AR models and the empirical data collected in area M1.

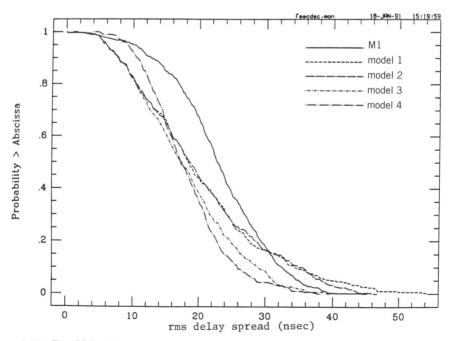

Figure 6.25 The CDFs of the rms multipath delay spread for the four statistical AR models and the empirical data collected in area M1.

Figure 6.20 shows a graph of (a) the CDF of the 3-dB width of the frequency correlation function obtained from 100 simulated measurements for each of the four models of experiment G1 and (b) the actual CDF determined from the 70 measurements for G1. Figure 6.21 shows corresponding results for the rms delay spreads. In the two figures it is seen that the first three models provide very similar results, in either the time or frequency domain. In these models, the magnitude of the first pole changes for each regenerated measurement. It can be seen that the simpler model 4, with its fixed poles, cannot accommodate the entire range of variations of 3-dB width or rms delay spread.

Figures 6.22 and 6.23, derived from the local experiment L1, are similar to Figs. 6.20 and 6.21, respectively. The results of model 4 with fixed poles are in better agreement with the actual data. This indicates that local movements do not significantly change the locations of the poles. Figures 6.24 and 6.25 present results similar to those in Figs. 6.20 and 6.21, for the mixed experiment M1. Again, the simple model 4 cannot adequately represent variations observed in the empirical data. Therefore, we conclude that for global and mixed measurements at least the magnitude of the first pole should be considered random, while local measurements can be regenerated using a model having fixed poles.

6.4 COMPARISON BETWEEN STATISTICAL MODELS

Here we draw comparisons between the time-domain and frequency-domain approaches to statistical channel modeling. The time-domain approach is by far the most popular statistical approach used for modeling indoor and urban radio channels. In time-domain modeling there is a straightforward relationship between multipath propagation and the results of measurements, and therefore a physical interpretation of the model is quite evident. The amplitudes and arrival delays of the paths are determined directly from the results of measurements. However, systematic identification of the path clusters from measurement data is typically not a simple task. With frequency-domain modeling, physical interpretation of a model is not so straightforward, because the path arrivals are not directly related to the results of frequency-domain measurements. The second-order frequency-domain model associates the arriving paths with two clusters and the arrival delays of the clusters are evaluated indirectly by determining the angular positions of the poles of the AR process.

Computation of the parameters of the frequency-domain AR model is straightforward and generally accurate. The frequency-domain AR model is defined by the statistics of the pole locations. First the pole locations are calculated by solving a set of linear equations involving the complex frequency autocorrelation function. Then the location of each pole is modeled as a random variable. These calculations are systematic and routine, and efficient algorithms are available for doing them. In contrast, computation of parameters for the time-domain model is rather tedious and prone to inaccuracy. In time-domain modeling, path amplitudes and arrival times should be statistically modeled. First each time response has to be processed with a peak detection algorithm to determine the path locations and amplitudes. Peak detection involves calculation of a threshold to distinguish real peaks from those created spuriously by noise, and the resulting statistical model is sensitive to these details. The arrivals of the detected peaks are used for calculation of the parameters of the modified Poisson process. This calculation is iterative and requires careful attention to ensure accurate results. The amplitudes of the individual paths are then analyzed for best fit to one of the alternative statistical distributions. The mean and variance of the selected statistical model are then determined by best fit to another exponential curve.

As far as the results of simulations are concerned, in time-domain modeling the CDF of the rms delay spread of the modified Poisson/lognormal model have been found to fit measurement data more closely than the Poisson/Rayleigh model. In frequency-domain AR modeling, simulation results show close agreement with the CDF of the 3-dB width of the frequency correlation function of the measurement data, but the agreement is not as close for the CDF of the rms delay spread of the channel.

The time-domain channel impulse responses obtained from frequency-domain measurements typically represent higher numbers of arriving paths than do the responses obtained from direct time-domain measurements. Therefore, the results of time-domain modeling using frequency-domain measurement systems have to be evaluated carefully.

In a "block-oriented" computer simulation environment, a simulation of an AR model uses a filter block and a Fast Fourier Transform (FFT) block to generate the frequency- and time-domain responses. These two blocks are included in all block-oriented simulation packages. The time-domain simulation involves generation of a modified Poisson process and amplitude and phase distributions that are not typically provided in block-oriented simulation packages. Therefore, the implementation of the time-domain method in a block-oriented simulation environment is more complicated.

6.5 RAY-TRACING ALGORITHMS

In Chapter 3 we introduced the two-dimensional ray tracing technique through a simple example of modeling radio propagation within a room using free-space transmission and multiple reflections from walls. We also showed how, given the paths connecting the transmitter and receiver, one can determine the response of the channel to narrow transmitted pulses. In addition to free-space transmission and simple reflection, more complex ray-tracing algorithms include mechanisms of diffraction, diffuse wall scattering, and transmission through various materials. In this section we provide a more detailed treatment of signal reflection and introduce three other mechanisms, while describing more complex examples of propagation in indoor and outdoor areas. Appropriate combinations of these mechanisms can be used for simulation in a particular coverage area. For example, in small indoor areas with soft-surfaced walls, reflection and transmission are the dominant mechanisms for radio propagation at frequencies around 1 GHz; whereas for macrocellular high-rise urban canyons with antennas installed above roof level, diffraction is the main mechanism for signal propagation.

6.5.1 Reflection and Transmission Mechanisms

Figure 6.26 shows reflected and transmitted paths in a simple propagation example. As a ray meets a wall two paths emerge, one reflected and the other transmitted through the wall. To determine the reflected and transmitted paths in a ray-tracing algorithm, we assume walls, ceilings, floors, and other objects to be dark mirrors, and we use the theory of geometric optics to identify the paths. There are two general approaches that can be employed to calculate these paths. The first uses optical images of the transmitter and receiver [Law92, Rus91, McK91]. In this approach, reflections of the transmitted signal by various reflecting objects in the floor plan are described by images of the transmitter, and these images are used with images of the receiver to find all the paths to the receiver.

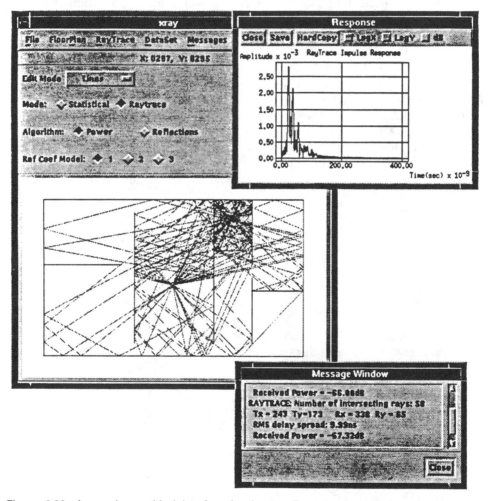

Figure 6.26 A sample graphical interface for the two-dimensional ray-tracing algorithm. The windows identify the trace of the paths, channel impulse response, channel parameters, and simulation parameters. (From [Hol92c].)

The path distance and direction are then used to determine the magnitude, phase and delay of each path arriving at the receiver. The two drawbacks in this method are as follows: (1) It is difficult to determine all image points if we must consider many reflections of the transmitted signal, and (2) it is difficult to automate the imaging procedure with a computer program. Thus to efficiently implement this approach, special procedures are employed to cancel the unnecessary images. This approach is best suited to environments, such as large hilly urban areas, where only a few dominant reflectors need be considered.

The second method for determining the reflected and transmitted paths is through the application of *ray-shooting* techniques [Des72, Ike91]. The ray-shooting algorithm is an intuitively simple approach to the problem of multipath propagation. A pincushion of rays is sent out from the transmitter, and the progress of each ray is traced through the environment until the ray has either intersected the receiver or has lost enough power that its contribution to the received signal is negligible. The time of arrival, intensity, phase, and direction of arrival are recorded for each ray that intersects the receiver. Once every ray has been traced to completion, the channel impulse response is formed. In the implementation of this algorithm, as shown in Fig. 6.27, the rays are actually treated as cones. This implementation has a few potential problems: (1) When a ray strikes near an edge of a wall it is difficult to trace the reflected and transmitted rays. (2) This method of ray tracing does not ensure that every signal path between the transmitter and receiver is considered. By taking special care in designing the algorithm, the harmful effects of the first problem can be avoided. The second problem is found to be minor in most applications. The ray-shooting approach is found to be suitable for areas having many irregular reflectors, such as small indoor areas.

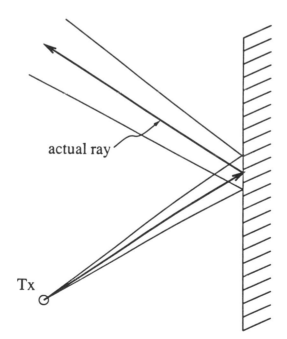

Figure 6.27 The cone structure used for the ray-shooting method.

Calculation of Reflection Coefficients. When a traced ray meets a wall or other surface, some of the energy is reflected, some is transmitted through the surface, and some is absorbed. There are three rays that can be defined at the wall: (1) the incident ray, (2) the reflected ray, and (3) the transmitted ray. Figures 6.28 and 6.29 represent all the electrical field components of the incident ray, reflected ray, and the transmitted or refracted rays. Figure 6.28 represents the horizontal polarization components, and Fig. 6.29 represents the vertical polarization field components. The angle of the transmitted ray is assumed to be the same as that of the incident ray, and the angle of the reflected ray is determined by the orientation of the reflecting surface. The amounts of energy reflected and transmitted are based on the structural material as well as the angle of incidence. The phase of the ray is also affected by a reflection or a transmission. We can assume a 180° phase reversal of the reflected ray and no phase change in the refracted ray. This assumption is valid as long as we consider high-frequency electromagnetic radiation. Over a smooth surface the reflection coefficients for the horizontal and vertical components are given by [And93]

$$R_{s,h} = \frac{E_{r,h}}{E_{i,h}} = \frac{\sin\psi - \sqrt{\varepsilon - \cos^2\psi}}{\sin\psi + \sqrt{\varepsilon - \cos^2\psi}} \qquad (6.5.1)$$

and

$$R_{s,v} = \frac{E_{r,v}}{E_{i,v}} = \frac{\varepsilon\sin\psi - \sqrt{\varepsilon - \cos^2\psi}}{\varepsilon\sin\psi + \sqrt{\varepsilon - \cos^2\psi}} \qquad (6.5.2)$$

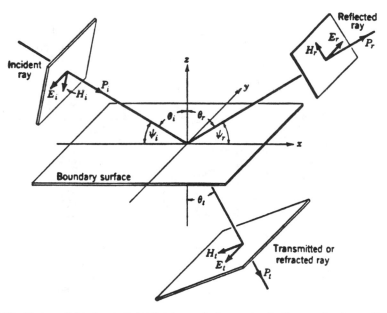

Figure 6.28 Various field elements for the transmission and reflection mechanism with horizontal polarization.

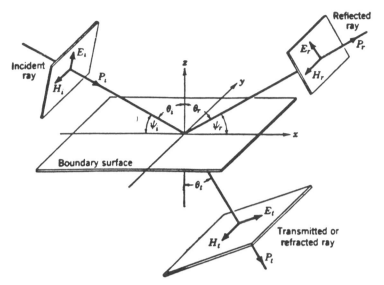

Figure 6.29 Various field elements for the transmission and reflection mechanism with vertical polarization.

where ψ is the angle of incidence and ε is the complex permittivity given by

$$\varepsilon = \varepsilon_r - j60\sigma\lambda \qquad (6.5.3)$$

where ε_r is the normalized relative dielectric constant of the reflecting surface, σ is the conductivity of the reflecting surface, and λ is the wavelength of the incident ray. Examples of the magnitude and phase of the reflection coefficients for different angles of incident are given in Figs. 6.30 and 6.31.

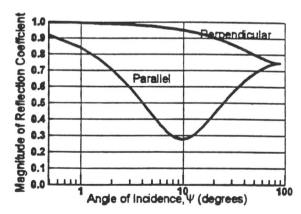

Figure 6.30 Magnitude of the reflection coefficient for $\sigma = 1.0$, $\varepsilon_r = 15$, and $f = 600$ MHz. (From [And93] © IEEE.)

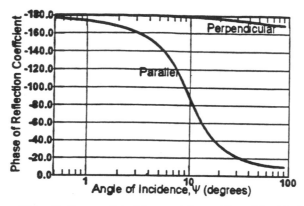

Figure 6.31   Phase of the reflection coefficient for $\sigma = 1.0$, $\varepsilon_r = 15$, and $f = 600$ MHz (From [And93] © IEEE.)

If the surface is rough, an additional surface roughness factor ρ is included in the equations and the reflection factors are then given by

$$R_h = \rho R_{s,h} \tag{6.5.4}$$

and

$$R_v = \rho R_{s,v} \tag{6.5.5}$$

where the roughness factor is given by [Cha82] as

$$\rho = e^{-\delta}$$

and

$$\delta = \frac{4\pi \Delta h}{\lambda} \sin \psi$$

where Δh is the standard deviation of the normal distribution assumed for the surface.

Calculation of the Wall-Transmission Coefficient. The *wall-transmission coefficient* is calculated from the reflection coefficient and the attenuation through the wall. A thick concrete wall has a very high transmission loss, almost totally blocking the transmission, while a glass wall may have a loss of only a fraction of a decibel. For outdoor applications where both transmitter and receiver are outside, but in the vicinity of one or more buildings, transmissions through building walls are not included in the ray-tracing models. Instead it is assumed that the most of the energy of the rays transmitted through the outside wall of a building is absorbed by other walls within that building and is thus rendered negligible. For outdoor-to-indoor communications, as we observed in the discussion of narrowband signal power loss, a path loss of 10–20 dB is usually included in the calculations. Propagation after the initial loss follows the same rules as described for indoor radio propagation. For indoor communications, the transmission through walls constitutes an important component of the received signal, especially for obstructed line-of-sight (OLOS) environments. In the remainder of this section we describe three simple models for transmission through walls on indoor radio channels, and then we introduce the slab model for transmission.

The three transmission models we describe here are simple and straightforward, though each is an approximation rather than an exact representation of an actual wall. In all three models, Eqs. (6.5.1) and (6.5.2) are used to determine the reflection coefficients of the structural material. The differences among the three models lie in the way that the transmission coefficients are determined.

For the first model it is assumed there is no energy lost in the wall itself, that is, all of the energy is either transmitted or reflected. The transmission coefficient for both horizontal and vertical components of this model is given by

$$T_{(h,v)} = \sqrt{1 + R_{(h,v)}^2} \tag{6.5.6}$$

This equation is derived from the fact that total power must be conserved in the system. Given that the reflection coefficient is determined by assuming a single planar boundary, this equation is not accurate. The second model assumes that there will always be some loss in the material and includes loss for the first and second layers of the wall. This is a simple *slab model* which includes the boundaries encountered at both sides of the wall. The horizontal and vertical transmission coefficients in this case are given by

$$T_{(h,v)} = \sqrt{\left(1 - R_{1,(h,v)}^2\right)\left(1 - R_{2,(h,v)}^2\right)} \tag{6.5.7}$$

The values of R_1 and R_2 are the reflection coefficients at the boundaries of the wall. R_1 is the transmission coefficient at the "front" boundary between the air and the wall, whereas R_2 is the transmission coefficient at the "back" boundary between the wall and the air. The two terms are multiplied together to obtain the total transmission through the wall. This model does not take into account the thickness of the material or the fact that some of the energy is reflected off the second boundary and back out of the material. A more elaborate model for the slab will be discussed later. The third and last model assumes a 3-dB loss through the wall, as defined by

$$T_{(h,v)} = \sqrt{0.5\left(1 - R_{(h,v)}^2\right)} \tag{6.5.8}$$

Figure 6.32 shows plots of the reflection coefficient and transmission coefficients for all three models, considering only the horizontal component. The plot shows the reflection coefficient from Eq. (6.5.1) as well as the transmission coefficient calculated using each of the three models. The permittivity of the wall is assumed to be 20 farads/m in this case.

The Slab Model. Figure 6.33 shows a model for reflection and transmission in a slab. The wave transmitted through the first layer of the slab is reflected several times, and each reflection from either side of the slab causes another transmission

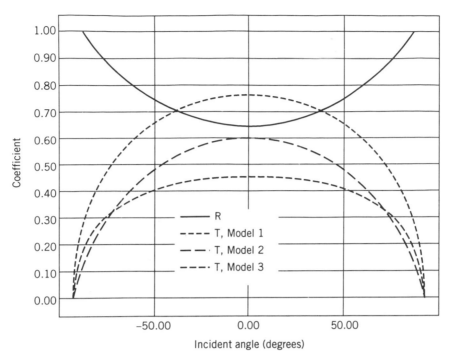

Figure 6.32 Reflection and transmission coefficients for three simple transmission models.

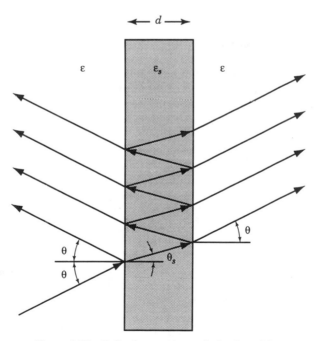

Figure 6.33 Reflection and transmission in a slab.

to the air. Therefore, each incident ray gives rise to multiple reflected and transmitted rays, and the strengths of these rays are functions of the material and the thickness of the slab as well as the signal frequency and the angle of the incident ray. If we assume that all the reflected and transmitted rays are added to form a single ray, it is possible to model the slab with a single reflection coefficient and a single transmission coefficient. These overall transmission and reflection coefficients are given by [Bur83] and [Bal82]:

$$R_{(h,v)} = \frac{R_{s,(h,v)}(1 - P_d^2 P_a)}{1 - R_{s,(h,v)}^2 P_d^2 P_a} \tag{6.5.9}$$

and

$$T_{(h,v)} = \frac{(1 - R_{s,(h,v)})P_d P_t}{1 - R_{s,(h,v)}^2 P_d^2 P_a} \tag{6.5.10}$$

where $R_{s,(h,v)}$ denotes the reflection coefficients for horizontal and vertical polarization of smooth surfaces as given by Eqs. (6.5.1) and (6.5.2). The other three parameters are given by

$$P_d = e^{-jk'l}$$

$$P_a = e^{j2kl\sin^2\theta}$$

$$P_t = e^{-jkl}$$

and $l = d/\cos\theta$, where d is the thickness of the slab and where $k = 2\pi\lambda$ and $k' = k\sqrt{\varepsilon}$ are the propagation constants in free space and the slab, respectively.

6.5.2 Diffraction Mechanism

Figure 6.34 shows a typical diffraction scenario. The ray striking the corner at angle ϕ is diffracted in all directions and one of the diffracted rays arrives at the receiver from angle ϕ'. The diffracted ray is attenuated by the *diffraction coefficient*. In outdoor radio applications, diffraction occurs at roof edges and the

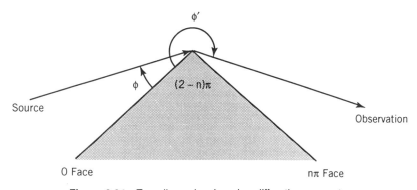

Figure 6.34 Two-dimensional wedge diffraction geometry.

corners of walls. In indoor radio propagation it usually occurs at the corners of rooms, at bent surfaces of walls, in corridors, and at the edges of windows. The energy contributions from diffracted paths can be strong if a first- or second-order diffracted path exists while other paths arriving with multiple reflections or transmissions are heavily attenuated. For example, if the transmitter and receiver are located at opposite sides of a building, it usually takes many reflections to establish a path connecting the transmitter and receiver, and transmission through the building involves many intervening walls. However, a second-order diffraction at two sides of the roof can provide a diffracted path of only second order.

The complete solution to the diffraction problem for perfectly conducting and finite-conducting wedges using the Uniform Theory of Diffraction (UTD) is treated in [Lue84a] and [Lue84b]. The diffraction coefficients given in [Lue84b] are as follows:

$$
\begin{aligned}
D_{h,v} = \frac{-e^{-j(\pi/4)}}{2n\sqrt{2\pi k}} \Bigg[&\cot\left(\frac{\pi + (\phi - \phi')}{2n} \right) F(kLa^{+}(\phi - \phi')) \\
+ &\cot\left(\frac{\pi - (\phi - \phi')}{2n} \right) F(kLa^{-}(\phi - \phi')) \\
+ R_{o(h,v)} &\cot\left(\frac{\pi - (\phi + \phi')}{2n} \right) F(kLa^{-}(\phi + \phi')) \\
+ R_{n(h,v)} &\cot\left(\frac{\pi + (\phi - \phi')}{2n} \right) F(kLa^{+}(\phi - \phi')) \Bigg] \quad (6.5.11)
\end{aligned}
$$

where $R_{o(h,v)}$ and $R_{n(h,v)}$ are the reflection coefficients of the incident wedge face and the opposite wedge face as determined from Eqs. (6.5.1) and (6.5.2). The function $F(x)$ is the Fresnel integral

$$
F(x) = 2j\sqrt{x}\, e^{jx} \int_{\sqrt{x}}^{\infty} e^{-j\tau^2}\, d\tau
$$

and the parameter L is defined by

$$
L = \frac{lr}{l + r}
$$

where l and r are the distances of the transmitter and the receiver, respectively, from the diffracting edge. The remaining parameters are given by

$$
a^{\pm}(\phi) = 2\cos^2\left(\frac{2n\pi N^{\pm} - \phi}{2} \right)
$$

and $N^{\pm}$ are the closest integers satisfying the equation

$$
N^{\pm} = \frac{\phi \pm \pi}{2n}
$$

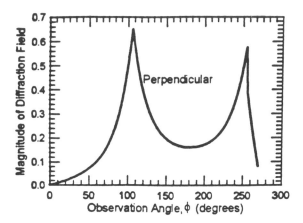

Figure 6.35   Magnitude of diffracted field for $\phi' = 75°$, $\sigma = 1.0$, and $\varepsilon_r = 15$. (From [And93] © IEEE.)

Figure 6.35 shows the magnitude of the diffracted field for a 90° wedge angle, $\phi' = 75°$, $\sigma = 1$, and $\varepsilon = 15$.

6.5.3   Diffused Wall Scattering

It is well known that when an electromagnetic wave meets a rough or nonuniform surface, it scatters. The scattering of radio signals has been studied extensively for many years, particularly in the field of radar [Eav87]. As the frequency of the impinging wave increases, the wavelength decreases and the roughness of the surface causes increased scattering. At optical frequencies scattering is the dominant source of electromagnetic wave propagation, as will be discussed in greater detail in Chapter 10. To model the scattering process, we may treat the intersection of the ray and the surface as a radiating point. The signal radiated from this point will have a special radiation pattern, and within the pattern the signal propagates just as it would from a radiating point in free space. The power is inversely proportional to the square of the distance from the radiating point. Inclusion of the effects of scattering in a transmission analysis can require substantial computational power. Optical signals are confined within a room; and as we will see later, inclusion of a detailed scattering model with Lambertian laws is numerically feasible. For radio signals, the structures are more complicated and direct inclusion of scattering is generally not computationally feasible. However, wavelengths are longer at radio frequencies than at optical frequencies, resulting in smaller amounts of scattering. In these cases it is possible to account for scattering by including an additional loss factor in the transmission analysis [And93].

6.5.4   Two-Dimensional Ray Tracing in Small Indoor Areas

In this section we compare the results obtained from a two-dimensional ray-tracing algorithm with the results of measurements made in a small indoor office area. Results of frequency-domain measurements described in Chapter 5 are used as a reference in evaluating the accuracy of the ray-tracing algorithm. The model uses the two-dimensional reflection and transmission mechanism to trace the rays using the ray-shooting technique. This model provides a low-cost means of doing propagation analysis for small indoor areas used for wireless local area network (WLAN) applications. As we discussed earlier, diffraction does not play a major role in most indoor radio propagation scenarios. The diffraction effect may influence propagation significantly in some locations in corridors when the LOS is blocked and the received signal involves multiple reflections and transmissions. However, this is not a likely situation for indoor WLAN applications, where terminals are typically used in reasonably open work areas. One of the major costs in ray tracing is the development of an accurate electronic map of the area. To minimize the cost of preparing the map, one may use the two-dimensional layout plans that are generally available for office, commercial, or residential buildings. Ideally, one would use a scanner to read the plan into the computer.

To examine the accuracy of the simple two-dimensional ray-tracing algorithm, we look at a typical indoor area, shown in Fig. 6.36, which was used for measurement and modeling in the last two chapters. The software and the interface graphics for this model were developed as part of the work reported in [Hol92a],

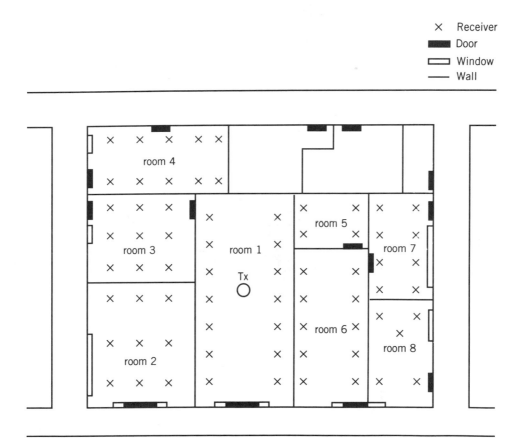

Figure 6.36 Floor plan of one section of the second floor of the Atwater Kent Laboratory. The dimension of the internal area is 12 m × 16 m.

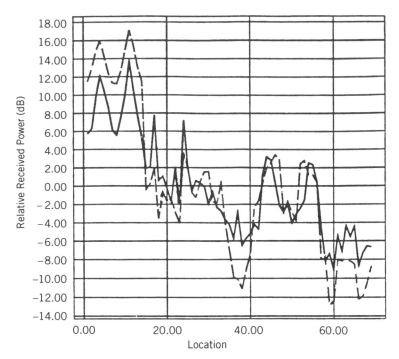

Figure 6.37 Average received power versus location for ray tracing, and the results of measurements. Solid line, Measurements; dashed line, simulation.

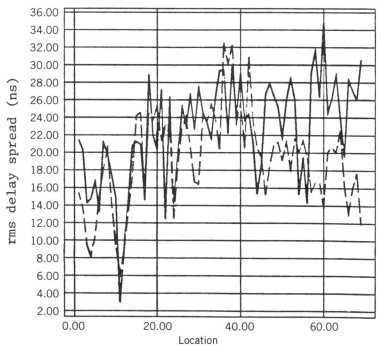

Figure 6.38 Average rms delay spread versus location for the ray tracing, and the results of measurements. Solid line, Measurements; dashed line, simulation.

[Hol92b], Figure 6.26 shows a sample measured ray tracing, the associated impulse response, and the message window seen on the computer screen. A total of 621 measurements were taken near the center of the building in an area comprising several labs and offices. The receiver was located at the center of room 1 (Fig. 6.36), and the transmitter was moved to different locations in several of the surrounding rooms as well as in room 1. A simplified model of the actual floor plan was entered into the ray-tracing program in order to simulate the environment. The model contains all of the walls, doors, and windows in the area being simulated. The simulation was performed for the same locations at which the measurements were taken.

The 621 measurements consisted of nine evenly spaced measurements in each of 69 locations distributed throughout the area shown in Fig. 6.36. The 69 locations are marked with "×" in the figure, and nine measurements were taken in a 2-ft × 2-ft area surrounding each of the marked locations. The average normalized received power and the average rms delay spread for each local area were calculated. The simulation consisted of calculating the average received power and rms delay spread in each of the 69 locations. Approximately 200 receiver locations, evenly distributed over the same 2-ft × 2-ft area, were simulated and the results

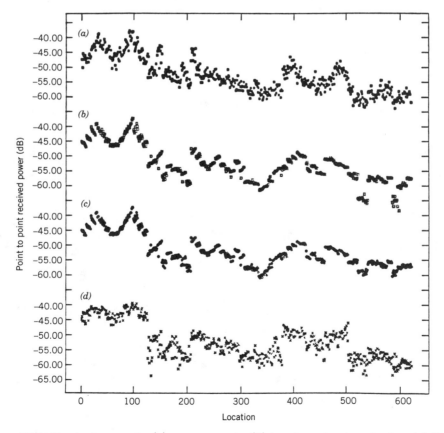

Figure 6.39 Received power for (a) measurements, (b) two-dimensional ray tracing, (c) three-dimensional ray tracing, and (d) two-dimensional FDTD modeling at 621 points in the area shown by Fig. 6.36. FDTD modeling is discussed in Section 6.6.

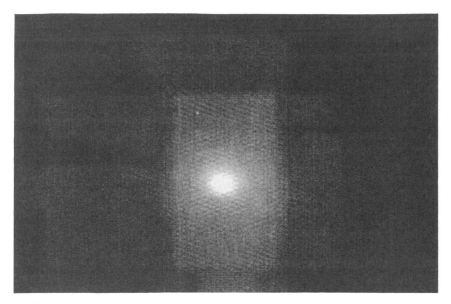

Figure 6.40 Power coverage prediction using two-dimensional ray tracing in the area shown by Fig. 6.36.

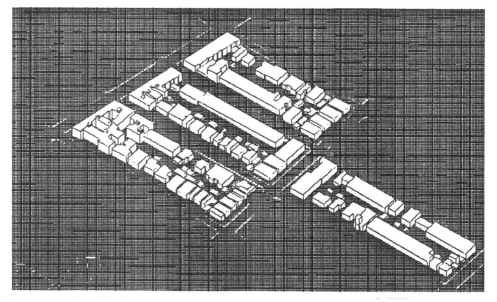

Figure 6.41 Four block area in Brooklyn, New York, displayed by AutoCAD. The area is approximately 400 m × 250 m.

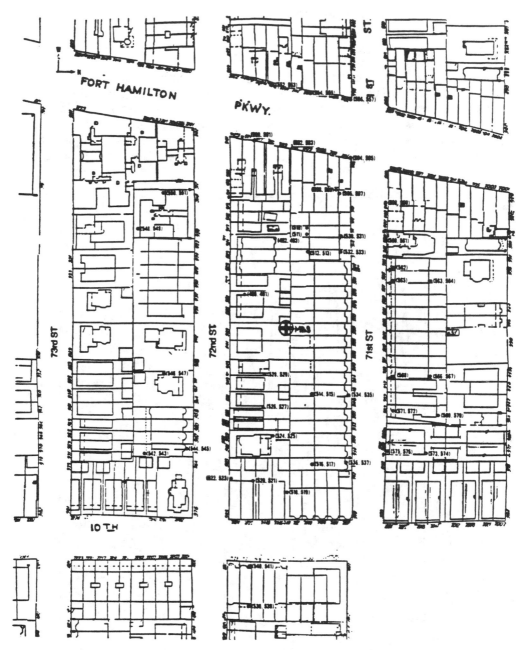

Figure 6.42 Location codes for measurements made with a mobile base station located in the backyard next to the telephone pole of 939 72nd Street.

from the 200 locations were averaged together. Figures 6.37 and 6.38 show the average received power and rms delay spread at each of the 69 locations for both the measurements and the simulation. These graphs show that the received power and rms delay spread follow essentially the same trends from one location to the other.

Based on these propagation models, received powers predicted by the models are examined by comparing them with measured data. Figure 6.39 shows the point-to-point comparison between predicted and measured power in all 621 locations for two-dimensional and three-dimensional ray-tracing models. It was determined that if two-dimensional and three-dimensional ray-tracing models are used, differences between the two predicted powers are less than 1 dB and, consequently, the two-dimensional model is judged to be valid and suitable for power prediction for this area. Judging from this figure, we see that the ray-tracing

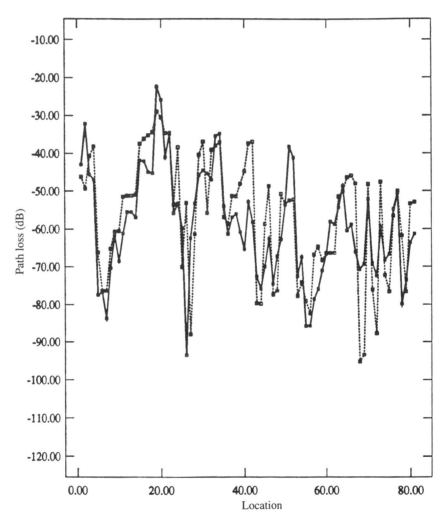

Figure 6.43 Path loss measurements and results of three-dimensional ray-tracing simulations. ●—, Measurement; ■—, simulation.

model is in good agreement with the measurements. The standard deviation of prediction error with the ray-tracing model is 2.3-dB. Figure 6.40 depicts the power coverage prediction for the area shown in Fig. 6.36.

6.5.5 Three-Dimensional Simulation in Urban Microcellular Environments

Here we consider the simulation of a typical residential area, located on 71st and 72nd Streets, between Fort Hamilton Parkway and 10th Avenue in Kings County, New York. This area consist mostly of two- and three-story private homes, though there are two four-story apartment buildings on the southern corner of 72nd Street and Fort Hamilton Parkway. Extensive measurements have been made in this area, and a three-dimensional description of these city blocks has been transferred to AutoCAD format [Yan94b]. Figure 6.41 depicts the overlays of the four blocks as displayed by AutoCAD. This simplified overlay provides a good model of realistic residential buildings, though it neglects complex roof structures, trees, and fences.

The simulation is carried out with locations of mobile base stations (MBSs) and personal subscriber units (PSUs), antenna patterns, and antenna heights kept the same as those used during the actual measurements. Figure 6.42 shows that the MBS receiver is located in the backyard of a residence next to a telephone pole while the PSU is positioned at more than 40 different locations. The MBS

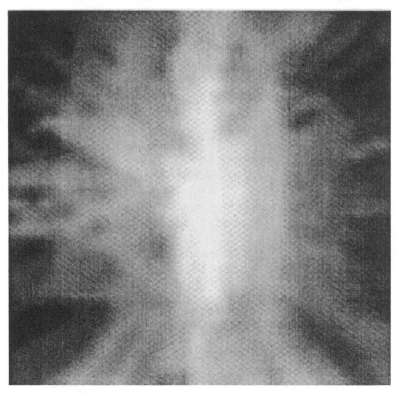

Figure 6.44 Power coverage for the area shown in Fig. 6.41.

receiving antenna is 18 ft high, and the PSU transmit antenna height is maintained at either 6 or 15 ft above the ground, providing two measurements for each location.

Figure 6.43 shows a comparison of path losses as determined by measurement and simulation in 80 receiver–transmitter configurations. The results show that power prediction errors are no more than 10 dB at 70% of the locations, while the remainder are in the range from 10 dB to 20 dB. The accuracy of these simulation results is comparable to the most accurate models utilized in the literature.

Realistic estimation of power coverage area is important in planning layouts of wireless communication systems. Figure 6.44 shows an example of power coverage predicted by simulations based on the simplified model described above. For more detailed treatments of ray-tracing, the reader may refer to [Leb89, Leb92, Ros92, Ros93, Law91, Law92, Rap92, Hol92a, Yan93a, Yam93b].

6.6 DIRECT SOLUTION OF RADIO PROPAGATION EQUATIONS

Although ray-tracing models can efficiently predict radio propagation characteristics for indoor and outdoor applications, these techniques are only approximations to the direct solution of electromagnetic wave propagation equations. The ideal method of simulating radio propagation is to solve Maxwell's equations numerically [Yee66, Taf75, Hol83, Fus90, Har92, Lee93]. The numerical solution of these differential equations over a designated area requires selection of a number of points at which the solution is to be determined iteratively. A systematic method of selecting these points is to draw a grid over the area, using certain specified rules, define meshes within the grid, and solve the equations at the nodes of the meshes. The dimension of the meshes within the grid is on the order of a wavelength at the carrier frequency. An increase in the carrier frequency or in the area for which the solution is to be obtained will increase the number of nodes exponentially. Therefore, numerical solution typically requires (a) large amounts of memory to keep the track of the solution at all locations and (b) extensive calculations to update the solution at successive instants of time. Given the steady advancements in high-performance computing, memory size and computational speed of computers are increasing rapidly, and this increases the feasibility of direct solution of Maxwell's equations for coverage areas of substantial size.

With today's computational capabilities, direct solution of the three-dimensional Maxwell's equations for areas with dimensions on the order of several meters and carrier frequencies around 1 GHz is generally not feasible. However, much work has been done with *finite-difference time-domain* (FDTD) numerical methods for performing direct solutions with reasonable computational complexity. The FDTD method is described in the next section.

6.6.1 Finite-Difference Time-Domain (FDTD) Model

The FDTD method is probably the most straightforward and most widely used method for numerical solution of Maxwell's equations. With this method, Maxwell's equations are approximated by a set of finite-difference equations. By placing the electric and magnetic fields on a staggered grid and defining appropriate initial conditions, the FDTD algorithm employs the central differences to approximate both spatial and temporal derivatives, and it solves Maxwell's equations directly. The distribution of electric and magnetic fields over the whole grid is calculated incrementally in time; and when the simulation is finished, the propagation characteristics are known at every location in the area under study.

In its original form, the FDTD algorithm used a rectangular grid [Yee66, Taf75], and curves and slanted lines were approximated by "staircases," which can introduce large errors unless the grid size is made very small. To circumvent this difficulty, a non-orthogonal quadrilateral grid was introduced by Holland [Hol83] in 1983. Later, Fusco [Fus90] successfully applied this approach to solving two-dimensional scattering problems. Recently, Lee and others [Har92, Lee93] have modified the approach to a more computationally efficient form that shows

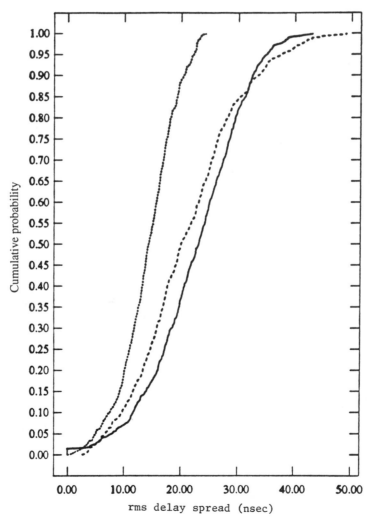

Figure 6.45 Comparison of the CDFs of the rms delay spreads for the measurements (———), three-dimensional ray-tracing (· · ·), and two-dimensional FDTD models (---).

significant improvement in accuracy over the rectangular FDTD algorithm. In this section we describe results obtained with a simple two-dimensional FDTD model of reasonable computational complexity, which was suggested in [Yan93] and is suitable for modeling areas with dimensions on the order of ten meters. A nonorthogonal mesh, rather than a rectangular mesh, is used in the FDTD model to reduce the mesh density required to achieve a certain precision and a further reduction of the computational load. The computational time needed with this method was comparable to that of a three-dimensional ray-tracing algorithm.

With this method, the indoor radio propagation problem is formulated as an E-field polarization scattering problem in two dimensions. Both the transmitting and receiving antennas are assumed to be ideally omnidirectional. The source electric field is specified as a function of time at the location of the transmitting antenna. By solving the Maxwell's equations numerically using the nonorthogonal FDTD algorithm, the electric field distribution in the area is calculated incrementally in time. Finally, from the spatial and temporal field distributions, the propagation characteristics are obtained. The area used for examination of this algorithm, shown in Fig. 6.36, is the portion of the Atwater Kent laboratory at WPI, which we had used previously for frequency-domain measurements, statistical modeling, and ray tracing.

Figure 6.39 shows the point-to-point comparison between predicted and measured power in all 621 measurement locations for two-dimensional and three-dimensional ray-tracing and for the FDTD models. From this figure we see that both ray-tracing models and the FDTD model are all in good agreement with the measurements. The standard deviations of prediction error with the ray-tracing and FDTD models are 2.3-dB and 2.7 dB, respectively, indicating that ray-tracing provides a more accurate estimate of the power. Figure 6.45 shows the cumulative distribution functions of rms delay spreads of measurements and simulations for all the locations. It is shown that the FDTD model can provide a more accurate prediction of rms delay spreads than the ray-tracing model.

6.7   COMPARISON OF DETERMINISTIC AND STATISTICAL MODELING

Development of a statistical model can involve a substantial effort to collect accurate databases for different environments of interest. However, implementation of statistical models on a computer requires minimal computational resources. In principle, development of deterministic models does not require that any measurements be made. In practice, some measurements are needed to check the accuracy of the model and to determine the values of model parameters such as conductivity and permittivity of structural materials. The implementation of both ray-tracing and FDTD models, however, always requires extensive computational resources.

Computation time required with the ray-tracing method grows exponentially with the complexity of the details included in the layout of the area, but the size of the area does not necessarily increase the computation time. As a result, the computation time required for modeling a small area with many walls might be as great as for a large area containing only a few buildings. As we noted earlier, application of the ray-tracing technique will in general involve different analytical approaches for different application areas. In indoor areas the transmission and reflections are the dominant mechanisms, whereas in outdoor applications the diffraction and reflections are the most important mechanisms. Computation time for the FDTD technique is proportional to the size of the area, and addition of structural details does not significantly affect the computation. However, the number of nodes used for computation is exponentially related to the size of the area and the frequency of operation. In general, the ray-tracing technique is suitable for radio propagation studies in large areas whereas the FDTD approach is appropriate only for smaller areas.

Statistical models have the advantage of being computationally efficient and easy to generalize. Furthermore, the development of a statistical model is independent of the layout and structural details of the coverage area, which removes the requirement for investing time and resources in surveying outdoor areas or building layouts for each individual application. The statistical time-domain models in particular are the popular choice of standards bodies, and their recommended models generally include selectable sets of parameters derived from earlier measurements in representative categories of coverage areas. However, a disadvantage of statistical models is that they cannot provide the relationship between the layout of a building or an outdoor area and the detailed channel response in a specific location. Building-specific radio propagation modeling is feasible only with deterministic approaches such as ray-tracing and FDTD. Applications such as the optimum siting of antennas, or the analysis of systems using sectored antennas, are best served by use of deterministic ray-tracing models, which can provide realistic estimates of the azimuthal distribution of rays received in a multipath environment.

APPENDIX 6A: GSM-RECOMMENDED MULTIPATH PROPAGATION MODELS

This appendix provides tables of parameters recommended by the GSM standards organization for modeling and simulation of mobile radio channels in various environments. The tables are taken from Annex 3 of GSM Recommendation 05.05. The sets of model parameters for outdoor areas are identified as RAx for rural areas, HTx for hilly terrain, and TUx for urban areas. A fourth set of parameters, EQx, is provided for standardized testing of adaptive equalizers. For each of the four cases, a profile of 6 tap settings is provided, and for the HTx and TUx cases, a 12-tap profile is also provided. The 6-tap models were defined by GSM to provide for situations in which a given simulator does not have the capability to implement a 12-tap model. For each model, two equivalent alternative profiles of tap settings are given, indicated respectively by (1) and (2) in appropriate columns. The Doppler spectrum to be applied at each tap, identified in the right-most column of each table, is either Rician (RICE) or the classical Doppler (CLASS) spectrum, as discussed in Section 6.2.

TABLE 6A.1   GSM Model Parameters for Hilly Terrain

Tap Number	Relative Time (μsec)		Average Relative Power (dB)		Doppler Spectrum
	(1)	(2)	(1)	(2)	

Typical case for hilly terrain (HTx): (12-tap setting):

Tap Number	(1)	(2)	(1)	(2)	
1	0.0	0.0	− 10.0	− 10.0	Class
2	0.1	0.2	− 8.0	− 8.0	Class
3	0.3	0.4	− 6.0	− 6.0	Class
4	0.5	0.6	− 4.0	− 4.0	Class
5	0.7	0.8	0.0	0.0	Class
6	1.0	2.0	0.0	0.0	Class
7	1.3	2.4	− 4.0	− 4.0	Class
8	15.0	15.0	− 8.0	− 8.0	Class
9	15.2	15.2	− 9.0	− 9.0	Class
10	15.7	15.8	− 10.0	− 10.0	Class
11	17.2	17.2	− 12.0	− 12.0	Class
12	20.0	20.0	− 14.0	− 14.0	Class

The reduced setting (6 taps) is defined hereunder:

Tap Number	(1)	(2)	(1)	(2)	
1	0.0	0.0	0.0	0.0	Class
2	0.1	0.2	− 1.5	− 2.0	Class
3	0.3	0.4	− 4.5	− 4.0	Class
4	0.5	0.6	− 7.5	− 7.0	Class
5	15.0	15.0	− 8.0	− 6.0	Class
6	17.2	17.2	− 17.7	− 12.0	Class

TABLE 6A.2   GSM Model Parameters for Urban Area

Tap Number	Relative Time (μsec)		Average Relative Power-(dB)		Doppler Spectrum
	(1)	(2)	(1)	(2)	

Typical case for urban area (TUx): (12-tap setting):

Tap Number	(1)	(2)	(1)	(2)	
1	0.0	0.0	− 4.0	− 4.0	Class
2	0.1	0.2	− 3.0	− 3.0	Class
3	0.3	0.4	0.0	0.0	Class
4	0.5	0.6	− 2.6	− 2.0	Class
5	0.8	0.8	− 3.0	− 3.0	Class
6	1.1	1.2	− 5.0	− 5.0	Class
7	1.3	1.4	− 7.0	− 7.0	Class
8	1.7	1.8	− 5.0	− 5.0	Class
9	2.3	2.4	− 6.5	− 6.0	Class
10	3.1	3.0	− 8.6	− 9.0	Class
11	3.2	3.2	− 11.0	− 11.0	Class
12	5.0	5.0	− 10.0	− 10.0	Class

The reduced TUx setting (6 taps) is defined hereunder:

Tap Number	(1)	(2)	(1)	(2)	
1	0.0	0.0	− 3.0	− 3.0	Class
2	0.2	0.2	0.0	0.0	Class
3	0.5	0.6	− 2.0	− 2.0	Class
4	1.6	1.6	− 6.0	− 6.0	Class
5	2.3	2.4	− 8.0	− 8.0	Class
6	5.0	5.0	− 10.0	− 10.0	Class

**TABLE 6A.3   GSM Parameters, for Testing Adaptive Equalizers,
Profile for Equalization Test (EQx), 6-tap setting**

Tap Number	Relative Time (μsec)	Average Relative Power-(dB)	Doppler Spectrum
1	0.0	0.0	Class
2	3.2	0.0	Class
3	6.4	0.0	Class
4	9.6	0.0	Class
5	12.8	0.0	Class
6	16.0	0.0	Class

APPENDIX 6B: WIDEBAND MULTIPATH PROPAGATION MODELS

This appendix provides tables of parameters, recommended by the Joint Technical Committee (JTC) for PCS Air Interface Standards, for time-domain simulation of PCS channels in various environments. The various categories of communication environments are identified in Table 6.3 and discussed briefly in Section 6.2.

TABLE 6B.1   JTC Channel Model Parameters for Indoor Office Areas

	Channel A		Channel B		Channel C		
Tap	Relative Delay (nsec)	Average Power (dB)	Relative Delay (nsec)	Average Power (dB)	Relative Delay (nsec)	Average Power (dB)	Doppler Spectrum
1	0	0	0	0	0	0	Flat
2	50	−3.6	50	−1.6	100	−0.9	Flat
3	100	−7.2	150	−4.7	150	−1.4	Flat
4			325	−10.1	500	−2.6	Flat
5			550	−17.1	550	−5.0	Flat
6			700	−21.7	1,125	−1.2	Flat
					1,650	−10.0	Flat
					2,375	−21.7	Flat

TABLE 6B.2   JTC Channel Model Parameters for Indoor Commercial Areas

	Channel A		Channel B		Channel C		
Tap	Relative Delay (nsec)	Average Power (dB)	Relative Delay (nsec)	Average Power (dB)	Relative Delay (nsec)	Average Power (dB)	Doppler Spectrum
1	0	0	0	−4.6	0	0	Flat
2	50	−2.9	50	0	50	−0.4	Flat
3	!00	−5.8	150	−4.3	250	−6.0	Flat
4	150	−8.7	225	−6.5	300	−2.5	Flat
5	200	−11.6	400	−3.0	550	−4.5	Flat
6			525	−15.2	800	−1.2	Flat
7			750	−21.7	2,050	−17.0	Flat
8					2,675	−10.0	Flat

TABLE 6B.3 JTC Channel Model Parameters for Outdoor Urban High-Rise Areas — Low Antenna

Tap	Channel A Relative Delay (nsec)	Channel A Average Power (dB)	Channel B Relative Delay (nsec)	Channel B Average Power (dB)	Channel C Relative Delay (nsec)	Channel C Average Power (dB)	Doppler Spectrum
1	0	0	0	0	0	−9.0	Class
2	50	−1.6	200	−1.2	50	0	Class
3	150	−4.7	250	−13.0	500	−1.1	Class
4	325	−10.1	800	−4.6	800	−11.2	Class
5	550	−17.1	1,250	−7.2	2,250	−4.9	Class
6	700	−21.7	2,100	−6.0	4,200	−9.1	Class
7			3,050	−13.0	6,300	−9.6	Class
8			3,750	−21.7	7,500	−16.3	Class
9					8,550	−18.6	Class
10					10,000	−21.7	Class

TABLE 6B.4 JTC Channel Model Parameters for Outdoor Urban Low-Rise Areas — Low Antenna

Tap	Channel A Relative Delay (nsec)	Channel A Average Power (dB)	Channel B Relative Delay (nsec)	Channel B Average Power (dB)	Channel C Relative Delay (nsec)	Channel C Average Power (dB)	Doppler Spectrum
1	0	0	0	0	0	0	Class
2	50	−1.6	50	−3.0	50	−0.1	Class
3	150	−4.7	200	−2.6	500	−6.0	Class
4	325	−10.1	475	−1.4	800	−1.7	Class
5	550	−17.1	1,000	−1.2	2,250	−4.9	Class
6	700	−21.7	1,650	−4.8	4,200	−9.1	Class
7			2,350	−5.2	6,300	−13.7	Class
8			2,800	−8.1	7,500	−7.0	Class
9			3,500	−10.1	8,550	−18.6	Class
10			5,100	−14.8	10,000	−21.7	Class

TABLE 6B.5 JTC Channel Model Parameters for Outdoor Residential Areas — Low Antenna

Tap	Channel A Relative Delay (nsec)	Channel A Average Power (dB)	Channel B Relative Delay (nsec)	Channel B Average Power (dB)	Channel C Relative Delay (nsec)	Channel C Average Power (dB)	Doppler Spectrum
1	0	0	0	0	0	0	Class
2	50	−2.9	50	−0.4	50	−3.0	Class
3	100	−5.8	250	−6.0	200	−2.6	Class
4	150	−8.7	300	−2.5	475	−1.4	Class
5	200	−11.6	550	−4.5	1,000	−1.2	Class
6	250	−14.5	800	−1.2	1,650	−4.8	Class
7	300	−17.4	2,050	−17.0	2,350	−5.2	Class
8	350	−20.3	2,675	−10.0	2,800	−8.1	Class
9					3,500	−10.1	Class
10					5100	−14.8	Class

**TABLE 6B.6   JTC Channel Model Parameters for
Outdoor Urban High-Rise Areas — Low Antenna**

	Channel A		Channel B		Channel C		
Tap	Relative Delay (nsec)	Average Power (dB)	Relative Delay (nsec)	Average Power (dB)	Relative Delay (nsec)	Average Power (dB)	Doppler Spectrum
1	0	0	0	−5.2	0	−4.6	Class
2	50	−0.4	50	−3.0	300	−0.1	Class
3	250	−6.0	300	0	350	0	Class
4	300	−2.5	750	−0.8	750	−0.3	Class
5	550	−4.5	1,250	−1.4	1,250	−0.5	Class
6	800	−1.2	5,000	−4.6	4,000	−7.0	Class
7	2,050	−17.0	8,900	−9.6	10,000	−4.3	Class
8	2,675	−10.0	13,000	−6.0	22,000	−4.0	Class
9			17,000	−18.5	29,000	−8.2	Class
10			20,000	−13.0	50,000	−16.0	Class

**TABLE 6B.7   JTC Channel Model Parameters for
Outdoor Urban / Suburban Low-Rise Areas — High Antenna**

	Channel A		Channel B		Channel C		
Tap	Relative Delay (nsec)	Average Power (dB)	Relative Delay (nsec)	Average Power (dB)	Relative Delay (nsec)	Average Power (dB)	Doppler Spectrum
1	0	−3.0	0	−1.2	0	−4.6	Class
2	50	−7.0	300	−6.0	300	−0.1	Class
3	200	0	700	0	350	−0.1	Class
4	500	−6.2	750	−0.7	750	−7.0	Class
5	1,200	−5.2	1,250	−1.1	2,250	−0.7	Class
6	1,525	−18.9	5,000	−5.2	8,000	0	Class
7	1,750	−21.7	8,900	−7.7	20,000	−5.8	Class
8			15,000	−3.0	32,000	−7.0	Class
9			21,000	−18.2	39,000	−7.0	Class
10			25,000	−16.0	55,000	−10.0	Class

TABLE 6B.8   JTC Channel Model Parameters for Outdoor Residential Areas — High Antenna

	Channel A		Channel B		Channel C		
Tap	Relative Delay (nsec)	Average Power (dB)	Relative Delay (nsec)	Average Power (dB)	Relative Delay (nsec)	Average Power (dB)	Doppler Spectrum
1	0	−6.0	0	−6.0	0	−4.6	Class
2	50	−3.0	450	−3.0	300	−0.1	Class
3	150	0	500	0	350	0	Class
4	500	−6.7	1,050	−1.5	750	−0.3	Class
5	850	−1.2	3,250	−4.7	1,250	−0.5	Class
6	1,325	−17.7	6,000	−3.0	4,000	−7.0	Class
7	1,750	−23.4	8,300	−12.0	10,000	−4.3	Class
8			10,000	−14.5	22,000	−4.0	Class
9			12,050	−17.4	29,000	−8.2	Class
10			15,000	−21.7	50,000	−16.0	Class

QUESTIONS

(a) Why do we need wideband channel simulations?

(b) Although the arrival times of signal paths are known to be random, the models recommended by standards bodies assume fixed arrival times for the paths. How are modem performance evaluations affected by the assumption of fixed arrival times in these models?

(c) Name three advantages of time-domain statistical modeling.

(d) Name three advantages of frequency-domain statistical modeling.

(e) What are the three major classes of environments in the JTC channel model? What are the three major categories within each environment?

(f) How many classes of channel profiles are considered in the GSM channel model?

(g) Why does the JTC model provide three classes of profiles for each specific area in an environment?

(h) What is the main advantage of deterministic modeling over statistical modeling?

(i) How does ray-tracing relate to the direct solution of Maxwell's equations?

(j) Discuss advantages and disadvantages of ray-tracing relative to statistical modeling.

(k) Name all the mechanisms considered in the ray-tracing algorithms.

(l) What are the major mechanisms used in ray-tracing of the indoor radio channel?

(m) What are the major mechanisms used in ray-tracing for microcellular outdoor areas?

(n) Which form of statistical modeling is commonly used by standards bodies to represent the radio propagation media?

PROBLEMS

Problem 1. If an exponential distribution with variance 2 is used for simulation of the interarrival times of a Poisson random process, what is the average arrival rate of the process? Using MatLAB or MathCAD, generate 10 samples of the Poisson arrivals and plot them versus time.

Problem 2. In the Poisson arrival process, if the time axis is divided into small intervals, referred to as *bins*, then the existence of arrivals in the bins forms a binomial distribution. This principle is commonly used for the simulation of the

Poisson path arrivals in computer simulation of wideband radio channel characteristics.

If we observe a Poisson process for T seconds and the average of path arrivals per second is μ, the probability of arrival of l paths in T seconds is given by

$$P(L = l, T) = \frac{(\mu T)^l}{l!} e^{-\mu T} \tag{P6.1}$$

Assume that T_m is the measurable delay spread of the channel (the span of time in which we are able to detect a path in background noise) and that Δt is the resolution of the measurement system (base of the pulses used for measurement). If we use the resolution as the bin interval and we have N bins, then $T_m = N\Delta t$.

(a) Comparing Eq. (P6.1) with Eq. (6.2.2), observe that $\lambda = \mu T_m$.
(b) From Eq. (P6.1), show that the probability of path occurrence in a bin is given by

$$r_i = 1 - P(L = 0, \Delta t) = \mu \Delta t \, e^{-\mu \Delta t}$$

(c) Using the exponential series expansion, show that for narrow bin widths where $\mu \Delta t$ is a very small number, $r_i = \mu \Delta t$.
(d) Show that Eq. (6.2.3) holds if the approximation in part (c) holds.
(e) Using the above discussions, explain a procedure for simulation of the Poisson path arrivals that uses only uniformly distributed random variables to determine the existence of paths in the bins.

Problem 3

(a) Use the parameters given in Table 6.1 to simulate the modified Poisson process and regenerate the path occurrence probabilities shown in Fig. 6.4.
(b) Use Table 6.1 to determine the arrival rate of the experiment in the manufacturing areas if the arrivals are assumed to be Poisson.
(c) Use the results of part (b) to regenerate Fig. 6.1.

Problem 4. For frequency-selective fading channels it is sometimes necessary to simulate a notch in the passband of the channel. A convenient model has been developed by Rummler for application in the microwave LOS channels. In this model the frequency response of the channel is represented by

$$H(j\omega) = a\left[1 - be^{-j(\omega - \omega_0)\tau}\right]$$

(a) Assume $a = 1$ and determine b so that the channel produces a 30-dB notch at the frequency f_0.
(b) For $\tau = 6.3$ nsec and $f_0 = 20$ MHz, sketch the amplitude and phase response of the transfer function of the channel for $0 < f < 100$ MHz.

Problem 5. Another approach to simulating a deep notch is to use the power series in the frequency domain. In the two-term power series model for microwave LOS, the channel frequency response is represented by

$$H(j\omega) = A_0 + (A_1 + jB_1)j\omega$$

(a) Assume $A_0 = 1$, and determine the values of A_1 and B_1 that produce an infinite decibel notch at $f_0 = 20$ MHz ($|H(j\omega_0)|^2 = 0$).

(b) Sketch the magnitude and phase of the frequency response for the parameters determined in part (a) in the range $0 < f < 100$ MHz.

(c) Describe a method to solve part (a) for any depth of the notch, where the notch depth is defined with respect to $H(0)$.

Problem 6

(a) Use MathCAD or MatLAB to simulate the Poisson/Rayleigh profiles used in Fig. 6.11.

(b) Generate the CDF of the rms delay spread using 100 samples, and compare the results with those presented in Fig. 6.11.

Problem 7

(a) Use MathCAD or MatLAB to simulate the modified Poisson/lognormal profiles used in Fig. 6.11.

(b) Generate the CDF of the rms delay spread using 100 samples, and compare the results with those presented in Fig. 6.11.

Problem 8

(a) Use MathCAD or MatLAB to simulate the channel A in indoor office areas with the JTC model described in Table 6B.1.

(b) Generate the CDF of the rms delay spread using 100 samples, and compare it with the results of Fig. 6.11.

(c) Calculate the rms delay spread of the channel from the table, and compare the result with average of the rms delay spreads used in the CDF curve.

Problem 9

(a) Use MathCAD or MatLAB to simulate the GSM model for hilly terrain, using 6 taps.

(b) Generate the CDF of the rms delay spread using 100 samples.

(c) Calculate the rms delay spread of the channel from the table, and compare the result with average of the rms delay spreads used in the CDF curve.

Problem 10

(a) Use MathCAD or MatLAB to simulate the frequency-domain AR model 2 for experiment M1.

(b) Generate the CDF of the rms delay spread using 100 samples, and compare the results with the results shown in Fig. 6.25.

(c)   Sketch the delay power spectrum of the channel using the locations of the poles. The inverse Fourier transform of a transfer function with poles located at the average location of the poles can be considered as the delay power spectrum of the channel.

(d)   Calculate the rms delay spread of the channel from the the delay power spectrum, and compare the results with the average of the rms delay spreads used in the CDF curve.

7

MODEM TECHNOLOGY

7.1 INTRODUCTION

In this chapter we describe various modulator and demodulator (modem) technologies. We begin with a discussion of basic modem techniques—that is, techniques widely employed in standard voiceband modems for use in the wired public telephone network. Then we discuss the application of the basic modem techniques to radio channels, where multipath and fading limit performance and give rise to certain design issues and enhancement techniques which are specific to the radio environment.

The organization of this chapter in a sense reflects the evolution of modem technology for application to wireline and radio systems. The development of modern wireline modem technology began in the late 1950s, when a market began to develop for transfer of data between computers over the public switched telephone network (PSTN). At that time, the PSTN, which had been designed for voice traffic, had characteristics which significantly limited its capability for carrying digital data. The frequency response characteristics of most telephone circuits exhibited considerable phase distortion, which did not necessarily impair voice quality but placed severe limitations on the data rates which could be sustained over the network. In addition, abrupt phase hits and noise impulses caused bursts of errors in transmitted data. Thus the early modems used low data rates and simple, robust modulation methods so as to provide reliable data delivery over arbitrary switched call connections across the network.

The early limitations of the PSTN for data transmission led to the development of adaptive equalization, by which modems could automatically (though not completely) compensate for the amplitude and phase distortion encountered in each new call connection. At the same time, the Bell System steadily improved the transmission characteristics of the PSTN, paving the way for increasingly sophisticated modem designs capable of achieving ever higher data rates over voice-bandwidth connections. The most sophisticated of today's modems are in fact wireline modems, which are approaching the ultimate limits set by the bandwidth and signal-to-noise ratio (SNR) provided by a voiceband PSTN channel. Pahlavan and Holsinger [Pah88c] provide a detailed survey of the development of modem technology for the voiceband telephone channel. A detailed description of the CCITT V-Series modem standards can be found in [Bla91].

The application of modem technology to radio systems has evolved along a path similar to that followed in PSTN data transmission, but has lagged several years behind the wireline developments. This has of course been due to the special characteristics of radio propagation, which create a much harsher environment for data transmission than we encounter in the telephone network. Location-dependent power variations, fading, and multipath act to limit the data rates and performance achievable over a radio channel. Consequently the earliest of modem applications in radio systems used only the simplest of wireline modem techniques, though much more advanced techniques were already in use in the PSTN. For example, while voiceband wireline modems incorporating adaptive equalization were coming into general use in the early 1970s, adaptive equalization was not successfully applied to radio systems until the late 1970s. Similarly, while the technique of trellis-coded modulation (TCM) has been used in standard commercial modems for several years now, the application of TCM to mobile radio systems is just beginning. Thus the transfer of evolving wireline modem technology from the wired network to wireless systems is a continual process, with the radio applications employing certain design features and enhancements which are specific to the special characteristics of that environment.

A number of the wireline modem techniques we describe here are being incorporated into standards for several new mobile wireless systems, including digital cellular systems, digital land-mobile radio systems, and emerging systems for personal communications services (PCS). In the wireless local area network (WLAN) market, the lack of an unlicensed frequency band without restriction on choice of modulation technique had held back the application of the more sophisticated modem design technologies in those systems. As the market for portable data terminals, notebook computers, and portable facsimile devices grows, the demand for a band specifically assigned to personal wireless data communications increases. Given the recent (September 1993) FCC designation of certain 2-GHz bands for data services, combined with the ever-growing demand for higher speeds and wider and more robust coverage for wireless data communications, there will undoubtedly be a migration of the more sophisticated wireline modem technologies into wireless data communication products.

Before we move on to the remainder of this chapter, it is worthwhile to review a very basic point—that is, why a radio system must include modulation for data transmission. While this discussion will be viewed by some readers as unnecessary, it will serve to highlight a few issues which are important in examining differences among different wireless systems designed for the same type of service or application.

In the case of data transmission over a telephone line, it is clear that we need a modulator to convert a stream of digital data to a line signal which is compatible with the passband characteristics of a voice circuit. A typical telephone voice circuit passes signals only in the frequency interval from about 300 Hz to about 3300 Hz, and thus a modem creates a line signal positioned appropriately in that band. Most of the existing modem standards utilize a carrier frequency of 1800 Hz, which is about at the center of the passband of this channel. In the case of a radio system, however, one might well ask why we do not simply apply the data stream to a power amplifier feeding a transmitting antenna; there are at least three reasons why we use modulation as an intermediate step.

First, we must consider the fact that for effective signal radiation, the length of the transmitting antenna must be comparable to the wavelength of the signal to be transmitted. Therefore, for example, a binary stream at 3 kbits/sec has a bandwidth of about 3 kHz, which implies an antenna of length around 100 km, a totally impractical design. But if we modulate the data stream onto a 3-GHz carrier, the appropriate antenna length will be around 10 cm. Furthermore, because the effective bandwidth of an antenna, normalized as a percentage of its center frequency, stays roughly constant from one wavelength to another, the absolute value of the available bandwidth increases proportionally with the center frequency. Thus we want to modulate the data stream onto an appropriate radio frequency (or onto some intermediate frequency, which is then up-converted to the radio frequency), where antenna size is practical and where antenna bandwidth is ample, or at least adequate, relative to the data rate we wish to transmit.

Second, by modulating the data stream onto a carrier, we can ensure the orderly coexistence of multiple signals in a given spectral band by arranging the carriers in a frequency division multiplex (FDM) format. In fact, FDM is the oldest and simplest method of providing multiple-user access to a shared segment of frequency spectrum.

Third, particular forms of modulation may provide an especially effective way of reducing interference among intended users or interference from unwanted sources. We refer here to spread-spectrum modulation, which was developed to defeat intentional jamming of military communication systems [Sch82, Pri83]. Spread-spectrum modulation is being used for both wireless voice and data communications. In wireless data communications, the availability of the frequency bands (ISM bands) for spread-spectrum modulation has motivated the implementation of a number of spread-spectrum WLAN products. In the wireless portable and mobile radio industries, spread-spectrum modulation is used with a multiuser access technique known as *code division multiple access* (CDMA) to increase the number of users supported in an allocated frequency band.

In Section 7.2 we describe basic modem techniques and show that a number of the techniques are closely related. We show the transmit and receive functions that are performed in all standard modems. We show how modem signal constellations of increasing complexity can provide corresponding increases in bandwidth efficiency— that is, increased data rate per unit of available bandwidth. Increasing bandwidth efficiency has been an important objective in the wireline modem industry, because the bandwidth of telephone channels is fixed by the design of the public network, while the demand for ever-higher data rates has increased unabated. The most sophisticated voiceband modems have been developed for use on leased lines, where the modems are part of a data network. Leased lines are more expensive than ordinary dial-up lines, and an increase in the achievable data rate will reduce the required number of leased lines, thus lowering the cost of operating the network. As a result, major users of these networks, such as banks and airline companies, have always been willing to make large investments to steadily update the data rate capabilities of their leased circuits. In the case of radio modems, an increase in the bandwidth efficiency will reduce the bandwidth required for a given data rate. For the wireless portable and mobile radio industry, where service is provided to a very large number of customers, increasing the bandwidth efficiency allows more simultaneous users in the allocated bandwidth, increasing the revenues of the operating company.

In Section 7.3 we review the theoretical limits on achievable data rate and communication efficiency in an ideal Gaussian-noise channel, as defined by Shannon's capacity formula. We then describe briefly the real-world channel impairments that constrain the data rates achievable in practice.

In Section 7.4 we examine the effects imposed on the performance of standard modem techniques by the characteristics of flat Rayleigh fading radio channels. We then discuss the effects of diversity combining methods and coding techniques on the performance of radio modems operating on these channels.

In Section 7.5 we begin by addressing the major issues which arise in the selection of a modulation technique for use on a radio channel. Then, we describe the standard radio modems which are being used in existing and emerging new wireless networks. Other modem technologies, utilizing spread-spectrum and infrared techniques, are discussed in Chapters 9 and 10 respectively.

7.2 BASIC MODULATION TECHNIQUES

In this section we outline basic modulation techniques used primarily in standard voice-band modems. It is not our intention here to provide an exhaustive treatment of modulation techniques but rather to explain the basic forms of modulation, to make comparisons among the different techniques, and to describe the principal functions which must be implemented in any modem. First we provide the framework for analysis of modem techniques and then we describe various approaches to implementing a modem. More detailed treatments of modulation techniques and modem design can be found in a number of references, including [Feh87], [Bin88], and [Pro89].

7.2.1 Framework for Analysis

In this subsection we discuss certain topics that underlie our subsequent discussions of modem techniques and achievable performance. Our discussion covers matched filtering, different measures and interpretations of SNR, and the criteria used for error rate analysis for different modems.

Matched Filtering. In establishing a digital communication link, the information digits arriving in sequence are mapped into symbols, which are in turn represented by pulses for transmission over the link. At the receiving end, each received pulse waveform is filtered and then sampled. The sampled signal is used for making a decision on the transmitted symbol, which is then mapped back to information digits. As a result, the basic elements of digital communication are built around pulse transmission and the design of the receiver filter. It is well known that on an additive noise channel, the optimum receiver filter, one that maximizes the SNR after sampling at the receiver, is a *matched filter*. Figure 7.1a shows a block diagram of the optimum receiver for pulse transmission. The transmitted pulse representing the ith symbol is $s_i(t)$, and it is disturbed by additive white Gaussian

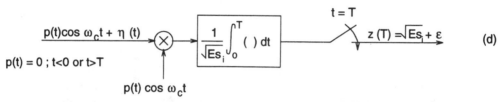

Figure 7.1 Matched filter. (*a*) Block diagram of the optimum receiver for pulse transmission. (*b*) Matched filter implemented with a correlator when the transmitted pulse is time-limited. (*c*) Matched filter implemented with a mixer followed by a filter matched to $p(t)$. (*d*) Matched filter implemented with a correlator when $p(t)$ is time-limited.

noise (AWGN) $\eta(t)$ with two-sided spectral density N_0. The received signal is passed through the filter and sampled at time $t = T$. The sampled signal $z(T)$ has a deterministic component $\sqrt{E_{si}}$, where

$$E_{si} = \int_{-\infty}^{\infty} |s_i(t)|^2 \, dt \qquad (7.2.1)$$

which is referred to as the *energy per symbol*; it also has a random component ε with variance

$$\overline{|\varepsilon|^2} = N_0$$

The impulse response of the matched filter is

$$f(t) = \frac{1}{\sqrt{E_{si}}} s_i(T - t) \qquad (7.2.2)$$

which is the transmitted pulse waveform reversed and shifted to time $t = T$. The SNR per symbol, γ_{si} at the output of the matched filter is then given by

$$\gamma_{si} = \frac{E_{si}}{N_0} \qquad (7.2.3)$$

which is the ratio of the square of the deterministic component of the sampled signal to the variance of the random component. The energy per symbol E_{si} is an indicator of the transmitted power, while N_0, the power spectral density of the AWGN, is an indicator of the noise power. Different modulation techniques can be based on different-size modulation symbol alphabets and employ different pulse waveforms. Thus to provide a more meaningful measure, the SNR and the energy per symbol are usually averaged over all symbols and are represented by γ_s and E_s, respectively. The error rate of any modulation technique is a function of γ_s. For different modulation techniques the mapping of the digits into symbols is different, resulting in different equations for calculation of the error rate.

If the transmitted pulse is time-limited, the matched filter can be equivalently implemented with a correlator, as shown in Fig. 7.1b. In this figure, $s_i(t) = 0$ if $t > T$ or $t < 0$. If the transmitted symbol is a radio-frequency (RF) pulse of the form $s_i(t) = p(t)\cos(\omega_c t)$, the matched filter will be implemented with a mixer followed by a filter matched to $p(t)$, as shown in Fig. 7.1c. As with the reception of baseband pulses, if $p(t)$ is time-limited, the matched filter can be implemented with a correlator, as shown in Figure 7.1d. The matched filter implementation is commonly used in voiceband modems, whereas the correlator implementation is used in direct-sequence, spread-spectrum modulation systems. The pulse $p(t)$ is a baseband pulse with energy $E_p = 2E_{si}$, so that the SNR at the baseband is given by $E_p/N_0 = 2E_{si}/N_0$.

Alternative Interpretations of the SNR. In order to have a basis for evaluating the communication performance achievable with various modulation and demodulation methods, we must carefully define our terminology for signals and noise. It is important to distinguish between two measures of signal energy. The measure of signal energy used to define γ_s in the previous section is the average signal energy per channel symbol, commonly denoted by E_s. In most cases of M-ary modulation, M is a power of 2, say $M = 2^m$, and thus each M-ary channel symbol carries m bits of information. Therefore we can define signal energy per bit as $E_b = E_s/m$ and the SNR per bit as $\gamma_b = \gamma_s/m$, which in effect normalizes the symbol energy in the channel to the individual bits in the reconstituted data stream appearing at the output of the demodulator. Both measures of signal energy are useful in making comparisons among the error rates of alternative communication techniques and systems.

Another approach is to define the SNR as the ratio of the received power to noise power in the communication channel. This method relates the SNR directly to the requirements on transmitted power. Here we are comparing modulation schemes based on the ratio of signal power S to noise power N in the transmission channel. We can relate channel S/N to the modulation parameters by assuming that the bandwidth is W and the symbol duration is T_s, which yields

$$\frac{S}{N} = \frac{E_s/T_s}{N_0 W} = \frac{R_s}{W} \times \gamma_s \qquad (7.2.4)$$

where $R_s = 1/T_s$ is the symbol transmission rate. This equation relates the received signal-to-noise power ratio S/N to the SNR per symbol γ_s usually employed for the calculation of error rate. Equation (7.2.4) is derived for the ideal case in which the receiver filter is matched to the transmit pulse-shaping filter. If the filters are not matched [for example, if the transmitter uses a filter with raised-cosine frequency roll off (to be discussed in Chapter 8) but the receiver uses a brick-wall filter], the required S/N for the same E_s/N_0 will change by 1–2 dB.

The symbol transmission rate is $R_s = R_b/m$, where R_b is the bit transmission rate of the system. Thus defining S/N in terms of bit rate we have

$$\frac{S}{N} = \frac{E_b R_b}{N_0 W} = \frac{R_b}{W} \times \gamma_b \qquad (7.2.5)$$

which relates the signal-to-noise power ratio to the SNR per bit, the spectral density of the additive noise, the data bit rate, and the occupied bandwidth. The ratio of the transmission data rate to the occupied bandwidth, $\eta = R_b/W$, is referred to as the *bandwidth efficiency* of the modulation technique. The ratio $\gamma_b = E_b/N_0$ is the SNR per bit, the quantity ordinarily used for calculation of the error rate of the system. (The quantity E_b/N_0 is sometimes referred to as the *energy contrast ratio*, because E_b and N_0 both have units of energy.) The ratio S/N is the received SNR parameter that provides a measure of the transmitted power. The parameters usually examined in evaluations of alternative modulation techniques are (a) the required minimum transmitted power for acceptable performance and (b) the bandwidth efficiency of the modulation technique. For a modulation technique with a bandwidth efficiency of $\eta = 1$ bit/sec/Hz, we have $S/N = E_b/N_0$; that is, the signal-to-noise power ratio is the same as the ratio of the energy per bit to the spectral density of the additive noise. In the communications literature, different combination of the above measures of SNR are used, and the reader must have an accurate understanding of the various parameters in order to make proper comparison among systems. For example, the power ratio S/N is usually used in the satellite communications literature and is sometimes referred to as the *carrier-to-noise ratio* (CNR or C/N). The energy ratio E_s/N_0 is the parameter ordinarily used in the literature on voiceband data communications.

Error Rate as a Performance Criterion. The standard performance criterion in digital communications is the probability of bit error or bit-error rate (BER) of a modem. Some voiceband modem applications, such as the transfer of financial data, permit error rates no greater than 10^{-5}, whereas other applications such as digitized voice in cellular or mobile radio systems will tolerate error rates as high as 10^{-2} to 10^{-3}. Meanwhile, high-fidelity digital audio systems (e.g., compact disk players) demand error rates on the order of 10^{-8}. From the design standpoint, for a given modulation and coding scheme there is a one-to-one correspondence between the BER and the received signal-to-noise power ratio S/N. From the user standpoint, S/N is not the favorite criterion for the performance evaluation of digital communication links, because the user measures the quality of a system by the number of errors in the received bits and prefers to avoid the technical details of modulation or coding. However, using received S/N rather than BER will allow us to relate our performance criteria to the required transmitted power, which is very important for battery-operated wireless operations. For analog communications the received S/N is the usual measure of performance quality. An 18-dB S/N is typically required for analog mobile radio systems, and 30-dB S/N is expected in FM broadcasting systems. In comparing analog and digital systems, we need to translate these performance criteria into a common basis that makes S/N the convenient criterion for comparing these systems. In comparing digital systems with one another, the BER is used most of the time.

The error rate for a digital modulation technique is almost always expressed in the form of an *exponential* function or a *complementary error function* (erfc), where the erfc function is defined as

$$\mathrm{erfc}(x) = \frac{2}{\sqrt{\pi}} \int_x^\infty e^{-t^2}\, dt.$$

For the exponential function we have the BER or probability of bit error, P_b, given by the general expression

$$P_b = \alpha e^{-\beta \gamma_b} \tag{7.2.6}$$

where α and β are given values appropriate to each specific modulation technique. The erfc function is bounded by an exponential function; as a result, for a large class of modulations we have

$$P_b = \alpha\, \mathrm{erfc}\,\sqrt{\beta \gamma_b} < \alpha e^{-\beta \gamma_b} \tag{7.2.7}$$

Figure 7.2 shows the two functions given in Eqs. (7.2.6) and (7.2.7), for $\alpha = 1/2$ and $\beta = 1$. With these parameter values, the two equations give the probability of error for DPSK and BPSK modulations, respectively, to be discussed later. The exponential approximation is an asymptotic bound providing a close approximation (less than 1-dB error in γ) for the low error rates required in most practical applications. It is convenient if we assume that the error rate is an exponential function with parameters α and $\beta \gamma_b$. The error rate is related linearly to the parameter α and exponentially to $\beta \gamma_b$. Furthermore, we observe that the value of α varies over a limited range from one modulation scheme to another, and only order-of-magnitude changes in the error rate are considered significant. As a result, we may ignore α and compare various modulation techniques on the basis of the value of $\beta \gamma_b$ needed to provide an acceptable error rate. This approach will allow us to use SNR as the basis for comparing modulation methods, rather than the precisely calculated error rate. For example, from Fig. 7.2 we see that either BPSK or DPSK modulation for steady signals in AWGN requires a γ_b of around 10 dB to provide an error rate of 10^{-5}. Using SNR rather than error rate has two

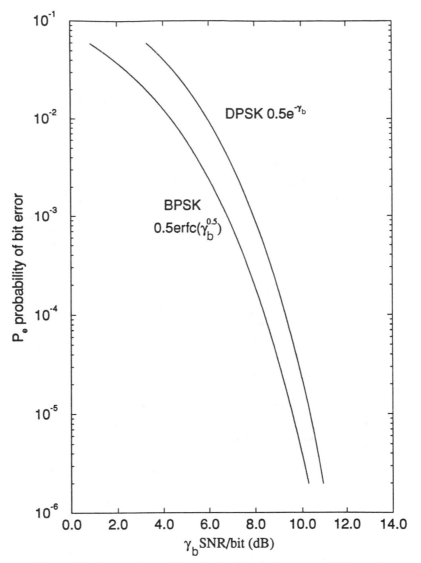

Figure 7.2 A comparison of the erfc function and its exponential bound.

advantages. First, SNR is the criterion used for assessing both digital and analog modulation techniques. Therefore using SNR we may compare, for example, the analog Advanced Mobile Phone Service (AMPS) cellular system with the new IS-54 TDMA digital cellular system. Second, SNR is directly related to the transmitted power, which is an important design parameter. As a rule of thumb, in the middle of the erfc curve each 3-dB change in γ_b will change the error rate by approximately two orders of magnitude.

In a modulation scheme based on a multiple-symbol alphabet, it is conventional to represent the symbols in a diagram known as a *signal constellation*, where the dimensions in the diagram are expressed in terms of the square root of the energy of the transmitted symbols. If we assume coherent symbol detection, the error rate in these cases is approximated by $0.5 \, \mathrm{erfc} \, d/2\sqrt{N_0}$, where d is the *minimum distance* between the points in the constellation. To determine the error rate of a modulation technique as represented by its signal constellation, the minimum distance is expressed as a function of average energy in the constellation. Then, by substituting d into $0.5 \, \mathrm{erfc} \, d/2\sqrt{N_0}$, one finds the error rate expressed in terms of average energy in the constellation.

In transmission on radio channels the signal is subject to fading over a wide dynamic range. In these cases the *average SNR* may be used as the performance parameter of interest. We shall see later in this chapter that the relationship between the average SNR and the average error rate over a fading channel does not follow the relatively steep exponential or erfc curves of the steady-signal AWGN channel. Therefore the average SNR is not a good indicator of performance in fading. Instead the average error rate provides a more meaningful performance parameter. On channels such as troposcatter or HF, where the system is subject to fading over time while the terminals are held fixed, the average error rate is defined as the average over time. In portable and mobile applications, where the error rate changes from one location to another, the average error rate is defined as the average over a range of locations. Thus we see that average SNR may imply either temporal or spatial averaging, depending upon the system under consideration and the physical mechanisms producing the signal fading.

Another important performance criterion for systems operating on fading channels is the *probability of outage*. The probability of outage is the percentage of time or locations at which the modem performance is unacceptable. The acceptable level of performance is defined by a required BER or SNR level, which we term the *performance threshold* or simply the *threshold*. In mobile applications a 1% outage probability is usually considered acceptable. This will subject the terminal to unacceptable performance in 1% of the locations in a service area.

In the remainder of this chapter we first review various modulation techniques designed for operation on steady-signal additive noise channels. We then examine the effects of fading and multipath, and finally we describe standard modem technologies used in the portable and mobile radio industries.

7.2.2 On-Off Keying

The simplest form of carrier modulation is *on–off keying* (OOK), in which the modulator simply turns a fixed-amplitude carrier signal on or off in accordance with the value of each information bit to be transmitted. Let us say that the carrier is turned on for a 1 and off for a 0, as shown in Fig. 7.3a. Demodulation of the OOK modulated signal can be done coherently with a carrier reference, as shown in Fig. 7.3b. To represent a 1, the symbol $p(t)\cos \omega_c t$ is transmitted, where $p(t)$ is the general pulse shape and ω_c is the radian carrier frequency. In Fig. 7.3a, $p(t)$ is a rectangular pulse spanning the symbol transmission time. Fig. 7.3d shows the signal constellation for the OOK signal. In this constellation the average energy per bit over all symbols is $E_b = E_s = E_{s_1}/2$. The minimum distance is expressed in terms of average energy in the constellation as $d = \sqrt{E_{s_1}} = \sqrt{2E_s}$. The probability

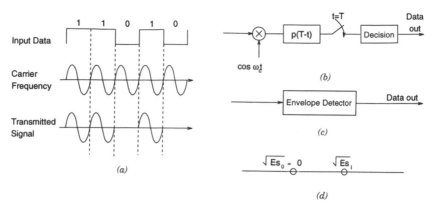

Figure 7.3 On–off keying (OOK). (*a*) Modulation. (*b*) Coherent matched filter. (*c*) Envelope detector. (*d*) Signal constellation.

of error for the coherent implementation is $0.5 \, \text{erfc} \, (d/2\sqrt{N_0})$, which yields

$$P_b = \frac{1}{2} \, \text{erfc} \left(\sqrt{\frac{E_b}{2N_0}} \right) = \frac{1}{2} \, \text{erfc} \left(\sqrt{\frac{\gamma_b}{2}} \right) < \frac{1}{2} e^{-\frac{\gamma_b}{2}} \qquad (7.2.8)$$

The received signal can also be detected noncoherently without a carrier reference by using a simple envelope detector as shown in Fig. 7.3c. Noncoherent reception provides a simpler implementation, but the output SNR of the noncoherent receiver is 3 dB lower than that of the coherent receiver. The bandwidth efficiency of OOK depends upon the chosen pulse shape $p(t)$. For the rectangular pulse shape, if we define the transmission bandwidth as the bandwidth between the first zero crossing of the spectrum, the bandwidth efficiency is $\eta = R_b/W = 0.5$. If ideal $\sin(x)/x$ pulses (having a rectangular spectrum) are used, the bandwidth efficiency increases to $\eta = 1$.

While this method of modulation is indeed very simple, the use of the scheme poses some nontrivial problems in the design of the demodulator. For efficient reception of the OOK signal in additive noise, the demodulator must set a detection threshold at a level which depends upon the received signal strength. Thus in a communications environment where received signal strength can vary with time or location or both, the detection threshold must be varied accordingly. Furthermore, long strings of 0s (carrier off in Fig. 7.3a) cannot be distinguished from the no-transmission state. Finally, the BER performance of OOK modulation is poorer than is achievable with other modulation techniques that are almost as simple.

The OOK modulation method is used in certain wireless information networks, (some optical WLANs in particular) where light emitting diodes (LEDs) and photo detectors offer practical and inexpensive transmitter and noncoherent receiver implementations. The transmitted light can be thought of as a carrier which is modulated by simply turning the LED on and off. The photo detector can be thought of as a noncoherent envelope detector demodulating the transmitted

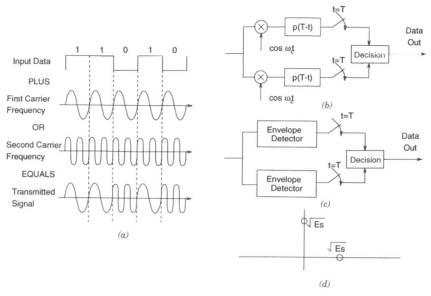

Figure 7.4 Binary frequency shift keying (BFSK). (*a*) Modulation. (*b*) Coherent detection. (*c*) Envelope detection. (*d*) Signal constellation.

signal by eliminating the optical carrier signal and detecting only the signal amplitude.

7.2.3 Frequency Shift Keying

The second simplest form of modulation is *frequency shift keying* (FSK), which uses two signal tones. In each bit interval, the modulator sends a pulse of one tone or the other in accordance with whether the information bit is 1 or 0. An FSK modulator implementation simply requires two oscillators; and it switches between the oscillators in accordance with the information bit to be transmitted, as shown in Fig. 7.4a. FSK signals can be demodulated coherently by correlating the received signal over each pulse interval with the two tones, sampling the result, and selecting the larger of the two outputs. Figure 7.4b shows a coherent receiver for binary FSK, where the two symbols are represented by $s_1(t) = p(t)\cos\omega_1 t$ and $s_0(t) = p(t)\cos\omega_0 t$. The receiver consists of two branches matched to the two transmitted symbols. The occupied bandwidth and consequently the bandwidth efficiency of the FSK scheme depends upon the separation between the center frequencies of the two tones. For proper operation of the system the two symbols must be orthogonal so that the signal intended for one detector branch does not cause interference (crosstalk) on the other branch. The orthogonality requirement is expressed mathematically as

$$\int_0^T s_1(t)s_0(t)\,dt = 0$$

where T is the pulse duration.

The signal constellation for FSK modulation is shown in Fig. 7.4d, where the orthogonality of the two signals is represented by placing the signals on orthogonal axes. The average energy over two symbols is given by $E_b = E_s = E_{si}$ and $d = \sqrt{2E_s}$. The relationship between average energy in the constellation and the minimum distance, and consequently the error rate, remains the same as for OOK. The symbol decision is made by comparing the outputs of the two matched filters. In FSK reception, one of the branches contains the signal plus the additive noise whereas the other branch contains only the additive noise. Therefore the variance of the noise involved in the symbol decision is twice the variance of the noise in each branch.

FSK can be implemented in either a coherent or noncoherent form; the noncoherent form is shown in Fig. 7.4c. The choice between the two affects the minimum frequency spacing between the tones required to achieve orthogonality. With coherent FSK the pulses are generated and demodulated with known phases. In this case, it can be shown that orthogonality is achieved if the two tones are separated by any integer multiple of $1/(2T)$ Hz, where T is duration in seconds of each FSK pulse. However, most applications use noncoherent FSK, in which the detector operates without knowledge of the received signal phase. Thus, it is necessary that the tones be spaced by an integer multiple of $1/T$ Hz in order to achieve orthogonality with arbitrary signal phases. As a practical matter, the tones may be spaced at any integer multiple of the minimum orthogonal spacing, but the most efficient use of bandwidth is achieved with the minimum tone spacing applicable to either coherent or noncoherent operation. With either form of binary FSK, implemented with minimum orthogonal spacing, we can say that the signal bandwidth is approximately equal to the channel signaling rate, $1/T$, which places the bandwidth efficiency at $\eta = 1$. The noncoherent implementation of FSK suffers a SNR disadvantage of about 3 dB relative to coherent FSK.

Coherent FSK with frequency spacing $1/(2T)$ Hz is referred to as *minimum shift keying* (MSK). The MSK scheme is the most bandwidth-efficient form of FSK, and a special version of this modulation, called *Gaussian-filtered MSK* (GMSK), is widely used in the portable and mobile radio industries. This topic is treated in greater detail in a later section of this chapter.

For M-ary FSK modulation we have $M = 2^m$ orthogonal signals, where m is the number of bits per symbol. The receiver for this case consists of M parallel matched filters. If all the signals have the same energy, the average signal energy remains the same as for binary FSK, but the noise involved in making each symbol decision is M times the noise in each branch, resulting in an increase in energy per bit by the factor $M/2m$ relative to binary FSK, to maintain the same error rate. The required bandwidth is M times that of OOK, while the bandwidth efficiency is $m/(M - 1)$ times that of OOK.

The 4-ary FSK modulation format is used in many wireless applications, including WLANs at 18–19 GHz and digital land mobile radios operating in VHF and UHF bands. A practical advantage of FSK is the availability of low-cost FM radios for analog voice applications such as AMPS and land mobile radio. In order to modify the system to accommodate data transmission, one need only organize the data into a stream of four-level pulses and use them as an input to the FM modulator. At the receiving end, the four-level symbol stream is extracted at the output of a simple frequency-discriminator detector. This approach provides for easy integration of voice and data services in a unit having low production cost.

7.2.4 Phase Shift Keying

In binary *phase shift keying* (PSK), there is only one signal oscillator with a constant known phase, and information is conveyed in each bit interval T by either leaving the signal phase unchanged or shifting the phase 180° relative to the oscillator phase, in accordance with the bit value to be transmitted. Binary PSK modulation, which is sometimes called *antipodal signaling*, is shown in Fig. 7.5a, where the two transmitted symbols are $\pm p(t)\cos \omega_c t$. Optimum coherent detection of PSK signals is done using a matched filter followed by a sampler, as shown in Fig. 7.5b; the sampled output of the matched filter is compared with a zero threshold to determine the polarity of the transmitted signal. The needed phase reference is extracted from the received waveform using a phase-locked loop. The signal constellation for PSK is shown in Fig. 7.5d. Similar to FSK, both symbols have the same energy and the average energy per bit is given by $E_b = E_s = E_{si}$. However, for the same average energy in the constellation the minimum distance is $d = 2\sqrt{E_s}$ and the squared minimum distance is twice that of coherent FSK or OOK. This increase in normalized distance gives a 3-dB SNR advantage to PSK when compared with FSK or OOK. In fact, it can be shown that the binary PSK format is the optimum binary signal set for communication in AWGN [Woz65]. Noncoherent implementation of PSK involves differential modulation at the transmitter and differential demodulation at the receiver, and it is referred to as *differential PSK* (DPSK). The differential demodulator is shown in Fig. 7.5c. DPSK modulation suffers a 1- to 2-dB performance disadvantage relative to PSK at levels of error rate required for most system applications. As with OOK and binary FSK, the bandwidth of a PSK signal is roughly $1/T$, where T is the PSK symbol duration, which results in a bandwidth efficiency of $\eta = 1$.

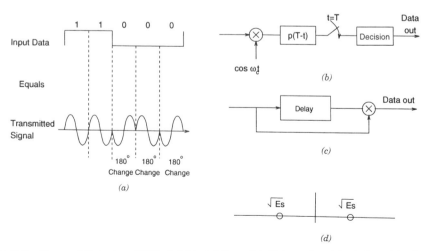

Figure 7.5 Phase shift keying (PSK). (*a*) Modulation. (*b*) Coherent matched filter detection. (*c*) Differential detection. (*d*) Signal constellation.

BPSK modulation is the building block for multiamplitude/phase modulation and coding techniques that are used in the most sophisticated voiceband modems. DPSK modulation is the building block for many of the radio modems designed to operate in harsh multipath fading environments.

7.2.5 Pulse Amplitude Modulation

Next we examine a form of nonbinary digital modulation, called *pulse amplitude modulation* (PAM), which can be viewed as an extension of binary PSK. This modulation technique was developed for use in telephone carrier systems. PAM uses one signal oscillator with known fixed phase, but allows the transmitted signal amplitude to have any of a set of discrete values (levels) $\{a_i\}$, where $i = 1, 2, 3, ..., M$ and $M = 2^m$ with m being the number of bits encoded into a symbol. In a PAM transmitter, shown in the upper half of Fig. 7.6a, information bits are buffered and encoded into a stream of pulse amplitudes $\{a_i\}$. If, for example, $M = 4$, two information bits at a time are buffered and encoded into one of four amplitude levels. If $M = 8$, three bits at a time are buffered and encoded into one of eight levels, and so forth. The encoding process is most easily done using a simple look-up table. The encoded amplitude is then applied to a fixed pulse shape, which we denote as $p(t)$. The amplitude-modulated pulse is next multiplied by the carrier signal $\cos \omega_c t$ and transmitted on the channel. The transmitted symbol in each symbol interval is given by $a_i p(t) \cos \omega_c t$, where one amplitude-modulated pulse is transmitted every T seconds. The choice of pulse shape is very important in the design of a PAM system, and we will say more about this when we discuss the demodulator. Note that because m information bits are encoded onto each transmitted pulse, the symbol rate in the channel is lower than the source information bit rate by the factor m and thus the bandwidth efficiency is $\eta = m$. Because the bandwidth of the transmitted signal is determined by the pulse shaping filter, regardless of the number of pulse amplitudes, PAM provides a very effective way of increasing the transmitted data rate within a fixed bandwidth.

In PAM, the pulse amplitudes are chosen with uniform spacing and are arranged symmetrically about zero. Figure 7.6b illustrates the signal constellation

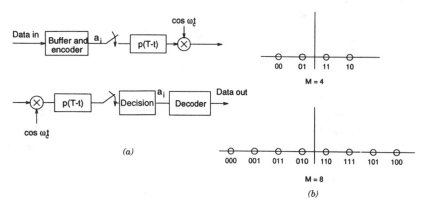

Figure 7.6 Pulse amplitude modulation (PAM). (*a*) Transmitter and receiver. (*b*) Signal constellations for $M = 4$ and $M = 8$.

for the cases $M = 4$ and $M = 8$, with allowed pulse amplitudes denoted by circles on the horizontal axis. It should be clear that if $M = 2$, this is equivalent to binary PSK modulation. The relationship between the minimum distance and the average energy per symbol in this constellation is given by

$$d^2 = \frac{12}{4^m - 1} E_s^m \qquad (7.2.9)$$

where m is the number of bits per symbol and E_s^m is the average energy per symbol for a constellation with 2^m symbols. The average energy in the constellation can be written in the following recursive form:

$$E_s^{m+1} = 4E_s^m + \frac{1}{3}d^2$$

In other words, the transmitted power must be increased by a factor of 4(6 dB) to compensate for the performance degradation caused by sending an additional bit per symbol.

At the receiver, shown in the lower part of Fig. 7.6a, the received signal is multiplied by the carrier signal $\cos \omega_c t$ and passed through a filter matched to the transmitted pulse shape. The multiplication by $\cos \omega_c t$ produces two realizations of the received signal: One is centered about zero frequency, which is the baseband signal, and the other is centered about $2 \times f_c$. We want the baseband signal, which is recovered by using the transmitter pulse shape $p(t)$ as the matched filter. Given that we sample at the optimum time, the output is the transmitted amplitude a_i. The detected amplitude is then decoded to the appropriate set of information bits, which can be done with the decoding version of the encoding look-up table.

7.2.6 Quadrature Amplitude Modulation

In the preceding discussion we saw that the use of M-level PAM provides a bandwidth efficiency proportional to m bits per channel symbol. We now describe a modulation scheme which doubles the bandwidth efficiency of PAM by simply applying the same amplitude levels on both the sine and cosine of the carrier, producing a transmitted signal of the form

$$s(t) = a_i p(t) \cos \omega_c t + b_i p(t) \sin \omega_c t \qquad (7.2.10)$$

Because the transmitted signal consists of two PAM pulse streams in phase quadrature, this modulation scheme is called *quadrature amplitude modulation* (QAM). This form of modulation was first used in a 9600-bit/sec commercial modem introduced to the market in the early 1970s [Pah88c].

Because the quadrature channels are orthogonal, the modulator can be designed with two signal branches, each configured exactly as in a PAM modem, one channel modulating the cosine of the carrier, the other the sine. At the receiver, the two channels are prevented from interfering with one another by their

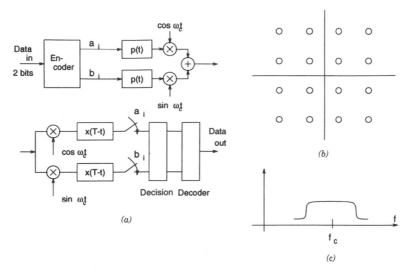

Figure 7.7 Quadrature amplitude modulation (QAM). (*a*) Transmitter and receiver. (*b*) Signal constellation for $M = 16$. (*c*) Power spectrum.

orthogonality. The data rate for the QAM modem is simply the sum of the data rates on the two channels, but the signal bandwidth, which is determined by the pulse shape $p(t)$, is unchanged from the single-channel PAM signal. Thus the bandwidth efficiency of the QAM design is twice that of the PAM design and is given by $\eta = m = 2n$, where $m = 2n$ is the number of bits for each point in the constellation and n the number of bits in each dimension.

Note that if we use antipodal amplitude values on each branch, we have two binary PSK signals in quadrature. This modulation is called *quadrature phase shift keying* (QPSK) and is commonly used, with a number of variations, in digital radio systems. QAM modulation is the predominant modulation technique in use in voiceband modems, and it is expected to eventually find adoption in the radio communications industry. There are a number of other modulation schemes that are implemented with versions of the two-branch modem structure. Important examples are minimum shift keying (MSK), Gaussian lowpass filtered MSK (GMSK), and $\pi/4$-shift QPSK, each of which is characterized by a particular form of pulse shaping used on the quadrature branches.

Figure 7.7a shows a block diagram of a QAM modulator and demodulator. Figure 7.7b shows an example of 16-ary QAM signal constellations that transmits two bits on each branch in each symbol interval, for a total of four bits per symbol interval. The voiceband modem mentioned in the previous paragraph uses the 16-ary QAM signal constellation and a symbol transmission rate of 2400 baud, yielding a data rate of 9600 bits/sec. The relationship between the minimum distance and the average energy over all QAM symbols is given by

$$d^2 = \frac{6}{2^m - 1} E_s^m \qquad (7.2.11)$$

This equation leads to the recursive equation

$$E_s^{m+1} = 2E_s^m + \frac{d^2}{6}$$

indicating a requirement of two times (3 dB) additional power for transmitting one additional bit in the constellation. This is 3 dB better than the 6 dB per added bit required with PAM.

In the implementation of digital modems, the real and imaginary parts of the transmitted symbols are modulated onto sine and cosine functions so as to achieve orthogonality. When the orthogonal channels are transmitted simultaneously, the peak in transmit power occurs at the peak of the pulse shaping filter. In many radio communication channels it is desirable to transmit at full power in order to maximize power efficiency, but at the same time, power amplifiers exhibit increasing nonlinearity as they are driven near their peak power limits. To deal with this problem, the drive level to the power amplifier is adjusted to keep the peak signal power at some specified margin relative to the peak power limit of the amplifier. Consequently the average power is kept at an even lower level, which reduces the overall power efficiency. Therefore, in many cases the modulation is staggered so that the transmitted pulses for the real and imaginary parts of the signals have a relative time delay of $T/2$ seconds. Staggering reduces the peak-to-average power ratio, allowing the average power to be set closer to the nonlinear range of the amplifier, achieving better overall power efficiency.

7.2.7 Multiphase Modulation

M-phase PSK modulation can also be implemented with the two-branch structure, when $M = 2^m$ with m the number of bits per symbol. All the transmitted M-PSK symbols have the same energy $E_s = mE_b$ and the average energy per symbol is the same as the energy of any individual symbol. As a result, the signals in the constellation are located on a circle with radius $\sqrt{E_s}$. The minimum distance for M-PSK modulation is given by

$$d = 2\sqrt{E_s}\,\sin\frac{\pi}{M} \tag{7.2.12}$$

For $M = 2$ and $M = 4$ we have BPSK and QPSK (4-QAM), respectively. An 8-ary PSK modulator can be structured as two quadrature branches, with three amplitude levels; 0, $\pm\sqrt{E_s/2}$, and $\pm\sqrt{E_s}$ transmitted.

Because all symbols have the same amplitude, PSK modulation is less sensitive to nonlinearities in the channel. As a result, PSK is widely used on power-limited radio channels such as satellite channels where the amplifiers are driven close to their nonlinear regions of operation in order to maximize power efficiency.

7.2.8 Partial-Response Signaling

The transmission of two symbols/sec/Hz with QAM requires ideal pulse shaping filters—that is, filters which completely eliminate intersymbol interference at the sampling instants. If we can remove the constraint of having no intersymbol interference at sampling instants, the system can be designed with physically realizable filters which arc generally easier to implement than the ideal pulse-shaping filters. Signaling techniques which allow symbol transmission at a rate equal to two symbols per hertz of bandwidth, with controlled amounts of intersymbol interference, are called *partial-response signaling* techniques. Partial response signaling was introduced in the 1960s for use in wireline modems and has sometimes been applied in radio modems. With partial-response signaling, the bandwidth efficiency of ideal QAM is achieved with a realizable filter. As will be shown in the section on pulse shaping, the pulse shape for partial response signaling is designed so that the information content of one transmitted symbol is distributed over two sample intervals. The partial-response form of the filter allows additional noise into the system, which reduces the SNR per bit by a factor of $(\pi/4)^2 = 2.1$ dB. Otherwise the error rates for one- or two-dimensional partial response signals are given by the same equations as for PAM and QAM signaling. However, the transmitted waveforms are different, and as we will see later, the performance analysis for frequency-selective fading channels is different from that of PAM or QAM. For more detailed discussions of partial response signaling see [Luc68], [Kab75], [Feh87], and [Pro89]. As shown in [Bel84] and [Pah85a], the performance of quadrature partial response (QPR) and staggered QPR (SQPR) signaling over frequency-selective fading channels is inferior to QPSK and staggered QPSK (SQPSK).

7.2.9 Trellis-Coded Modulation

In classical communication systems, error control is provided by coding the input data bits and then modulating a carrier with the coded signal. To keep the data rate unchanged, one should compensate for the error-correction parity bits by increasing the transmission rate. In bandlimited channels, such as voiceband channels, an increase in transmission rate requires an increase in the number of points in the constellation, resulting in a higher symbol error rate. For many years, it was believed that if the data rate remains the same, practical error-control codes cannot compensate for the performance loss caused by increasing the number of points in the signal constellation. As a result, coding techniques were not employed in voiceband modems.

About a decade ago, renewed attention was given to the concept of coding for band-limited channels, spurred by the development of a combined modulation and coding technique now referred to as *trellis-coded modulation* (TCM) [Ung82, Ung87]. The principal advantage of TCM over modulation schemes used with traditional error-correction coding is its ability to achieve improved power efficiency without the customary bandwidth expansion introduced by the use of coding. Various versions of TCM can improve the performance of a modem by 3–6 dB on steady-signal channels. The 8-state trellis code with nominal gain of 4 dB is perhaps the most attractive, because more complex trellis codes offer little additional improvement with extensive additional implementation complexity. A version of TCM which can resolve 90° phase ambiguity [Wei84a, Wei84b] has been adopted by CCITT as a standard for QAM voiceband modems [CCI84a]. A comprehensive treatment of coded and uncoded signal constellations for band-limited channels is available in [For84]. A historical overview of the development of the TCM for wireline modems is available in [Pah88c]. Standard coded and uncoded QAM signal constellations can be modified to improve performance of the wireline modem in the presence of "nonuniform" noises, such as those arising from phase jitter or nonlinear quantization [Pah91].

The TCM technique is an extension of QAM in which the number of points in the constellation is increased to create redundancy. These extra symbols enable the transmitter to create dependency between successive transmitted symbols, and in this way, only certain sequences of symbols are valid. The received sequence of symbols is compared with all valid sequences, and the sequence with maximum likelihood is chosen. The efficient search method under the maximum likelihood criterion is the Viterbi algorithm [Vit67]. Implementation of the Viterbi algorithm for TCM is computationally complex. The CCITT-recommended TCM technique almost doubles the required processing power for the implementation of a modem.

In addition to its applications in wireline modems, TCM has also been studied for application to fading channels, particularly mobile satellite channels [Div87, Div88a, Div88b, McL88, Sch89, Moh89, Big91]. An important point to be noted in the application of TCM to fading channels is that the criteria for designing optimum trellis codes for fading channels are different from the design criterion for steady-signal AWGN channels. This point is discussed in detail in [Big91].

7.2.10 Comparison of Modulation Methods

The probabilities of bit error for the most common binary modulation techniques are given by

$$
\begin{aligned}
\text{FSK or OOK} - \text{CD:} \quad & P_b = \frac{1}{2}\text{erfc}\sqrt{\frac{\gamma_b}{2}} \\
\text{BPSK} - \text{CD:} \quad & P_b = \frac{1}{2}\text{erfc}\sqrt{\gamma_b} \\
\text{DPSK} - \text{NCD:} \quad & P_b = \frac{1}{2}e^{-\gamma_b} \\
\text{FSK} - \text{NCD:} \quad & P_b = \frac{1}{2}e^{-\frac{\gamma_b}{2}}
\end{aligned}
\tag{7.2.13}
$$

where CD denotes coherent demodulation, NCD denotes noncoherent demodulation, and in each case steady-signal reception in AWGN is assumed. The four formulas in Eq. (7.2.13) are plotted in Fig. 7.8. As is seen in the figure, the best BER performance is achieved with coherent BPSK. The BER performance achieved by coherent FSK (or OOK) is exactly 3 dB poorer than coherent PSK, simply reflecting the doubled noise level associated with the detection of two orthogonal signals in FSK, as contrasted with detection of antipodal signals in PSK.

It can be seen in Fig. 7.8 that DPSK provides somewhat poorer performance than does coherent PSK, but at high SNR the curves are very close together. The relationship between the PSK and DPSK curves is in fact given by the analytical bound shown earlier in Eq. (7.2.7). A simple heuristic explanation of the near-

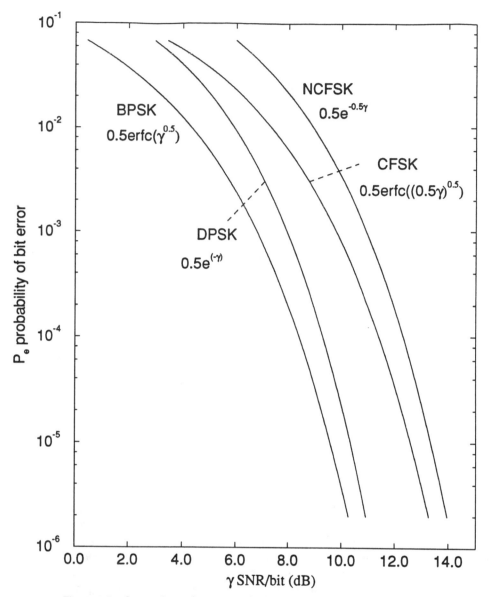

Figure 7.8 Comparison of error rates for four binary modulation techniques.

identical performance of PSK and DPSK at high SNR is as follows: While a PSK pulse is demodulated with an ideal noiseless phase reference, each DPSK pulse is in effect demodulated using the (noisy) previous pulse as its phase reference. As the SNR increases, the previous pulse becomes steadily less noisy and thus becomes more like the ideal noiseless phase reference.

Noncoherent FSK is the highest of the four BER curves in Fig. 7.8, and it is in fact exactly 3 dB poorer than DPSK. (It is left as an exercise for the reader to explain the exact 3-dB difference between FSK and DPSK BER performance.)

For nonbinary signal constellations, we assume $M = 2^m$ symbols with m the number of bits conveyed in each symbol. The approximate equations for the probability of symbol error with coherent detection in AWGN are:

$$M - PSK: \quad P_s \simeq \text{erfc}\sqrt{\sin^2\frac{\pi}{M}m\gamma_b}$$

$$M - FSK: \quad P_s \simeq M\text{erfc}\sqrt{m\gamma_b/M} \qquad (7.2.14)$$

$$M - QAM: \quad P_s \simeq 2\,\text{erfc}\sqrt{\frac{3}{2 \times 2^m - 1}m\gamma_b}$$

The bandwidth efficiency of M-PSK and QAM for pulse-shaping with ideal $\sin(x)/x$ pulses is $\eta = m$, whereas the bandwidth efficiency of M-FSK with tones spaced $1/T$ Hz apart is $\eta = m/M$-1.

Writing the previous equations in terms of signal-to-noise power ratio S/N we have

$$M - PSK: \quad P_s \simeq \text{erfc}\sqrt{\sin^2\frac{\pi}{M}\frac{S}{N}}$$

$$M - FSK: \quad P_s \simeq M\text{erfc}\sqrt{\frac{M-1}{M}\frac{S}{N}} \qquad (7.2.15)$$

$$M - QAM: \quad P_s \simeq M\text{erfc}\sqrt{\frac{3}{2M-1}\frac{S}{N}}$$

For M-PSK and M-QAM modulations, the bit error probability calculation depends on the encoding scheme for the symbols in the constellation—that is, the mapping of information bits onto modulation symbols. If the symbols are Gray-coded, each symbol error that is a transition to an adjacent symbol in the constellation causes only one bit error. Thus, given a reasonably high SNR, we may assume that the bit error probability is m times smaller than the symbol error probability. Generally m is a small number and the symbol error rate provides a reasonable approximation to the bit error rate. For M-FSK modulation, the symbol error probability can be converted to a bit error probability in the corresponding m-bit groups by assuming that when an M-ary symbol is in error, each of the $2^m - 1$ incorrect symbols is equally likely. Then, each bit in the erroneous symbol has 2^{m-1} chances out of the $M - 1$ possibilities to be in error. This leads to the relationship

$$P_b = \frac{2^{m-1}}{M-1}P_s$$

Curves of P_s versus γ_b for M-ary PSK with coherent demodulation are shown in Fig. 7.9. In this presentation, it can be seen that at very low values of symbol error probability P_s, binary and 4-ary PSK operate at essentially the same levels of E_b/N_0. In the approximation given by Eq. (7.2.14) the probability of symbol error

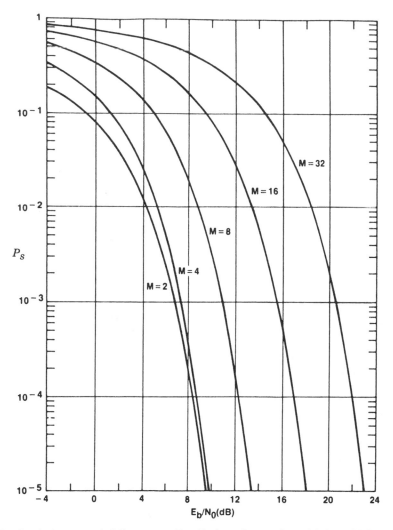

Figure 7.9 Symbol error probability versus E_b / N_0 for coherent demodulation of PSK and M-ary PSK. $E_b / N_0 = (S / N)(W / R_b)$.

versus γ_b is the same for binary PSK (BPSK) and 4-ary or quadrature PSK (QPSK). If we were to plot the exact bit error probability P_b instead of P_s, the BPSK and QPSK curves would be identical, because coherent QPSK is equivalent to two orthogonal BPSK channels. This means that QPSK allows us to double the data rate of BPSK with no increase in bandwidth and no penalty in communication efficiency. However, as the M-ary phase constellation changes from 4 to 8 phases, there is a loss of communication efficiency of nearly 4 dB. For each further doubling of the PSK signal constellation, there is a steady growth in the corresponding loss in communication efficiency.

Figure 7.10 shows symbol-error probability versus γ_b for a selection of M-ary PSK and M-ary QAM modulation systems. Note that 4-ary PSK and 4-ary QAM give exactly the same symbol error probability performance; this is to be expected,

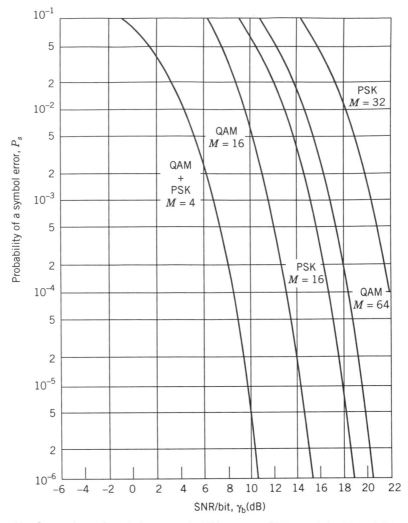

Figure 7.10 Comparison of symbol error probabilities versus SNR per bit for *M*-ary PSK and *M*-ary QAM signal constellations.

because they are really the same scheme, as we have pointed out earlier. Note, however, that as M is increased to larger values, the QAM constellations have better communication efficiency than do the PSK constellations for the same number M of signal points. The reason for this is the relatively more efficient "packing" of signal points in a rectangular QAM constellation. For a thorough treatment of the effects of various packing strategies on the performance of the multidimensional modulation techniques, the reader is referred to [For84].

Next let us compare performance curves on the basis of signal-to-noise power ratio, S/N, where we recall from Section 7.2.1 that S/N is related to γ_b, the SNR per bit, by $S/N = (R_b/W)\gamma_b$ with W the signal bandwidth and R_b the data rate in bits/sec. As we noted in Section 7.2.1, the required signal-to-noise power ratio

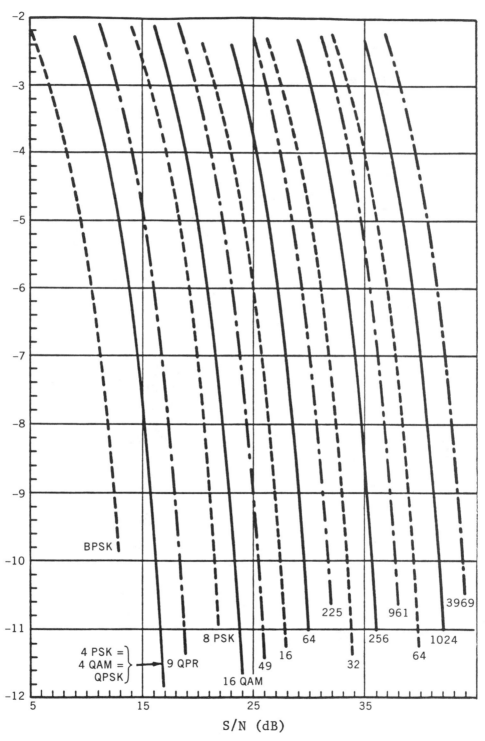

Figure 7.11 Symbol error probability versus S / N for coherent demodulation of M-ary PSK (---), M-ary QAM (—), and N-ary QPRS (—·-). (From [Kuc85] © IEEE.)

is a measure of the received power before processing and is directly related to the power requirements for the transmitter.

Figure 7.11 shows a set of performance curves calculated for a selection of modulation schemes operating on a steady-signal channel with AWGN. The curves show probability of symbol error P_s as a function of received signal-to-noise power ratio, S/N, for BPSK, QPSK, M-ary PSK for M up to 64, QAM for M up to 1024, and a selection of N-ary quadrature partial response (QPRS) systems for N up to $63 \times 63 = 3969$.

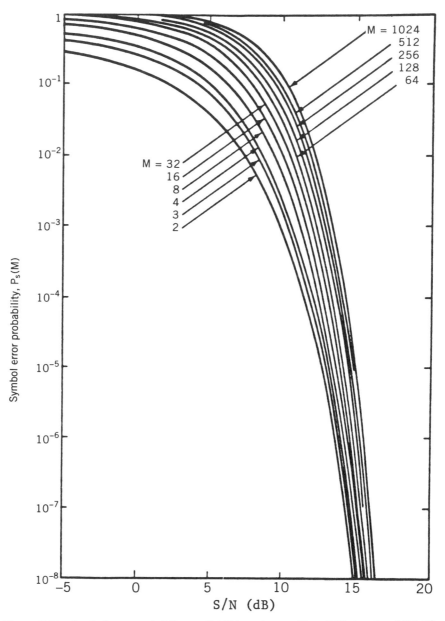

Figure 7.12 Symbol error probability vs. S/N for coherent M-ary FSK signaling [NBS63].

To compare the performances, let us consider a BER of 10^{-5} for each of the modulations schemes. For the nonbinary modulations, though Fig. 7.11 gives probability of symbol error, we can assume the use of Gray coding, as discussed above, and use the approximation for the bit error probability $P_b = (1/m)P_s$, where $m = \log_2 M$ and P_s is the M-ary symbol error probability.

Let us compare BPSK with QPSK. For BPSK the S/N required at 10^{-5} is somewhat under 10 dB whereas QPSK requires just under 13 dB. A comparison of required SNR per symbol is really a comparison of required signal power, if we assume that the background noise is the same in each case. Therefore we see that QPSK requires 3 dB, or two times more signal power than does BPSK, to achieve the same error rate in the same noise background. This is exactly what we should expect when we recall the earlier description of QPSK as two BPSK channels operating on orthogonal phases of the same carrier. For the higher-order M-ary PSK and M-ary QAM curves, it can be seen that steadily more signal power is required as the number of points in the signal constellation grows. In particular, observe that as we discussed in Section 7.2.6, we pay a 3-dB penalty for each additional bit contained in a QAM signal.

Figure 7.12 gives curves of P_s versus SNR for M-ary FSK with coherent demodulation. Here we see that the performance becomes poorer as the number of tones increases. Increasing the number of FSK tones does not affect the minimum distance between signals, because the tones are always chosen to be orthogonal over the symbol interval T. But increasing M does increase the number of "competitors" for the correct signal in the demodulation process, and this is exhibited in the multiplicative factor M seen in the M-FSK formulas in Eq. (7.2.15).

7.3 THEORETICAL LIMITS AND PRACTICAL IMPAIRMENTS

In this section we review the theoretical limits on achievable data rate and communication efficiency as defined by the Shannon capacity for steady-signal Gaussian noise channels. We then review briefly the real-world channel impairments encountered in practical situations.

7.3.1 Theoretical Limits of Communication Performance

In our discussions thus far, we have shown how the design of the modem signal constellation relates to the achievable data rate and the energy efficiency of the modem design. Therefore it is useful to consider the ultimate limits on data rate and efficiency that are theoretically achievable. This is best done by examining Shannon's well-known formula for the channel of a band-limited continuous AWGN channel:

$$C = W \log_2(1 + S/N) \qquad \text{bits/sec}$$

where C is the maximum achievable information transfer rate of the channel, W is the channel bandwidth in Hertz, and S/N is the signal-to-noise power ratio in the bandwidth W [Sha48]. Stated succinctly, the essence of Shannon's work on channel capacity is as follows:

> If we take increasingly long sequences of source information bits and map them into correspondingly long transmission waveforms, the error rate in the delivered data can be brought arbitrarily close to zero, as long as we do not attempt to transmit data at a rate higher than C. Therefore, at any nonzero level of channel signal-to-noise ratio S/N, there is some nonzero information transfer rate below which arbitrarily accurate communication can in principle be achieved.

The significance of Shannon's result, which is called the *Channel Coding Theorem*, is that channel noise does not inherently limit the accuracy with which communication can be achieved but only the rate at which information can be reliably transmitted [Sha49, Sha59].

We can readily apply the capacity formula to the case of a voiceband telephone channel, which has a bandwidth of 3–4 kHz and a typical S/N equal to 28 dB for conditioned leased lines. This yields a theoretical channel capacity of about 30 kbits/sec. It will therefore be useful to determine how closely this limit can be approached with various signal constellations. In comparing various modulation schemes with the Shannon limit, it is instructive to rewrite the capacity formula in the form

$$\frac{C}{W} = \log_2\left[1 + \frac{S}{N_0 C}\frac{C}{W}\right] \quad \text{bits/sec/Hz}$$

where N_0 is the one-sided power spectral density of the white Gaussian noise. We can further rewrite the formula as

$$\frac{S}{N_0 C} = \frac{W}{C}[2^{C/W} - 1] \tag{7.3.1}$$

For transmission at capacity, the signal power is $S = CE_{b_{\min}}$ and the left side of Eq. (7.3.1) becomes

$$\frac{S}{N_0 C} = \frac{E_{b_{\min}}}{N_0}$$

where $E_{b_{\min}}$ is the minimum transmitted energy per source information bit required for reliable communication. Finally we rewrite Eq. (7.3.1) as

$$\frac{E_{b_{\min}}}{N_0} = \frac{W}{C}[2^{C/W} - 1] \tag{7.3.2}$$

This equation describes channel capacity in terms of two convenient normalized parameters, $E_{b\min}/N_0$ and C/W. The first parameter is the minimum value of SNR per source information bit required for reliable transmission of data at capacity over an AWGN channel of bandwidth W. The second parameter, C/W, simply normalizes the channel capacity to an arbitrary bandwidth and represents the maximum achievable value of bandwidth efficiency; its reciprocal W/C is the

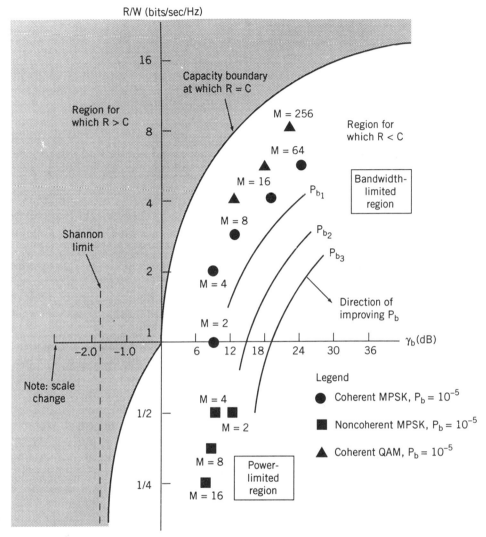

Figure 7.13 Channel capacity, and a comparison of several modulation methods at bit error probability equal to 10^{-5}.

bandwidth expansion factor for operation at capacity. Therefore Eq (7.3.2) expresses channel capacity in terms of two parameters defining the achievable limits of communications efficiency as measured by SNR-per-bit and bandwidth utilization. This now provides us with a convenient framework for assessing the communications efficiency of any chosen modulation scheme.

In Fig. 7.13 we show the capacity formula as a plot of R/W versus E_b/N_0, where R is the information rate in the channel, with $R = C$ at channel capacity. Note that the lower portion of the scale is expanded for convenience in drawing the figure. This figure essentially represents a bandwidth-versus-efficiency plane, and the capacity curve divides the plane into two regions. The shaded area to the left of the curve defines the region in which reliable communication cannot be achieved; that is, no modulation or coding scheme can be devised to operate in that region with low BER in delivered data. In the right-hand area of the figure, which defines the region of achievable signal designs, design points are shown for several modulation methods which we have discussed earlier. For all the cases shown, the delivered BER is 10^{-5}. The displacement of each design point from the capacity boundary indicates how close the communication efficiency of the corresponding modulation scheme comes to the capacity limit. The horizontal displacement measures the shortfall in terms of SNR per bit, while the vertical displacement measures the shortfall in terms of bandwidth utilization. Note that if we were to plot the modem design points for a lower level of delivered BER, the points would all move to the right (i.e., further away from the capacity boundary), whereas if we used a higher BER, they would move closer to the capacity boundary.

It is conventional to call the region of $R/W > 1$ the *bandwidth-limited region* of operation and to call the region of $R/W < 1$ the *power-limited region* of operation. The bandwidth-limited region includes all the modulation schemes we have described for use on voiceband telephone circuits, where rigid channel bandwidth limitations are imposed by the existing design of the public network. There we see that the M-ary modem signal constellations provide steadily increasing bandwidth utilization as M is increased. It can be seen from the figure that the QAM schemes are closest to the capacity boundary.

As can be seen from the figure, the Shannon capacity formula shows that the greatest energy efficiency is achieved in the power-limited region, where the bandwidth must be made very large relative to the information rate. In the limiting case of very large bandwidth and C/W approaching zero, E_b/N_0, approaches $\ln 2$ or -1.6 dB, which is called the *Shannon limit*. In the power-limited region, we show design points for noncoherent binary and M-ary FSK modulations. Because these modulations are bandwidth-expansive, they are not used in modern voiceband modems. However, binary FSK was used in the earliest of the voiceband modems, as was binary PSK. As modem technology developed, the industry evolved modems with roughly the succession of modulation methods seen as we progress to the right and upward in the bandwidth-limited region of Fig. 7.13. The current state of the art in high-rate modem technology for the PSTN is high-order QAM with TCM, and data rates up to 28.8 kbits/sec are now being achieved.

Many radio systems must also be treated as bandwidth-limited media, due in most cases to ever-increasing demands for access to the available spectrum in various frequency bands. Consequently, increasingly sophisticated modem techniques are migrating from the voiceband applications, for which they were developed, to various radio applications. As examples, M-ary PSK modulation with adaptive equalization is now a proven technique for radio applications, and there is recent work on the use of 16-ary QAM with trellis coding in some frequency bands.

7.3.2 Transmission Channel Impairments

In reality, most data communication channels are not accurately described by the steady-signal AWGN model. To describe real-world channel impairments, it is again helpful to use signal constellation diagrams. This is done in Fig. 7.14, where

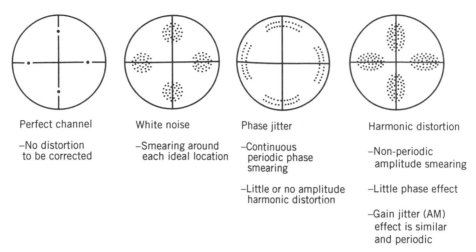

Perfect channel	White noise	Phase jitter	Harmonic distortion
–No distortion to be corrected	–Smearing around each ideal location	–Continuous periodic phase smearing	–Non-periodic amplitude smearing
		–Little or no amplitude harmonic distortion	–Little phase effect
			–Gain jitter (AM) effect is similar and periodic

Figure 7.14 Transmission channel impairments and their effects on a four-phase modem signal constellation.

several common forms of channel impairment are illustrated by their effects on a 4-ary PSK signal constellation. The first diagram shows an undisturbed constellation, representing a distortion-free channel with no measurable noise. The second diagram shows the constellation with each of the four points smeared into a circular "cloud" caused by additive Gaussian noise. The third diagram shows the effects of *phase jitter*, a prevalent effect on most channels. The phase jitter effect is a continuous periodic smearing of the signal phase, with little or no effect on the signal amplitude. The last diagram shows the effects of harmonic distortion, again common on channels with nonlinearities, which results in a nonperiodic smearing of the signal amplitude, with a somewhat smaller effect on signal phase. This appears as an elliptical cloud around each signal point in the constellation. Another impairment observed on some channels, termed *gain jitter*, is a random-appearing amplitude modulation similar in its effect to harmonic distortion. While the impairments depicted in Fig. 7.14 are shown individually, they will in general appear in various combinations on different channels. To visualize the effects of these impairments on modem performance, one can pass decision boundaries through the origin at 45° to the right and left of the vertical axis. Received signal points crossing one or more decision boundaries result in demodulated symbol errors.

All of the channel impairments shown in Fig. 7.14 are found on both wired and wireless channels. The wireless channels also suffer from large amplitude fluctuations caused by signal fading—We treat this topic in detail in Section 7.4. Another category of transmission impairment is one that can be described equally well for wireline and radio systems as distortion due to nonideal channel frequency-response characteristics. That is, when the channel has nonflat amplitude and delay response over the bandwidth occupied by the transmitted signal, the channel acts as a nonideal filter causing intersymbol interference in the received symbol stream. Intersymbol interference imposes the principal limitation on achievable data rates on band-limited channels. The difference between the amplitude and phase distortion in wireline and wireless channels is that the distortions in the wireline channels are on the edges of the band while a radio channel may by subjected to frequency-selective fading even in the mid-region of the band.

7.4 RADIO COMMUNICATION ON FLAT RAYLEIGH FADING CHANNELS

Having discussed the principal modem techniques used in steady-signal wireline systems, we will next consider the limitations encountered in applying these techniques to fading-multipath radio channels. The fundamental problem to be dealt with here is multipath, which causes fluctuations in the received signal power, frequency-selective fading, and multipath delay spread. The signal fluctuations cause an increase in the signal power required, relative to steady-signal operation, to achieve the same overall BER performance. Frequency-selective fading, if it occurs in the mid-region of the band, can disable the proper operation of the modem. The time dispersion of the signal due to multipath puts a limit on the speed at which modulated symbols can be transmitted in the channel. Here we discuss the effects of power fluctuations and then we explain diversity combining and coding as approaches to counteracting the effects of power fluctuations. In Chapter 8 we will discuss the effects of frequency-selective fading and the methods used to improve performance in the presence of frequency-selective fading.

7.4.1 Effects of Flat Fading

Here we consider the case of frequency-nonselective or flat fading channels. As the name implies, flat fading is a form of fading in which all the frequency components of the transmitted signal rise and fall in exact unison. Let us recall the discussion in Chapter 3 of the classical uncorrelated scattering model of a multipath fading channel, where we defined the root mean square (rms) delay spread of the channel, τ_{rms}, and referred to its reciprocal as the coherence bandwidth of the channel. We noted there that the symbol transmission rate should be much smaller than the coherence bandwidth of the channel if multipath distortion of transmitted symbols is to be made negligible. The assumption of flat fading, therefore, is simply the assumption that the transmission bandwidth is significantly small relative to the coherence bandwidth of the channel. If the contrary were true, and the transmission bandwidth were comparable to or wider than the coherence bandwidth, we would describe the fading model as frequency-selective or nonflat fading.

In pulse transmission under the assumption of flat fading, we can characterize the received sampled signal at the output of the matched filter, shown in Fig. 7.1, as

$$z(T) = a\sqrt{E_{si}} + \varepsilon \tag{7.4.1}$$

where a is the channel gain factor imposed on the signal by flat fading. Note in this expression that the additive noise ε is assumed to be unaffected by channel fading. In the flat fading model, the channel gain factor a is a random variable which is described completely by a probability density function $f_A(a)$. As was noted in earlier chapters, many fading radio channels are accurately characterized by the

Rayleigh model of fading. This is analytically convenient, because closed-form solutions for calculation of average error rates are readily available for a number of common modulation techniques. With Rayleigh fading, the probability density function of a is given by the Rayleigh distribution

$$f_A(a) = \frac{a}{\Gamma} e^{-a^2/2\Gamma} \tag{7.4.2}$$

where Γ is the mean-square amplitude of the channel gain factor. In this situation, the signal-to-noise ratio per bit is given by

$$\gamma_b = \frac{a^2 E_b}{N_0}$$

In contrast with data communication over wireline circuits, here the γ_b is a random variable that changes with time or spatial movements of the transmitter or the receiver. Because E_b and N_0 have fixed values, the probability density function of γ_b will follow the probability density function of a^2. Given that the channel gain factor a has a Rayleigh distribution, a^2 has an exponential distribution, the chi-square distribution with two degrees of freedom [Pro89]. The probability density function of γ_b is then given by

$$f_\Gamma(\gamma_b) = \frac{1}{\bar{\gamma}_b} e^{-\gamma_b/\bar{\gamma}_b} \tag{7.4.3}$$

As we discussed earlier, there are two performance criteria for digital communication over fading channels: (1) probability of outage (2) and average probability of error. Probability of outage is the probability that the modem performs more poorly than a specified threshold. The threshold for most digital communications applications is usually defined by a certain error rate, which we will call $P_{e\text{-th}}$. For a given modulation technique, we may use the error probability formula or the error rate curve for a nonfading channel to determine the corresponding value of γ_b, which we refer to as γ_{out}. If the error probability for the modulation technique over a nonfading channel is described by the general exponential function

$$P_e = \alpha e^{-\beta\gamma_b} \tag{7.4.4}$$

where $\gamma_b = E_b/N_0$ is the received SNR per bit if the channel is nonfading, then we have

$$\gamma_{\text{out}} = -\frac{1}{\beta} \ln\left(\frac{P_{e\text{-th}}}{\alpha}\right) \tag{7.4.5}$$

The probability of outage P_{out} is the probability that γ_b, having the probability density function given by Eq (7.4.3), is less than γ_{out}:

$$P_{\text{out}} = \int_0^{\gamma_{\text{out}}} f_\Gamma(\gamma_b)\, d\gamma_b = 1 - e^{-\gamma_{\text{out}}/\bar{\gamma}_b} = 1 - \left(\frac{P_{e\text{-th}}}{\alpha}\right)^{1/\beta\bar{\gamma}_b} \tag{7.4.6}$$

If the error probability for the chosen modulation method on a nonfading channel is given by

$$P_e = \alpha \operatorname{erfc} \sqrt{\beta \gamma_b} \qquad (7.4.7)$$

the inverse mapping to P_{out}, similar to Eq. (7.4.5), is not analytically feasible. In this case either the asymptotic exponential bound for erfc can be used in conjunction with the above equation, or plotted curves of the error probability can be used for numerical inverse mapping. The value γ_{out} determined from a plotted curve is then substituted into the integral over $f_\Gamma(\gamma_b)$ in Eq. (7.4.6). Figure 7.15 shows the probability of outage versus threshold for several modulation techniques.

Now let us consider the average probability of error. Given our assumption of the flat Rayleigh fading model and fixed noise level, the received SNR per bit is a random variable with the exponential probability density function of Eq. (7.4.3). Therefore we can find the average probability of error in fading by averaging P_e

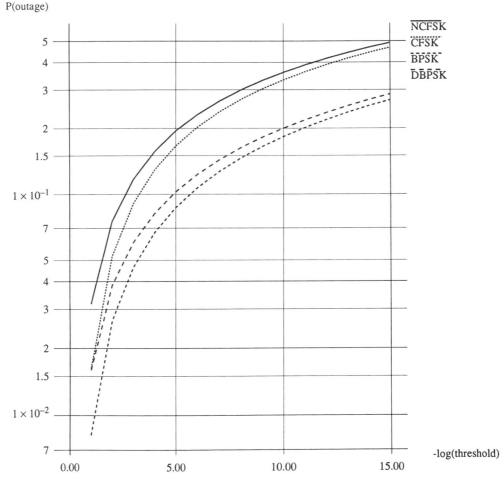

Figure 7.15 Probability of outage versus threshold for basic modulation methods.

given by Eq. (7.4.4) or Eq. (7.4.7) over the probability density function of γ_b, given by Eq. (7.4.3). Both integrals have closed-form solutions. For bit error probability of the form of Eq. (7.4.7), this averaging is given by the integration

$$\bar{P}_b = \frac{\alpha}{\bar{\gamma}_b} \int_0^\infty \mathrm{erfc}\,\sqrt{\beta\gamma_b}\,e^{-\gamma_b/\bar{\gamma}_b}d\gamma_b = \alpha\left[1 - \sqrt{\frac{\beta\bar{\gamma}_b}{1 + \beta\bar{\gamma}_b}}\right] \simeq \frac{\alpha}{2\beta\bar{\gamma}_b} \quad (7.4.8)$$

For bit error probability of the form of Eq. (7.4.4), the averaging is given by the integration

$$\bar{P}_b = \frac{\alpha}{\bar{\gamma}} \int_0^\infty e^{-\beta\gamma_b}e^{-\gamma_b/\bar{\gamma}_b}d\gamma_b = \frac{\alpha}{1 + \beta\bar{\gamma}_b} \simeq \frac{\alpha}{\beta\bar{\gamma}_b} \quad (7.4.9)$$

which gives performance 3 dB poorer than that of Eq. (7.4.8). In both cases the average error probability is reduced by only one decade per 10 dB of increase in $\bar{\gamma}_b$. This is to be contrasted with the exponential reduction of error probability with increasing $\bar{\gamma}_b$ on nonfading channels. This clearly indicates the need for substantial additional power to provide the same average error probability on fading versus nonfading channels. This increase in required signal power is referred to as the *fade margin*.

Figure 7.16 shows the probability of bit error versus SNR per bit for coherent BPSK modulation, for both the nonfading and flat Rayleigh fading cases. As can

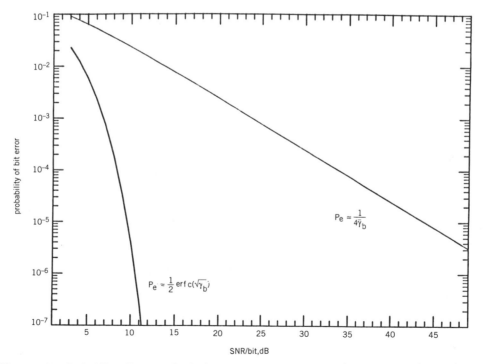

Figure 7.16 Probability of bit error for BPSK modulation on nonfading (*left-hand curve*) and fading (*right-hand curve*) channels.

be seen from the figure, the presence of signal fading causes a large increase in the SNR required to achieve reasonable levels of BER. For example, to achieve BER equal to 10^{-5} we require about 10 dB on a nonfading channel but require nearly 45 dB SNR in fading, a penalty of 35 dB. The reason for this large SNR penalty can be understood by examining the probability density function for the received signal power in fading, given by Eq. (7.4.3), where the exponential form of the distribution places some of the signal power at very low levels, where the instantaneous BER is near 0.5. For higher levels of signal power, the error rates are negligible. Therefore the average BER is dominated by the intervals of time in which the SNR is low and the BER is high. As a result, a large increase in average signal power, which greatly widens the distribution $f_\Gamma(\gamma)$, is needed in order to reduce the probability that the instantaneous signal power lies in the region of high BER.

Because averaging over Rayleigh fading involves integration over an exponential power distribution function, Laplace transform tables can be used for the evaluation of the integral. Laplace transform tables are bountiful, and a wide variety of closed forms can be found for application to many different modulation techniques. As a result, most of the closed-form solutions available in the communications literature have been derived for Rayleigh fading channels. In the remainder of this section we provide a selection of these derivations that are widely used in different applications to derive formulas for performance of various modulation techniques over fading channels.

7.4.2 Diversity Combining

As we observed in previous chapters, multipath fading is manifested as signal amplitude fluctuations over a wide dynamic range. In particular, during short periods of time, the channel goes into deep fades causing significant number of errors that virtually dominate the overall average error rate of the system. In order to compensate for the effects of fading when operating with a fixed-power transmitter, the power must typically bc increased by several orders of magnitude relative to nonfading operation. This increase of power protects the system during the short intervals of time when the channel is deeply faded. A more effective method of counteracting the effects of fading is to use diversity techniques in transmission and reception of the signal. The concept here is to provide multiple received signals whose fading patterns are different. With the use of diversity, the probability that all the received signals are in a fade at the same time reduces significantly, which in turn can yield a large reduction in the average error rate of the system.

Diversity can be provided spatially by using multiple antennas, in frequency by providing signal replicas at different carrier frequencies, or in time by providing signal replicas with different arrival times. It is conventional to refer to the diversity components as *diversity branches*. We assume that the same symbol is received from different branches, with each branch exposed to a separate random fluctuation. This has the effect of reducing the probability that the received signal will be faded simultaneously on all the branches; this in turn reduces the overall outage probability, as well as the average BER.

A variety of techniques are available for reception of the diversity signals. With *selection diversity*, one signal is chosen from the set of diversity branches, usually on the basis of received signal strength. With *linear combining*, as the name suggests, the diversity branches are simply summed together before demodulation. In the optimum method of combining, called *maximal-ratio combining*, the diversity branches are weighted prior to summing them, each weight being proportional to the received branch signal amplitude. The maximal-ratio combiner for the diversity channel can be considered equivalent to a discrete matched filter receiver, in the sense that it provides the optimum post-demodulation SNR for the received signal, which in this case is made up of diversity components.

Performance Evaluation. This section provides an analytical framework for calculation of the error probability achieved with the use of diversity reception. The channel is assumed to be a Rayleigh fading channel, and most of the derivations are based on the use of maximal-ratio combining. In later chapters, we will describe various innovative techniques for providing diversity with different modulation methods. However, the equations used for performance calculations will be those introduced in this section. As a basis for discussing the performance improvements obtained through diversity, let us assume that the we have a Rayleigh fading channel and are using diversity of order D; that is, the transmitted signal is arriving from D independent diversity branches each equipped with a matched filter. The set of sampled signals received at one instant of time from the diversity branches, after sampling at the output of the matched filters, is given by the following vector equation:

$$z_j(T) = a_j\sqrt{E_{si}} + \varepsilon_j, \qquad 0 < j \le D \qquad (7.4.10)$$

where E_{si} is the energy of the transmitted symbol, a_j is the amplitude fluctuation in the jth branch caused by flat fading in that branch, and ε_j is the AWGN associated with the jth branch. For a maximum-ratio combiner the received signal in each branch is scaled by the amplitude of that branch a_j and the scaled signals are added to form $z(T)$, the sample used for decision-making:

$$z(T) = \sum_{j=1}^{D} |a_j|^2 \sqrt{E_{si}} + \sum_{j=1}^{D} a_j \varepsilon_j$$

The signal-to-noise ratio per bit is then given by

$$\gamma_b = \sum_{j=1}^{D} |a_j|^2 \frac{E_b}{N_0} \qquad (7.4.11)$$

which is a random variable following the distribution function of $\sum_{j=1}^{D} |a_j|^2$ in which $|a_j|^2$ is exponentially distributed. If we assume that the average of $\sum_{j=1}^{D} |a_j|^2$ is normalized to 1, the average signal-to-noise ratio per bit is then given by

$$\overline{\gamma_b} = \frac{E_b}{N_0}$$

It can be shown [Pah 79], [Pah 80] that the most general form for the probability distribution function of γ_b is given by

$$f_{\Gamma}(\gamma_b) = \sum_{n=1}^{D} \frac{A_n}{2\lambda_n} e^{-\gamma_b/2\lambda_n} \qquad (7.4.12)$$

where $\{\lambda_n\}$ are eigenvalues of the $D \times D$ branch amplitude covariance matrix whose elements are defined by $r_{ij} = E\{a_j a_i^*\}$ and $\{A_n\}$ are defined by

$$A_n = \prod_{\substack{k=1 \\ k \neq n}}^{D} \frac{1}{1 - \lambda_k/\lambda_n}$$

Given the probability distribution function of the SNR per bit, as given in Eq. (7.4.12), the probability of outage with diversity reception of order D is given by

$$P_{\text{out}} = \int_0^{\gamma_{\text{out}}} f_{\Gamma}(\gamma_b) d\gamma_b = 1 - \sum_{n=1}^{D} A_n e^{-\gamma_{\text{out}}/2\lambda_n} \qquad (7.4.13)$$

where γ_{out}, the threshold SNR for a given modulation technique, is determined by the same method used for flat fading and described in Section 7.4.1. The average probability of error for coherent PSK, expressed in terms of $\bar{\gamma}_b$, and determined from Eq. (7.4.8) using the density function in Eq. (7.4.12), is given by [Pah 79], [Pah 90c]

$$\overline{P_b} = \sum_{n=1}^{D} \frac{A_n}{2} \left\{ 1 - \left[\frac{2\overline{\gamma_b}\lambda_n}{1 + 2\overline{\gamma_b}\lambda_n} \right]^{1/2} \right\} \qquad (7.4.14)$$

This equation incorporates the correlation among all diversity branches and assumes that the values of all eigenvalues are different.

Special Cases. Let us now assume that the amplitudes of the signals received on different branches are all uncorrelated Rayleigh-distributed random variables. For our first case, we also assume that the same average signal power is received on each diversity branch and that the average SNR on each branch is denoted by $\bar{\gamma}_b$. The probability distribution function of the post-combining SNR is then given by

$$f_{\Gamma}(\gamma_b) = \frac{1}{(D-1)!\,\overline{\gamma_b}^D} \gamma_b^{D-1} e^{-\gamma_b/\overline{\gamma_b}} \qquad (7.4.15)$$

The probability of outage at the post-combining SNR level γ_{out} is then given by

$$P_{\text{out}} = \int_0^{\gamma_{\text{out}}} f_{\Gamma}(\gamma_b) d\gamma_b = 1 - \frac{1}{e^{\gamma_{\text{out}}/\overline{\gamma_b}}} \sum_{j=1}^{D} \frac{1}{(j-1)!} \frac{\gamma_{\text{out}}}{\overline{\gamma_b}} \qquad (7.4.16)$$

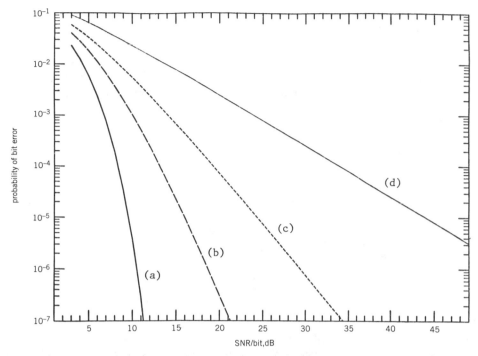

Figure 7.17 Probability of bit error for BPSK modulation for no fading and for Rayleigh fading with various orders of diversity. (a) no fading; (b) $L = 4$ in fading, (c) $L = 2$ in fading, (d) $L = 1$ in fading.

The average probability of error for coherent PSK demodulation at the maximal-ratio combiner output is given by

$$\overline{P_b} = [P(\overline{\gamma_b})]^D \sum_{l=1}^{D} \binom{D-2+l}{l-1} [1 + P(\overline{\gamma_b})]^{l-1} \approx \left(\frac{1}{4\overline{\gamma_b}}\right)^D \binom{2D-1}{D}$$

(7.4.17)

where $P(\overline{\gamma_b})$ is found from Eq. (7.4.8) or (7.4.9) and is the equation giving the average error probability for a specific modulation technique used in the system. We use the standard notation for a binomial coefficient:

$$\binom{N}{k} = \frac{N!}{(N-k)!k!}$$

The expression in Eq. (7.4.17) shows that average BER performance at the maximal-ratio combiner output improves exponentially with increasing D, the order of diversity. Figure 7.17 shows the average BER $\overline{P_b}$ versus average SNR per bit $\overline{\gamma_b}$, for different orders of diversity, with an assumption of independent equal-power Rayleigh fading on diversity branches. Included in the figure is the BER curve for steady-signal reception. As we saw earlier, with a single antenna, we lose 30–35 dB in performance relative to steady-signal reception at reasonable levels of BER. With two independent diversity branches, the performance loss is reduced to about 25 dB, and with four orders of diversity the SNR penalty is reduced to around 10 dB. With additional orders of diversity the penalty relative to nonfading can be further reduced. There will of course be a practical limit to the order of diversity implemented because, for example, one cannot put an arbitrarily large number of antennas into a communications terminal.

If the average received powers from different branches are unequal, but the amplitudes are uncorrelated Rayleigh-distributed random variables, the average error probability for maximum ratio combining with D paths is given by [Pro89]

$$\overline{P_b} = \frac{1}{2} \sum_{i=1}^{D} \pi_i \left[1 - \sqrt{\frac{\overline{\gamma_{bi}}}{1 + \overline{\gamma_{bi}}}} \right] \tag{7.4.18}$$

where

$$\pi_i = \prod_{\substack{j=1 \\ j \neq i}}^{D} \frac{\overline{\gamma_{bi}}}{\overline{\gamma_{bi}} - \overline{\gamma_{bj}}}$$

and $\overline{\gamma}_{bi}$ is the average signal power received on path i.

For selection diversity, where the strongest path is selected for making the symbol decision, the average probability of PSK error for equal-power uncorrelated branches is given by the following closed-form equation [Sch66, Pro89]:

$$\overline{P_b} = D \sum_{k=1}^{D} \binom{D-2}{k-1} \frac{(-1)^{k-1}}{k} P\left(\frac{\overline{\gamma_b}}{k} \right) \tag{7.4.19}$$

Calculation of average probability of error for equal-gain combining is relatively difficult. However, the results given above for BER with maximal-ratio combining and selective combining can be used as lower and upper bounds, respectively, on the BER achieved with equal gain combining. The actual BER performance for equal-gain combining in most cases of interest is closer to the performance of maximal-ratio combining.

The treatment of diversity combining techniques in this chapter is somewhat abstract. In the following chapters we discuss practical methods for providing diversity in real systems.

7.4.3 Error-Control Coding for Fading Channels

A fundamental problem in digital communication over wireless networks is that the wireless environment typically produces levels of BER in delivered data which are several orders of magnitude poorer than usually found on wireline channels. Added to this is the fact that in the wireless environment channel errors often occur in bursts, essentially coinciding with the incidence of deep fades or even signal blockage on the link. The result of these effects is that average untreated error rates on wireless channels will often be at a relatively high level (perhaps around 10^{-2} on outdoor cellular and mobile radio channels) and will occur as mixtures of random-appearing and bursty patterns, where within the bursts the local BER will approach 50%. Such error characteristics may be marginally acceptable for some user applications such as digital voice service, but may be completely unacceptable for other applications such as message or data services or for critical system functions relying on over-the-air exchange of control signaling. Because of this, most digital wireless systems utilize one or more forms of *error-control coding* as one means of mitigating the effects of multipath fading on received bit quality.

From a broad perspective, error-control coding involves the addition of systematic redundancy to the transmitted data and the use of that redundancy at the receiver to improve the quality of the delivered data, relative to uncoded transmission. As described in these terms, coding has some similarity to diversity transmission and reception. In fact, diversity transmission of information symbols, say on multiple frequencies or in multiple time slots, is essentially the same as repetition coding, a simple form of error-control coding. However, an extensive array of coding techniques are available which make much more efficient use of redundancy than can be obtained with simple diversity transmission. In this subsection, we briefly describe the principal coding techniques used in existing and planned wireless systems and discuss the issues arising in the application of coding to fading channels.

Error-control coding techniques used in wireless systems can be grouped broadly into three categories as follows:

1. Error-detection coding using block codes.
2. Forward-error correction (FEC) coding, using both block and convolutional codes.
3. Automatic repeat request (ARQ) schemes.

Error-Detection and FEC Block Coding. In block coding for error detection or error correction, k information digits are used with a prescribed encoding rule to calculate a set of $n - k$ *parity-check digits*, which are transmitted along with the information digits on the channel as an n-digit *code block* or *codeword*. It is customary to refer to a parity-check block code as an (n, k) code, where n is the *block length* and k is the number of information digits encoded in each code block. The *code rate* is defined as the ratio $R = k/n$. The encoding rule, defined by a *parity-check matrix* or an equivalent *generator polynomial*, determines the mathematical structure of the code and hence its capabilities for detecting and correcting errors. Note that in this description we have used the term "digits"; this is to denote the fact that block codes can be constructed on various nonbinary alphabets as well as on the binary alphabet {0, 1}. In all cases, encoding and decoding are performed with *finite-field arithmetic*, with *modulo-2 arithmetic* as the special case. The mathematical details of coding theory, along with discussions of widely-used coding techniques, can be found in a number of texts, including [Pet72], [Mac77], [Lin83], and [Mic85]. The application of coding techniques to fading channels is discussed in [Cla81], [Pro89], and other references cited therein.

The received version of the transmitted codeword is called simply the *received word* to allow for the possibility that it may not be the intended word or even a valid codeword (as defined by the encoding rule), if errors occurred during transmission. The *decoder* follows the demodulator and operates on the received word to determine whether the received information and check digits satisfy the encoding rule, and it uses any observed discrepancy to detect and possibly correct transmission errors.

A given block code may be used for error detection only or, given a sufficient amount of redundancy in the design, for error detection and correction, commonly referred to by the acronym *EDAC*. In error-detection decoding, the parity check bits are recomputed from the received k information bits and compared with the check bits received from the channel. If the received and recomputed check bits are in agreement, the k information bits contained in that block are accepted as correct. Cyclic redundancy check (CRC) codes are widely used error-detection codes which can be applied to information-bit strings of arbitrary length [Cas 93]. Each CRC code is defined by its *generator polynomial* which establishes its encoding rule. A number of CRC codes are now standardized, and microchips are available for implementing some of the more popular codes.

It is important to note that no error-detection code can be designed to successfully detect all received error patterns. In particular, an error pattern that changes the transmitted codeword into a different valid codeword is always undetectable. Thus an apparently successful outcome of the error-detection decoding procedure does not guarantee that the code block was received correctly, because some error patterns will produce undetectable false decodings. For many error-detection block codes, the probability P_{FD} of false decoding is upper-bounded by a function of the number of parity bits n-k [Wol82]:

$$P_{FD} \leq 2^{-(n-k)}$$

It should be noted here that the bound given above does not necessarily hold for an arbitrarily chosen error-detection code. For some codes, in certain ranges of channel BER, the probability of false decoding rises above the $2^{-(n-k)}$ "bound." This issue is dealt with at some length in a number of papers, including [Wol82] and [Cas93].

Error-detection codes are incorporated into the transmitted signal structure for some of the wireless standards. For example, the IS-54 TDMA digital cellular standard includes error-detection coding on selected bits in each frame of digitized speech. At the receiving end, where speech frames are being synthesized into analog speech, techniques such as frame interpolation can be employed to "mask" the effects of an occasional segment of detected error bits. Error-detection codes are also used in retransmission protocols, as we shall discuss in a later paragraph.

The mathematical treatment of the structure and performance capabilities of block codes is based upon the concepts of Hamming weight and Hamming distance. The *Hamming weight* or simply *weight* of a codeword is the number of nonzero elements in the codeword. Similarly, the weight of an error pattern is the number of errors in the pattern. The error-correction capability of a block code is determined by its *minimum Hamming distance*, or simply *minimum distance*, where the Hamming distance between two codewords is the number of symbol positions in which two codewords differ. A block code with minimum distance d_{min} can correct any error pattern having weight up to t_{max}, where

$$t_{max} = \left\lfloor \frac{d_{min} - 1}{2} \right\rfloor$$

where $\lfloor x \rfloor$ denotes the largest integer not greater than x.

FEC coding techniques can be implemented to perform either *hard-decision* or *soft-decision* decoding, depending on the amount of information conveyed to the decoder with each demodulated symbol. In the simplest implementations, the demodulator makes a definite decision on each received symbol and passes the bits or symbols to a hard-decision decoder. Hard-decision decoding algorithms are essentially efficient algebraic equation-solving routines, although simple look-up tables are sometimes used for decoding very short block codes. Single-error-correcting codes are sometimes implemented with simple shift-register encoders and decoders. Here the codewords are represented as polynomials, and encoding and decoding are done using polynomial multiplication and division operations [Pet72, Lin83].

Soft-decision FEC decoding begins with soft-decision demodulation, in which the demodulator output is quantized to Q levels, where Q is greater than the size of the transmission alphabet. Quantization incurs loss of information, and thus soft-decision demodulation preserves information that can be profitably utilized with appropriate decoding algorithms. Soft-decision decoding algorithms more nearly resemble signal correlation or matched-filtering operations than equation-solving routines. A number of efficient soft-decision techniques have been devised for decoding block codes, and several are described in [Mic85]. It has been shown that the soft-decision Viterbi decoding algorithm, widely used for decoding convolutional codes, can also be used to perform optimum soft-decision decoding for some block codes [Wol78].

Soft-decision decoding provides better performance than hard-decision decoding at a cost of increased demodulator and decoder complexity. The range of performance improvement achievable will depend to a great extent on the characteristics of the transmission channel. On steady-signal AWGN channels, the theoretical limit on SNR improvement achievable with soft-decision decoding is 3 dB [Cha72]. However, practical experience shows that actual improvements of 1–2 dB are feasible with algorithms of reasonable complexity [Mic85]. Much greater SNR improvements are achievable on fading channels, as we show later.

When a coding technique is specified in a wireless standard, no particular method of decoding is mandated, because manufacturers can choose to implement various receiver techniques without affecting interoperability with other manufacturers' products. Whether the use of soft-decision decoding is justified in any given application is a matter of trading off between improved performance, which may translate into widened signal coverage, and increased complexity, which may be reflected in size, weight, and cost of a communication terminal. A great many choices are available to the manufacturer. For example, a mobile terminal manufacturer may choose to implement hard-decision decoding in a mobile terminal, in order to minimize complexity and cost of the subscriber equipment. At the same time, a base-station equipment manufacturer may choose to implement soft-decision decoding in order to maximize the geographic range over which mobile subscribers can operate.

Convolutional FEC Coding. In encoding a convolutional code, the encoder accepts information bits as a continuous stream and generates a continuous stream of encoded bits at a higher bit rate. The information stream is fed to the encoder b bits at a time, where b is usually in the range 1–6. The encoder operates on the current b-bit input and some number of immediately preceding b-bit inputs to produce V output bits with $V > b$. Thus the code rate is $R = b / V$. The number of successive b-bit segments of information bits over which each encoding step operates is called the *constraint length* of the code, which we also denote by k. The encoder for a convolutional code might be thought of as a form of digital filter with memory extending $k - 1$ symbols into the past. A typical binary convolutional code is one having $b = 1$, $V = 2$ or 3, and k in the range 4–7.

As with block codes, convolutional codes may be decoded with either hard- or soft-decision decoding, and the performance advantage for soft-decision decoding will again vary with channel characteristics. Though a number of different algorithms are available for decoding convolutional codes, the most frequently used is the Viterbi algorithm [Vit67], which is in fact a *maximum likelihood decoding* algorithm for the steady-signal AWGN channel [For73a]. Microchip Viterbi decoders are now available from a number of manufacturers, and some of the chips provide both hard- and soft-decision decoding options.

We shall not delve any further into the details of code design here, because these details are amply treated in a number of texts, including those cited earlier in this section. Our brief discussion has served our primary purpose, which is to introduce the parameters: code rate, block length, and constraint length. These are key parameters in the selection of a coding technique, because the reciprocal of code rate provides a measure of required bandwidth expansion, while the block length or constraint length gives a measure of complexity of the required encoding and (more important) decoding operations.

TABLE 7.1 Examples of Coding Techniques Used in Wireless Systems

Type of Code	Comments
Hamming codes	Length $= 2^m - 1$, $m = 2, 3, 4, \ldots$ Minimum distance $d_{min} = 3$
BCH codes	Length $= 2^m - 1$, $m = 2, 3, 4, \ldots$ $d_{min} \geq 2t - 1$, t any integer Number of parity checks: $n - k \leq mt$
Golay (23, 12) code	$n = 23$, $k = 12$, $d_{min} = 7$, $t = 3$
Reed–Solomon (RS) codes (q-ary)	$q = p^m$, p prime, m integer $N = q - 1$, $K = 1, 2, 3, \ldots, N - 1$ $d_{min} = N - K + 1$
Walsh / Hadamard codes	$n = 2^m$, $d_{min} = n / 2$
Binary convolutional codes	Typically used code rates: $1/2$, $1/4$, $1/8$ Typical constraint lengths: $k = 5, 6, 7$

Table 7.1 lists some of the coding schemes used in wireless systems. The first row in the table shows the Hamming codes, an infinite class of single-error-correcting ($d_{min} = 3$) codes. The "natural" length of a Hamming code is $n = 2^m - 1$, where m can be any integer greater than 1, and we use the term *natural length* to mean the block length prescribed by the formal mathematical definition of the code. Any Hamming code can be shortened by replacing some of the information bits with assumed zeros. Hamming codes and shortened versions thereof are used in the link-layer coding specifications for a number of wireless standards, such as IS-54 and IS-95.

The second row in the table shows Bose–Chaudhuri–Hocquenghem (BCH) codes, an infinite class of binary multiple-error-correcting codes. For any positive integers m and $t < n/2$, there exists a binary BCH code with natural block length $n = 2^m - 1$ and minimum distance $d \geq 2t - 1$ having no more than mt parity check bits. Each such code can correct up to t random errors per codeword and thus is a t-error-correcting code. BCH codes with $t = 1$ are identical to Hamming codes. BCH codes have been studied and utilized extensively, and much work has been done on developing efficient decoding algorithms for the codes. As with the Hamming codes, the BCH codes can be shortened from their natural lengths as needed, and thus the BCH class provides a rich assortment of codes with various block lengths and degrees of error-correction capability. Various BCH codes, including shortened versions, have been incorporated into wireless system specifications.

The next row in the table shows the 3-error-correcting $(23, 12)$ Golay code. This code has the special property that with hard-decision decoding, all received error patterns will be decoded, that is, no received word will be declared as having a "detectable but uncorrectable" error pattern. This property defines the Golay code as a *perfect code*. A frequently used variation of the Golay code is the $(24, 12)$ distance-8 code, often called the *extended Golay code*, which is obtained by appending an overall parity check to the distance-7 $(23, 12)$ code. The extended code is found to be attractive in some applications because its code rate k/n is exactly equal to 0.5. Both the basic and extended Golay codes have been widely used for a number of years, and decoders are available as commercial chips. Both versions of the Golay code can be shortened as well, and several versions of both codes have been incorporated into wireless system specifications, such as the APCO/TIA standard for digital Land Mobile Radio [TIA 93a].

The next row in the table shows Reed–Solomon (RS) codes, which are an important class of nonbinary block codes. The symbols in RS codewords are drawn from an alphabet of size $q = p^m$, where p can be any prime and m is an integer. In most applications $p = 2$, so that $q = 2^m$, and m bits are mapped into each q-ary symbol. An (N, K) RS code has block length $N = q-1$, and the number of symbols in a codeword can have any value $K = 1, 2, 3, \ldots, N - 1$. As with the Hamming codes and binary BCH codes, the RS codes can be shortened to lengths smaller that the natural length. Thus a wide range of code designs is available within this class of codes. An important property of the RS codes is that any (N, K) RS code has the largest minimum distance achievable with the given values of N and K,—specifically, $d_{min} = N - K + 1$. For this reason, RS codes are described as *maximum-distance separable codes*. RS codes have proved to be very effective for use on channels where errors occur in bursts, or as mixtures of random errors and bursts. The RS codes can be defined as a special subclass of nonbinary BCH codes, and several of the efficient decoding algorithms developed for BCH codes can be applied to RS codes as well [Mic85]. RS codes are used in the link-layer coding structure specified for the CDPD cellular data standard [CDPD93] and have also been proposed for use in the planned new APCO/TIA standard for digital Land Mobile Radio [TIA93a].

Walsh codes, also called Hadamard codes, are constructed by selecting as codewords rows of special square matrices called *Hadamard matrices*. A Hadamard matrix of order n is an $n \times n$ matrix of $+1$s and -1s, where all pairs of rows are orthogonal. Hadamard matrices exist only for orders 1 and 2 and multiples of 4. (That they exist for all multiples of 4 has been conjectured but not yet proved.) If the matrix elements $\{\pm 1\}$ are replaced by ones and zeros, respectively, a Hadamard matrix has one row of all zeros, and the remaining rows each have an equal number of ones and zeros. Walsh or Hadamard codes can be constructed for block lengths $n = 2^m$, where m is an integer, all nonzero codewords having Hamming weight $n/2$, and all pairs of codewords being separated by Hamming distance $n/2$. The IS-95 CDMA cellular system uses Walsh codes of order 64.

The last row in Table 7.1 shows binary convolutional codes, which are included in several wireless standards. Though convolutional codes are designed to encode continuous streams of data, they are readily adapted to a block-structured transmission format simply by truncating the information bit stream and inserting known *tail bits* into the encoder, which facilitate decoding the information bits at the end of the block. The tail bits represent overhead in the transmission format and thus impact overall bandwidth efficiency. However, in most practical cases the impact is small. Convolutional codes are part of the link-layer protocols for both the IS-54 and IS-95 digital cellular standards.

ARQ Schemes. A long-standing and widely used method of error control combines error detection block coding with retransmission-on-request in a technique called *automatic repeat request* or simply *ARQ*. If forward-error correction coding is used in conjunction with an ARQ protocol, the technique is referred to as *hybrid ARQ*. Over the years, many variations of the ARQ technique have been studied, and detailed treatments can be found in a number of references, including [Lin83], [Lin84], [Com84], [Tan88], and [Dos92]. ARQ techniques are particularly well-suited to any channel where errors tend to occur in bursts, and to fading radio channels in particular. In essence, ARQ is a method of adapting the effective information transmission rate to the conditions of the channel. That is, when the channel transmission quality is high, most of the code blocks are received correctly on the first try, and information is carried over the channel at a rate at or near the maximum rate allowed by the transmission format. When channel quality degrades due to fading, signal blockage, or other temporary signal disruption, code blocks are received with detected errors, and transmission of new data is slowed down or even halted ("flow-controlled") while erroneous blocks are retransmitted, perhaps multiple times, until the channel returns to a state of good transmission quality. Effective applications of ARQ are not limited to fading channels, of course. Forms of ARQ are incorporated into all of the common contention-based multiuser access protocols, such as ALOHA, CSMA and others, where the principal source of errors is collision between different users' packets (see Chapter 11). ARQ protocols are also part of the design of standard data network protocols, such as BISYNC, X.25, and TCP/IP.

A key figure of merit for an ARQ system is its *throughput efficiency* or simply *throughput*, which is defined as the ratio of the average number of information bits accepted at the receiver to the maximum data transmission rate on the channel. The achievable throughput is determined to a large extent by the chosen retransmission strategy, of which several are available. The relative advantages and disadvantages of one retransmission strategy relative to another is influenced somewhat by the detailed error-clustering characteristics of the channel at hand. Here we briefly define the principal retransmission strategies and comment upon their effectiveness on fading channels.

The three basic types of ARQ strategies are: stop-and-wait, go-back-N, and selective repeat. The simplest strategy is *stop-and-wait*, in which the transmitter stops after transmitting each data block and waits until an acknowledgment (ACK) or retransmission request (NAK) is sent back from the receiver, or a timer expires. In full-duplex transmission, ACKs and NACKs are sent along with data blocks on the return channel. Typically, only error detection rather than FEC is implemented at the receiving end, though hybrid forms of stop-and-wait have been proposed for some applications [Com84]. An obvious potential problem with stop-and-wait ARQ is that if the transmitter must be idle while waiting for acknowledgments, throughput will suffer, and if round-trip delays are long, throughput can suffer appreciably.

The problem of idling with stop-and-wait ARQ is alleviated with the use of a slightly more complex strategy called *continuous ARQ* or *go-back-N*. This is the retransmission protocol in predominant use in packet-switching networks. Here the transmitter does not wait for ACKs or NAKs, but instead continuously transmits code blocks until receipt of a NAK or expiration of a timer. Then, the transmitter stops, backs up to the code block that was not successfully received, and restarts the transmission with that block. All N blocks that were transmitted in the time interval between the original transmission and the receipt of the NAK receipt or timer expiration are sent again in sequence. The throughput enhancement achieved with the pipelining nature of go-back-N can be pronounced. However, many of the blocks that are retransmitted may have already been successfully received, as many as all $N - 1$ following the one received with detected errors. Thus additional throughput gains can be realized if only those blocks found to be in error are retransmitted. This is the essence of the strategy we describe next.

The throughput inefficiency caused by retransmission of error-free blocks can be overcome by *continuous ARQ with selective repeats*, commonly termed *selective repeat* (SR) or sometimes termed *selective reject*. Here only a NAK'd frame, or a frame for which the timer has expired without receiving an ACK, need be retransmitted. In most applications the SR scheme provides the best throughput performance of the three basic ARQ strategies, however, its implementation is appreciably more complex than that of the other two. In particular, buffer management for the SR scheme is rather involved, because a reordering of blocks is required at the receiving end before releasing data to the user interface. Because of the relative complexity and cost of this protocol, it has not received wide commercial adoption. However, with modern developments in VLSI technology, the SR scheme is increasingly viewed as a cost-effective ARQ strategy. Recent studies of ARQ protocols for application to mobile and cellular radio networks have shown the SR strategy to be superior to go-back-N. The margin of performance is small for slow-fading conditions, but becomes significantly greater in fast fading [Cha91, TIA93b]. The performance improvement achievable with one protocol over another can be very important in networks where maximum signal coverage and information throughput are needed. A version of SR (based on Tannenbaum's Protocol 6 [Tan88]) has been incorporated into the asynchronous data service standard in the IS-54 TDMA digital cellular standard [TIA94].

Effects of Fading on Coding Performance. As we have noted in earlier discussion, the principal issue arising in the application of error-control coding techniques on wireless channels is that of error clustering,—that is, the occurrence of errors in bursts or clusters of varying density as a direct result of the nonuniformity of signal propagation in the wireless environment. The key point here is that error-control codes generally perform better against statistically independent errors than against clustered errors. One consequence of this is that the performance of the code will vary with the temporal characteristics of the fading. Let us say, for example, that we transmit codewords with their symbols appearing in direct consecutive order in the channel. If the fading rate is very slow relative to the duration of a code block, so that a deep fade in effect "submerges" the entire block, then clearly the code will be rendered ineffective and only retransmission (if it is being used) can provide successful delivery of the contained information. However, if the fading rate in the channel is speeded up, with fade durations made comparable to the symbol interval (in this discussion we are ignoring associated phase distortion effects), errors in adjacent symbols may appear nearly independent, and the code may yield satisfactory performance. This variation in performance with fading rate can be seen in the results of a number of investigations coding applied to mobile and cellular radio channels, where coding performance improves as vehicle velocity increases [Iye93]. Figures 7.18 and 7.19 show coding performance measured in computer simulations of a rate-1/2 convolutional code operating on a mobile radio channel with various mobile vehicle speeds. The code in each case is the rate-1/2 code specified in the IS-54 digital cellular standard. Figure 7.18 gives output BER versus channel SNR for hard-decision decoding, whereas Fig. 7.19 gives the corresponding results for soft-decision decoding of the convolutional code. It can be seen from the figures that BER performance improves as vehicle speed increases and channel errors become less correlated, and that greater improvements are achieved with soft-decision decoding.

A technique which can be used to reduce the statistical dependence of errors in a code block is *interleaving*. With interleaving, the symbols contained in one code block are not transmitted in consecutive order but instead are interspersed among other transmitted symbols so that a signal fade is less likely to impose a dense burst of errors on an individual code block. If the system design allows interleaving of code blocks over a sufficiently long time span, errors affecting individual blocks can be made to appear essentially independent, enhancing code performance. However, practical system considerations often rule out long interleaving spans.

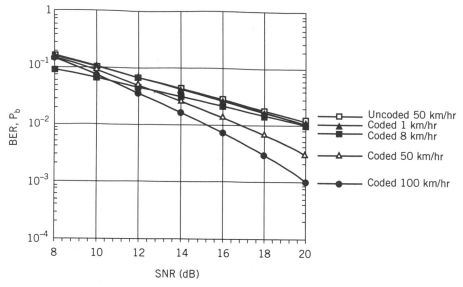

Figure 7.18 BER performance for the IS-54 rate-1/2 convolutional code on a simulated mobile radio channel (hard-decision decoding). [Iye93]

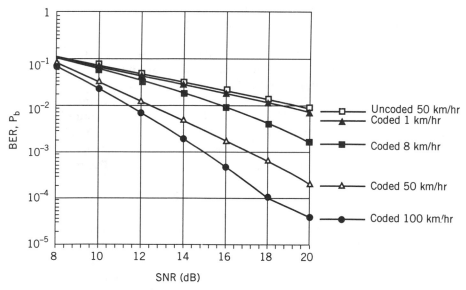

Figure 7.19 BER performance for the IS-54 rate-1/2 convolutional code on a simulated mobile radio channel (soft-decision decoding). [Iye93].

For example, if the system is providing digital voice service, interleaving over multiple voice frames may impose an unacceptable time delay in a conversation. The IS-54 standard, for example, provides for interleaving over only two 20-msec voice frames, in the interest of minimizing time delay. Even in a digital data service, the use of large interleaving spans in turn requires large data buffers, which have a complexity and cost impact on the user terminal. Thus the benefits of interleaving must be balanced against the need to properly satisfy user requirements and to achieve a cost-effective product design.

There is one type of system in which independence between consecutive errors can be achieved automatically; this is a system using *frequency hopping*. As we noted in an earlier chapter, fading can be made independent from one frequency to another, provided that the frequencies are sufficiently separated. In the ideal case, if the carrier is hopped to a new frequency in each consecutive symbol interval, the error-control code will be dealing with independent errors and will exhibit a corresponding improvement in performance. The Pan-European digital cellular system GSM is designed to operate with frequency hopping, though this feature has not yet been implemented in commercial operation [Rah93].

The use of coding redundancy with multifrequency operation is very suggestive of frequency diversity operation, and on this point we return to the topic of the connections between diversity and coding as it is employed on fading channels. We do this in the context of examining soft-decision decoding, which we have already noted offers significant performance gains relative to hard-decision decoding on fading channels. The reason why soft-decision decoding is more beneficial on fading channels than on steady-signal channels is that in the fading case the soft-decision demodulator output conveys information about the instantaneous level of signal fading imposed on that particular symbol, and this in turn represents a "quality metric" for that symbol. By using the symbol quality metrics appropriately in decoding, one can achieve more reliable decoding than is achieved with use of only hard-decision demodulator outputs. Furthermore, one can show that given the use of coding on a Rayleigh fading channel, there is a direct connection between soft-decision decoding and diversity combining. Specifically, one can show the performance achieved with optimum soft-decision decoding of a block code having minimum distance $d_{\min}$ is equivalent to optimum diversity combining with order of diversity equal to $d_{\min}$ [Pro89]. The significance of this result can be appreciated by considering the example of using the (24, 12) extended Golay code on a Rayleigh fading channel. This is a rate-1/2 code with minimum distance 8. Therefore the bandwidth expansion factor needed for use of this code is 2, as with dual-diversity, but optimum soft-decision decoding, with independent bit- to-bit fading, yields performance equivalent to eighth-order diversity. The results obtained for this example are shown in Fig. 7.20 [Pro89]. The figure shows average probability of bit error versus average SNR per bit $\bar{\gamma}_b$ for binary and 4-ary FSK, both with dual-diversity, and Golay coding used with binary FSK and optimum soft-decision decoding. At $\bar{P}_b = 10^{-5}$ the Golay code outperforms dual-diversity 4-ary FSK by about 10 dB. Therefore rate-1/2 coding can be used to make much more effective use of available bandwidth than does dual diversity, at the cost of the greater complexity required for a decoding implementation, and a factor-of-2 reduction in bandwidth efficiency of the system.

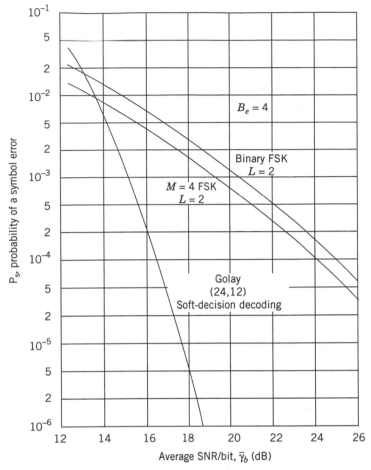

Figure 7.20 BER performance obtained with conventional dual-diversity as compared with rate-1/2 coding for bandwidth expansion factor $B_c = 4$. (From [Pro89] © Mc-Graw-Hill with permission.)

7.5 STANDARD RADIO MODEMS FOR WIRELESS NETWORKS

In our earlier discussion of modem technology, we described the progression of modem techniques, ranging from simple OOK and FSK through PSK, PAM, QAM, partial-response, and TCM. In this section we describe modulation techniques which have been adopted in most of the developing standards for second-generation wireless information networks. In principle, the modulation techniques discussed earlier are applicable to all wireline and wireless modems. That is, there are basic design issues that are common to both wireline and wireless systems. In general we would like to transmit data with the highest achievable data rate and with the least expenditure of signal power. In other words, we usually want to maximize both bandwidth efficiency and power efficiency. However, the emphasis on these two objectives varies from one application to another, and there are certain details that are specific to particular applications.

In voiceband telephone channels, high bandwidth efficiency has a direct economic advantage to the user, because it can reduce connect time or avoid the necessity of leasing additional circuits to support the application at hand. The typical telephone channel is less hostile than a typical radio channel, providing a fertile environment for examining complex modulation techniques and signal processing algorithms. Specific impairments seen on telephone channels are amplitude and delay distortion, phase jitter, frequency offset, and effects of nonlinearities. Many of the practical design elements of wireline modems have been developed to efficiently deal with these categories of impairments.

In radio systems, bandwidth efficiency is also an important consideration, because the radio spectrum is limited and many operational bands are becoming increasingly crowded. Radio channels are characterized by multipath fading and Doppler spread, and a key impediment in the radio environment is the relatively high levels of average signal power needed to overcome fading. However, there are other considerations which impact on the selection of a modem technique for a wireless application. In the next subsection we discuss requirements for radio modems in greater detail before going on to a description of the modem techniques that are in most widespread use in the evolving wireless networks.

7.5.1 Requirements for Radio Modems

There are a number of considerations that enter into the choice of a modulation technique for use in a wireless application, and here we briefly review the key requirements. These requirements can vary somewhat from one system to another, depending upon type of system, the requirements for delivered service, and the users' equipment constraints.

Bandwidth Efficiency. Most wireless networks that support mobile users have a need for bandwidth-efficient modulation, and this requirement steadily grows in importance each year. For example, land-mobile radio (LMR) communications in licensed VHF and UHF bands, which has long provided service for public safety organizations and commercial fleet-dispatch users, is a rapidly expanding market. While these networks were originally designed for FM voice service, they are seeing growing use for digital voice and data services. The growth in demand for mobile data services, combined with the need to provide ever more channels, makes bandwidth efficiency a crucial issue. Standard channel spacing in most LMR bands is 25 kHz, and most mobile data applications are supported by data devices operating at data rates of 2400 bits/sec or lower. However, a joint Public Safety and U.S. Government LMR standard is currently under evaluation which will specify a digital transmission format for use with 12.5-kHz channel spacing and data rates as high as 9600 bits/sec [FS1102].

Other important developments are the initiatives in North America, Europe, and the Far East to define new standards for digital cellular telephone services to replace the existing analog cellular systems (see Chapter 2). The driving force for these developments has been the rapid growth in the market for cellular services, which has resulted in loading analog systems to full capacity during busy hours in some large metropolitan areas. A cellular carrier company is assigned a specified amount of licensed bandwidth in which to operate their system, and therefore an increase in system capacity leads directly to increased revenues. This defines another clear need for modulation techniques that provide efficient utilization of available bandwidth.

An area of wireless communications network development where the modulation bandwidth efficiency is not yet as critical is that of WLANs. Unlike cellular systems, which to date have been used to support circuit-mode services (services, like standard telephone service, in which a user has access to the full user-channel capacity for the duration of the call connection), WLANs are typically used for burst-mode traffic. Because of the bursty nature of the user data, the aggregate data traffic on a WLAN rarely approaches system capacity. Furthermore, many WLAN products operate in the unlicensed ISM bands, where the same frequencies are reused again and again even in relatively close geographic areas. It is for these reasons that the WLAN industry has placed relatively little emphasis on bandwidth-efficient modulation techniques.

Power Efficiency. Power efficiency is another parameter which may not be of major importance in some of the wireless applications such as WLANs used to interconnect stationary workstations, because these types of equipment are typically powered from the AC power sources already available in the office or factory environment. However, in most other applications such as digital cellular, cordless phones, or mobile data services power translates into battery size and recharging intervals and, even more important to the mobile user, into size and weight of the portable terminals. Thus power efficiency is important in most personal and mobile communication systems, and this will become increasingly important as consumers become accustomed to the convenience of small hand-held communication devices.

There are two facets of the power requirement: One is the power needed to operate the electronics in the terminal, and the other is the amount of power needed at the input to the power amplifier in order to radiate a given amount of signal power from the antenna. The radiated signal power, of course, translates directly into signal coverage, and it is a function of the data rate and the complexity of the receiver. Higher data rates require higher operating levels of SNR. More complex systems using TCM or adaptive equalization require less transmission power. However, more complex receiver design increases the power consumed by the electronics and consequently reduces battery life. In some applications a compromise has to be made between the complexity of the receiver and the electronic power consumption. For example, for handheld local communication devices some manufacturers avoid the use of complex speech coding techniques in order to hold down battery consumption. Also, in the design of high-speed data communication networks for laptop or pen-pad computers, some designers find it difficult to justify the additional electronic power consumption required for inclusion of adaptive algorithms.

In spread-spectrum CDMA systems, power efficiency and overall system bandwidth efficiency are closely related. The use of a more power-efficient modulation method allows a system to operate at lower SNR. The performance of a CDMA system is limited by the interference from other users on the system, and an improvement in power efficiency in turn increases the bandwidth efficiency of the system. The discussions that follow will address only the issue of efficiency through the power amplifier, and they will not deal with the issue of power consumption in the electronics of the wireless terminal.

Out-of-Band Radiation. An important issue in the selection of a modulation technique for a radio modem is the amount of transmitted signal energy lying outside the main lobe of the signal spectrum. This issue requires more careful definition than we shall undertake here, but the key point is that in many multiuser radio systems the performance is limited by adjacent-channel interference rather than additive noise. For example, in the design of VHF/UHF land-mobile radio systems, the major design parameter is adjacent-channel interference (ACI), which is the interference that a transmitting radio presents to the user channels immediately above and below the transmitting user's channel. The ACI will determine the geographic area over which mobile users can be served by a single base station. This is because a low level of ACI will permit a distant mobile transmitter to reach the base station with a weak signal while another mobile much closer to the base station is transmitting in an adjacent channel. Thus ACI specifications will indirectly influence system capacity and cost. Evaluation of the ACI involves the characteristics of the transmit and receive channel filters, nonlinearities in the transmitter, and of course the height and rolloff characteristics of the skirts of the transmitted signal spectrum. Radio manufacturers strive to design radios that keep ACI below a specified level, typically -60 dB, and the out-of-band spectral power of the modulation scheme is the principal ingredient in achieving that goal. In contrast, the out-of-band signal power in voiceband modems is not as critical and a voiceband modem manufacturer would be satisfied with out-of-band power of around -40 dB.

Resistance to Multipath. Another important issue in the design of a radio modem is sensitivity to multipath. Various modulation techniques have different degrees of resistance to multipath. This was a major issue in the development of the digital cellular standards, where it was necessary that each standard be written to accommodate the worst-case multipath conditions likely to be encountered by users over the entire geographic region of usage for that standard. Considerable attention is also being given to multipath specifications as part of the standardization of the air interface for 2-GHz PCS systems in the United States [JTC94].

Constant Envelope Modulation. Most mobile radio products are designed with Class-C power amplifiers, which provide the highest power efficiency among the common types of power amplifiers. However, Class-C amplifiers are highly nonlinear, and therefore it is necessary that the signal to be amplified is constant-envelope or as nearly so as is practical. The reason for this is that any amplitude fluctuations in the input signal will give rise to a spectral widening of the output signal, in turn causing increased ACI. It is because of these considerations that frequency modulation has remained in widespread use in the mobile radio industry. Though analog FM mobile radio systems were originally designed for analog voice, they have been extended to data service simply by feeding baseband digital data streams to the frequency modulator. This in effect is a method of FSK, where input amplitude levels correspond to transmitted tones. An FM signal is by its very nature constant-envelope, however, it is not spectrally efficient due to its high sidelobes. Thus as the needs for greater bandwidth efficiency have grown, efforts have been made to design modulation schemes that are less wasteful of bandwidth while preserving (or nearly so) the constant-envelope nature of FM. In order to conform to spectrum constraints, it is necessary in some systems to apply filtering to the modulated waveform before power amplification, and this filtering produces amplitude variations. In order that undesirable out-of- band spectral components not be generated, it is then necessary that the amplification be linear. Consequently such non-constant-envelope filtered signals are commonly referred to as *linear modulation* systems.

In the subsections that follow, we describe three modulation techniques that are specified in prominent standards for radio modems. These are four-level-FM, GMSK, and $\pi/4$-shift QPSK. We shall see that the first two are constant-envelope modulations, while the third, which is specified in a filtered version in the IS-54 TDMA digital cellular standard, is a form of linear modulation.

While today's implementations of linear modulation schemes require linear amplifiers, and thus suffer a loss in efficiency relative to Class-C amplification, much research work is in progress to develop new methods of power amplification that combine near-linearity with power efficiency approaching Class-C characteristics.

7.5.2 Digital Frequency Modulation

As we noted earlier in this section, FM is the predominant form of modulation used in the mobile radio industry. While FM has long been used for carrying analog voice over radio systems, newer digital systems have also been based on FM, specifically multilevel digital FM. A widely used format is four-level FM, which is equivalent to 4-ary FSK. A typical modulator implementation will use direct modulation, in which the four-level baseband digital signal is applied directly to the voltage-controlled oscillator (VCO). This provides a relatively simple design which is compatible with Class-C power amplification, and which permits demodulation with a simple frequency discriminator followed by a sampler. A disadvantage of digital FM is that the spectral skirts are relatively high, and therefore it becomes difficult to move carrier frequencies closer together while complying with limits on ACI. However, it is possible to reduce the spectral skirts by filtering the baseband signal before it is applied to the VCO. Because this filtering affects the frequency excursions, and not the amplitude of the signal, the constant-envelope nature of the transmitted waveform is preserved. Nevertheless, in development of some of the new wireless networks, multilevel FM has not been judged to provide sufficient bandwidth efficiency, and other modulation techniques have been adopted. We shall describe these newer techniques in the following subsections.

7.5.3 OQPSK, MSK, and GMSK

Here we briefly describe a set of modulation techniques that are closely related to one another: offset quadrature phase shift keying (OQPSK), minimum shift keying

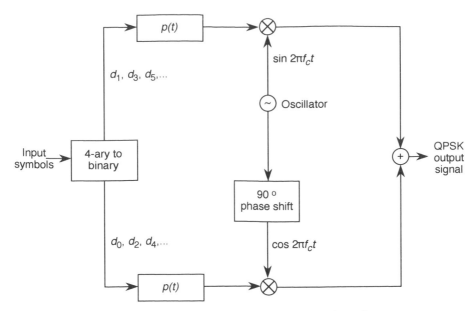

Figure 7.21 Quaternary phase shift keying (QPSK).

(MSK), and Gaussian MSK (GMSK). Two of these modulations, MSK and GMSK, can be explained by direct reference to FSK modulation, but it is useful to begin our discussion instead with quadrature phase shift keying (QPSK), which we mentioned briefly in Section 7.2.6.

As shown in Fig. 7.21, the transmitted QPSK signal can be written as

$$x_t(t) = d_k p(t) \cos \omega_c t + d_{k+1} p(t) \sin \omega_c t, \qquad k = 0, 2, 4, \ldots$$

where $\{d_k\}$ is the stream of data bits, which we have divided into even-numbered bits $d_0, d_2, d_4, \ldots$, and odd-numbered bits $d_1, d_3, d_5, \ldots$.

On the right-hand side of the equation above, the first term is called the *in-phase signal* and the second term is called the *quadrature signal*. The spectrum of $x_t(t)$ is simply the sum of the spectra of the two terms; and because each term represents binary PSK signaling, the overall spectrum has the same shape as does binary PSK. Therefore the bandwidth efficiency of QPSK is twice that of BPSK. Modems using QPSK modulation, and variations of this technique, have been used in many different radio systems. One common variation of basic QPSK is differential QPSK, in which the information to be transmitted is sent as successive changes in the signal phase. Another important variation of this modulation technique is offset QPSK, which we discuss next.

OQPSK. Figure 7.22 shows the basic structure of offset QPSK (OQPSK), which is also referred to as staggered QPSK. In this figure we denote the streams of even-numbered and odd-numbered data bits as $d_I(t)$ and $d_Q(t)$, respectively. In an

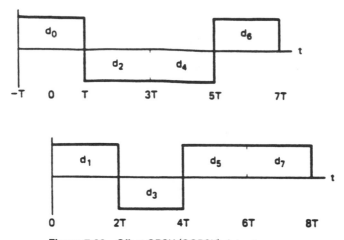

Figure 7.22   Offset QPSK (OQPSK) data streams.

OQPSK modulator, instead of applying the source data bits to the two branches simultaneously every T seconds (Fig. 7.21), the two branches are offset by $T/2$ seconds. The QPSK and OQPSK transmitted waveforms are compared in Fig. 7.23. The benefit of this scheme can be seen by referring to the QPSK modulator shown in Figure 7.21. In ordinary QPSK modulation, the data bit polarity can change sign simultaneously on the two branches, which results in a 180° phase shift in the transmitted waveform. (In a QPSK stream carrying random data,

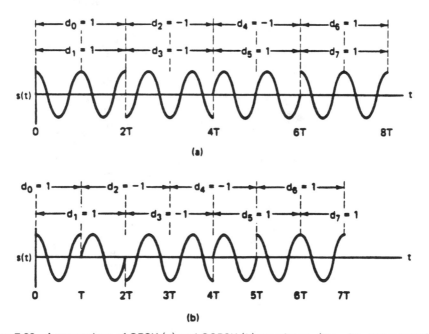

Figure 7.23   A comparison of QPSK (*a*) and OQPSK (*b*) waveforms. (From [Pas79] © IEEE.)

one-fourth of the symbol-to-symbol phase transitions will be 180° phase shifts.) When the QPSK-modulated signal is filtered, as must typically be done to constrain its radiated spectrum, the resulting signal exhibits significant amplitude variations. The presence of these amplitude variations in turn rules out the use of highly nonlinear, but power-efficient, amplifiers. Nonlinear amplification would simply regenerate the out-of-band spectral components which the filtering is meant to suppress.

However, we can see, by referring to the diagrams in Fig. 7.22, that with OQPSK the data bits can change polarity on only one branch at a time, every $T/2$ seconds, and thus the phase of the modulated signal can change by at most 90° from one symbol to the next. The end result of this modification is much lower variations in the envelope of the modulated signal after passing through a band-limiting filter. Thus the filtered signal can then be put through a nonlinear amplifier with very little growth in the out-of-band spectral components. Note that because the offset alignment does not change the spectra of the branch signals, each is the spectrum of BPSK modulation with keying interval T, and therefore the spectrum of OQPSK is identical to the spectrum of basic QPSK. However, the spectra of the two signals are very different after band-limiting and nonlinear amplification.

Because the offset time alignment does not affect the orthogonality of the two branches of the modulator, the theoretical BER performance of OQPSK is identical to that of QPSK for the same received signal and noise power. But because OQPSK can be used with Class-C power amplification, it can be implemented with considerably less prime power in the transmitter. OQPSK was first used in satellite systems because of its compatibility with the highly nonlinear amplifiers used in satellite repeaters and because of the critical importance of minimizing power consumption onboard the satellite. Power consumption is also important in mobile communications systems, where low weight and long battery life are important to users of hand-held radios. As will be shown in the next chapter, the equalizers used for OQPSK modulation and their performance on multipath channels are slightly different from standard equalizers used for QPSK.

Thus we see that modulation techniques which constrain the instantaneous phase transitions in the transmitted waveform can yield important benefits in the design of power-efficient systems. A further improvement of this nature is provided by the modulation technique we examine next, which is closely related to OQPSK.

MSK. Minimum shift keying (MSK) can be described either as a special case of frequency modulation or as a variation of OQPSK. Perhaps the simplest definition of MSK is that it is phase-continuous coherent binary FSK with modulation index $m = 0.5$. The modulation index is defined by $m = T\Delta f$, where Δf is the frequency spacing between the FSK tones; and therefore the tone spacing is $1/2T$, the minimum spacing for which orthogonality over the symbol interval T can be achieved. This is what gives MSK its name. Analytically, we can write the transmitted MSK signal as

$$x_t(t) = \cos\left[2\pi\left(f_c + \frac{d_k}{4T}\right)t + x_k\right], \qquad kT < t < (k+1)T \qquad (7.5.1)$$

where f_c is the carrier frequency, $\{d_k\}$ $(k = 0, 1, 2, 3, \dots)$ is the sequence of data

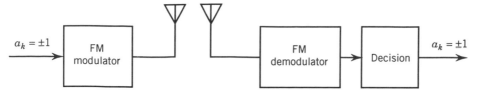

Figure 7.24 MSK implemented as phase-continuous FSK.

bits, and x_k is a value of phase which is constant over the kth successive T-second symbol interval. During each interval, x_k is either 0 or π, to meet the requirement that the phase be continuous from the end of one symbol interval to the start of the next. This requirement is satisfied if x_k is given by the following recursion:

$$x_k = \left[x_{k-1} + \frac{\pi k}{2}(d_{k-1} - d_k) \right] \bmod 2\pi \qquad (7.5.2)$$

Examining Eq. (7.5.1), we see that the frequency in each symbol interval is either $f_c + 1/(4T)$ or $f_c - 1/(4T)$, in accordance with the data bit value to be transmitted in that interval. Therefore the tone spacing is $1/2T$, the minimum tone spacing for signal orthogonality over the interval T.

One way of implementing MSK is to use phase-continuous FSK with a modulation index of 0.5, as represented by Eq. (7.5.1) and as shown in Fig. 7.24. At the receiving end, for best BER performance, the signal is demodulated by coherent FSK demodulation. Alternatively, the MSK signal can be detected with a frequency discriminator followed by a slicer, a simpler implementation which yields somewhat poorer BER performance than does coherent FSK demodulation. In practice in radio modems, MSK has typically not been implemented using the FM approach depicted in Fig. 7.24, because it requires the modulation index to be implemented very precisely in order for phase coherence to be preserved. (This of course is not important if frequency discriminator detection is to be used.) Consequently, a quadrature modulator structure is ordinarily used, as described next. Our presentation follows those of Pasupathy [Pas79] and Sklar [Skl88].

Using standard trigonometric identities, we can rewrite Eq. (7.5.1) in the form

$$x_t(t) = a_k \cos\frac{\pi t}{2T}\cos 2\pi f_c t + b_k \sin\frac{\pi t}{2T}\sin 2\pi f_c t, \qquad iT < t < (i+1)T$$

$$(7.5.3)$$

where

$$a_k = \cos x_k = \pm 1$$

$$b_k = d_k \cos x_k = \pm 1 \qquad (7.5.4)$$

In this quadrature form, the in-phase signal component is represented by $a_k \cos(\pi t/2T)\cos 2\pi f_c t$, where $\cos(\pi t/2T)$ is described as sinusoidal symbol weighting, and a_k depends upon the data through Eq. (7.5.4). Similarly, the

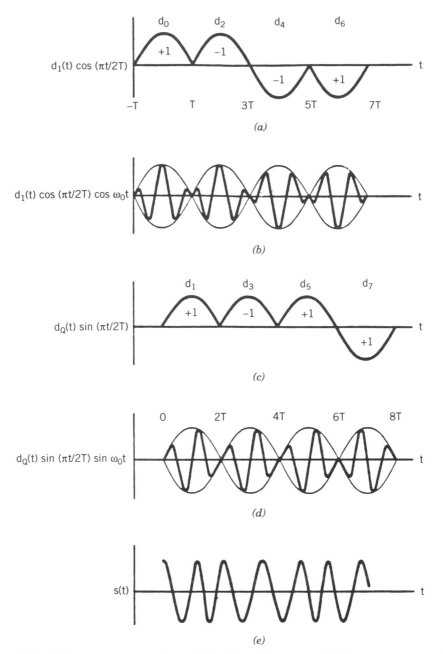

Figure 7.25 MSK waveform composition. (a) Modified *I* bit stream. (b) *I* bit stream times carrier. (c) Modified *Q* bit stream. (d) *Q* bit stream times carrier. (e) MSK waveforms. (From [Pas79] © IEEE.)

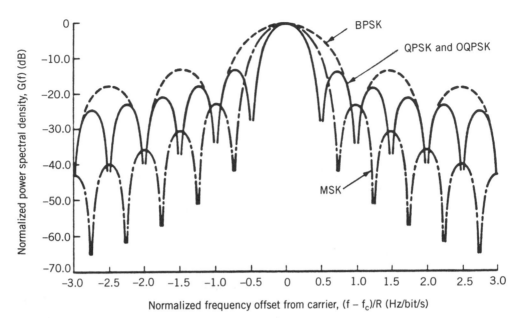

Figure 7.26 Normalized power density spectra for BPSK, QPSK, OQPSK, and MSK. (From Amoroso © 1980 IEEE.)

quadrature signal component is given by $b_k \sin(\pi t/2T) \sin 2\pi f_c t$, where $\sin(\pi t/2T)$ is a sinusoidal symbol weighting and b_k also depends upon the data through Eq. (7.5.4). At first examination of Eq. (7.5.4), it might appear that a_k and b_k can change every T seconds, because the data bit value d_k can certainly change every T seconds. However, because of the continuous-phase constraint, as ensured by Eq. (7.5.2), a_k can change sign only at the zero crossings of $\cos(\pi t/2T)$ and b_k can change sign only at the zero crossings of $\sin(\pi t/2T)$. Thus the symbol weighting in either the inphase or quadrature channel is a half-cycle sinusoidal pulse of duration $2T$ seconds with alternating sign. As with OQPSK, the two channels of this modulator are offset by T seconds. Therefore we can regard the MSK modulation as a form of OQPSK having sinusoidal symbol weighting on the quadrature channels and a special relationship between the source data stream d_k and the binary values applied to the quadrature channels. For steady-signal reception in AWGN, the BER performance for MSK is exactly the same as for OQPSK and QPSK. The composition of the MSK waveform is shown in Fig. 7.25.

As with OQPSK, MSK is an offset constant-envelope modulation, but MSK has the further advantage that it is phase-continuous, and therefore even the $\pm 90°$ shifts of OQPSK are eliminated. In fact, it is readily seen from Eqs. (7.5.1) and (7.5.2) that in each T-second interval, the phase of the signal moves ahead of or behind the carrier phase at a constant rate for a total excursion of exactly $\pi/2$ radians over the interval. The phase continuity of MSK results in a signal spectrum with tails somewhat lower than those of OQPSK. Figure 7.26 shows the normalized power density spectra for BPSK, QPSK, OQPSK, and MSK. As can be seen in the figure, MSK has lower sidelobes than QPSK or OQPSK, but the main lobe of the MSK spectrum is broader. At the 3-dB power points, the MSK main lobe is about

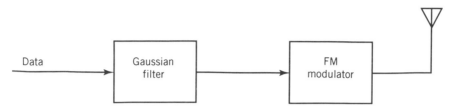

Figure 7.27 Gaussian MSK (GMSK) modulation.

30% wider than that of QPSK and OQPSK. Considerations of how one might further improve upon MSK led to the development of GMSK, which we describe next.

GMSK. Murota and Hirade [Mur81] observed that the MSK signal spectrum could be made even more compact, and the spectral skirts further lowered, by implementing the modulation in its direct FM form with a low-pass filter applied to the data stream before modulation. This technique provides a smoothing of the phase transitions at symbol boundaries; and because the filtering is done before modulation, the constant-envelope property of the signal is preserved. The specific filter characteristic proposed by Murota and Hirade is the Gaussian low-pass filter. This modulation technique, shown in Fig. 7.27, is called *Gaussian-filtered MSK* or

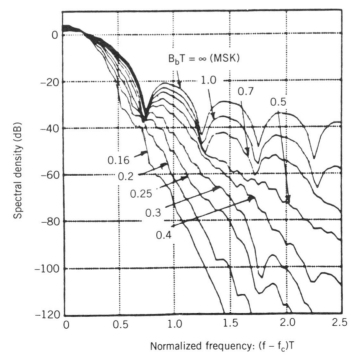

Figure 7.28 Power Spectral density of GMSK. $B_b T$ is the normalized 3-dB bandwidth of the premodulation Gaussian LPF; T is the unit bit duration. (From [Mur81] © IEEE.)

simply *Gaussian MSK* (GMSK). The rapid rolloff achievable with Gaussian filters provides spectra which are more compact than that of strict-sense MSK. Because the transform of a Gaussian function is also Gaussian, the impulse response of the premodulation filter is Gaussian. The tails of the Gaussian time-domain function remain above zero, and this results in some amount of intersymbol interference. The choice of rolloff characteristic for the Gaussian filter involves a tradeoff between spectral confinement and performance loss. Figure 7.28 shows power spectra for GMSK versus the normalized frequency difference from the carrier center frequency, $(f - f_c)T$, with normalized 3-dB filter bandwidth B_bT as a parameter. The parameter selection $B_bT = \infty$ effectively removes the filter, and thus corresponds to pure MSK. Decreasing values of B_bT produce corresponding narrowing of the power spectrum, while producing increasing degradation in BER performance relative to unfiltered MSK. The spectral plots presented here represent a practical range of bandwidth selections. In their paper, Murota and Hirade suggest $B_bT = 0.25$ as an optimum choice and indicate that this produces a performance degradation of no more than 0.7 dB relative to MSK. The GMSK modulation, with bandwidth parameter $B_bT = 0.3$, has been adopted for GSM, the Pan-European Digital Cellular Standard [Rah93, Hau94].

In can be seen from the spectral plots that as the Gaussian filter bandwidth is reduced, both the sidelobes and the width of the main lobe are reduced. The narrowing of the main lobe relative to a fixed data rate yields an improvement in

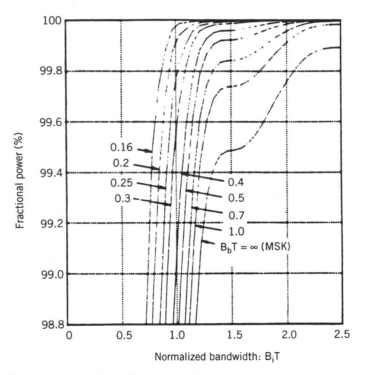

Figure 7.29 Fractional power ratio of GMSK. B_i is the bandwidth of an ideal rectangular predetection BPF; B_bT is the normalized 3-dB bandwidth of a premodulation Gaussian LPF; T is the unit bit duration. (From [Mur81] © IEEE.)

TABLE 7.2 Occupied Bandwidth Normalized by Bit Rate for Specified Percentage Power

B_bT	Power%			
	90	99	99.9	99.99
0.2	0.52	0.79	0.99	1.22
0.25	0.57	0.86	1.09	1.37
0.5	0.69	1.04	1.33	2.08
MSK	0.78	1.20	2.76	6.00
TFM	0.52	0.79	1.02	1.37
	Occupied bandwidth			

Source: [Mur81] © IEEE.

bandwidth efficiency, as measured by bits/sec/Hz. While GMSK provides some improvement in bandwidth efficiency relative to MSK, about 1.6 rather than 1.4, the bandwidth efficiency of QPSK or OQPSK is still better, at about 1.8 bits/sec/Hz. However, GMSK does provide an important advantage in the low level of its spectral sidelobes, which in turn translates into superior ACI performance.

Figure 7.29 shows the fractional power in the desired bandwidth as a function of the normalized bandwidth B_iT of a rectangular predetection bandpass filter, with normalized Gaussian filter bandwidth B_bT as a parameter. Table 7.2 gives occupied bandwidth for several values of percentage of total signal power, with B_bT also a variable parameter. The table provides a convenient means of assessing how the selection of pre-modulation filter bandwidth affects the distribution of transmitted signal power between the main lobe and the sidelobes of the spectrum.

As with MSK, GMSK can be viewed as a form of digital FM, and therefore can be demodulated with a simple limiter–discriminator detector [Eln86, Hir84]. However, because of its close relationship to ordinary MSK, best performance is obtained by demodulating the received GMSK signal with a two-branch coherent demodulator very much the same as used with MSK. In GMSK reception, extra care must be taken in the design of the demodulator to ensure reliable carrier and timing recovery, given the use of premodulation filtering. These issues are discussed in some detail in [Mur81].

In a signal fading environment, coherent detection of GMSK has been shown to exhibit high error floors, or irreducible error rates [Hir79]. Thus there has been much work on developing various forms of differential detection for GMSK [Sim84, Ogo81, Yon86a, Yon86b, Yon88].

Though we have described GMSK modulation in terms of direct frequency modulation with prefiltering, the reader may well ask whether some form of the two-branch MSK modulator discussed earlier can be used instead. The answer is that a two-branch modulator can be used; but its structure is not as simple as in the case of unfiltered MSK, because the Gaussian filtered input signal does not have a simplified equivalent in-phase and quadrature representation. However, Murota and Hirade [Mur81] suggest that a quadrature modulator with digital waveform generators might be used. With this form of modulator the input data stream would be fed to a digital-waveform-generating function which would produce two signals, $\sin[f(t)]$ and $\cos[f(t)]$, which would then modulate the sine

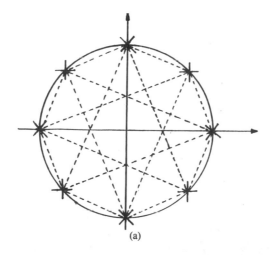

(a)

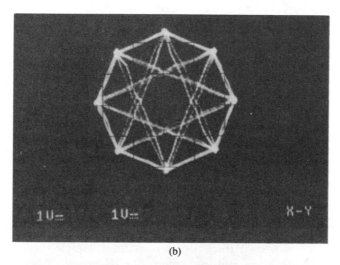

(b)

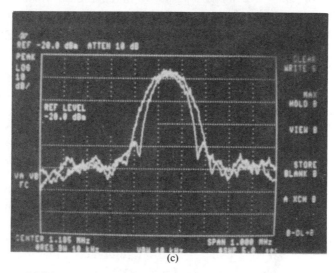

(c)

Figure 7.30 $\pi/4$-QPSK modulation. (*a*) Possible phase states of the $\pi/4$-QPSK modulated carrier at sampling instants. (*b*) The signal constellation with sinusoidal shaping. (*c*) Spectrum of the $\pi/4$-QPSK signal (*upper trace*) compared with that of an SQAM signal (*lower trace*). (From [Feh91] © IEEE.)

and cosine branches, respectively, of the two-branch modulator. While use of this form of modulator for GMSK has not appeared to date in the communications literature, Davarian and Sumida [Dav89] have described a multipurpose digital modulator which has been designed to implement several different modulation methods with a single I–Q structure.

7.5.4 $\pi/4$-Shift QPSK

In our previous discussions, we noted that multilevel FM, which has been widely used in mobile radio systems, in large part due to its amenability to nonlinear amplification and limiter–discriminator detection, does not provide the bandwidth efficiency required in the newly emerging digital radio systems. We saw that QPSK provides much better bandwidth efficiency, and the staggered version, OQPSK, also provides greater compatibility with nonlinear amplification due to its lower amplitude fluctuations after filtering. However, a drawback of OQPSK modulation is that it may require coherent demodulation and may suffer performance degradations on radio channels with large Doppler shifts [Feh91]. These limitations relate directly to the $T/2$-staggered structure of the OQPSK signal. The search for nonstaggered modulation schemes having low postfiltering amplitude variations led to work by Akaiwa and Nagata [Aka87] and others, including [Liu89], [Goo90], and [Liu90], on $\pi/4$-shift QPSK modulation.

Simply described, $\pi/4$-shift QPSK is a form of QPSK modulation in which the QPSK signal constellation is shifted by $45°$ each symbol interval T. This means that the phase transitions from one symbol to the next are restricted to $\pm\pi/4$ and $\pm3\pi/4$. By eliminating the $\pm\pi$ transitions of QPSK, the amplitude variations after filtering are significantly reduced. Figure 7.30a shows the eight possible phase states of the $\pi/4$-shift QPSK modulated carrier at the sampling instants. The eight phases represent the 4-phase QPSK constellation in its two shifted positions. The four dashed lines radiating from each point on the circle indicate the allowed phase transitions. Figure 7.30b shows the signal constellation and the transitions as displayed on an oscilloscope. In the implementation depicted, the modulation is implemented with sine-wave pulse shaping [Feh91]. Figure 7.30c shows spectra measured with two versions of $\pi/4$-shift QPSK, nonlinearly amplified. The upper trace is strict-sense $\pi/4$-shift QPSK, while the lower trace is the spectrum for sine-wave pulse-shaped $\pi/4$-shift QPSK.

Thus the $\pi/4$-shift QPSK modulation provides the bandwidth efficiency of QPSK together with a diminished range of amplitude fluctuations. Furthermore, the $\pi/4$-shift QPSK modulation has the advantage that it can be implemented with coherent, differentially coherent, or discriminator detection [Liu90]. These multiple advantages of $\pi/4$-shift QPSK led to its adoption for the North American Digital Cellular TDMA standard [TIA92], as well as the Japanese Digital Cellular standard [Nak90] and the planned new standards for Trans-European Trunked Radio (TETRA) [Hai92].

Although it is not essential, $\pi/4$-shift QPSK modulation is frequently implemented with differential encoding, because this permits the use of differential detection in the receiver, though coherent detection may also be used to achieve optimum performance. The use of differential detection avoids the complexity required to reliably extract a coherent carrier reference under multipath fading

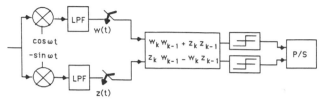

Figure 7.31 Baseband differential detector for $\pi/4$-DQPSK. The low-pass filters are assumed to be square-root raised cosine. (From [Feh91] © IEEE.)

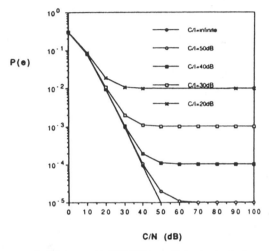

C/N (dB)

Figure 7.32 $P(e)$ versus C/N for $\pi/4$ DQPSK in a flat, slowly fading channel, with Gaussian noise and co-channel interference. x, $C/I = 20$ dB; $\Box$, $c/I = 30$ dB; $\blacksquare$, $c/I = 40$ dB; $\bigcirc$, $C/I = 50$ dB; $\bullet$ $C/I = \infty$ (From [Feh91] © IEEE.)

conditions. This scheme is termed *differential $\pi/4$-shift QPSK*, denoted simply as $\pi/4$-DQPSK. Figure 7.31 shows a block diagram of a baseband differential detector for $\pi/4$-DQPSK. Figure 7.32 shows the BER performance of $\pi/4$-DQPSK in a flat, slowly fading channel corrupted by AWGN and co-channel interference.

QUESTIONS

(a) Explain why block codes are typically preferred in data-oriented WIN systems whereas convolutional codes are preferred in voice-oriented WIN systems.

(b) Give an equation for calculation of the error rate of the $\pi/4$–QPSK modulation scheme in AWGN.

(c) Why are constant-envelope modulation techniques preferred for use on radio channels?

(d)   Why is bandwidth efficiency important for WIN systems?

(e)   Why is power efficiency important for WIN systems?

(f)   Name three major design requirements that are more important for radio modems than for wireline modems, and explain why these issues are important.

(g)   Why was GMSK modulation adopted for several of the leading WIN systems?

(h)   What is the major advantage of OQPSK modulation relative to QPSK?

(i)   Explain why the signal constellation for $\pi/4$–QPSK modulation, shown in Fig. 7.30, has eight points rather than the four points of standard QPSK. How many points do we expect in the signal constellation of OQPSK?

(j)   Compare the effectiveness of Golay coding with that of diversity. The improvement using Golay coding approximates that of how many orders of diversity?

(k)   Discuss the relative advantages and disadvantages of $\pi/4$–QPSK and GMSK modulation.

(l)   Why are multiamplitude and multiphase modulation techniques not very popular in mobile radio systems?

PROBLEMS

Problem 1. The standard pulse shape used in most short-distance cable communication applications such as RS232 is a rectangular pulse.

(a)   The matched filter receiver for the rectangular pulse transmission is usually implemented with an integrator and dump circuit. Give a block diagram for this receiver, and explain how it works and why it is a matched filter.

(b)   If the voltage used for the amplitude of the pulses is $\pm A$ volts and the variance of the received noise is N_0, determine the signal-to-noise ratio after sampling at the receiver.

(c)   The integrate-and-dump circuit can be implemented with a simple RC low-pass filter and a switch. Give the circuit diagram for this matched filter receiver.

(d)   Determine the signal-to-noise ratio after the sampler in terms of the RC constant of the low-pass filter and the variance of the received noise.

Problem 2. Assume that we have a BPSK modem operating on a voiceband channel with a symbol transmission rate of 2400 symbols per second and a bandwidth efficiency of 1. We want to increase the data rate to 19.2 kbits/sec.

(a)   If we increase the number of points in the constellation until the data rate becomes 19.2 kbits/sec while the baud rate remains at 2400 symbols per second, what is the number of points in the constellation? What is the bandwidth efficiency of the modulation technique? What is the additional power requirement for the transmitter to keep the quality the same as before (the approximations used in Section 7.2 can be applied)?

(b)   Repeat (a) if we use a trellis-coded modulation that doubles the points in the constellation and has an overall coding gain of 3 dB.

(c)   If we increase the symbol transmission rate to 19,200 and use the same BPSK modulation, what is the additional power requirements for the transmitter to maintain the same quality as a 2400-bit/sec modem?

Problem 3. Consider the 16-PSK constellation and the rectangular 16-QAM constellation shown in Fig. P7.1. We have two transmitters, each using one of these constellations. In order to have approximately the same performance (error rates) at the receiver, what should be the difference (in decibels) of the transmitted power for the two transmitters. What are the advantages of a 16-PSK modem?

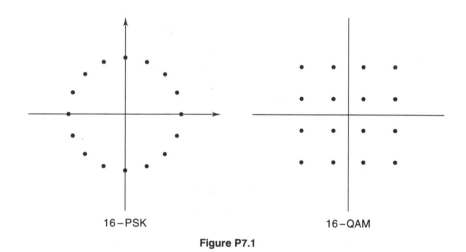

16-PSK   16-QAM

Figure P7.1

Problem 4. Suppose you are asked to design a 16.8-kbit/sec constellation for a voiceband modem. The symbol rate is 2400 symbols/sec and the received signal-to-noise ratio is 28 dB.

(a)   For an uncoded constellation, find the alphabet size and select a constellation. Give your reasoning for the selection. Using the asymptotic bound, give the symbol error rate for the modem.

(b)   Repeat part (a) for a trellis-coded constellation with 3-dB coding gain.

Problem 5

(a)   Use MathCAD or MatLAB to plot Eq. (7.2.14) for $M = 16$. The format of the plot should be similar to Fig 7.2. Find the γ_b's for which each plot provides an error rate of 10^{-5}.

(b) Plot Eq. (7.2.15) for $M = 16$, and determine the signal-to-noise ratios for which each plot provides an error rate of 10^{-5}.

(c) Compare your results with those given in Figs. 7.10 and 7.11. If there are discrepancies, explain them.

Problem 6. Sketch the probability of outage versus error rate threshold for BPSK modulation with one and two orders of diversity and average signal-to-noise ratios of 10, 20, 30 and 40 dB. Assume that the channel is flat Rayleigh fading and that the two diversity branches are independent with the same average power. Use logarithmic scales for both axes.

Problem 7

(a) Give an equation for approximate calculation of the probability of outage of M-ary coherent PSK modulation over flat Rayleigh fading channels. The equation should be written in terms of the average received signal-to-noise ratio per bit γ_b and the threshold signal-to-noise ratio γ_{th}.

(b) For $M = 16$, plot the probability of outage versus threshold signal-to-noise ratio γ_{th} for the average signal-to-noise ratios of $\gamma_b = 10$ dB, 20 dB, and 30 dB. For an acceptable error rate threshold of 10^{-5}, compare the outage rates for the three different plots.

(c) For $M = 16$, plot the probability of outage versus average signal-to-noise ratio γ_b for the threshold signal-to-noise ratios of $\gamma_{th} = 10$ dB, 20 dB, and 30 dB. For an average error rate of 10^{-5}, compare the outage rates for the three different plots.

Problem 8

(a) Give an equation for approximate calculation of the probability of outage of M-ary coherent QAM modulation over flat Rayleigh fading channels. The equation should be written in terms of the average received signal-to-noise ratio per bit γ_b and the threshold signal-to-noise ratio γ_{th}.

(b) For $M = 16$, plot the probability of outage versus threshold signal-to-noise ratio γ_{th} for the average signal-to-noise ratios of $\gamma_b = 10$ dB, 20 dB, and 30 dB. For an acceptable error rate threshold of 10^{-5}, compare the outage rates for the three different plots.

(c) For $M = 16$, plot the probability of outage versus average signal-to-noise ratio γ_b for the threshold signal-to-noise ratios of $\gamma_{th} = 10$ dB, 20 dB, and 30 dB. For an average error rate of 10^{-5}, compare the outage rates for the three different plots.

Problem 9

(a) What is the required average received signal-to-noise ratio for a QPSK modem operating over a flat fading radio channel to have an outage rate of 10^{-2} relative to the threshold error rate of 10^{-4}?

(b) If we reduce the acceptable threshold level in part (a) to 10^{-2}, what improvement in the outage rate would we see?

(c) If we increase the transmitted power in part (a) four times (6 dB), how much improvement would be seen in the outage rate?

(**d**) If we operate the modem over a fixed wireline channel, what would be the required signal-to-noise ratio to maintain an error rate of 10^{-2}? How much improvement in the error rate would occur if we increase the power for 6 dB?

Problem 10. A coherent BPSK receiver operates in a Rayleigh fading channel with two antennas. The signals received from the two antennas are correlated, and the eigenvalues of the covariance matrix are 0.4 and 0.6.

(**a**) Plot the average probability of error on a logarithmic scale versus the average signal-to-noise ratio in decibels.

(**b**) Plot the probability of outage on a logarithmic scale versus the average signal-to-noise ratio in decibels.

(**c**) Repeat (a) and (b) when the two diversity branches are uncorrelated and the received power in both branches is the same.

Hint: Note that the calculations involve subtraction of two exponentials with values very close to each other. For an accurate computation, either very-high-precision computation or series expansions may be needed.

Problem 11. Use MathCAD or MatLAB to plot the probability of error versus average signal-to-noise ratio of a binary DPSK-NCD modem over a Rician fading channel with $k = 6$ dB.

Problem 12. Use MathCAD or MatLAB to plot the probability of error versus average signal-to-noise ratio of a binary DPSK-NCD modem over a zero-mean lognormal fading channel.

SIGNAL PROCESSING
FOR WIRELESS APPLICATIONS

8.1 INTRODUCTION

In this chapter we describe signal processing techniques commonly implemented in wireless systems. Figure 8.1 shows a block diagram of a transmitter and a receiver for a typical modem. The transmitter consists of a coder, a pulse-shaping filter, a modulator and a power control element. The receiver consists of a decoder, the second part of the pulse shaping filter, an automatic gain (power) control circuit, a timing recovery circuit, a phase recovery circuit, a demodulator, and a box called *digital processor*. The encoder will perform both source and channel coding plus data scrambling, if utilized.

There are numerous applications of signal processing identified in Fig. 8.1. These applications can be divided into three categories: signal processing needed for proper operation of the modem, signal processing used to mitigate the effects of frequency selective fading or cancel interference in the channel, and source-coding algorithms used for information compression.

Proper operation of a receiver requires power control normally done with automatic gain control (AGC) circuits, recovery of a timing reference for the transmitted pulses, recovery of the transmitted carrier phase for coherent modulation, and pulse shaping to optimize the use of the channel bandwidth and minimize the *intersymbol interference* (ISI). In many of the modern wireless information networks, transmitter power control is implemented as well.

The purpose of source coding is to minimize the volume of transferred information to save transmission time and expense and to accommodate a larger number of users in a given bandwidth. Earlier we discussed various modulation techniques to increase the information transmission rate for a given bandwidth. Source coding is another approach to increasing the bandwidth efficiency of the system, one which is independent of the transmission channel or modulation technique. The future direction of wireless information networks is toward multimedia systems integrating data, voice, and image/video in a unified network. Speech- and image-coding techniques are usually treated in the signal processing and communication literature, while data compression is regarded as a strict-sense communication problem. Data compression techniques are very popular in voiceband data

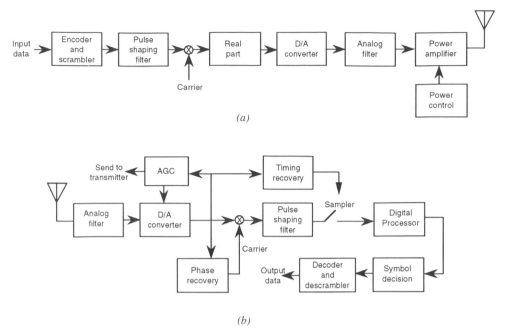

Figure 8.1   Block diagrams for a typical modem. (*a*) Transmitter. (*b*) Receiver.

communications and some of the new wireless data service standards provide for the use of compression techniques.

Speech coding is very important in digital cellular communications where the support of large number of users requires transmitting digitized speech at the lowest possible rates. Sophisticated model-based speech coding algorithms are used in this industry to encode the speech to rates as low as 7 kbits/sec. The systems being developed for personal communications services (PCS) are intended to extend the standard telephone services, which have higher quality than the typical mobile radio serviccs. Thc PCS systems will also require low levels of electronic power consumption in order to maximize battery life, and will have smaller number of users per cell, which makes bandwidth utilization efficiency less critical. As a result, this industry will use simple speech coding algorithms such as adaptive differential PCM (ADPCM) with speech-coding rates around 32 kbits/sec.

Low-bit-rate moving image coding and fixed-image coding algorithms for com- munication over fading channels are under investigation in the wireless informa- tion network industry. Low-bit-rate image coding is suitable for video-telephony applications, while fixed-image coding applications are of interest for wireless multi media networks. For further information on speech-coding techniques we refer the reader to [Rab78] and [Cro83]; for voiceband waveform-encoding tech- niques, to [O'Ne80] and [Jay84]; for image-coding techniques, to [Net80] and [Kan80]; and for data compression techniques, to [Hel84]. Later in this chapter we provide a brief discussion of the most widely used speech- and image-coding algorithms.

Frequency-selective fading channels deliver a distorted form of the transmitted symbol stream to the receiver. As the data rate is increased, the channel appears more frequency-selective and modem performance further degrades. To recover the transmitted symbols and improve the quality of the detected information at higher data rates, adaptive signal processing algorithms are implemented in modems. The emphasis in this chapter is on the applications of signal processing algorithms in the receiver to improve the performance of the modem on fading multipath wireless media.

Section 8.2 provides an overview of the speech and image coding algorithms used in wireless information networks. The rest of the chapter is devoted to signal processing applications to modem design technology. Section 8.3 describes the basic signal processing elements necessary for proper operation of a modem. In Section 8.4 we discuss the degradations caused by frequency-selective multipath fading on the performance of typical modems. In Section 8.5 we introduce the concept of in-band diversity. The remainder of the chapter describes various methods that can be used to take advantage of the in-band diversity to increase the data rate limitations of modems operating on radio channels. Adaptive discrete matched filtering, maximum likelihood sequence estimation, and equalization are studied first. Then application of sectored antennas, multicarrier systems, multi-amplitude and multiphase modulations, and multirate transmission are described. Because of its importance, particular attention is paid in this chapter to the application of adaptive equalization techniques.

8.2 SOURCE CODING FOR WIRELESS APPLICATIONS

In this section we describe briefly the principal techniques used for source coding of speech signals and images. We identify the speech-coding techniques that have been adapted in various standards and comment on recent work in the application of image coding to wireless systems.

8.2.1 Speech Coding

In any wireless system that provides digital voice service, there is clearly a need for a speech-coding process which will encode analog speech into a digital stream for transmission over the air interface and then reconstitute the speech with a fidelity that the listener will find acceptable. The field of speech coding is one in which an enormous amount of research has been done, leading to a wide variety of coding and synthesis algorithms. The interested reader can refer to a number books [Fla72, Rab78, Jay84] and survey articles [Fla79, Cro83, IEE88b, Jay90, IEE92] on this interesting and wide-ranging subject. An overview of CCITT activities on speech processing standards up to 1988 is given in [Dec88]. Here we shall not attempt to cover this subject in any great depth, but instead will focus on the application of speech coding techniques in the wireless environment. Other treat-ments of speech coding for mobile radio applications can be found in a number of articles, including [Cox91b] and [Ste93], and in the text by Steele [Ste92].

Speech Coding in the Wireless Environment. The key parameters of concern in the choice of a speech-coding technique for application in a wireless system are: (1) transmitted bit rate, (2) delivered speech quality, (3) robustness in the presence of transmission errors, and (4) complexity of implementation. The relative

TABLE 8.1   Speech Coders Used in Second-Generation Wireless Systems

System	Application	Voice Coder	Uncoded Rate (kbit / sec)	Overall Rate (kbits / sec)
APCO-25	Public Safety LMR, United States	IMBE	4.4	7.2
IS-95	CDMA Digital Cellular, North America	QCELP	9.6, 4.8, 2.4, 1.2	28.8, 19.2 (FEC + repetition)
IS-54	TDMA Digital Cellular, North America	VSELP	8	13
JDC	Digital Cellular, Japan	VSELP	8	13
GSM	Digital Cellular, Europe	RPE-LTP	13	22.8
DCS-1800	PCN, United Kingdom	RPE-LTP	13	22.8
PCS-2 GHz	PCS, United States	ADPCM	32	32
Handi-Phone	PCS, Japan	ADPCM	32	32
CT2	Cordless, Europe and Asia	ADPCM (G721)	32	32
DECT	Cordless & WPBX, Europe	ADPCM (G721)	32	32

importance of each of these parameters will vary from one category of system to another. Furthermore, these parameters are all interrelated, as we shall see in the ensuing discussion.

The transmitted bit rate for a candidate speech-coding technique is an important element in determining the overall spectral efficiency for a wireless system. This is a particularly important issue in the design of mobile and cellular systems, which are intended to serve large numbers of users. Thus we see use being made in these systems of some of the most complex of the speech-coding algorithms, techniques which operate at bit rates of 13 kbits/sec and below. In some systems, where spectral efficiency is less important than voice quality and implementation complexity, 32-kbit/sec coder–decoders (codecs) are used. Table 8.1 shows the speech-coding techniques, with transmission rates, used in a number of second-generation cellular and cordless systems. There are two bit rates of concern here: the basic bit rate of the speech coder, termed uncoded in the table, and aggregate bit rate when error-protection coding is applied to the coded voice bits. We shall say more about the issue of error protection, as well as the speech coders themselves, in later paragraphs.

While it can be assumed that high quality in reconstructed speech is important in any communication system, the requirements for voice quality do vary from one system to another, based in part on perceived levels of expectation by the user of the service. For example, in the new digital cellular telephone systems, the standard of voice quality which has been adopted for selection of voice coders has been the level of voice quality typically provided by the existing analog cellular FM systems. While FM generally provides a very satisfactory quality of voice service under most conditions, it does exhibit the characteristics of the mobile radio environment, with occasional fading and signal dropouts. Users of cellular telephone service do not expect to hear the voice quality typical of a typical toll-quality PSTN call connection. Thus in mobile and cellular systems, the priority given to system capacity, which leads to the adoption of low-rate coders, ranks above the requirement for voice quality. In contrast, PCS is seen as the next step in evolution of cordless telephone, and as the specifications for PCS are being developed, emphasis is being placed on providing near-toll-quality voice service, which has led to the adoption of 32-kbit/sec voice coding.

The robustness of a speech coding scheme in the presence of channel errors is a very important consideration for applications in wireless networks. While resistance to channel errors is an aspect of the overall quality characteristics of a candidate speech-coding technique, it is important to know how different voice coders respond to the error mechanisms of the wireless environment. Voice coders that perform very well on telephone circuits, where BERs of 10^{-5} and better can be expected on most call connections, may perform very poorly on wireless connections, where average BERs as high as 5% will often be experienced, and where errors will occur in bursts associated with signal fades and with call handoffs in mobile and cellular systems. As we shall see in later discussion, this problem is dealt with by the judicious application of error-control coding. Thus, in a sense, we use speech coding to compress the voice signal by removing its natural and somewhat unstructured redundancy, and then insert structured redundancy in the form of channel coding to protect the relatively fragile digital voice bit stream against transmission errors. Also, because the error-control decoder can also provide short-term information about the quality of the received data stream, it can also play a role in controlling the operation of the network—for example, by providing an indication of when a cellular call should be handed off from one base station to another having better transmission quality [Ste93]. In some cases, interleaving is used in conjunction with channel coding to combat burst errors, but interleaving buffers introduce time delay, and long delays become objectionable to talkers and listeners, and thus interleaving must be used sparingly. The appropriate balance among speech compression, channel coding, and interleaving will be different for each speech-coding technique.

The complexity incurred in implementing various speech-coding techniques is closely related to the information compression achieved by various candidate techniques. Broadly speaking, the coding algorithms capable of operating at the lowest transmission rates require the most complex signal processing; in some instances, for example, a DSP chip must be devoted entirely to the speech-encoding and -decoding functions. As with the other speech-coding parameters, the importance of implementation complexity varies somewhat from one system application to another. In cellular and other mobile radio systems, where many user terminals are used in vehicles, the added power consumption is not critical, and the same may be the case with respect to the size and weight of the mobile terminal. However, in PCS applications, where users expect to have the convenience of pocket-portable devices, and where their expectations have been conditioned by the small size, light weight, and low cost of cordless phones, it will be very important to minimize the complexity of the electronics in user terminals. As we shall see in the following paragraphs, this requirement can be met with relatively simple voice coders that also provide near-toll quality speech.

Overview of Speech-Coding Techniques. There are three basic categories of speech-encoding techniques: (1) waveform encoding, (2) model-based source coding, and (3) hybrid techniques [Fla79]. Within each category we can find a wide array of different techniques providing different levels of fidelity and transmission rates. Here we shall describe briefly the key concepts utilized in the three categories and comment on the issues involved in selecting among the available techniques.

Waveform coders are by far the simplest category of speech-coding techniques, and both time-domain and frequency-domain versions have been developed. A waveform encoder simply accepts a continuous analog speech signal and encodes it into a digital stream for transmission. At the receiving side of the link or network, a decoder reverses the encoding process to provide an accurate reconstruction of the original speech signal. The objective in designing a waveform-encoding and -decoding technique is faithful reproduction of the speech waveform, under the assumption that this will result in good voice quality as perceived by the listener. The encoding process introduces *quantization noise*; but given a sufficient number of quantization steps, this can be brought to such a low level that it is not noticeable to the listener. Given the fundamental premise of waveform coding, which is accurate reproduction of the original analog signal, the logical figure of merit for a candidate encoding and decoding technique is signal-to-noise ratio (SNR) in the reconstructed waveform. Waveform-coding techniques can be implemented with relatively little complexity; thus encoder–decoders, commonly called *codecs*, are relatively inexpensive and consume little power. Also encoding delay is typically very low, which is very attractive for application in a digital voice service. It should be clear from this description that waveform coding would find useful application to many other types of signals, and in fact waveform codecs are used for digitization of a wide variety of signals, including telephone signaling tones, modem line signals, imagery, and telemetry signals [Jay84].

Perhaps the simplest and best-known waveform-encoding technique is *pulse code modulation* (PCM), a time-domain technique in which an analog signal is sampled at the Nyquist rate (2 times bandwidth); each sample is quantized and transmitted as a binary number. Other examples of time-domain waveform coding include differential PCM (DPCM), delta modulation (DM), and adaptive predictive coding (APC). We shall say more about time-domain waveform coding in later paragraphs.

Examples of frequency-domain waveform coding include sub-band coding (SBC) and adaptive transform coding (ATC). With SBC [Cro76, Cro77], the speech signal is filtered into contiguous nonoverlapping sub-bands and each sub-band is then encoded independently using time-domain techniques. A different number of bits is used in encoding each sub-band, with the bits assigned according to the perceptual value of information in different sub-bands. Special filtering techniques must be used to ensure that the aliasing error at each sub-band edge is canceled by the aliasing error of the adjacent sub-band. Detailed treatments of SBC can be found in the references cited above as well as in [Jay84]. The SBC technique has been considered for use on mobile radio channels in several investigations, including [Cox91] and [Ste92]. Several versions of SBC were among the prime candidates considered for inclusion in the Pan-European GSM digital cellular standard [Ste92]. A related frequency-domain technique is ATC, in which a block of speech samples is transformed into a corresponding set of frequency-domain coefficients [Zel77] using, for example, the cosine transform [Ahm74]. As in the SBC technique, different numbers of bits are assigned to the ATC frequency-domain coefficients.

It is generally accepted that waveform coders cannot produce high-quality reconstructed speech at transmission rates below about 16 kbits/sec. At the lower rates, more effective techniques are those referred to by the name *model-based speech coding* or the more general term *source modeling*. This category comprises a wide variety of techniques based on a model for the production of speech sounds in the human vocal tract. Model-based speech coders are among the earliest types of speech compression systems and have long been known by the name *vocoder* [Cro83]. With all model-based techniques, signal processing algorithms are used to extract from the analog speech waveform certain parameters that correspond to time-varying physical parameters of the speech production mechanism, and these extracted parameters are transmitted as indirect representations of the speech sounds. At the receiving end, the parameters are used to "drive" an algorithmic representation of the speech generation model, thereby generating an approximation to the speech waveform. The underlying concept here is that if the transmission rate is constrained to rates too low for the effective use of waveform coding, the available bits can be made much more effective by utilizing them in conjunction with knowledge of the speech production process. While waveform-coder performance is usually evaluated using output SNR as a figure of merit, model-based vocoders must be evaluated using more subjective criteria. The reason for this is that there is no SNR metric which correlates well with the differences in speech quality perceived with different vocoders. Thus in order to compare the performance of different vocoders, standardized tests are used to provide indicators of the subjective intelligibility and quality of the speech produced by each vocoder [Gib93]. The term *intelligibility* refers to whether spoken words

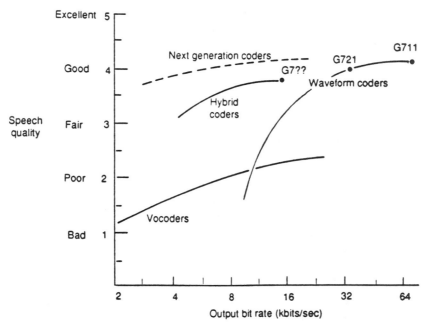

Figure 8.2 Hypothetical quality versus bit rate for various speech-coding techniques. (Adapted from [Mac91] © Peregrinus Ltd, with permission.)

and phrases are easily understood. The term *quality* usually refers to the "naturalness" of the speech, as perceived by the listener. The commonly used measures of intelligibility and quality include the Mean Opinion Score (MOS) [Jay84], the Diagnostic Rhyme Test (DRT) [Voi77a], and the Diagnostic Acceptability Measure (DAM) [Voi77b].

Figure 8.2, adapted from [Mac91], compares a selection of voice-coding techniques using a hypothetical quality measure plotted against transmitted bit rate. At the highest transmission rates, the best performance is provided by waveform coders; and, in fact, 32- and 64-kbit/sec coders, based on refinements of PCM, are widely used in the public telephone network. The voice coders adopted for the CT2 and DECT standards are 32-kbit/sec waveform coders. At the lowest rates, model-based vocoders are clearly preferred, and the voice coders adopted for the emerging digital cellular and digital mobile radio systems are all model-based vocoders operating at rates of 13 kbits/sec and lower (see Table 8.1). The parametric and hybrid techniques referred to in the figure are a number of techniques which combine waveform coding with model-based coding and are shown to provide better speech quality in a certain range of transmission rates than either technique alone can provide. We shall not delve into hybrid techniques, but refer the interested reader to [Cro83] and [Fla79], where they are discussed.

PCM, DPCM, and ADPCM Waveform Coding. Here we will briefly describe PCM and its principal variations, including differential PCM (DPCM) and adaptive DPCM (ADPCM) techniques which are widely used for digitization of speech signals in both telephone and wireless networks. These topics are treated in detail in a number of standard communications texts, including [Jay84], [Pro89], and [Bel83]. Our description follows [Pro89].

Let us consider an analog signal $x(t)$ band-limited to bandwidth W, which we sample at the Nyquist rate $1/T_s = 2W$, where T_s is the sampling interval. In PCM each signal sample x_n is quantized to one of 2^b amplitude levels, where b is the number of bits used to represent each sample. The transmitted bit rate from the PCM encoder is therefore b/T_s bits/sec.

Speech signals have the characteristic that small signal amplitudes occur more frequently than large amplitudes. However, the uniform quantizer assumed above provides the same spacing between successive levels over the entire amplitude range of the signal. A better quantizer is one that has more closely spaced levels at low signal amplitudes and more widely spaced levels at large amplitudes. The nonuniform quantization technique yields a mean-square quantization error smaller than is achievable with uniform quantization. In practice, a nonuniform quantizer characteristic is obtained by passing the input signal through a nonlinear device that compresses the signal amplitude, followed by a uniform quantizer and binary encoder. At the receiving end, the binary decoding is performed, and the inverse nonlinear function is used to expand and recover the speech signal. The combined compressor–expander pair is referred to as a *compandor*. A version of PCM called *logarithmically companded PCM* or *log-PCM*, operating at 64 kbits/sec (8000 samples/sec ×8 bits/sample), is used throughout the public telephone network for digitization of voiceband signals. Studies of transmission of log-PCM signals over mobile radio channels are described in [Sun87] and other references cited therein.

In PCM each sample of the input signal is quantized and encoded independently of all other samples. However, most speech signals exhibit considerable correlation from one sample to the next, due to both (a) the quasi-periodic vibrations of the vocal cords and the constrained movements of components of the vocal tract and (b) the inherent structure of language. In other words, the average change in amplitude between successive samples is relatively small. Consequently an encoding technique that represents the signal in terms of appropriate sample differences can achieve a reduction in quantization noise or, equivalently, a reduced bit rate for a given requirement on quantization noise. This can be appreciated qualitatively by considering a case in which the successive samples are highly correlated. For this case let us assume that the encoder first takes successive adjacent sample differences $\{x(n) - x(n - 1)\}$ and then quantizes each difference value for transmission. The decoder will then construct an approximation to the original signal by integrating over the stream of quantized differences. Given that the original signal samples are highly correlated, the variance of the quantizer input is much smaller than that of the encoder input samples $\{x(n)\}$, and this leads directly to a reduction in the quantization noise power for a given value of b.

A straightforward extension of this basic idea is to predict the current sample based on a weighted sum of the previous p samples. To be specific, let x_n denote

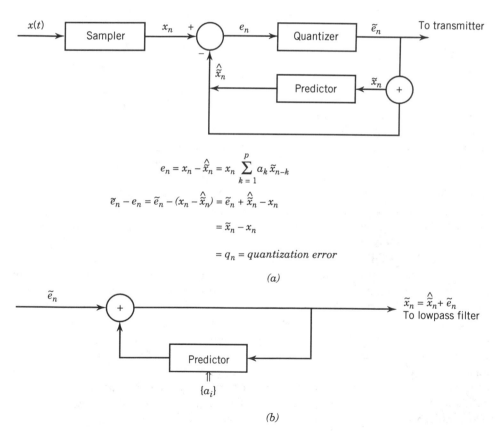

$$e_n = x_n - \hat{\tilde{x}}_n = x_n \sum_{k=1}^{p} a_k \tilde{x}_{n-k}$$

$$\tilde{e}_n - e_n = \tilde{e}_n - (x_n - \hat{\tilde{x}}_n) = \tilde{e}_n + \hat{\tilde{x}}_n - x_n$$

$$= \tilde{x}_n - x_n$$

$$= q_n = \text{quantization error}$$

(a)

$$\tilde{x}_n = \hat{\tilde{x}}_n + \tilde{e}_n$$
To lowpass filter

(b)

Figure 8.3 Differential pulse code modulation (DPCM). (a) Encoder. (b) Decoder. (From [Pro89] © McGraw-Hill, with permission.)

the current sample of the input speech signal and let $\hat{x}_n$ denote the predicted value of x_n, defined as

$$\hat{x}_n = \sum_{i=1}^{p} a_i x_{n-i}$$

Thus $\hat{x}_n$ is a weighted linear combination of the previous p samples and the $\{a_i\}$ are called the *predictor coefficients*. The predictor coefficients are selected so as to minimize the mean-square error between x_n and $\hat{x}_n$. The details of this minimization problem are discussed in a number of texts, including [Jay84] and [Pro89]. This technique is a simple example of *linear prediction*, a concept which underlies many of the commonly used source-coding and signal estimation techniques. A practical implementation of this technique, called *differential pulse code modulation* (DPCM), is shown in Fig. 8.3. In this configuration the linear predictor in the encoder, Fig. 8.3a, does not operate directly on the input speech signal but instead is placed in a feedback loop around the quantizer. The input to the predictor, denoted by $\tilde{x}_n$, represents the signal sample x_n modified by the quantization process, while the output of the predictor is given by

$$\hat{\tilde{x}}_n = \sum_{i=1}^{p} a_i \tilde{x}_{n-i}$$

The difference

$$e_n = x_n - \hat{\tilde{x}}_n$$

is the input to the quantizer and $\tilde{e}_n$ denotes the output. Each value of the quantized prediction error $\tilde{e}_n$ is encoded for transmission. The quantized error $\tilde{e}_n$ is also added to the predicted value $\hat{\tilde{x}}_n$ to yield $\tilde{x}_n$. At the receiving end, Fig. 8.3b, the same predictor is used again and its output $\hat{\tilde{x}}_n$ is added to $\tilde{e}_n$ to yield $\tilde{x}_n$. The signal $\tilde{x}_n$ is both the input to the predictor and the desired output sequence from which the recovered signal $\tilde{x}(t)$ is obtained by filtering. The feedback configuration for the DPCM encoder has the important advantage of eliminating the accumulation of quantization errors in the decoder. This point is explained in the references cited earlier.

The PCM and DPCM techniques described above are based on the assumption that the input signal is stationary—that is, that the variance and autocorrelation function of the signal are constant with time. However, many real signal sources, including speech signals, are only quasi-stationary and have statistics which vary slowly with time. Consequently, the efficiency of these encoders can be improved by having them adapt to the time-varying signal statistics. Adaptive versions of DPCM vary the quantizer or the predictor, or both. The quantizer is made adaptive by relating the quantization steps to an estimate of the signal power. The predictor is made adaptive by changing the prediction coefficients in accordance with the changing spectrum of the input signal. In some adaptive speech coders, two predictors are used, one to follow the variations in the vocal tract, the other to follow the pitch of vocal cord vibrations. When adaptive quantizers or predictors are employed, the technique is called *adaptive DPCM* (ADPCM). A 32-kbit/sec

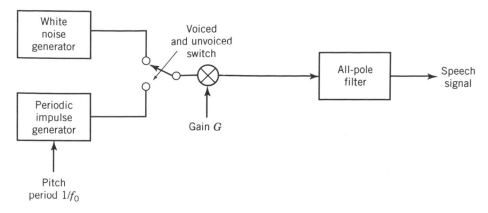

Figure 8.4 A simple model of speech production. ([Cro83] © IEEE.)

ADPCM codec has been standardized by the CCITT as Recommendation G.721 [Dec88]. This CCITT ADPCM scheme is specified in the CT2 and DECT wireless standards in Europe.

There are several other widely used waveform coding techniques that are closely related to DPCM and ADPCM. Delta modulation (DM) can be described as a simplified form of DPCM in which a two-level (1-bit) quantizer is used in conjunction with a fixed first-order predictor. Various forms of adaptive DM have been implemented; and one of these techniques, *continuously variable slope delta modulation* (CVSD), is widely used as a speech codec. CVSD at 16 and 32 kbits/sec is widely used in many military wireline and radio communication systems, and 12-kbit/sec CVSD is used in some digital land-mobile radio systems.

Model-Based Speech-Coding Techniques. As their name implies, model-based speech-coding techniques are based upon a simple representation or model of the speech production process. This model, shown in Fig. 8.4, represents speech as being generated by either of two acoustic sources, one a periodic pulse source and the other a noise source. The pulse source is intended to represent *voiced* sounds, whereas the noise source represents *unvoiced* sounds. Voiced sounds (e.g., clear vowel sounds) occur when the vocal cords are vibrating at an essentially periodic rate with a fundamental frequency f_0 or pitch period $1/f_0$. Unvoiced sounds are generated by forming a constriction somewhere in the vocal tract (usually in the forward part of the mouth), thereby causing turbulence and thus a random source of acoustic noise. In this model, one source at a time excites a filter that represents the vocal tract and provides the speech signal. This two-source model is a very simplified representation of the speech generation process, and more complete models have been described in the literature. However, most of the model-based speech-coding techniques which have been developed are based upon the simple two-source model. The reader is referred to [Rab78] for a detailed treatment of this subject.

Depending upon the observation interval, the speech signal can be described as either locally stationary on nonstationary [Cro83, Fla79]. Over intervals of a second or more, there is considerable variation in the speech signal, due to the sequential occurrence of voiced and unvoiced sounds, silent intervals, and variations in the speech amplitude. However, over intervals in the range of 20–40 msec, the speech signal is locally stationary. In such short intervals, the speech signal can be accurately described as voiced or unvoiced and will have a well-defined spectrum. Because of this, model-based voice coders are based upon the processing of *frames* of the sampled voice signal, with the frame duration usually chosen to be in the range 20–40 msec.

Essentially all the model-based voice coders which have been studied in recent years are based upon some form of *linear predictive coding* (LPC), which we shall describe briefly. Linear prediction methods permit the representation of speech signals in terms of a small number of slowly time-varying parameters. We discussed a common application of linear prediction in our description of differential PCM earlier in this subsection.

In linear predictive coding of speech, we begin by assembling a frame of signal samples taken at the Nyquist rate. A set of predictor coefficients is then found by minimizing the mean-square error between the actual and predicted values of the speech samples. Determination of the predictor coefficients is generally the most computationally burdensome part of the LPC analysis, and a great deal of work has been devoted to devising efficient algorithms for doing this [Rab78, Gio85]. Additional processing of the speech samples provides classification of the speech segment as voiced or unvoiced, and if voiced, the pitch period is estimated. An additional parameter required in LPC analysis is the root mean square (rms) level or *gain* of the error signal. The resulting set of parameters—the predictor coefficients, the pitch period, a binary indicator of voiced or unvoiced excitation, and the gain of the error signal—are assembled into a data frame and transmitted. At the receiver, the transmitted parameters are used to synthesize the predictive filter, which is then excited by a stream of pitch pulses or by pseudorandom noise, in accordance with the voiced/unvoiced indicator.

Because of the time-varying nature of speech, it is necessary to update the parameters frequently. Typically the 20- to 40-msec speech frame is divided into several subframes with interval of about 5 to 10 msec, and the parameters are updated in each subframe. In almost all applications of LPC to speech coding, the vocal-tract filter is represented by an all-poles filter of order p between 4 and 10. The assumption of an all-poles filter incurs less computational complexity than would be required for a more general filter model having both poles and zeros. Furthermore, the all-poles model is very well suited to modeling the acoustic characteristics of the vocal tract, which result in most of the signal energy being concentrated several resonance frequencies called *formants*. Human speech typically exhibits 3–6 formants in the 0- to 4-kHz bandwidth of interest, and two real-valued predictor coefficients are required to represent each formant. Thus 6–10 predictor coefficients are appropriate for an accurate description of the vocal-tract filter. A 10th-order 2400-bit/sec LPC vocoder, known as LPC-10 [Ata79, Tre82], is established as Federal Standard FS-1015 and is widely used in government-standard secure telephone terminals.

One of the characteristics of LPC-processed speech is that some of the derived parameters are more sensitive than others to transmission errors in the encoded bits. Operational experience with the standard LPC-10 algorithm, for example, has shown that the intelligibility and quality of the delivered speech are particularly sensitive to errors in the pitch values. In view of this problem of error sensitivity, LPC-based speech coding designs generally include error-correction coding for certain of the model parameters. This issue becomes very important in the application of coded speech to wireless channels, where channel error rates can be quite high and errors tend to occur in bursts. For this reason, all of the emerging wireless standards for digital voice service include error-control coding of transmitted voice bits.

LPC has proved to be a highly effective means of coding speech signals where very low transmission rates, say 2400 bits/sec and lower, are needed. However, the fidelity provided by basic LPC vocoding is generally regarded as not satisfactory for widespread commercial application, and thus many refinements of LPC, requiring higher transmission rates, have been developed. Next we briefly describe two schemes—multipulse excitation (MPE) and regular-pulse excitation (RPE)—which are direct extensions of the basic LPC model. In the MPE and RPE schemes, the simple two-source representation of voice excitation is replaced by a more general model in which a specific number of pulses is assumed to occur in short subframe interval of about 5–10 msec. No explicit distinction is made between voiced and unvoiced segments, and instead least-square estimation techniques are then used to determine the best pulse pattern for each subframe. With MPE [Ata82] the pulse positions and amplitudes are determined one at a time in a recursive procedure, and the resulting pulses are in general not uniformly spaced. In RPE [Kro86], as the name implies, the pulses are assumed to be uniformly spaced by a specified distance in the speech segment, and an estimation algorithm is used to determine the best choice of first-pulse position and the individual pulse amplitudes.

A version of RPE, called RPE-LTP, was adopted for use in the full-rate GSM cellular system. The designation LTP refers to the *long-term predictor*, or pitch predictor, which considerably enhances the performance of the basic RPE scheme. The GSM full-rate vocoder was selected after extensive comparative testing of six candidate vocoders originally submitted for evaluation [Nat88]. In the course of the evaluation, two systems were withdrawn and the final stage of evaluation considered two sub-band coders, one MPE-LTP coder, and one RPE coder. In this final evaluation, no single coder was found to be superior in all respects. However, the RPE coder showed best speech quality and had acceptable complexity and delay, and thus was singled out for further improvement. This led to the RPE-LTP scheme, which was ultimately selected for the standard. The vocoder operates at 13 kbits/sec and utilizes a speech frame of 20-sec duration [GSM89].

Given the importance of achieving the greatest possible bandwidth efficiency in the new wireless networks, there has been much research activity directed toward designing voice coders that will deliver high-quality speech with data rates below about 8 kbits/sec. The MPE and RPE coders outlined in the previous paragraphs provide good speech quality at bit rates down to about 9.6 kbits/sec, but are generally considered to give unacceptable quality at lower rates, because lower rates cannot support the information needed to accurately represent the optimized pulse patterns. This led to the development of a technique called *code-excited linear predictive* (CELP) coding, first introduced in 1984 [Ata84, Sch85]. Broadly stated, CELP coding is based on yet another approach to selecting the form of the excitation for the linear predictive filter. A block diagram of a CELP encoder is

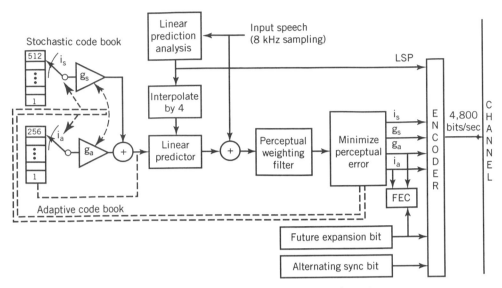

Figure 8.5 Block diagram of a codebook-excited linear predictive (CELP) speech coder [FS 1016].

shown in Fig. 8.5. The CELP encoder is an example of the *analysis-by-synthesis* technique, and the speech decoding or synthesis procedure, shown within the figure, is an integral part of the encoder. The reconstructed speech is produced by passing the excitation signal through a filter defined with short-term predictor coefficients. The parameters of this filter are determined from processing the input speech signal and are updated at regular intervals. As the figure shows, the excitation signal is formed as a weighted sum of vectors drawn from two codebooks, one adaptive and one fixed. The adaptive codebook represents the periodicity of the speech signal and actually follows the speaker's pitch in voiced sounds. The fixed codebook, also called the *stochastic codebook*, contains vectors that are characteristic of the speech signals. The vectors can be determined in a variety of ways. In some CELP implementations, the vectors are generated as noise-like sequences, but vectors with more deterministic structure can have computational advantages [Cam91]. The filter output is compared with the actual speech signal to produce an error signal which is then passed through a *perceptual weighting filter* designed to emulate the way in which the human ear perceives speech. Minimization of the weighted error is then used as the criterion for optimum selection of code vectors and their weights. The computational requirements for CELP encoding, which are dominated by the codebook searches, are considerable. Computations can be reduced by constraining the search to a subset of either codebook, at the expense of speech quality, and this becomes an implementation tradeoff issue. Since its introduction, the CELP technique has been the subject of a great deal of research, much of it devoted to reducing the computational complexity of the technique while preserving speech quality [Kro92].

A 4800-bit/sec CELP coder has been standardized as Federal Standard 1016 [Cam91]. The FS-1016 coder specifies a 10th-order linear prediction filter. The fixed codebook contains 512 codewords, whereas the adaptive codebook has 256 codewords. The encoder operates at an 8-kHz sample rate and a 30-msec frame size with four 7.5-msec subframes. The adaptive codebook is updated with every new subframe. The FS-1016 standard has been implemented in government standard secure telephones and is being considered for application in military radio communication systems.

A technique closely related to CELP, called *vector-sum excitation linear prediction* (VSELP), provides an efficient means of forming the excitation signal [Ger90]. In this technique, the excitation is formed as a linear combination of basis vectors weighted with values ± 1. This approach yields a significant reduction in the amount of computation required to determine the optimum excitation vector. An 8-kbit/sec VSELP coder was selected as the full-rate voice coder for the IS-54 TDMA digital cellular standard [TIA92]. In the IS-54 specification, the VSELP-coded bits in each 20-msec frame are designated as either Class I or Class II bits. The Class I bits are those bits judged to be most critical to the preservation of speech quality, and they are encoded for transmission with a rate-1/2 binary convolutional FEC code. The Class II bits are transmitted without FEC coding. The inclusion of FEC coding raises the transmission rate from 8 to 13 kbits/sec.

Before leaving the topic of speech processing, we briefly describe another model-based voice-coding technique which has been adopted for several wireless applications, called *improved multiband excitation* (IMBE) [Har89]. The IMBE technique, which is based upon the multiband excitation speech generation model, was first described in 1984 [Gri85]. As with traditional LPC, the IMBE operates on speech segments short enough, (e.g., 20 msec) that the spectral characteristics are essentially stationary over each segment. However, while traditional LPC classifies each segment as either entirely voiced or entirely unvoiced, the IMBE system divides the excitation spectrum into a number of nonoverlapping sub-bands and makes a voice versus unvoiced decision on each sub-band. This technique allows the excitation signal for any speech segment to be a mixture of periodic and nonperiodic signals. A real-time implementation of the IMBE speech coder is described in [Bra90]. The IMBE system has been investigated extensively for application in various wireless systems. A 6.4-kbit/sec IMBE coder was adopted as the voice coding standard for the INMARSAT-M and AUSSAT mobile satellite communication systems. This IMBE system uses a 20-msec speech frame, so that there are a total of 128 bits per frame, of which 45 bits (2250 bits/sec) are used for FEC coding using Golay and Hamming block codes [Har91]. Prior to transmission, the 128 bits contained in a speech segment are interleaved in such a way that at least five bits separate any two bits from the same codeword. Another version of the IMBE system has been incorporated into the GTE Genstar airliner-to-ground telephone system. In that implementation the voice coder operates at 4.8 kbits/sec, and 1.7 kbits/sec of FEC coding brings the aggregate transmission rate to 6.5 kbits/sec. A third version of IMBE is specified in the proposed APCO-25 standard for narrowband digital land mobile radio [TIA93a]. There, the voice coder operates at 4.4 kbits/sec, and 2.8 kbits/sec of FEC coding brings the transmission rate to 7.2 kbits/sec.

8.2.2 Image Coding

In existing wireless networks, the provision of image transmission is limited to facsimile transmission between portable fax terminals and standard office fax machines. However, as wireless networks become more widely deployed, and as the demand for wireless services continues to grow, the need will inevitably arise for a variety of wireless imagery and video transmission capabilities similar to those which are becoming available in the existing public switched network and in wired office environments. In the planning of new mobile data and digital cellular networks, needs are arising for transmission of multitone still images and slow-scan or freeze-frame video images. The emerging vision of universal personal communications includes the concept of a portable multimedia terminal incorporating video services [She92]. In the office environment, as wireless local area networks (WLANs) come into more widespread use, users will eventually expect to have access to such services as video conferencing, already becoming available in many business enterprises. Therefore it is of interest to examine the issues involved in migrating imagery and video transmission to the wireless environment.

Issues for the Wireless Environment. Key issues which arise here are (1) data rate requirements, (2) complexity and power requirements for portable terminals, and (3) impact of wireless channel characteristics. Even a cursory examination of data transmission requirements for imagery will reveal that there are wide discrepancies between requirements and network capabilities for some networks. The least demanding requirement is for transmission of still images, and this requirement can reasonably be met by many wireless systems. For example, a single VGA-resolution 640- $\times$ 480-pixel full-color image with an uncompressed format of 24 bits/pixel (8 bits each for red, green, and blue color components) creates a file of over 700 kbits, requiring almost 13 minutes of transmission time at 9600 bits/sec. Here we choose 9600 bits/sec as a data rate representative of the capabilities of cellular and mobile data systems. Clearly, data compression is in order here, and, fortunately, techniques have been developed which will encode a still image of this size and deliver good picture quality at 0.25 bits/pixel and excellent quality at 1 bit/pixel [Jay92], reducing the transmission time in the above example to 8 or 2 seconds, respectively. This would certainly be regarded as an acceptable grade of service for most users. As we examine higher data rate networks, such as WLANs with data rate capabilities of a few Mbits/sec and the newly developing PCS networks which will provide rates of tens of Mbits/sec, it becomes reasonable to consider real-time video transmission. The current state of compression technology is such that full-motion digital video can require data rates exceeding 2 Mbits/sec. Thus it is clear that spectrum utilization is major concern if a network is to serve multiple simultaneous users with full-motion video.

The issues of implementation complexity and power consumption are especially important in the design of portable multimedia terminals, and the image processing functions have the potential of dominating the signal processing requirements, given the high data rates involved. Therefore it is important to utilize processing techniques that can be realized in low-complexity, low-power VLSI implementations.

The characteristics of wireless channels present the same issue here as they do in the operation of any data service over wireless networks—that is, the relatively high error rates experienced in comparison with transmission in wired networks. However, the problem is made more severe for the transmission of compressed imagery. In order to achieve highly efficient image compression, many systems use *variable-rate coding* techniques such as *entropy coding* and *run-length coding* [Rab78]. Variable-rate coding techniques provide much better compression ratios than do fixed-rate techniques, but are degraded more severely by channel errors. Error control can be applied using FEC coding, but the coding redundancy expands the data rate unnecessarily under benign conditions and may still not be able to overcome the effects of errors under severe channel degradation. Thus the wireless environment presents a difficult tradeoff between the need to constrain transmission rates and the need to provide acceptable picture quality in the presence of channel errors.

Next we briefly describe the most widely used image-coding techniques, including techniques which have been established as standards.

Image- and Video-Coding Standards. Many of the coding techniques that we described in relation to speech coding can be applied to image processing as well. For example, the PCM, DPCM, and ADPCM waveform-coding techniques can be adapted to the processing of two-dimensional images. In the simplest scheme, PCM, the sampled image is quantized and encoded one pixel at a time. It is generally found that to provide decoded images of acceptable quality, PCM requires 6–8 bits/pixel. (For a full-color image we must triple the number of bits.) However, DPCM and ADPCM can provide the same quality with fewer bits per pixel by exploiting the spatial correlation of adjacent pixel values, just as these techniques exploit the temporal correlation between speech samples. With DPCM, good picture quality can be achieved with 2–3 bits/pixel, whereas with ADPCM, good quality can be achieved with as little as 1 bit/pixel [Mod81, Dau83, Gio85]. The model-based LPC approach discussed earlier has also been applied to image coding [Rab78].

The image-coding techniques which have enjoyed widest use are sub-band coding and transform coding for compression of both still and moving images [Jay92]. Sub-band and transform coding have both been used effectively in speech processing, and they can be carried over to the compression of two-dimensional images with our accustomed concept of frequency being interpreted as *spatial frequency* in the processing of an image. Compression or redundancy removal is accomplished by first representing the image in the frequency or transform domain and observing that components in different regions of the frequency or transform domain have different variances or energy levels. The image features most discernible to the eye lie in low spatial frequencies and have the greatest variance. The higher spatial frequencies, which represent edges and fine details, have the lowest variance. Then, by allocating more bits to components of larger variance, and fewer bits to components of smaller variance, the overall mean-square error can be minimized for a given constraint on total bit rate.

An especially useful transform for image coding is the two-dimensional discrete cosine transform (2-D DCT) [Ahm74, Jay84]. The DCT is an orthonormal transform having certain properties of near-optimality for representation of most images (and speech signals). This transform has the further important advantage that fast algorithms have been developed for its implementation. We shall not delve any further into the properties of the DCT, but shall simply say that the 2-D DCT provides an effective and computationally efficient means of decomposing an image into a spatial-frequency representation. The 2-D DCT is at the heart of several image- and video-coding standards, which we shall describe below.

In discussing the compression of still and moving images, we distinguish between *intraframe coding* and *interframe coding*. In intraframe coding, as the name implies, redundancy is removed from a single image frame by exploiting the spatial correlation within that frame. Of course, a moving image can be compressed by only applying intraframe coding separately to each successive frame. However, much greater compression of moving images is achieved by exploiting the similarity between successive frames, and this is termed *interframe coding*. There are two types of interframe coding: predictive and interpolative. Predictive coding of image frames is much like predictive coding of speech signals. First one picture frame is encoded using intraframe techniques, and then the differences between the reference frame and successive frames are encoded. To prevent error propagation, the process is periodically restarted with a new reference frame. With interpolative coding, reference frames are again used, but some frames between reference frames are simply not transmitted and are restored during decompression by interpolating between reference frames. We shall next briefly describe three international standards for compression of still and moving images. The three standards are commonly known as *JPEG* (Joint Photographic Experts Group) [ISO92], *MPEG* (Moving Pictures Experts Group) [ISO91], and *CCITT H.261*, commonly referred to as $p*64$ [CCI91].

The JPEG standard is the simplest of the three standards and is intended for compression of still images, though it is sometimes applied frame by frame to video images. With JPEG the image is converted to a spatial frequency representation using the 2-D DCT technique, and different transform components are quantized with different granularity, in accordance with the spatial-frequency sensitivity of the human visual system. The 2-D DCT operates on 8×8 blocks of 64 pixels, with spatial frequency ranging from "dc" $(0, 0)$ to a maximum frequency $(7, 7)$, and the spatial frequency components are quantized and coded by *entropy coding* [Wal91]. Entropy coding represents more frequently occurring values with short code sequences and less frequent values with longer code sequences. The well-known *Huffman coding* technique is an example of entropy coding [Jay84]. JPEG achieves compression ratios of about 10:1 with little or no loss of picture quality. The JPEG standard accommodates image sizes as large as $65,536 \times 65,536$ pixels.

The MPEG standard is intended for addressable video in multimedia applications, and its encoder is significantly more complex than the JPEG encoder. The MPEG encoder operates on groups of frames rather than on individual frames. Each group typically begins with a reference frame compressed by intraframe coding as in JPEG. The encoder then uses motion-compensated prediction of successive image frames and applies 2-D DCT and entropy coding to the difference between each prediction and the actual frame [LeG91]. MPEG then creates a third type of frame by calculating macroblock motion between reference and predicted frames, using both forward and backward motion compensation. As a consequence, MPEG uses a considerable amount of buffering, and the frames in a group are transmitted in an order different from the display order. The MPEG standard limits images to 4095×4095 pixels and recommends a "constrained parameters" image of 768 times 768 pixels or smaller. The interframe coding in

MPEG provides 10–100 times the compression achievable with transform coding alone. MPEG-coded maximum size video images require a transmission rate of about 100 Mbits/sec, while the constrained-parameter option requires about 2 Mbits/sec.

The CCITT H.261 standard was developed for "video telephony" and is called $p*64$ ($p = 1$–24) because it operates with data rates that are multiples of the 64 kbits/sec transmission rate of a basic digital telephone channel. Consequently $p*64$ operates with smaller image sizes (352×288 or 176×144 pixels) than does either JPEG or MPEG. The $p*64$ encoder is much like the MPEG encoder, using prediction and (optional) motion estimation to remove the frame-to-frame redundancy in moving images. However, in order to minimize processing and buffering delay for real-time video telephony, the encoder operates on 16×16 pixel blocks instead of entire image frames. The $p*64$ encoder first encodes a frame using only intraframe transform coding and transmits that frame as a reference frame. It then compares each block in the next frame with the version transmitted in the previous frame. If sufficient change is observed, the encoder transform-codes and transmits the difference between the new block and the previous block. If the change is not sufficient, the block is not transmitted. The definition of "sufficient change" can be varied from block to block [Lio91].

Future Directions. The three compression standards described above are all widely used, and a number of proprietary algorithms, available through licensing agreements, have also been developed. The particular technique one chooses depends upon the application, the image quality required, and the bandwidth available for transmission of the coded images. However, it is important to note that essentially all application of image compression to date have been for low-BER environments. In the wireless environment, one must closely examine the effects of channel errors on the quality of the reconstructed images. Generally speaking, the coding techniques offering the greatest amounts of compression are the most vulnerable to errors. In the three examples we described above, the JPEG technique is the least vulnerable to errors, because image frames are encoded individually. However, MPEG-encoded video images are highly vulnerable to channel errors, due to the extensive use of interframe coding, which is susceptible to error accumulation upon decoding. Of course, error-control channel coding can be applied to the compressed data streams, introducing structured redundancy which somewhat offsets the redundancy removal achieved by compression. However, research is being done to find other image compression techniques which are less error-prone than the techniques described above. We shall briefly describe on such approach.

One category of image-coding technique which is being investigated specifically for application in wireless environments is *vector quantization* (VQ). The VQ technique has been employed very effectively in a number of speech and image processing systems [Ger83, Jay92], though its encoding requirements are sometimes regarded as overly complex. The basic concept of VQ coding is to group image samples into a vector and then quantize the vector using a codebook [Jay84]. At the encoder, a vector of image samples is formed, say a 4×4 block giving 16 bytes. The vector is then compared against all the vectors in the codebook with the goal of minimizing the distortion between the input image vector and its representative codeword. The codeword providing the closest match is transmitted. If, let us say, the codebook contains $256 (= 2^8)$ codewords, an index consisting of a single byte is transmitted, for a 16:1 compression ratio. The decoder is very simple, because a simple table look-up is all that is needed to retrieve the image vector. The problems of designing VQ codebooks and the distortion measure have been treated extensively in the literature, and algorithms have been proposed that will reduce encoder complexity with little loss in image quality. An important advantage for VQ coding in the wireless environment is that, in its basic form, it can be provide significant compression ratios while encoding image frames individually. A recent paper addresses the problem of video compression for wireless communications and proposes an approach based on a variant of VQ coding that provides good compression efficiency and error recovery with low-power implementations [Men94].

8.3 BASIC SIGNAL PROCESSING

The most basic signal processing functions in a modem are power control, timing, and pulse shaping. Here we provide an overview of these operations.

8.3.1 Automatic Power Control

The traditional method of controlling received power is the technique known as *automatic gain control* (AGC). Most modern radio modems for wireless network applications also include transmitter power control. With any modem using multiple signal amplitudes (e.g., versions of QAM), some form of gain control is required as part of demodulation. This function is required because, at a minimum, received levels cannot in general be expected to be the same as the transmitted levels. Therefore some form of gain adjustment is required in order to establish correct decision threshold levels relative to the received signal levels. This problem is compounded in modems operating on radio channels, where there can be wide variations, including fading to depths of 30 dB, in received signal power. In practice, this is accomplished with an AGC circuit. We will say more about AGC in a later subsection.

Figure 8.6 represents an AGC for a digitally implemented modem. The received signal is passed through an anti-aliasing filter to eliminate out-of-band noise and then is sampled with an A/D converter. The samples are squared and then low-pass-filtered to provide an estimate of the power. The power is compared with a reference, and the difference is used to adjust the step size of the A/D converter. In this way, the range of received signal is adjusted in accordance with average received power, and the digital representation of the received signal remains essentially independent of power fluctuations in the channel.

Transmitter power control is the simplest method for counteracting the effects of fading. As the signal goes into a fade, the transmitted power is increased to compensate for the performance degradation due to the fading. With power control, the average transmitted power is reduced, which results in significant extension of battery life in a mobile terminal. In an interference-limited environment such as mobile radio, reduction in the average transmitted power will reduce

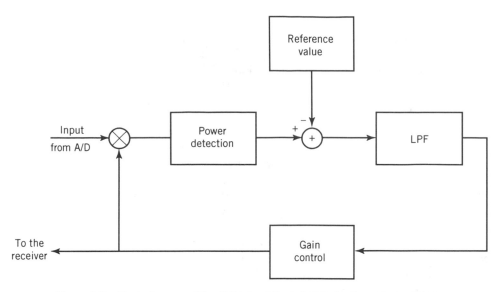

Figure 8.6 Block diagram of the AGC circuit for a digitally implemented modem.

the interference, which will in turn increase the number of simultaneous users in a service area. The capacity increase achieved with power control is more significant in code-division multiple access (CDMA) spread-spectrum communication systems. The common methods of power control are open-loop and closed-loop approaches. In the open loop scheme, the sum of the transmitted and received power in a two-way communication is always kept constant. Assuming a reciprocal channel and two-way communication, a weak received signal is an indicator of deep fading. An increase in the transmitted power compensates for the fading. With closed-loop power control the receiver informs the transmitter about the level of the received signal, which allows the transmitter to adjust its power accordingly. The open-loop method has a faster speed of convergence, but as a practical matter, not all channels are reciprocal. We will provide further discussion of power control in Chapters 9 and 11.

8.3.2 Timing Recovery

The most important ingredient in accomplishing accurate sampling in a digital demodulator is provision of a reference signal at the received channel symbol rate. This is done with various forms of timing recovery circuits that "learn" the symbol timing by processing the received signal. For coherent demodulation of the signals, the carrier frequency must also be acquired; this is ordinarily done with a phase-locked loop (PLL).

To achieve lower error rates, modems usually adopt coherent modulation techniques, which require a phase reference to be derived from the carrier. The easiest way to provide a reference phase is to transmit some part of the carrier, termed a *pilot tone*, and detect it at the receiver with a narrowband filter. However, this method wastes a portion of the transmitted signal power. To avoid this inefficiency, the received signal should be used for recovering the phase reference. A PLL determines the carrier phase and corrects it either before the decision directly on the received point in the signal constellation or at the demodulator. In the first approach the loop is shorter, which results in a faster tracking of the phase variations. Currently, the decision-aided phase-locked loop (DAPLL) technique is used in most modems. The principle of operation of a DAPLL is explained in [Pah88b].

In synchronous modems, proper recovery and tracking of symbol timing are critically important for proper operation. Several techniques are suggested and examined for this purpose in [Pah88b]. One approach is to adopt the minimum mean-square error at the equalizer output as the criterion for a decision aided algorithm for finding the optimum sampling phase [Kob71a, Kob71b]. This method generally yields a relatively complicated implementation and slow convergence; consequently, it has not been widely adopted.

A second approach is to use narrowband filters tuned to half the baud rate, followed by a square-law device and a bandpass filter [Fra74]. Analysis of the relationship of this method to the band-edge component maximization which yields the optimum sampler phase for infinite tap equalizers is given in [Lyo75] and [Maz75]. As compared with the method suggested earlier, this technique yields more economical implementation while giving comparable performance. Extracting timing from band-edge components of the received signal is widely used in modems employing standard baud rate equalizers.

Figure 8.7a shows a block diagram of the timing recovery circuit described above. The received signal is passed through two filters centered at $f_c - 1/(2T)$ and $f_c + 1/(2T)$; the outputs of the filters are multiplied, and the frequency component at $1/T$ is filtered out and used for extracting timing information. Figure 8.7b represents a digital implementation of the above circuit. The clock used for sampling is driven from the output of a long counter. The output of the filter at $1/T$ is quantized to one bit to cancel fluctuations of the amplitude due to the multilevel received symbols. The quantized output is integrated and compared with a reference value, and the difference is used to adjust the size of the counter. The size of the counter controls the sampling of the A/D converter.

As discussed earlier, as an alternative to coherent modulation and demodulation, one can use differentially coherent modulation and demodulation. In differentially coherent forms of modulation, it is not necessary that the carrier phase be known. This is because the data bits to be transmitted are mapped into successive differences in the transmitted channel symbols differential PSK, or DPSK. A direct extension of DPSK is differential PAM. In demodulating the DPSK signal or the differential PAM signal, it is not necessary to recover the carrier phase, and thus the PLL is not needed. Rather, each received symbol is used as the reference for the succeeding symbol. This simplifies the receiver design in systems (e.g., some radio systems) where it is difficult to reliably recover the received carrier phase. However, the simplification of the receiver design comes with a cost in performance. Differential detection will typically result in a performance loss of about 1–2 dB relative to coherent demodulation of the same signal. We also note that symbol timing recovery is still required for demodulation of the differentially modulated signals. DPSK and other forms of differential modulation have been used for a number of years in both wireline and radio modems. Most modems

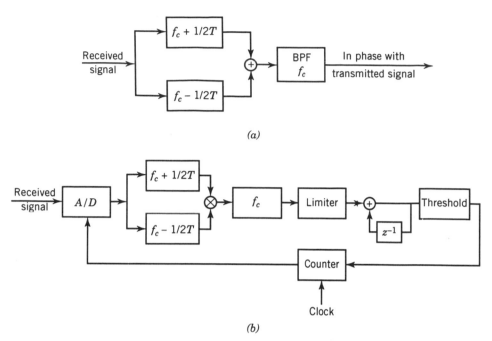

Figure 8.7 Block diagram of a band-edge timing recovery circuit: (*a*) Conceptual structure. (*b*) Digital implementation of the circuit.

currently in use in radio systems use differentially coherent forms of modulation. However, there have been important advances in recent years in the use of coherent modulation techniques on radio channels.

8.3.3 Pulse Shaping

Next we consider the choice of pulse shaping filter. If we were to choose an arbitrary shape for $f(t)$, then a stream of PAM pulses might overlap one another in such a way that a sample of an individual received pulse will be interfered with by many of the neighboring pulses. This effect is called *intersymbol interference* (ISI). Ideally we would like $f(t)$ to be rectangular over the pulse interval $(0, T)$. However, the sidelobes of the spectrum of this pulse shape, which has the functional form $(\sin f)/f$, decrease slowly with frequency, and this can lead to unacceptable signal interference in adjacent user channels. What is preferred is a pulse shape $f(t)$ that has its peak at time 0 and value zero at sampling times $T, 2T, 3T, \ldots$. Pulses having this characteristic are referred to as *Nyquist pulses*, and filters having impulse responses with this characteristic are called *Nyquist filters*. With transmission of Nyquist pulses at uniform intervals of T seconds, each received pulse can be sampled free of intersymbol interference from other transmitted pulses. A widely studied class of Nyquist filters is the class of *raised-cosine filters*. The frequency response of a raised-cosine filter has a flat amplitude portion and sinusoidal rolloff to zero. The raised-cosine rolloff characteristic is one that can be realized without difficulty in a practical design. Raised-cosine filters have been used extensively in modems designed for both wireline and radio systems.

The spectral characteristic of a raised-cosine filter is given by

$$\mathscr{F}\{f(t)\} = \begin{cases} T, & 0 \le |f| \le \dfrac{1-\beta}{2T} \\ \dfrac{T}{2}\left[1 - \sin\dfrac{\pi T}{\beta}\left(|f| - \dfrac{1}{2T}\right)\right], & \dfrac{1-\beta}{2T} \le |f| \le \dfrac{1+\beta}{2T} \end{cases} \quad (8.3.1)$$

where β is called the *rolloff factor* and can range between 0 and 1. The corresponding time domain Nyquist pulse is

$$f(t) = \frac{\sin \pi t/T}{\pi t/T} \times \frac{\cos \beta \pi t/T}{1 - 4\beta^2 t^2/T^2} \quad (8.3.2)$$

A few examples of raised-cosine frequency responses and their corresponding impulse responses are shown in Fig. 8.8 for selected values of the rolloff parameter β. As can be seen in the figure, small values of β yield the sharpest spectral rolloff characteristics, with $\beta = 0$ corresponding to a rectangular spectrum. The rolloff value $\beta = 1$ eliminates the flat portion of the spectrum and yields a pure raised-cosine shape. Though it is the frequency-domain characteristic which has the raised-cosine characteristic, the corresponding time-domain impulse responses are often called *raised-cosine pulses*.

Figure 8.9 [Feh87] shows oscilloscope displays of a measured power spectrum and so-called *eye patterns* or *eye diagrams* for symbol streams having two and four amplitude levels. The measured spectrum shows sidelobes 55 dB below the maximum value, which is considered a very good reduction of sidelobe level with Nyquist filtering. Eye patterns are readily produced on an oscilloscope by feeding the received pulse stream to the vertical input of the oscilloscope and a clock signal at the correct sampling rate to the external trigger. The horizontal time scale should be set approximately equal to the symbol duration. The display shows the superposition of many segments of the received pulse stream. The center of each eye, where the vertical extent of the eye opening is greatest, is an optimum sampling instant. A tight bundle of traces coming together at the top and bottom of each eye opening is an indication of very low intersymbol interference and highly reliable detection of pulse amplitudes when there is little or no noise present. An increasing level of intersymbol interference will be evident in a corresponding closing of the eye pattern.

In practice, a pair of matched filters are used to implement the raised cosine spectrum. The spectrum of the pulse-shaping matched filters used at the transmitter and the receiver is the square root of the raised cosine spectrum. The corresponding impulse response for individual filters is no longer represented by Eq. (8.3.2). This equation represents the overall impulse response used for analysis of the system. For design purposes the impulse response of the square root of the

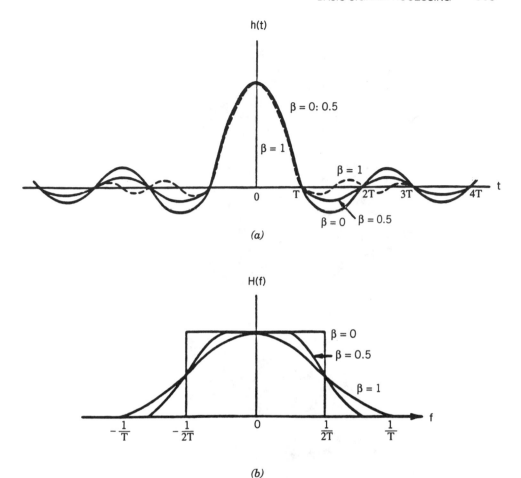

Figure 8.8 Time- and frequency-domain representations of raised cosine pulses.

raised cosine function is needed, which is given by

$$f_2(-t) = f_1(t) = \frac{\sin\left[\pi(1 - \beta)t\right] + 4\beta t \cos\left[\pi(1 + \beta)t\right]}{\pi\left[1 - (4\beta t)^2\right]t} \qquad (8.3.3)$$

The easiest method for digital implementation of this filter is to use the samples of this function as the discrete impulse response of a finite impulse response (FIR), discrete, pulse-shaping filter. To make the samples finite and to control the sidelobes of the spectrum, a window is applied to the time-domain representation. In voiceband modems a rolloff factor of around 0.2 with 20–30 taps is typical, and the sidelobes are designed to be more than 40 dB below the main lobe. For radio modems, analog filters with higher rolloff factors and lower sidelobes are used.

The maximum one symbol/sec/Hz for QAM modulation corresponds to the case of $\beta = 0$, which is not practically feasible. In order to preserve the same

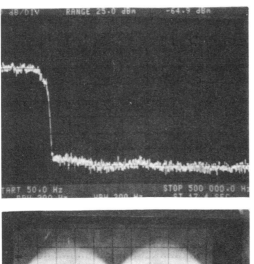

Measured power spectrum

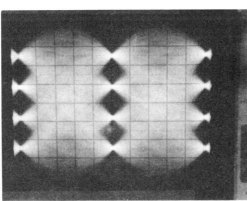

Two-level eye diagram

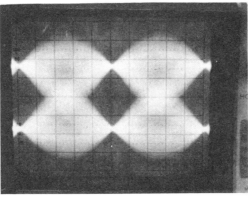

Four-level eye diagram

Figure 8.9 Oscilloscope displays of a measured power spectrum and eye patterns for symbol streams having two and four amplitude levels. (From [Feh87] © 1987 Prentice – Hall, reprinted with permission.)

bandwidth efficiency with a realizable filter, one may use partial response signaling. The channel impulse response of a partial response system has value one for two consecutive samples spaced by T seconds, and zero at all other samples. As a result, in the absence of channel distortions the received samples are $\{a_k + a_{k-1}\}$ rather than $\{a_k\}$. This can be viewed as known "forced" ISI, which creates no problem for detection, as long as the decisions are based on the last two received sampled pulses. The frequency-domain representation of the half cosine pulse-shaping filters used for pulse shaping in partial-response systems is given by

$$\mathscr{F}\{f(t)\} = \begin{cases} 2T\cos\pi Tf, & |f| \le 1/2T \\ 0, & |f| \ge 1/2T \end{cases} \tag{8.3.4a}$$

The time-domain representation for this waveform is

$$f(t) = \frac{\pi}{4}\frac{\cos\pi t/T}{1 - 4t^2/T^2} \tag{8.3.4b}$$

Figure 8.10 shows time- and frequency-domain representations of half cosine pulses used in partial response signaling. More extensive discussion of partial response signaling, with pertinent references, can be found in [Feh87, Pro89].

Several approaches have been used for the optimal design of digital pulse-shaping filters. Digital linear-phase FIR filters are discussed in [Mul73] with special attention to zero ISI and minimum stopband attenuation. An iterative technique using the steepest descent algorithm to design a pair of zero ISI matched filters with maximum spectral power in the passband is available in [Che82]. Other methods using linear programming [Sal82, Lia85] and modified Remez exchange algorithm [Sar86] are also available in the literature.

8.4 PERFORMANCE IN FREQUENCY-SELECTIVE MULTIPATH FADING

As the data rate of a modem increases beyond fractions of the rms multipath spread of the channel, the channel becomes frequency selective. In a frequency-selective fading multipath channel a null may occur in the passband of the channel. Equivalently, we can say that the data rate is high enough with respect to the multipath spread that the multipath causes performance degradation due to ISI [Bel63b]. The performance degradation caused by ISI forces the performance curves into flat areas where any increase in the SNR does not improve the error rate performance of the modem. The error rate obtained at these values is sometimes referred to as the *irreducible error rate* of the system.

To quantitatively represent the effects of frequency-selective multipath fading, we provide two examples in the following two sections. The first example shows the effects of frequency-selective fading on the performance of typical radio modems

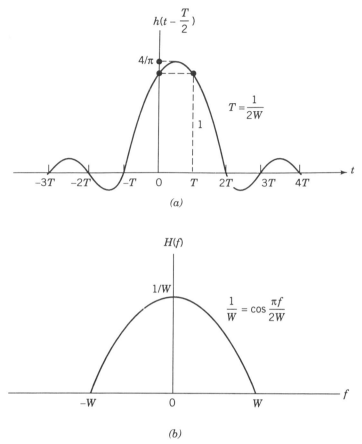

Figure 8.10 Time-domain (*a*) and frequency-domain (*b*) representation of half cosine pulses used in partial-response signaling.

by evaluating the performance of the modem with a deep null in various locations in the passband of the channel. In the following section we consider a hypothetical two path model and we examine the effects of the distance and relative amplitude of the paths on the error rate of the modem.

8.4.1 Effects of Frequency Selectivity

Figure 8.11 shows the results of simulations performed to model a microwave LOS link exhibiting a spectral null in its amplitude frequency response [Bel84, Pah85a]. In these simulations, the spectral null was placed at various positions in the frequency band, ranging from the center to the edge of the band. Results are shown for four different modulation techniques. It can be seen from the figure that the performance degradation is greatest when the spectral null is at the center of the band, resulting in a BER for PSK modulation approaching 0.5, a very poor level of quality. With the null at the band edge, the BER is around 10^{-5}, an

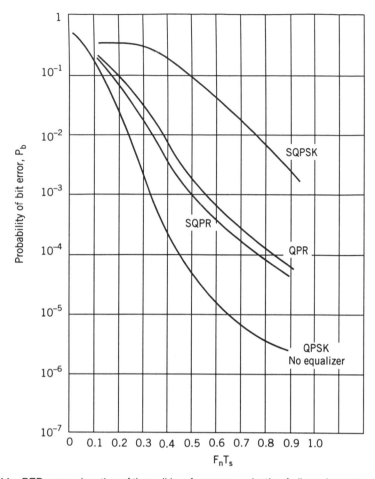

Figure 8.11 BER versus location of the null in a frequency-selective fading microwave channel.

acceptable level of quality. The four modulations simulated are quadrature partial response (QPR), staggered QPR, QPSK, and staggered QPSK. It can be seen from the figure that QPSK is more resistant to the effects of the spectral null than either QPR or staggered QPR, but the degradation of QPSK performance is nevertheless considerable when the spectral null is at the center of the band.

8.4.2 Multipath Effects and Data Rate Limitations

While signal fading, which arises from channel multipath, penalizes performance by greatly increasing the required SNR, the multipath also has a direct effect on performance in the form of intersymbol interference (ISI), as shown in Fig. 8.12. The top part of this figure shows the arrival of three consecutive pulses in a steady channel with no multipath. At the sampling time 0 only the sample associated with symbol "1" has a nonzero value. At following sampling times of $T, 2T, 3T, \ldots,$ the

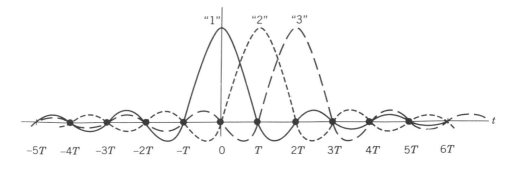

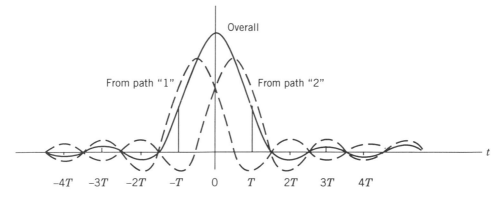

Figure 8.12 Intersymbol interference caused by multipath.

samples from symbol "1" have a zero value. As a result, symbol "1" has no contribution in decisions made for other symbols. The lower part of Fig. 8.12 shows the received pulses in a two-path channel and the overall received pulse associated with one transmitted symbol, which is the summation of the two pulses received from the two paths. In this channel samples of the received overall pulse have nonzero values at $T, 2T, 3T, \ldots$; this causes interference in decisions made on other symbols. Therefore, multipath causes intersymbol interference.

As we increase the symbol signaling rate in a multipath channel, the received symbols increasingly flow into one another, and this places an upper limit on rate at which symbols can be transmitted. Let us consider a theoretical example which will serve to illustrate this point. Figure 8.13 shows performance results obtained with a two-path channel model, where a_1 is the amplitude of one path, and the amplitude of the second path, a_2, is allowed to take on various values relative to a_1. The curves show received bit-error probability for different values of separation between the two paths normalized by the symbol duration. The figure shows that for values of a_2 up to about 20% of a_1, the ISI has very little effect on performance. However, for higher relative values of a_2 and when $t = T$, we find orders of magnitude of degradation in BER performance. These results give a general indication of how multipath affects performance as a function of data rate. For very low data rates the multipath has very little effect, while as the data rate increases, the performance can degrade markedly.

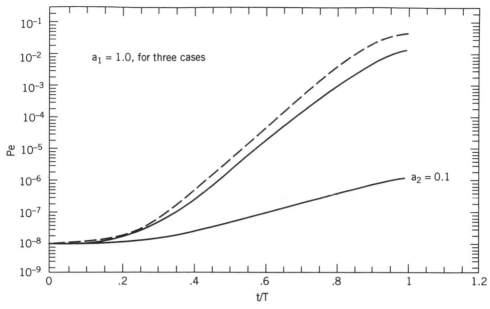

Figure 8.13 Probability of error versus normalized delay for a two-path channel model and three cases of relative path strengths. SNR = 15 dB for all three cases when $t/T = 0$.

In communications environments where the received BER can vary widely with time or with location of the transmitter or receiver, the overall average received BER is not necessarily a useful measure of system performance. A performance measure which is often more useful is the *outage probability*, which is defined as the probability that the received BER lies above some preselected threshold of acceptable performance. In Table 8.2 we show outage probabilities which have been computed from signal measurements made in five factory locations (results presented earlier, in Chapter 5). The third, fourth, and fifth columns in the table give outage probabilities computed for BER thresholds of 10^{-2}, 10^{-4} and 10^{-8}, respectively. These BER values represent performance thresholds that might be deemed acceptable for different data applications. The outage probabilities are calculated for BPSK modulation and are given for various data rates in the five manufacturing areas. The BER for each individual location was calculated using the measured channel impulse response in that location. All the points were calculated for 40-dB received SNR, in order to eliminate the effects of additive noise and include only the effects of multipath. All points were measured for transmitter-to-receiver distances less than about 50 m. If we accept outage rates around 1% and set 10^{-4} as the outage threshold on BER, a data rate on the order of 1 Mbit/sec is supported in all five areas.

These results, which can be regarded as typical of many indoor wireless communications environments, indicate that a data rate of 1 Mbit/sec is often feasible in these situations. However, for applications such as WLANs we wish to provide higher data rates in the same environments. We can of course replace

TABLE 8.2 Outage Probabilities Computed from Signal Measurements Made in Five Factory Locations [How91]

Area	Data Rate (MHz)	Outage for BER		
		10^{-2}	10^{-4}	10^{-8}
A	0.1	0.00093	0.00241	0.00444
	1.0	0.00130	0.00296	0.00630
	4.0	0.00741	0.01204	0.01778
	8.0	0.02667	0.03426	0.04333
	16.0	0.09778	0.10907	0.11926
B	0.1	0.00311	0.00867	0.02289
	1.0	0.00400	0.01089	0.02556
	4.0	0.02333	0.03333	0.05378
	8.0	0.07378	0.09044	0.11733
	16.0	0.23667	0.26733	0.30378
C	0.1	0.00067	0.00173	0.00440
	1.0	0.00680	0.00867	0.01387
	4.0	0.08347	0.09040	0.10227
	8.0	0.25293	0.26680	0.28453
	16.0	0.60373	0.61867	0.63600
D	0.1	0.00067	0.00200	0.00556
	1.0	0.01311	0.01800	0.02400
	4.0	0.13867	0.15178	0.16444
	8.0	0.39222	0.40889	0.42400
	16.0	0.67178	0.68289	0.69333
E	0.1	0.00015	0.00076	0.00182
	1.0	0.00273	0.00424	0.00606
	4.0	0.04045	0.04333	0.04788
	8.0	0.12121	0.12742	0.13439
	16.0	0.29136	0.29682	0.30515

BPSK modulation with QPSK, which doubles the data rate; but we may want even greater increases for our intended application in order to make the data rate comparable to those of WLANs. In the remainder of this chapter we shall describe several ways in which this can be achieved. Analysis of this situation requires an understanding of the implicit in-band diversity in frequency-selective fading channels, which we describe first.

8.5 MULTIPATH FADING AND TIME (IN-BAND) DIVERSITY

Multipath has the harmful effect of causing ISI, but at the same time the signals arriving from different paths are exposed to different fading patterns. The multipath signals can be regarded as a form of diversity, and a smart receiver can use this diversity to improve its performance. In this section we assume that we have a wide-sense stationary uncorrelated scattering (WSSUS) frequency-selective fading multipath channel and we show how we can take advantage of the multipath to provide diversity for the received signal. This diversity is not provided explicitly

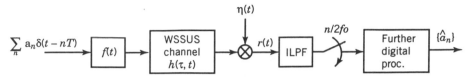

Figure 8.14 Overall baseband model of digital communication over a WSSUS fading multipath channel.

with multiple antennas or frequencies or with repeated transmissions, and thus it is referred to as *internal* or *implicit*-diversity.

The baseband model which we consider, shown in Fig. 8.14, is a general model of a high-speed digital communication system operating over a fading multipath channel. Closely following Pahlavan and Matthews [Pah90c], in this discussion the model, extending from the information source in the transmitter to the input of the digital processor in the receiver, will be recast as a discrete tapped delay line.

The data sequence $\{a_k\}$ modulates an impulse train $\delta(t - kT)$, $k = 1, 2, 3, \ldots,$ where T is the symbol period. The impulse train acts as an input to a filter whose impulse response, $f(t)$, is the fundamental transmitted symbol. The output of the filter,

$$\sum_k a_k f(t - kT)$$

is the transmitted signal. It is assumed that $f(t)$ is band-limited to $f_0 = W/2$ Hz.

The signal passes through a fading multipath channel with time-variant impulse response $h(\tau, t)$. The function $h(\tau, t)$ represents the path gain associated with delay τ at time t. White complex Gaussian noise $\eta(t)$ is added to the information-bearing signal to form the received signal

$$r(t) = \sum_k a_k g(t, t - kT) + \eta(t) \tag{8.5.1}$$

where

$$g(u, t) = \int h(\tau, t) f(u - \tau)\, d\tau$$

The receiver would normally perform matched filtering by pulse-shaping filtering followed by a sampler at the output, operating at the symbol rate of $1/T$ samples/sec. This is an optimal way to process $r(t)$ in the absence of frequency selective fading. In frequency-selective fading the received waveform is continually changing, and the matched filtering with the pulse-shaping filter does not provide optimum filtering. Because $f(t)$ and hence $g(t)$ are band-limited, it is also possible to pass $r(t)$ through an ideal low-pass filter followed by a sampler operating at the Nyquist rate. In this manner the matched filtering can be deferred to a subsequent digital processor.

The sampled output is

$$r(t)\big|_{t=n/2f_0} = \sum_k a_k g(t - kT, t)\bigg|_{t=n/2f_0} + \eta(t)\big|_{t=n/2f_0}$$

or

$$r_n = \sum_k a_k g(t - kT, t)\bigg|_{t=n/2f_0} + \eta_n \qquad (8.5.2)$$

Equation (8.5.2) is not in a convenient form because it represents sampling at one rate and transmission at another rate. Therefore we define

$$M = 2f_0 T$$

and increase f_0 somewhat, if necessary, to ensure that M is an integer. We now define the discrete sequence

$$x_n = \begin{cases} a_n/M, & n/M \text{ an integer} \\ 0, & \text{otherwise} \end{cases} \qquad (8.5.3)$$

which consists of the data interleaved with $M - 1$ zeros. Then the sampled output may be written compactly as

$$r_n = \sum_k x_k g_{n-k}(n) + \eta_n = \sum_k x_{n-k} g_k(n) + \eta_n \qquad (8.5.4a)$$

where

$$g_k(n) = g(u, t)\big|_{u=k/2f_0,\, t=n/2f_0}$$

Because $g(u, t)$ is band-limited, it cannot also be time-limited. However, in all practical situations $g(u, t)$ will have a finite duration by some suitable engineering definition. Let L denote the number of samples in that duration.[‡] The model corresponding to Eq. (8.5.4a) is shown in Fig. 8.15, where sufficient delay is added to make the model causal.

This is a multirate discrete system, because the rate at which information enters the system is not the same as the sampling rate before digital processing. The transmitted symbol x_n appears at the input of the equivalent discrete time channel model every M samples and we have L samples of r_n for each transmitted information symbol x_n. If the transmission system is designed so that $M \geq L$, there is no ISI and there is a vector of sampled received sequences with length L corresponding to each single symbol a_n. Therefore, Eq. (8.5.4a) reduces to

$$r_{n-k} = a_n g_k(n) + \eta_{n-k}, \qquad 0 \leq k < L \qquad (8.5.4b)$$

[†]Because $E\{|g_k(n)|^2\}$ is the average power associated with a tap weight, one method of choosing L is to retain all taps whose power is within "x dB" of the largest tap power. The criterion of 10 dB was found to be a good choice for the numerical results presented later.

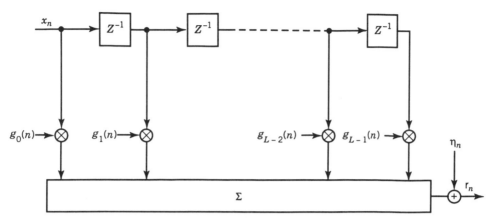

Figure 8.15 Discrete channel model representing transmitter filtering, the WSSUS channel, additive noise, and prefiltering at the receiver.

at the baud rate.[‡] In vector form, this equation is written as $\mathbf{R} = a_n\mathbf{G} + \mathbf{\eta}$. For each transmitted symbol a vector of length L is received, resulting in L-order implicit diversity. To normalize Eqs. (8.5.4a) and (8.5.4b), throughout the analysis it is assumed that $a_n = \pm \sqrt{E_b}$ where E_b is the transmitted energy per bit,

$$\sum_{k=0}^{L-1} E\left\{\left| g_k(n) \right|^2\right\} = 1,$$

and $|\eta_n|^2 = N_0$. Then the average SNR per bit is $\bar{\gamma}_b = E_b/N_0$. In the absence of ISI the optimum receiver is a discrete matched filter (DMF) which is a maximal-ratio combiner for the implicit diversity of the channel. In the presence of ISI, we have $M < L$ and additional signal processing is needed to eliminate the ISI. Maximum likelihood sequence estimation (MLSE) and decision feedback equalization (DFE) are the two methods considered in this case.

8.6 ADAPTIVE CHANNEL MEASUREMENT

The channel measurement techniques that we studied earlier were used for the development of statistical models for radio propagation. In this section we introduce another concept in channel measurement. Here we measure the taps of the discrete time-domain channel model developed in the previous section. The results of these measurements are used to implement adaptive receivers that take advantage of the in-band diversity. In the next section we introduce two specific receivers, the *adaptive-discrete-matched-filter* (ADMF) and the adaptive MLSE, using the adaptive channel estimator introduced in this section as a part of the receiver.

[‡]Note that during the transmission time of several symbols the channel is assumed to be fixed; that is, $g_k(n) = g_k(n + k)$ for small values of k.

Estimation of the taps of the overall channel impulse response requires a reference signal. The reference signal can be a transmitted probe sequence known to the receiver, or it can be some sequence of detected symbols. The probe signal can be time-multiplexed with the transmitted information symbols, or it can be overlaid onto the traffic channel as a weak spread-spectrum signal. The effects of the transmitted reference signal on the overall performance of an ADMF are analyzed in [Pah90c].

8.6.1 The MMSE Solution for Channel Estimation

The input to the channel estimator is given by

$$r_n = \sum_{k=0}^{L-1} g_k(n) x_{n-k} + \eta_n \tag{8.6.1}$$

where $g_k(n)$ is the equivalent discrete channel impulse response, x_n is the reference signal, and η_n is the additive noise. The channel estimator forms estimates of the channel impulse response $\hat{g}_k(n)$ and the received signal $\hat{r}_n$, which are related as follows:

$$\hat{r}_n = \sum_{k=0}^{L-1} \hat{g}_k(n) x_{n-k}$$

It is well known [Pah79, Pro89] that the minimum mean-square error (MMSE) estimate of the channel impulse response that minimizes $E\{|r_n - \hat{r}_n|^2\}$ satisfies the following linear set of simultaneous equations:

$$\mathbf{A}\hat{\mathbf{G}} = \mathbf{b} \tag{8.6.2a}$$

where $\hat{\mathbf{G}}$ is the vector of MMSE estimates of the tap gains, and the elements of matrix $\mathbf{A}$ and the vector $\mathbf{b}$ are defined by

$$
\begin{aligned}
A_{ij} &= E\{x_{n-j} x_{n-i}\}, \qquad 0 \le i, j < L \\
b_i &= E\{r_n x_{n-i}\}, \qquad 0 \le i < L
\end{aligned}
\tag{8.6.2b}
$$

For a reference signal with statistically independent symbols, such as a pseudonoise (PN) sequence, $\mathbf{A}$ is a diagonal matrix with E_b the value of all diagonal elements. The solution to Eq. (8.6.2a) is then

$$\hat{g}_k(n) = \frac{1}{E_b} E\{x_{n-k} r_n\} = g_k(n), \qquad 0 \le k < L \tag{8.6.3}$$

8.6.2 Implementation Using Cross-Correlation, and the LMS Algorithm

There are two ways to implement the channel estimator discussed above. One is to use cross-correlation to calculate Eq. (8.6.3) directly. The second is to use an equalizer-like channel estimator described in [Mag73]. This estimator is implemented by applying the steepest-descent algorithm to minimization of mean-square

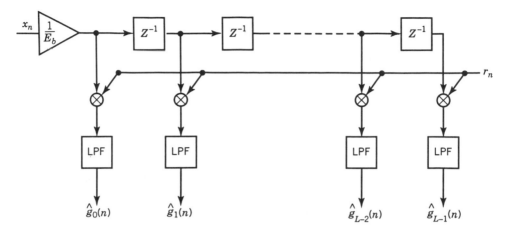

Figure 8.16   Discrete channel estimation using cross-correlation.

error. The steepest-descent algorithm was first introduced and named the *least-mean-square-(LMS)-algorithm* by Widrow [Wid66, Pro89].

Figure 8.16 shows the implementation of discrete channel estimation using cross-correlation [Pah90c]. The reference signal is passed through a tapped-delay line (TDL), and the tapped signals are multiplied with the received signal and then passed through a bank of low-pass filters. To normalize the output of the filters to actual samples of the channel impulse response, one of the signals is scaled to $1/E_b$, as indicated in Eq. (8.6.3).

Figure 8.17 shows equalizer-like channel estimation using the LMS algorithm. Here the reference signal is passed through a TDL, and the channel tap gains are

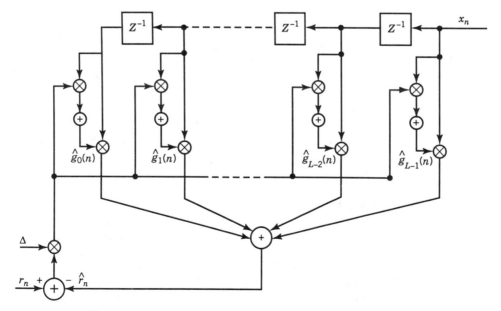

Figure 8.17   Adaptive channel estimator using the LMS algorithm.

adjusted according to

$$\hat{g}_j(n+1) = \hat{g}_j(n) + \Delta e_n x_{n-j} \tag{8.6.4}$$

where Δ is the adjustment rate parameter and

$$e_n = r_n - \hat{r}_n = r_n - \sum_{k=0}^{L-1} \hat{g}_k(n) x_{n-k} \tag{8.6.5}$$

The adaptive channel measurement techniques used with MLSE are described in [Mag73] and [Ung74], with application to slowly time varying channels. Analysis of adaptation techniques used for fading multipath radio channels is available in [Pah79].

8.7 ADAPTIVE RECEIVERS FOR TIME (IN-BAND) DIVERSITY

Since the mid-1950s many different types of adaptive receiving techniques have been studied for a variety of fading multipath channels. These techniques will increase achievable modem data rates by overcoming to varying degrees the multipath fading characteristics of the channel. These techniques are in effect intended to take advantage of the in-band diversity provided by the multipath nature of the received signal. If the modem is not designed to take advantage of the multipath structure, as we saw earlier, the multipath causes destructive ISI limiting the maximum data rate achievable with reliable communication over the channel. Historically, the most important adaptive receiver for fading multipath channels is the Rake system invented by Robert Price and Paul Green of MIT Lincoln Laboratory [Pri58]. The Rake system was originally designed for teletype communication with a baud rate of 90 chips/sec operating over an ionospheric channel. The envelope of the transmitted binary FSK signal was a 10-kHz PN sequence. The received signal was passed through two tapped delay lines—the matched filters—for mark and space frequencies, and the outputs were compared for decision-making. The tap gains of the delay lines were adaptively adjusted by cross-correlating the received signal with both mark and space reference signals at the receiver. Because the data rate was considerably smaller than the transmission bandwidth, the ISI was negligible. More recently, other versions of Rake receivers have been examined for urban radio [Kam81], HF radio [Bel88], and indoor radio [Cha93] channels.

In the past two decades other adaptive techniques have emerged in the development of a new generation of radios for fading multipath channels. Time-gating of the transmitted pulse to avoid ISI, with adaptive discrete matched filtering (DMF) of each received pulse, was the approach taken for a family of military troposcatter radios [Pah80, Con78]. The adaptive decision feedback equalizer (DFE) was another approach investigated for application to troposcatter [Grz75, Mon84], microwave line-of-sight [Bel84, Pah85a], HF [Fal85], and, more recently, indoor radio [Sex89a] channels. Finally, an adaptive version of the MLSE technique [For72] was investigated for troposcatter in [Bel69] and for HF in

[Cha75]. In the remainder of this section we discuss DMF and MLSE and the next section is devoted to adaptive equalization with emphasis on DFE. The RAKE receiver will be addressed in the next chapter, where we discuss spread spectrum communications.

8.7.1 Adaptive DMF

The simplest method of avoiding ISI is to reduce the symbol rate relative to the transmission bandwidth, which also reduces the bandwidth efficiency of the system. There are two approaches to implementing this technique. One approach is to provide a time guard between the transmitted symbols so that the signals arriving along different paths and associated with different transmitted symbols do not interfere with one another. The time guard should be equivalent to the maximum multipath delay spread of the channel so as to completely avoid the ISI, as shown in Fig. 8.18. In this case the bandwidth efficiency of the system is reduced $T/T + \tau_m$, where T is the duration of the pulse and τ_m is the maximum multipath spread of the channel. As will be explained in the next chapter, a second approach to avoiding ISI is to use direct sequence spread-spectrum transmission. The adaptive matched filter in this case is referred to as the *Rake receiver*. The advantage of spread spectrum for multiuser applications is that the loss in the bandwidth efficiency can be overcome by the use of code-division multiple access. Time gating has been used in less bandwidth-sensitive applications such as troposcatter communication, whereas the spread-spectrum technique is used in a number of recently developed wireless networks.

In the absence of ISI the optimum receiver is an adaptive discrete matched filter consisting of two parts: an adaptive channel estimator and a DMF which is a maximal ratio combiner of the implicit (in-band) diversity. Figure 8.19*a* shows a general block diagram for the adaptive DMF (ADMF). The adaptive channel estimator, described in the previous section, evaluates the instantaneous values of the taps in the equivalent discrete channel model. The DMF uses the estimated values of the channel impulse response to perform maximal ratio combining of the

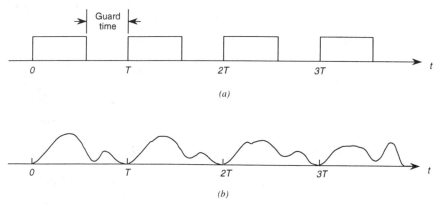

Figure 8.18 Time gating: a technique for avoiding intersymbol interference. (*a*) Transmitted symbols. (*b*) Received symbols.

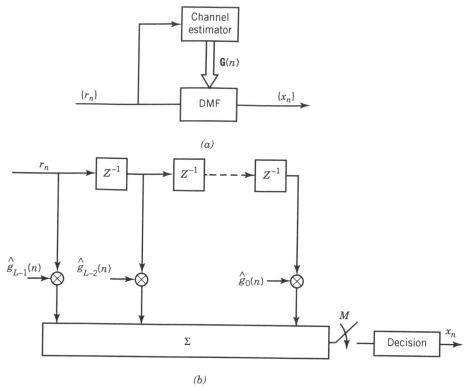

Figure 8.19 Adaptive matched filtering. (a) General block diagram for the adaptive discrete matched filter (ADMF). (b) Implementation of the DMF using a finite impulse response digital filter.

diversity signal represented by the tapped channel model. Figure 8.19b shows the implementation of the DMF using a finite impulse response digital filter. The taps of the DMF are the complex conjugates of the taps of the discrete channel impulse response in reverse time order. If we consider the overall discrete model, this is the discrete match filter for the overall discrete channel impulse response, and thus is the optimum receiver in the absence of ISI. Another way of looking at the DMF is to consider each tap of the overall channel impulse response as a branch of the inband diversity. Then the DMF multiplies each branch with the complex conjugate of the amplitude of that branch, which constitutes maximal-ratio combining under the assumption of equal noise powers on the diversity branches.

Performance Prediction for the DMF. If we assume that the inband diversity branches are all Rayleigh fading signals, Eq. (7.4.14) presented in Chapter 7 for calculation of maximal-ratio combining performance are applicable to performance analysis of the DMF with order of in-band diversity L replacing the order of external diversity D. With these equations the average probability of error for coherent PSK as a function of average signal-to-noise ratio $\overline{\gamma}_b$ was given by

$$\overline{P}_e = \sum_{n=0}^{L-1} \frac{A_n}{2} \left\{ 1 - \left[\frac{2\overline{\gamma}_b \lambda_n}{1 + 2\overline{\gamma}_b \lambda_n} \right]^{1/2} \right\} \tag{8.7.1}$$

where $\{\lambda_n\}$ are the eigenvalues of the covariance matrix of tap gains and

$$A_n = \prod_{\substack{k=0 \\ k \neq n}}^{L-1} \frac{1}{1 - \lambda_k/\lambda_n}$$

In terms of the parameters of the model, the elements of the covariance matrix are given by

$$r_{ij} = E\{g_i g_j^*\} = \int_{-\infty}^{\infty} f\left(\frac{i}{2f_0} - u\right) Q(u) f^*\left(\frac{j}{2f_0} - u\right) du$$

where $Q(\tau)$ is the delay power spectrum of the channel and $f(t)$ is the impulse response of the pulse-shaping filter.

Example 8.1. For a numerical example let us assume that the delay power spectrum is given by

$$Q(\tau) = \begin{cases} b^2 \tau e^{-b\tau}, & \tau \geq 0 \\ 0, & \tau < 0 \end{cases}$$

where b is related to the channel rms multipath spread τ_{rms}, through the expression

$$b = \frac{2\sqrt{2}}{\tau_{\text{rms}}}$$

This delay power spectrum is an approximation found by curve-fitting to typical measurements made on troposcatter links [She75] and is consistent with theoretical predictions given in [Bel69].

The system considered here for performance prediction uses a 100% raised-cosine pulse with a bandwidth of 2 MHz. The samples at the receiver are taken at the Nyquist rate, 4 Mbits/sec. With the ratio of the rms multipath spread to symbol interval set at $\tau_{\text{rms}}/T = 0.11$, the equivalent tapped delay line model for this example contains four significant tap gains. The average received power for each tap gain and the eigenvalues of the covariance matrix of the tap gains are given in [Pah79]. Using these values in Eq. (8.7.1), a plot of average PSK error probability versus $\bar{\gamma}_b$ is generated, as shown in Fig. 8.20. The figure also includes the performance calculated for a one-path case. The performance gain achieved through optimum exploitation of the in-band diversity is shown to be about 10 dB at $\bar{P}_e = 10^{-5}$.

In the above example for the DMF we assumed that the channel estimator makes perfect estimates of the tap weights in the channel model. In practice the adaptive channel estimator introduces measurement noise which somewhat degrades the performance of the system from the theoretical optimum. The effect of the noise depends on the approach used in providing the reference signal, on the Doppler spread of the channel, and on the parameters of the adaptive filters. The influence of these factors on performance of the ADMF is discussed in [Pah90c].

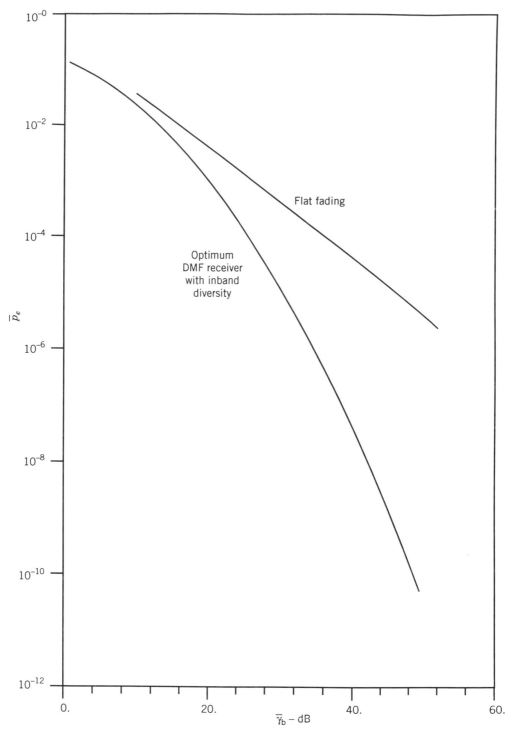

Figure 8.20 Performance of the discrete matched filter receiver operating on a troposcatter channel. The single-path flat fading case is included for comparison.

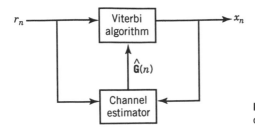

Figure 8.21 Adaptive maximum likelihood sequence estimation (MLSE).

8.7.2 Adaptive MLSE

The MLSE is the optimum receiver in the presence of ISI. Given the estimates of the channel impulse response, an MLSE receiver uses a trellis diagram with the Viterbi algorithm to obtain maximum-likelihood estimates of the transmitted symbols. The adaptive MLSE receiver, shown in Fig 8.21, is similar to the ADMF. It consists of two parts, the adaptive channel estimator and the MLSE algorithm. The sampled channel impulse response is measured with the adaptive channel estimator described in the previous section. Given the samples of channel impulse response, the sequence of the sampled received signal is compared with all possible received sequences. The maximum likelihood procedure [For72] is to determine the squared distance between the sequence of the sampled received signal and all possible received sequences and to determine the transmitted symbol sequence with minimal distance. The Viterbi algorithm [Vit67] is used as the computationally optimal and efficient solution to this search. This application of the Viterbi algorithm is similar to other well-known applications, including decoding of convolutional error-control codes and demodulation of TCM signals. If the ISI is negligible, the performance of the MLSE is the same as that of the DMF. With ISI, the performance of the ADMF provides a tight upper bound to the performance of the adaptive MLSE and can be used as an approximation for purposes of performance prediction.

The MLSE is the optimal method of canceling the ISI; however, the complexity of this receiver grows exponentially with the length of the channel impulse response, whereas the complexity of the equalizer grows only linearly with the length of the impulse response. For this reason, MLSE is an attractive option for channels with short impulse responses, but for longer impulse responses, equalizers are more practical. In the radio communication literature, MLSE is usually compared with decision feedback equalization (DFE), which is the equalizer of choice for frequency-selective fading multipath channels and will be discussed in more detail below. Comparisons of MLSE versus DFE performance are given for telephone line modems in [Fal76b] and [Fal76c], for HF radio in [Fal85], and for troposcatter radio links in [Mon77, Cha75].

8.8. ADAPTIVE EQUALIZATION

A third technique for increasing the data rate in multipath fading environments is adaptive equalization. Frequency-selective fading channels produce ISI in the received signal. In these channels an increase in the power does not improve performance, because additional power amplifies the ISI in step with the desired signal. The traditional method of compensating for ISI is to equalize the channel impairment by applying a filter at the receiver. In general, channel characteristics are subject to variations in time, which leads to the need for adaptive equalizers. Tapped delay line equalizers are the most commonly used equalizers, and their detailed analysis and theory of operation will be given in this section. For more detailed treatments of equalization techniques the reader can refer to [Pro89], [Qur85].

The principles of operation for TDL equalizers are the same as those for the MMSE channel estimator discussed earlier, and all of these techniques are considered special cases of adaptive filtering. Equalizers have wider applications in telecommunications, and they have been introduced and applied earlier than the other techniques. For these reasons, more detailed attention in this chapter is devoted to equalizers. Here, issues arising in the application of TDL equalizers to radio modems, such as sensitivity to frequency-selective fading, sensitivity to timing recovery, variations related to choice of modulation technique, and bounds for performance evaluations, are discussed. We start with a review of various equalizer architectures, and then we examine the effects of equalization on the performance of radio modems over frequency-selective fading channels. This discussion will be followed by analysis of performance over measured indoor radio channels. Here we provide a bound for calculation of the probability of outage and average error rate on radio channels.

8.8.1 Equalizer Architectures

Equalizer architectures considered for modems are: linear transversal equalizers (LTEs), linear fractionally spaced equalizers (FSEs), decision feedback equalizers (DFEs), pass-band equalizers, and blind equalizers. We next discuss these architectures, with detailed emphasis on LTE, FSE, and DFE equalizers.

Linear Transversal Equalizer (LTE). The LTE, shown in Fig. 8.22, is the earliest of the TDL equalizers. The received signal is passed through a tapped delay line with tap spacing $\Delta = T$. The tapped signals are weighted and added to form the

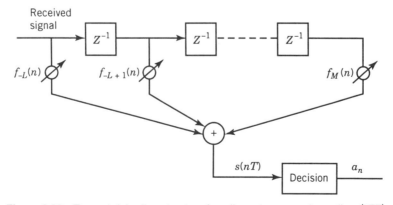

Figure 8.22 Tapped-delay-line structure for a linear transversal equalizer (LTE).

equalized output signal. The optimum tap gains are determined using either the zero-forcing or the MMSE criterion. In a zero-forcing algorithm, the tap gains are adjusted to force the overall sampled impulse response after equalization to have minimum ISI [Luc65, Luc66, Pro89]. For MMSE equalization, tap gains are adjusted to minimize the mean square of the error signal between equalized signal and the actual transmitted symbols [Pro89].

The equalizer is a discrete-time filter intended to compensate for the amplitude and phase distortions of the channel. One can see intuitively that for an infinite-tap equalizer, the sampled frequency response of the equalizer should be the inverse of the frequency response of the channel, which is a correct characterization of zero-forcing equalization. In MMSE equalization, the tap gains are set so as to minimize both channel distortions and additive noise, and therefore the frequency response function of the equalizer depends in part on the variance of the noise.

Today, MMSE is the dominant criterion used in the design of equalization algorithms, and various versions of the equalizer architectures are developed with this criterion. As a result, we focus primarily on MMSE equalization.

Fractionally Spaced Equalizer (FSE). The structure of the FSE is the same as that of the LTE, shown in Fig. 8.22, except that in FSE the tap spacing is $\Delta = kT/n$, where k and n integers and $k < n$. This minor difference in spacing of the taps results in three basic advantages for a modem using FSE over a modem using LTE. (1) A modem with a FSE is relatively insensitive to timing error, (2) it does not require accurate pulse shaping, and (3) it is more effective in treating the channel phase distortions at the edges of the band. For these reasons, FSEs are widely used both for voiceband and radio modems.

It has been found that while an FSE will perform well in computer simulation with floating point arithmetic, in real-time implementation with fixed-point arithmetic and limited word length, the tap gains tends to diverge. An investigation of tap gain divergence and its relationship to characteristics of the channel is given in [Git82]. In general, the simple solution to the tap gain divergence problem is the so-called *tap-leakage-algorithm* [Ung76, Git82]. This *ad hoc* solution introduces small corrections in the tap gains in order to prevent them from growing excessively.

Another issue concerning the comparison between LTE and FSE equalizers is that with the same amount of hardware complexity, an LTE spans a longer time interval, which intuitively suggests superior performance for an LTE. However, results of computer simulations reveal that with the same number of taps, an FSE performs at least as well as an LTE, and the FSE provides significantly better performance on channels with severe phase distortions at the edges of the band [Qur77].

Decision Feedback Equalizer (DFE). A DFE [Bel79b, Sal73], shown in Fig. 8.23, consists of two tapped-delay-line filters, referred to as the *forward* and the *backward* equalizers. The input to the forward equalizer is the received signal, and it operates similarly to the linear equalizers discussed earlier. The input to the backward equalizer section is the stream of detected symbols. The tap gains of this section are the estimates of the channel sampled impulse response including the forward equalizer, and this section cancels the ISI due to past samples. For

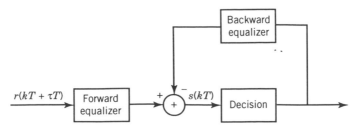

Figure 8.23 The decision feedback equalizer (DFE). The forward and backward equalizers are both tapped-delay-line equalizers.

finite-tap equalizers the backward equalizer of length M eliminates the ISI due to M past symbols.

A linear equalizer is unable to properly equalize a channel having a deep null in the pass band. For such a channel, the linear equalizer applies high gain in its frequency response to compensate for the null, which in turn causes noise enhancement, as we noted in an earlier subsection. However, the backward filter of a DFE does not suffer from the noise enhancement problem, because it estimates the channel rather than its inverse. As a result, for channels with deep nulls in the passband, DFEs are superior to linear equalizers.

In frequency-selective fading radio channels, channels occasionally experience deep nulls in the pass band, resulting in unsatisfactory performance for linear equalizers. For these channels, the DFE is the preferred design, and DFEs have been examined for troposcatter [Mon77], HF [Fal85], and microwave line-of-sight [Bel84, Pah85a] channels. On telephone channels, the most significant amplitude and delay distortion is found at the edges of the passband. As a result, a DFE, which is more effective against nulls in the middle of the passband, has little to offer over a linear equalizer with a large number of taps. Thus linear equalization is the predominant design choice for voiceband data modems operating over telephone channels.

Blind Equalization. Standard equalizers are trained with a known sequence at the start of data transmission. After training, with a high probability, the decisions at the output of the equalizer are the same as the transmitted symbols and are used as the reference for the tap gain adjustment. In many cases, both radio and telephone line modems need to be retrained without interruption of normal data transmission. The retraining algorithm should be independent of any particular information about the sequence of transmitted symbols. This equalization is usually referred to as *blind-equalization*. A simple algorithm for blind equalization, which uses the square of the error signal for adjustment of equalizer tap weights, is given in [Fal76a], and today this algorithm is widely used to retrain voiceband modems after an interruption of data. Because in these algorithms reference to transmitted symbols is unavailable at the receiver, the convergence is about one order of magnitude slower than with standard LMS equalization algorithms. Additional analysis of the transient behavior of blind equalizers, applied to radio communications, is available in [God80] and [Fos85].

Bandpass Equalization. Bandpass equalization [Git73, Fal76a] is performed before demodulation—in contrast with standard baseband equalization, which is done after demodulation. When the phase reference of the oscillator in the demodulator is obtained from the detected data stream, the delay between demodulation and phase recovery is smaller in passband equalizers, resulting in faster tracking of phase variations and, consequently, a more robust modem. However, in modern digital designs it is possible to demodulate without a phase reference and to adjust the phase by multiplying the demodulated symbol with a numerical phasor which shifts the point in the constellation into its proper place. For these implementations, there is no difference between baseband or bandpass equalization. In radio modems operating at high carrier frequencies, SAW (surface acoustic wave) devices sometimes provide a cost-effective solution for implementation of the equalizer. The SAW devices can operate at the required high frequencies; and given their cost advantage, passband equalization at either IF or RF is preferred to baseband equalization.

8.8.2 Analysis of the Equalizers

Calculation of the optimum tap gains for equalizers is treated thoroughly in the literature [Pro89, Pah88b]. Normally the tap values that minimize the mean-square error can be found from the solution of a set of linear equations. The following set of equations derived in [Pah88b] provides a unified solution for both the LTE and FSE as a function of tap spacing and timing error:

$$\sum_{j=-L}^{N} \left[f_j \overline{|a_k|^2} \sum_p g(pT + \tau T - j\Delta)g^*(pT + \tau T - l\Delta) + N_0 \delta_{jl} \right]$$
$$= \overline{|a_k|^2} g^*(\tau T - l\Delta), \qquad -L \le l \le N \qquad (8.8.1)$$

In the equation above, the $\{f_i\}$ are the optimum tap gains of the equalizer, Δ is the tap spacing, $g(t) = f(t) * h(t)$ is the overall channel impulse response including the modem filtering $f(t)$ and channel characteristic $h(t)$, τ is the normalized sampling-time error, and $\delta_{jl} = 1$ for $j = l$ and zero otherwise. The minimum mean square error calculated for this system is given by

$$\xi_{\min} = \overline{|a_k^2|} \left[1 - \sum_{i=-L}^{N} f_i g^*(\tau T - i\Delta) \right] \qquad (8.8.2)$$

where the $\{f_i\}$ are the optimal values of the tap gains found from Eq. (8.8.1).

As shown in [Pah88b], for the DFE with $N + 1$ forward taps and M feedback taps the optimum forward tap gains $\{f_i\}$ and feedback tap gains $\{b_i\}$ are the results of the solution of the following sets of linear equations:

$$\sum_{j=-L}^{N} \left[f_j \overline{|a_k|^2} \sum_p g(pT + \tau T - j\Delta)g^*(pT + \tau T - l\Delta) + N_0 \delta_{jl} \right]$$
$$- \overline{|a_k|^2} \sum_{j=1}^{M} b_j g^*(jT + \tau T - l\Delta)$$
$$= \overline{|a_k|^2} g^*(\tau T - l\Delta), \qquad -N \le l \le 0 \qquad (8.8.3)$$

and

$$b_l = \sum_{j=-N}^{0} f_j g(lT - j\Delta + \tau T), \qquad 1 \le l \le M \qquad (8.8.4)$$

The right-hand side of the equation above is a convolution of the sampled channel impulse response from the transmitter to the input of the forward equalizer taken at time spacing of Δ, with the discrete time impulse response of the forward equalizer. The results are sampled at time T, and only values on the right-hand side of the center sample are calculated; these samples are associated with ISI due to the past M transmitted symbols. Therefore, the optimum tap gains completely eliminate the ISI due to the past M samples. The value of the minimum MSE is

$$\xi_{min} = \overline{|a_k^2|} \left[1 - \sum_{i=-L}^{0} f_i g^*(\tau T - i\Delta) \right] \qquad (8.8.5)$$

At first glance, this equation suggests that the MMSE is independent of the backward tap gains, however this conclusion is incorrect. In Eq. (8.8.3) when the optimum tap gains are determined, one minimizes the error jointly for both sets of tap gains. As a result, optimum values of forward tap gains are affected by the optimum values of backward tap gains, which indirectly influence the MMSE.

The equalizers introduced thus far are suitable for PAM, PSK, and QAM modulation techniques. For other modulation techniques the setup for optimum tap gains and MMSE can be different. The equations for the calculation of the tap gains of the equalizer for SQPSK and SQPR are given in [Bel84]. A comparison between the performance of SQPSK, SQPR, QPS, and QPR over frequency-selective microwave LOS channel is given in [Pah85a].

Example 8.2. Here we provide some numerical examples to compare the effectiveness of different equalizers on frequency-selective fading channels. As an example applicable to radio communications, consider a frequency-selective fading channel whose frequency response is given by

$$H(j\omega) = A + B(j\omega)$$

where A and B are complex numbers. This model is used to represent frequency-selective fading in microwave line-of-sight (LOS) channels.

The transfer function of this channel has one zero at $s = -A/B$, which results in a null at the frequency $\omega = \text{Im}[-A/B]$. The depth of the notch is determined by $\text{Real}[-A/B]$.

Figure 8.24 shows plots of the inverse of the MMSE versus normalized timing error for $A = 1$, $B = 0.4$ and various numbers of taps for a DFE working with an SQPR modem. The inverse of the MMSE is a measure of SNR after equalization, which can be viewed as a performance criterion for an equalizer. The SNR before equalization is 23 dB. The forward tap gains are associated with $\Delta = T/2$, and plots include various numbers of forward (N) and backward (M) tap gains. As the

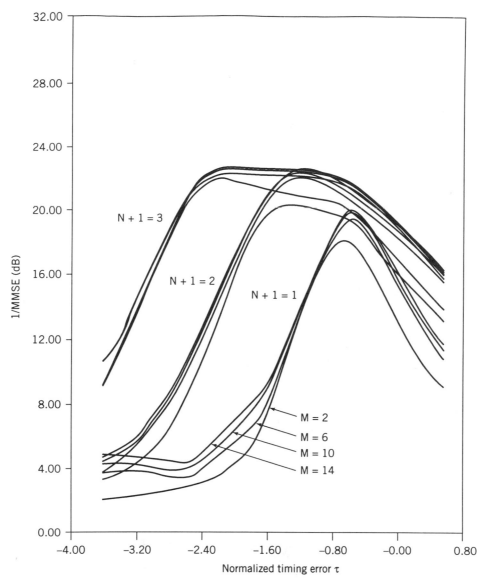

Figure 8.24 Inverse MMSE versus normalized timing error for a staggered quadrature partial response DFE operating over a frequency-selective microwave LOS channel. The inverse of the MMSE is a measure of SNR after equalization. $N + 1 =$ number of forward taps; $M =$ number of feedback taps; decision variable SNR $= 23$ dB with no distortion [Bel84].

number of forward tap gains increases, the flat region of the curve widens, indicating a larger span of time in which the sampling can take place without significant performance degradation. The insensitivity to sampling time will significantly reduce the complexity of the timing recovery circuit. An increase in the number of backward taps will increase the SNR. For this particular example, three forward and three feedback taps are adequate to provide a wide region of allowed sampling times and a maximum SNR.

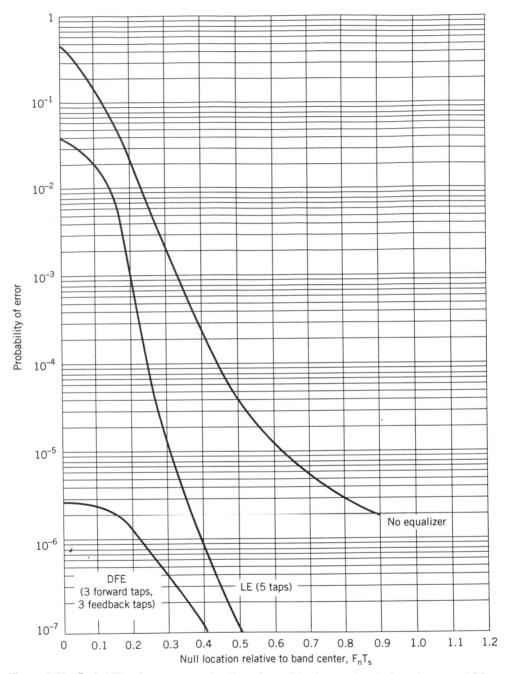

Figure 8.25 Probability of error versus location of a null in the passband of a microwave LOS channel for a DFE with three forward and three backward taps, an LTE with five taps, and the QPSK modem without equalization [Pah85a].

Figure 8.25 shows the probability of error versus location of a deep null in the passband of the channel for a QPSK modem. The channel model is the same as above, but the parameters are adjusted to force a null on frequency axis. The probability of error is calculated by determining the overall sampled impulse response after equalization and calculating the average error rate for all possible combinations of the ISI [Pah85a]. Plots include: the modem without an equalizer; an FSE with five $T/2$-spaced taps; and a DFE modem with three forward and three feedback taps. For the deep null near the center ($FT \approx 0$), only the DFE shows an acceptable level of performance. As the null moves toward the edges of the passband, the performance of the other two modems improves rapidly. As a result, one can conclude that for robust performance over a range of channel conditions, a DFE is required.

8.8.3 Adaptive Algorithms for Equalizers

In the previous section we analyzed the effectiveness of various equalizers under the MMSE criterion and found that the optimal tap gains in each case are determined from a set of linear equations. Direct solution of these equations requires measurement of the instantaneous overall channel impulse response, measurement of the variance of the noise, formation of the covariance matrix, and solution of a large set of linear equations. This method requires extensive numerical calculations, which is by no means economically attractive for real-time applications. For real-time computations we need a numerically efficient method of solution. Fortunately, both the peak distortion function used in zero-forcing algorithm and the MSE function used in MMSE equalizers are convex functions of the tap gains, which allows application of the LMS algorithm. The LMS algorithm is computationally attractive and can easily adapt to channel variations. Application of the LMS algorithm to adaptive equalization was first done for the peak distortion criterion [Luc65, Luc66] and was subsequently investigated for adaptive MMSE equalization [Pro89].

The LMS algorithm implementation for the forward tap gains $\{f_l\}$ of the equalizer is given by

$$f_l(k + 1) = f_l(k) + \mu e(kT) r(kT + \tau T - l\Delta) \qquad (8.8.6)$$

where μ is a small constant known as the step size, $e(kT)$ is the error between the equalizer output and the transmitted symbol, and $r(t)$ is the received signal at the input of the equalizer. For the feedback taps of the equalizer we have

$$b_l(k + 1) = b_l(k) + \mu e(kT) d_{k-l} \qquad (8.8.7)$$

where d_k is the transmitted symbol. The LMS algorithms was found to be slow for some applications, and rapidly converging algorithms for fast start-up equalization and fast tracking of rapidly changing channels were studied extensively. For an overview of these algorithms see [Pah88b]. For discussion of the methods applied in voiceband modems see [Qur85], and for detailed derivation of specific algorithms see [Pro89].

8.8.4 Equalization of the Indoor Radio Channel

Data rates on the order of 10 Mbits/sec are desirable for WLANs in order to make them compatible with existing wired or cabled LANs. Severe multipath fading, which is characteristic of indoor radio channels, limits the achievable data rates of BPSK modems to less than 1 Mbit/sec.

In this section we closely follow Pahlavan, Howard, and Sexton [Pah93], who used measured channel profiles from several indoor sites to determine the probability of outage and the average probability of error of BPSK/DFE modems. It is shown that a BPSK/DFE modem with three forward and three feedback taps provides a data rate an order of magnitude higher than a BPSK modem operating without equalization. The results of simulation based on the measured channel profiles are compared with performance predictions obtained from computer simulations and analytical bounds.

Performance Prediction Based on Channel Measurements. Here we use the time-domain measurement database described in Chapter 5 to predict the performance of the modem. These measurements provide the time profile of the channel at 288 locations in five areas (identified as A through E) on manufacturing floors of three companies. The results of these measurements were summarized in Table 5.1 of Chapter 5. For each profile, 1000 impulse responses are formed by associating 1000 independent uniform random phases with each measured path amplitude. Using the method described in previous sections, the bit error rate of a BPSK modem is calculated for each of the 288,000 impulse responses. Using the BERs calculated for all locations, the average error rate and the probability of outage for each area are determined.

Figure 8.26 shows the average BER for the BPSK/DFE modem with three forward and one, three, and five feedback taps in area C. Three feedback taps are adequate to provide an average error rate better than 10^{-4} for data rates around 5 Mbits/sec. We use this value for the rest of our computations. Figure 8.27 shows the average probability of error versus SNR in area D for the BPSK modem without equalizer and the BPSK/DFE modem at various data rates. For the 0.1-Mbit/sec rate, where the multipath spread is negligible relative to the symbol duration, the performance of the BPSK/DFE modem is slightly better than the performance of the BPSK modem. As the data rate is increased, performance of the BPSK modem is degraded due to the additional ISI. For the BPSK/DFE modem, initially, the performance gain due to the in-band (implicit) diversity is more than the performance loss caused by ISI. Eventually, the data rate of the BPSK/DFE modem increases to a point where the harmful effects of increasing ISI exceed the implicit diversity gain, and the average probability of error increases.

For 1 Mbit/sec and high SNR, the BPSK/DFE modem performs three orders of magnitude better than the BPSK modem at the same data rate and one order of magnitude better than the BPSK modem at 0.1 Mbits/sec. The BPSK/DFE modem at 10 Mbits/sec performs at nearly the same level as the BPSK modem at 1 Mbits/sec. These results indicate that for area D maximum data rates with equalization are at least 10 times greater than the data rates achievable without equalization. Data collected in other measurement areas provide similar results.

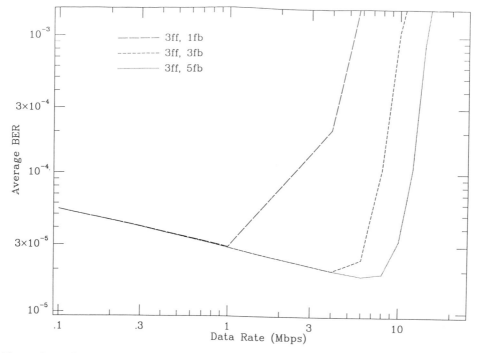

Figure 8.26 Average probability of error of the BPSK / DFE modem with three forward taps and one, three, and five feedback taps in area C. The SNR is 40 dB.

Figure 8.28 shows the probability of outage versus the BER outage threshold for BPSK and BPSK/DFE modems in area D with SNR = 40 dB and various data rates. The observations made regarding Fig. 8.27 are also supported by the results given in this figure.

Performance Prediction Based on Computer Simulation of the Channel.
Here we use computer-enerated profiles and channel modeling for performance prediction in order to determine the sensitivity of the results to the accuracy of the profiles. The profiles are generated from the model suggested in [Sal87b], which is based on the measurements made in a research laboratory. The performance evaluation methods are identical to those discussed in the previous section, except that computer-generated profiles are used instead of measured channel profiles. The average BER and the probability of outage are calculated for 50,000 channel impulse responses generated by associating 1000 independent phases with 50 computer-generated channel profiles. The results for 50,000 independent channel impulse responses are compared with the previous results to verify the accuracy of the modeling.

Figure 8.29 shows the probability of outage versus outage threshold in all five areas and the outage results obtained with the computer-simulated channel model, all for a BPSK modem operating at 1 Mbit/sec and SNR = 40 dB. Figure 8.30 shows a similar plot for the BPSK/DFE modem at 8 Mbits/sec. For both BPSK

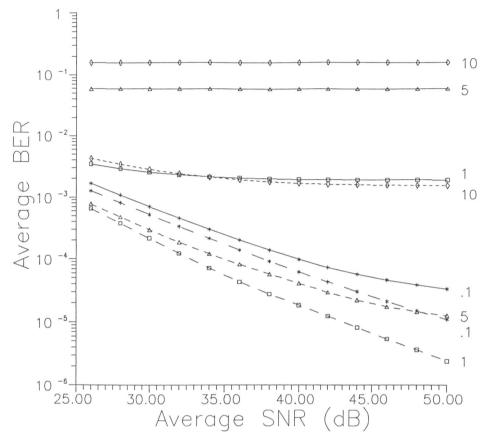

Figure 8.27 Average probability of error versus average SNR for the BPSK modem without equalizer (solid lines) and the BPSK / DFE modem (dashed lines) operating in area D.

and BPSK/DFE modems, beyond certain outage thresholds, a change of 10 orders of magnitude in the threshold results in at most one order of magnitude change in the probability of outage. This shows that when the channel is in a fade, the performance is unacceptable regardless of the chosen outage threshold. For both BPSK and BPSK/DFE modems, the performance in area B with metallic screens and in area D (the "jungle") are worse than the performance in areas A and E with ample open areas providing many locations with LOS propagation between transmitter and receiver. The results obtained from the computer simulations are seen to lie roughly in the middle of all the performance curves. Note also that performance predicted for area C, with its large pieces of machinery, benefits from the use of equalization relatively more than the other areas.

Analytical Performance Prediction. As we saw in the analysis of the DMF and the specific example in that section, if the channel is assumed WSSUS and the delay power spectrum is known, it is possible to develop an analytical framework

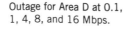

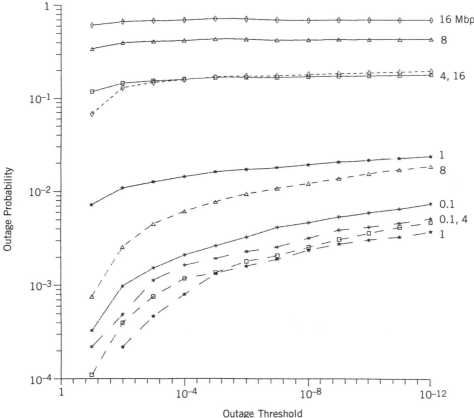

Figure 8.28 Probability of outage versus outage threshold (outage BER) for BPSK (solid lines) and BPSK / DFE (dashed lines) modems operating at various data rates in area D with SNR = 40 dB.

for performance prediction. This analysis can be further extended to performance evaluation of the DFE with an infinite number of feedback taps [Mon77]. If we wish to apply these analytical results to portable and mobile radio channels, we need to determine the delay power spectrum based on measured data. For fading multipath channels such as troposcatter, the multipath is caused by the reflection from small scattering elements in the atmosphere, which are changing continuously. In these channels, for a fixed location of the transmitter and receiver, the channel impulse response is a continuous-delay random process $h(\tau, t)$. As we explained in Chapter 3, if we assume that the channel is WSSUS, we have

$$E\{h(\tau, t)h^*(\tau_1, t_1)\} = Q(\tau, t - t_1)\delta(\tau - \tau_1)$$

where $Q(\tau, t)$ represents the received power as a function of multipath delay τ, measured at time t, and $\delta(\tau)$ is the delta function. For a slowly fading channel,

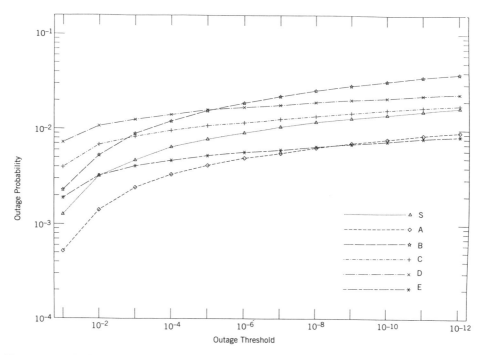

Figure 8.29 Probability of outage versus outage threshold for a BPSK modem in all five areas and for the computer-simulated channel model (*S*). The data rate is 1 Mbit / sec and the SNR is 40 dB.

where the Doppler spread of the channel is much lower than the data rate, $Q(\tau, t)$ is approximated by

$$Q(\tau, t) \simeq Q(\tau, 0) = Q(\tau)$$

in which $Q(\tau)$ is referred to as the delay power spectrum of the channel [Bel69]. The delay power spectrum $Q(\tau)$ represents the time-averaged received power as a function of the delay between arrivals of different paths. It can be viewed as the time average of the squared magnitude of the channel impulse response.

For the portable and mobile radio channels, multipath is caused by reflections from walls, ceiling, floor, furniture, and objects moving in the vicinity of the transmitter and receiver. As a result, the channel impulse response is a function of location of the transmitter and receiver, and the delay associated with the arriving paths, which we denote by writing the impulse response as $h(\tau, x)$. Because the portable and mobile radio channels are represented by a set of channel impulse responses $h(\tau, x)$ measured in different locations in an area, we may define an equivalent delay power spectrum as the average of the channel impulse responses over all locations in an area. Then, similar to the troposcatter channel, if we assume that the channel is WSSUS over all locations, we have

$$E_x\{h(\tau, x)h^*(\tau_1, x)\} = C(\tau)\delta(\tau - \tau_1)$$

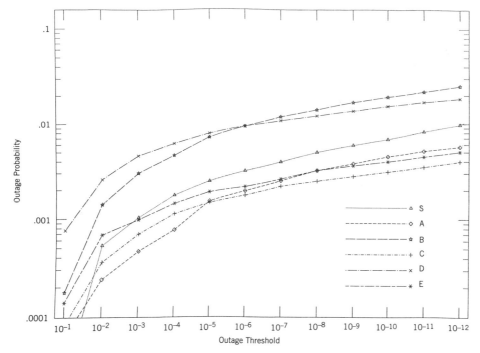

Figure 8.30 Probability of outage versus outage threshold for the BPSK modem in all five areas and for the computer-simulated channel model (*S*). The data rate is 8 Mbits / sec and the SNR is 40 dB.

where $E_x\{\cdot\}$ denotes the average over the ensemble of all channel profiles observed in an area, and $C(\tau)$ is the equivalent delay power spectrum of the channel.

For a set of measured profiles $h(\tau, x)$ from different locations in an environment, it is possible to determine the equivalent delay power spectrum, $C(\tau)$, experimentally. Starting from the arrival of the first path, the time axis is divided into bins of width equal to the transmitted pulse duration. The average received power over all profiles is calculated for each bin and plotted against the arrival time of the paths. Figure 8.31 shows the equivalent delay power spectrum measured from the indoor measurement database used in this section, and it also shows an exponential curve obtained as best-fit to the empirical data. The decay rate of the exponential curve is 15 nsec. The exponential fit to the equivalent delay power spectrum is used for performance predictions.

Calculations of the BER based on the equivalent delay power spectrum are done for an infinite number of feedback taps, which provides a lower bound to the calculations based on real data or on the computer-generated channel impulse responses. The theoretical results presented for the BPSK/DFE modem are based on the analysis provided in [Mon77]. This work assumes an infinite number of feedback taps, which results in complete removal of the ISI due to the past decisions. In this work, the probability density function of the SNR per bit after

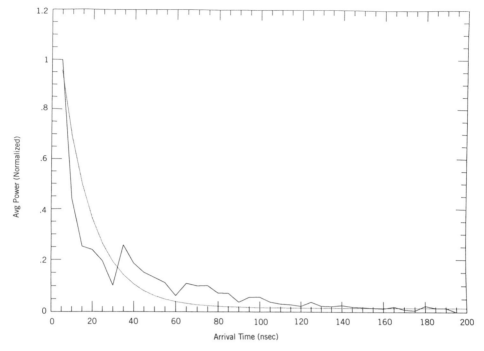

Figure 8.31 Equivalent delay power spectrum obtained from the database used in this chapter, along with an exponential fit to the empirical data.

equalization, γ_b, is given by

$$f_\Gamma(\gamma_b) = \sum_{i=1}^{D} \sum_{k=0}^{K} \frac{A_{ik} \gamma_b^{k-1}}{(\Gamma_j)^k (k-1)!} \exp\left[\frac{-\gamma_b}{\Gamma_j}\right] \tag{8.8.8}$$

where $\Gamma_j = \lambda_j \overline{\gamma_b}$ ($\overline{\gamma_b}$ is the average SNR per bit) and λ_j are the eigenvalues of the matrix $\mathbf{G}^{-1}\mathbf{C}(t_0)$. The elements of the matrix $\mathbf{C}(t_0)$ are given by

$$C_{kl}(t_0) = \int_{-\infty}^{\infty} g(t_0 - k\tau_s - u) g(t_0 - l\tau_s - u) Q(u)\, du$$

where τ_s is the equalizer tap spacing, $g(\tau)$ is the overall impulse response of the pulse shaping filters (raised cosine, 50% rolloff), and $Q(u)$ is the delay power spectrum of the channel. The elements of the matrix $\mathbf{G}$ are given by

$$G_{kl} = g(k\tau_s - l\tau_s) + \overline{\gamma_b} S^2 \sum_{i=1}^{I} C_{kl}(t_0 - iT)$$

where S^2 is the variance of the ISI, I is the number of future symbols with significant ISI, t_0 is the sampling instant, and T is the symbol period.

In our calculations these equations are used with the equivalent delay power spectrum $C(\tau)$ instead of the delay power spectrum $Q(\tau)$. The shape of the equivalent delay power spectrum is exponential with the decay rate of 15 nsec, as shown in Fig. 8.31, and the equalizer uses three $T/2$-spaced forward taps. Then the probability of outage is determined from [Sex89b] to be

$$P_{\text{out}} = \int_0^{\gamma_{\text{out}}} f_\Gamma(\gamma)\, d\gamma$$

$$= 1 - \sum_{i-1}^{D} \sum_{k-0}^{N} \frac{A_{ik}}{\Gamma_j^k} \exp\left[-\frac{\gamma_{\text{out}}}{\Gamma_j} \sum_{i=0}^{k-1} \frac{\gamma_{\text{out}}^i}{i!} \Gamma_j^{k-i} \right] \qquad (8.8.9)$$

where γ_{out} is the SNR which would produce the threshold error rate if the channel were nonfading. Figure 8.32 shows the probability of outage versus data rate using an equivalent delay power spectrum with one and two orders of diversity. It can be seen from the figure that the system with two orders of diversity performs more than three orders of magnitude better than the system with single diversity.

Figure 8.33 shows the probability of outage versus data rate at SNR = 40 dB and BER threshold 10^{-4} for the theoretical bound using the equivalent delay

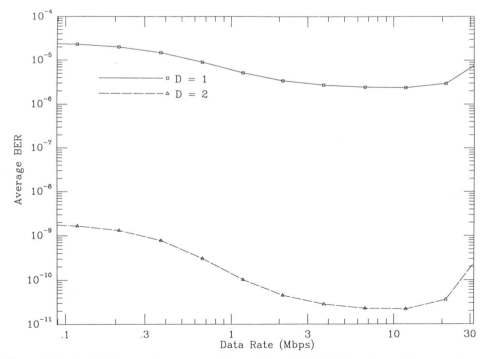

Figure 8.32 Probability of outage versus data rate for a BPSK / DFE modem with one and two orders of diversity, for SNR equal to 40 dB. Theoretical results obtained using an equivalent delay power spectrum.

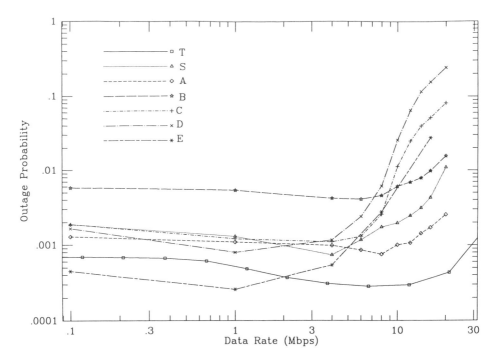

Figure 8.33 Probability of outage versus data rate for a BPSK / DFE modem in each of the five areas, for the computer-simulated channel model (S), and theoretical bound results (T). The SNR is 40 dB and the outage threshold is 10^{-4}. The theoretical bound uses the equivalent delay power spectrum.

power spectrum, together with results from computer-generated channel profiles and the results for the measured channel profiles in the same five areas, A through E, described earlier. The simplified calculation using the equivalent delay power spectrum and an infinite number of feedback taps provides a close lower-bound approximation to the actual probability of outage with three forward and three feedback taps in four out of the five areas. The results of performance analysis also show close agreement with the results obtained from the computer-simulated channel. At the lower data rates, the worst performance is observed in area B, containing a storage area partitioned with metallic screens. The best low data rate performance is observed for area E, which is a large open area used for final inspection of automobiles coming off the assembly line. At high data rates, area D, the "jungle," yields worst performance, whereas area A with its considerable open space shows the highest data rates.

The average probability of error for a BPSK/DFE modem [Mon 77] is given by

$$
\begin{aligned}
\overline{P_e} &= \frac{1}{2} \int_0^\infty \mathrm{erfc}\!\left(\sqrt{\gamma_b}\right) f_\Gamma(\gamma_b)\, d\gamma_b \\
&= \sum_{i=1}^{D} \sum_{k=0}^{N} A_{ik} \left(\frac{1 - u_j}{2}\right)^k \sum_{i=0}^{k-1} \binom{k-1+i}{i} \left(\frac{1 + u_j}{2}\right)^k
\end{aligned} \tag{8.8.10}
$$

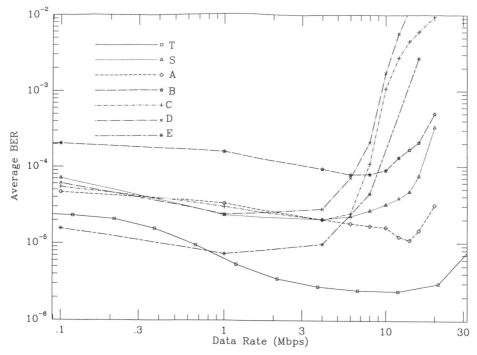

Figure 8.34 Average probability of error versus data rate for a BPSK / DFE modem in each of the five areas, for the computer-simulated channel model (*S*), and for theoretical bound results (*T*). The SNR is 40 dB and the theoretical bound uses the equivalent delay power spectrum.

where

$$u_j = \left(\frac{\Gamma_j}{1 + \Gamma_j} \right)^{1/2}$$

Figure 8.34 shows the average BER versus data rate at SNR = 40 dB for the theoretical bound using the equivalent delay power spectrum, together with results from computer-generated channel profiles and the results found for the measured channel profiles in the same five areas. For low data rates in area E, which is the most open area, the performance is better than the values predicted using the lower bound.

8.9 OTHER METHODS TO INCREASE THE DATA RATE

Thus far in this chapter we have described three techniques—DMF, MLSE, and DFE—capable of increasing the data rates achievable with modems operating on fading multipath radio channels. All three methods use in-band diversity to exploit the multipath condition and improve modem performance at high data rates. In this section we introduce additional methods for increasing the data rate, and we compare them with the methods described earlier. We begin with a detailed analysis of the effects of using sectored antennas and the use of multicarrier systems. Then we examine briefly multiamplitude and multiphase modulation and coding, and multirate transmission.

8.9.1 Sectored Antennas

One means of increasing the data rate is to use some form of antenna diversity, and three approaches are depicted in Fig. 8.35. Spatial diversity can be implemented using multiple separate antennas of the same design, or polarization diversity can be used, which has been employed in an NCR WLAN product. A

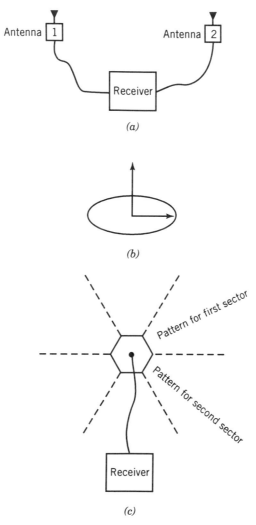

Figure 8.35 Three approaches to providing antenna diversity. (*a*) Physical antenna diversity. (*b*) Polarization diversity. (*c*) Sectored antenna diversity.

third approach is to use a sectored antenna, as is done in the ALTAIR WLAN product. The spatial diversity provided by multiple antennas, and the polarization diversity provided by orthogonally polarized antennas, will each reduce the power margin required to overcome the effects of power fluctuations caused by fading. The diversity analysis presented earlier in this chapter applies directly to the multiple-antenna and polarization diversity techniques. Therefore in this subsection we give primary attention to the sectored-antenna technique.

A sectored antenna observes the signal arriving from different directions (paths) and selects the sector with maximum power. The use of sectored antennas has two advantages relative to the first two techniques: (1) A sectored antenna reduces the multipath in each sector, increasing the maximum data rate achievable on the channel, and (2) effective diversity can be provided without requiring wide physical separation between antennas, making compact product packaging feasible. As an example, the ALTAIR WLAN product uses two six-sector antennas, one each at the transmitter and receiver, which provides a total of 36 effective sectors. With such a design, the array of all multiply reflected signal paths arriving at the receiver is divided into 36 subsets, and with appropriate signal processing (say selection diversity), the receiver can extract the subset producing the least signal degradation, discarding all the other signal arrivals.

Performance analysis given in the last section for decision feedback equalization (DFE) with omnidirectional antennas over measured and modeled indoor radio channels showed that data rates on the order of 10 Mbits/sec are attainable with BPSK/DFE modems operating in indoor areas. It has been demonstrated that sectored antennas can also provide similar enhancements for high speed wireless data communications in indoor areas [Fre91]. From one perspective, sectored antennas and DFE techniques represent two different approaches to exploiting diversity. A DFE modem uses the signals arriving from different paths to provide implicit or in-band diversity which is, in effect, combined with an adaptively optimized FIR digital filter. In contrast, sectored antennas serve to create signal components of reduced multipath dispersion, from which the receiving system extracts the component of best quality.

In the last section, measured channel profiles and statistical models based on measured channel profiles were used for performance prediction of the DFE modems. These measurements and modeling techniques cannot be applied to the analysis of the modems using sectored antennas, because all of the available wideband channel measurements and related statistical models have been developed from measurements made with omnidirectional antennas. These measurements identify all the arriving paths in the time domain, but do not identify the directions from which the individual paths arrive. To evaluate the performance of a modem with a sectored antenna, the ray-tracing method, which provides the direction of the arriving paths, should be used to model the propagation characteristics. In the remainder of this section, following Yang, Pahlavan, and Holt [Yan94c], we use the ray-tracing algorithm for the performance analysis of BPSK and BPSK/DFE modems operating with a sectored receiver antenna in a typical indoor area used for WLANs.

In analyzing the performance of sectored antennas, we assume that the receiver is equipped with a six-sector directional antenna whose polarizations are vertical.

The ith idealized antenna pattern is defined by the function

$$g_i(\phi_k) = \begin{cases} f\left(\phi_k - \dfrac{\pi i}{3}\right), & \dfrac{\pi i}{3} - \dfrac{\theta}{2} \le \phi_k < \dfrac{\pi i}{3} + \dfrac{\theta}{2} \\ 0, & \text{otherwise} \end{cases}$$

where $g_i(\phi_k)$ is the normalized power gain, ϕ_k is the antenna orientation angle, and θ is the 3-dB beamwidth of the antenna which is around $\pi/3$ for a six-sector antenna. The channel impulse response used for each sector of the antenna includes all paths arriving in that direction with appropriate attenuation given by the antenna pattern. The overall impulse response for an omnidirectional antenna includes all paths. The technique used in the previous section will be used for performance evaluation of systems with omnidirectional and sectored antennas, using the appropriate channel impulse response in each case. For the sectored antenna the impulse response associated with the sector receiving maximum power is used in the calculations. In another words, ideal selection combining is assumed to be used to process the diversity branches provided by the sectors of the antenna. Other diversity combining techniques or selection criteria may result in better performance in some cases, but selection diversity is considered the most attractive technique due to its relative simplicity of implementation.

To compare the performance of omnidirectional and sectored antennas, the received unfaded SNR at 1-m from the transmitter is assumed to be the same for both antennas. As an example for calculation of the unfaded SNR, assuming that the received bandwidth B is 10 MHz, the noise temperature T is 290° K (17° C), transmitter power is 100 mW, and the received front end noise figure is 9 dB. Then the thermal noise, kTB, is -104 dBm [Lee89] at the receiving antenna and thus the net noise level is -95 dBm. If we assume the attenuation at 1 m from the transmitter is 35 dB, then the unfaded SNR will be 80 dB for both omnidirectional and sectored antennas.

Performance Prediction in an LOS Environment. Office environments often include large, open inner areas. A simple square-room LOS environment will now be used to compare the performance of the four types of modems being examined. The receiver is assumed to be located at the center of the room, and the transmitter is located at many different locations throughout the room.

Figure 8.36 illustrates outage probabilities versus data rate for systems using different modems in a 30-m × 30-m room. The transmitter power is assumed to be 100 mW. This figure illustrates that if a system with an omnidirectional antenna is used in a 30-m × 30-m room, the data rate can reach only 3 Mbits/sec with an outage rate of 0.01 for an acceptable error rate of 10^{-5}. If a six-sector antenna system is used, and the selection criterion is to choose the sector with the highest received power, a data rate of 20 Mbits/sec can be achieved, whereas a system with a DFE can only reach a data rate of less than 15 Mbits/sec. The results in the figure indicate that a DFE modem with three forward taps and three feedback taps can operate successfully at a data rate of around 10 Mbits/sec but shows very high outage probabilities at data rates beyond about 15 Mbits/sec. However, it can be seen that a DFE modem operating with a sectored antenna can operate successfully at data rates up to about 40 Mbits/sec.

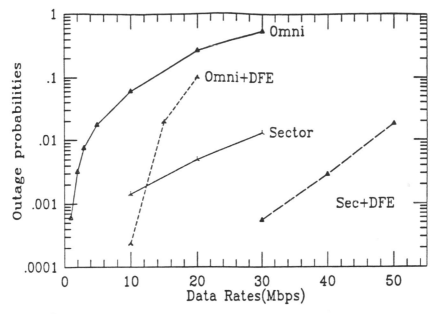

Figure 8.36 Outage probabilities versus data rate for systems with four different modems operating in a 30-m × 30-m room.

Figure 8.37 shows outage probabilities versus unfaded SNR at 1 m from the transmitter for a BPSK modem operating at three different data rates with sectored antennas in a 30-m × 30-m room. It can be seen that the probabilities of outage show little sensitivity with respect to unfaded SNR when the transmitter power level is in the region of high unfaded SNR.

Figure 8.38 shows the probabilities of outage versus the room size (square room with one wall length given) for different modems with a transmitter power of 100 mW at a data rate of 20 Mbits/sec. For a room smaller than 20 m × 20 m, it is difficult to conclude from these results whether the sectored antenna or the DFE delivers better performance. However, for room sizes larger than 30 m × 30 m, significantly better performance will be achieved with the sectored antenna.

Figure 8.39 shows outage probabilities versus data rate for a BPSK modem operating at two different power levels with three different sectored antennas (see Fig. 8.40): (I) a nonoverlapped antenna span pattern, (II) an optimum antenna span pattern, and (III) the mathematically ideal pattern. The results in Fig. 8.39 show that the sectored antenna patterns provide significant benefits for modem performance, and that a 10-dB change in the transmitted power causes little change in the outage probability. These results also provide information on the performance gains achievable with different antenna patterns.

Performance Prediction in an Obstructed LOS (OLOS) Environment. The simulation results discussed above show that using a sectored antenna is an effective technique for counteracting the harmful effects of multipath. In this section we examine the sectored antenna and the DFE in an OLOS environment. The floor plan that was analyzed in Chapter 6 for 2 D ray tracing is used again

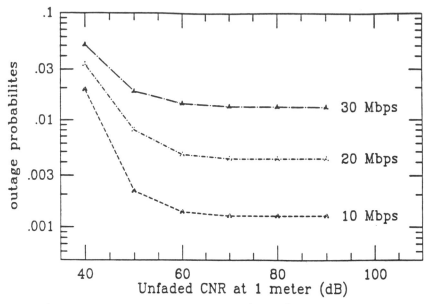

Figure 8.37 Outage probabilities versus unfaded CNR (or SNR) at 1 m from the transmitter given different data rates for a BPSK modem with sectored antennas in a 30-m × 30-m room.

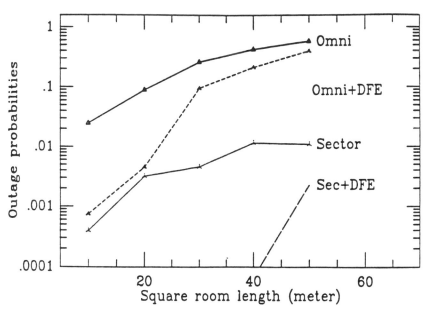

Figure 8.38 Outage probabilities versus room size for different modems with transmitted power 100 mW at a data rate of 20 Mbits / sec.

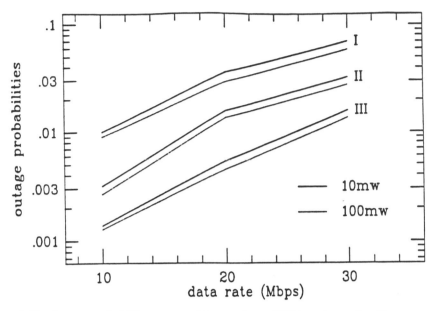

Figure 8.39 Outage probabilities versus data rate for a BPSK modem with different sectored antennas. (I) Nonoverlapped span pattern. (II) Optimum span pattern. (III) Ideal pattern. (See Fig. 8.40.)

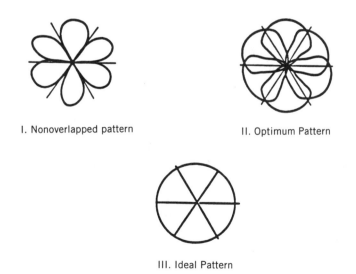

I. Nonoverlapped pattern   II. Optimum Pattern

III. Ideal Pattern

Figure 8.40 Three sectored antenna patterns. (I) Nonoverlapped antenna-span pattern. (II) The optimum antenna-span pattern. (III) Ideal pattern.

Figure 8.41 Floor plan of one section of the second floor of the Atwater Kent Laboratory. This is the site of the indoor propagation experiments described earlier.

here in Fig. 8.41, with rooms numbered. The receiver is located at the center of room 1 and the transmitter is moved to various locations in the floor plan. In order to obtain the outage probabilities, the number of simulations run for each room is around 5000. The maximum data rates are obtained for four different types of modems in each room, and an average maximum data rate is obtained for the overall floor plan.

From Table 8.3, it can be seen that the average maximum data rate over the entire floor plan is 12 Mbits/sec for a BPSK/DFE modem with an omnidirectional antenna and 10 Mbits/sec for a BPSK modem with a sectored antenna. The DFEs with three forward taps and three feedback taps perform slightly better than the nonequalized modems with six-sector antennas. In a small LOS environment such as room 1, high data rates can be achieved using either technique. If the receiver moves to one of the adjoining rooms, such as room 2, 3, 5 or 6, the BPSK/DFE modem with omnidirectional antennas can still attain data rates above 15 Mbits/sec. However, the maximum achievable data rate for the BPSK modem with the sectored antennas in room 3 drops to 12 Mbits/sec. In rooms 4, 7, and 8, only the BPSK/DFE modem with sectored antennas can achieve a data rate of above 10 Mbits/sec, and sectored antennas provide slightly better performance than DFE.

The results of Table 8.3 might seem to be overly optimistic but are in fact quite reasonable for the case examined here. The entire floor plan is confined to a

TABLE 8.3 Maximum Data Rates for BPSK and BPSK / DFE Modems with and without Adaptive Antenna Arrays in Different Rooms

Room	Maximum Data Rate (Mbit / sec)			
	Omni	Omni + DFE	Sector	Sector + DFE
1	8	40	50	50
2	5	20	12	25
3	4	15	13	20
4	2	7	9	13
5	4	20	20	40
6	7	25	30	50
7	2	6	8	12
8	2	7	8	15
Overall	3	12	10	18

relatively small area in which the maximum radius is about 10 m, the rms delay spread is less than 30 nsec, and the received power is not exceptionally low in any location. The results shown for transmitter and receiver both in room 1 show that any modem can achieve high date rates. In reality, of course the achievable data rates will be somewhat lower than are indicated here, due to the effects of phase and timing jitter.

The poorest performance is achieved in room 7. Figure 8.42 shows that if the data rate is below 12 Mbits/sec, the performance of the BPSK modem with sectored antennas is superior to that of a BPSK/DFE modem with an omnidirectional antenna. However, the sectored antenna seems to be less effective for data rates higher than 15 Mbits/sec in this worst case. The BPSK/DFE

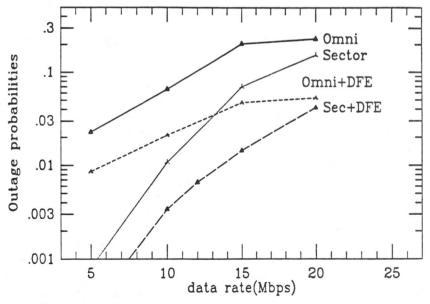

Figure 8.42 Outage probability versus data rate for four types of modems operating in room 7.

modem with sectored antennas offers relatively little advantage over the BPSK/DFE modem with omnidirectional antennas at very high data rates.

8.9.2 Multicarrier Modulation

Another approach to dealing with the problem of multipath distortion is one which has been used in radio modems for more than 30 years, known by several names, including *multicarrier-modulation* or *multitone-modulation* [Bin90]. The concept here is very simple: Instead of modulating a single carrier at rate R symbols/sec, we use n carriers spaced by about R/n Hz and modulate each of the carriers at the rate R/n symbols/sec. The advantage of this scheme is that on a multipath channel having a given value of multipath spread, say τ_m sec, the multipath is in effect reduced relative to a symbol interval by the ratio $1/n$ and thus imposes less distortion in each demodulated symbol. If the symbols are made sufficiently long relative to the multipath spread, reliable demodulation performance can be achieved without the need for adaptive equalization. Therefore, on multipath (frequency-selective) fading channels, multicarrier modulation provides a simpler alternative to single-carrier modulation with adaptive equalization. A further advantage of multicarrier modulation is that in frequency-selective fading, the subchannels provide a form of frequency diversity, which can be exploited by applying error-control coding across symbols in different subchannels.

Example 8.3. Common examples of multicarrier systems include several HF modems widely used by the U.S. military that operate with phase or frequency modulation on "parallel tones," each tone or carrier typically modulated at 75 baud. As an example, let us postulate a 16-tone modem using QPSK modulation at 75 baud on each tone, for an overall data rate of 2400 bits/sec. Consider a typical HF channel with multipath spread of $\tau_m = 2$ msec. If a single carrier were modulated at 1200 ($= 16 \times 75$) baud, the symbol duration would be 0.42 msec, for which the 2-msec multipath spread would cause enough intersymbol interference that adaptive equalization would be required for acceptable BER performance. However, with 16 channels modulated at 75 baud, the symbol interval in each channel is 13.3 msec and the multipath distortion can be essentially eliminated by providing "time guards" between successive symbols in each channel. (Frequency guard bands may also be required in order to avoid interference between adjacent subchannels.) Using exactly this approach, so-called "parallel-tone" modems have been used very effectively on long-haul HF circuits for a number of years. Several papers describing the theoretical and measured performance of parallel-tone HF modems are included in [Bra75].

In recent years, the multicarrier transmission approach has been investigated for application to microwave radio [Cho90], land-mobile [Sas93], and indoor wireless [Har93, Yee93, Yan94] communications. A brief discussion of some of the results presented in [Har93] will serve to illustrate the principles of designing a multicarrier system for a fading channel. In that paper the authors derive theoretical BER performance for a multicarrier system using DPSK modulation and differential detection on an indoor frequency-selective Rayleigh channel. Figure 8.43 shows block diagrams of the N-carrier transmitter and receiver that were

(a)

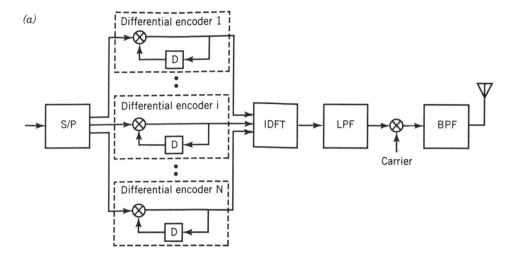

(b)

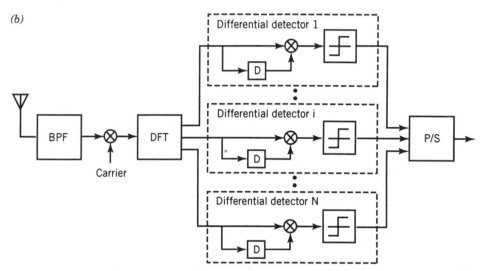

Figure 8.43 A multicarrier transmission system. (a) Transmitter. (b) Receiver. S / P, serial / parallel converter; D, unit delay; IDFT, inverse Fourier transformer; LPF, low-pass filter; BPF, bandpass filter; DFT, Fourier transformer; P / S, parallel / serial converter; □, decision blocks. (From [Har93] © IEEE.)

analyzed. The modulator and demodulator are implemented with an inverse discrete Fourier transformer (IDFT) and discrete Fourier transformer (DFT), respectively, instead of with N oscillators. The DFT technique for multicarrier transmission allows minimum frequency separation between carriers and efficient demodulation with computationally efficient DFT algorithms [Wei71]. The transmitted signal $x(t)$ can be written as

$$x(t) = \sum_{i=-\infty}^{\infty} \sum_{k=0}^{N-1} \text{Real}\left[c_{ki} e^{j2\pi f_k(t-iT)}\right] f(t-iT) \qquad (8.9.1)$$

In the equation above, f_k is the frequency of the kth carrier,

$$f_k = f_0 + \frac{k}{t_S} \tag{8.9.2}$$

where f_0 is the lowest carrier frequency. The function $f(t)$ is the pulse waveform for each transmitted symbol, defined as

$$f(t) = \begin{cases} 1, & -\Delta \le t \le t_S \\ 0, & t < -\Delta, t > t_S \end{cases} \tag{8.9.3}$$

where Δ is the time-guard interval, t_S is the observation interval, $T = \Delta + t_S$ is the total symbol duration, and c_{ki} is the output of the kth differential encoder in the time interval $(iT - \Delta, iT + t_S)$. In [Har93], M-ary DPSK modulation is used on each of the subcarriers. Therefore, the transmitted signal $x(t)$ is the sum of N M-ary DPSK signals with symbol duration T and with carriers separated by $1/t_S$ Hz. The $1/t_S$-Hz separation of carriers insures orthogonality (when the receiver is correctly time-synchronized) of symbols demodulated on different carriers.

The guard-time implementation described above is referred to as a *receiver-gated time guard*, in which length T symbols are transmitted in continuous succession, but the receiver is time-gated "on" only during the observation interval t_S. An alternative implementation is a *transmitter-gated time guard*, in which the transmitter is actually turned off for the interval Δ in each symbol interval T in order to keep successive symbols separated in the multipath channel. These two time-guard approaches result in somewhat different synchronization characteristics, an issue which is treated in detail in [Bel65].

Continuing with our discussion of the receiver-gated multicarrier implementation, as treated in [Har93], we briefly discuss two system issues: (1) BER performance in multipath fading and (2) the overall bandwidth utilization efficiency achieved by the system design. System performance is directly related to the choice of time-guard interval Δ in relationship to the multipath characteristics of the channel. As a specific example, [Har93] considered a system using $N = 32$ carriers operating on an indoor wireless channel with the symbol duration on each carrier chosen as $T = 1/128 \times 10^{-3}$ sec or 7.8125 μsec. With QDPSK modulation, this design yields a data rate of 256 kbits/sec per carrier for an overall data rate of 8.192 Mbits/sec. The indoor channel was characterized by a nine-ray multipath model, with rays fluctuating independently with a Rayleigh distribution, and overall rms delay spreads of $\tau_{rms} = 50$ and 100 nsec, typical values for such a channel. Figure 8.44, from [Har93], shows the theoretical BER performance versus normalized guard time (Δ/T) for the two assumed values of τ_{rms} and three values of $\bar{\gamma}_b$. These results were calculated for a case of frequency-selective, time-non-selective fading, which means that the multipath rays are fixed in time. Note that for each combination of rms multipath spread and SNR there is an optimum value of normalized guard time for which BER is minimized. With shorter guard times, the BER increases due to the multipath distortion at the edges of the received symbols. With longer guard times, the BER increases due to the nonutilization of signal energy transmitted during the guard time. It can be seen from Fig. 8.44 that

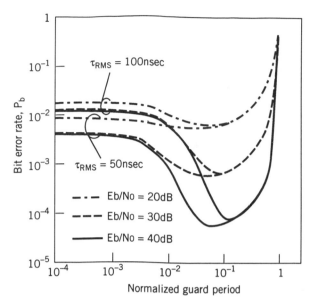

Figure 8.44 BER versus normalized guard time Δ/T for a multicarrier system operating on an indoor wireless channel. The number of carriers is 32, the modulation is QPSK, and the overall data rate is 8.192 Mbits / sec. (From [Har93] © IEEE.)

the optimum values of normalized guard time are in the range from 0.03 to 0.1 or about 250 to 750 nsec for the system considered here. As one would expect, smaller values of τ_{rms} result in shorter optimum guard times and lower levels of achievable BER.

Figure 8.45, also from [Har93], shows BER performance versus normalized delay spread (τ_{rms}/T) for three different modulation techniques used with the same 32-carrier system, and for a single-carrier QDPSK system operating at 8.192 Mbits/sec without equalization. The normalized guard time for each of the multicarrier cases is $\Delta/T = 3.03 \times 10^{-2}$, and the SNR is 40 dB for all cases. The results given in the figure clearly show the greater robustness of the multicarrier systems relative to the non-equalized single-carrier system in the indoor multipath environment.

It is possible to combine the basic multicarrier transmission method with various modulation and coding techniques to achieve various system objectives. As one example, a recent paper [Yee93] describes a technique given the name *multicarrier code-division multiple access* (MC-CDMA) and analyzes its performance on indoor wireless channels. With this technique, symbols are transmitted in each time interval on multiple subcarriers, where each subcarrier is modulated with a $\pm\pi$ offset in accordance with a pseudonoise (PN) sequence. Thus the signal structure is akin to conventional CDMA designs except that in MC-CDMA the transmitted signal has a PN structure in the frequency domain rather than the time domain. Multiple access is provided for in this system by allowing different users to transmit in the same time interval with orthogonal PN codes. As with other multicarrier systems, the subchannel symbol duration is chosen to be long with respect to the

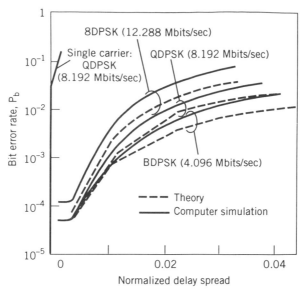

Figure 8.45 BER versus normalized delay spread τ_{RMS} / T_S for multicarrier systems and a single-carrier system operating on a wireless indoor channel. The normalized guard time is $1 / 33$ for all cases. $E_b / N_0 = 40$ dB; $D / T_S = 1 / 33$; $T_S = 7.8 / 25$ msec. (From [Har93] © IEEE.)

expected multipath spread. The advantage for this technique over a more conventional CDMA technique using single-carrier transmission and PN coding in the time domain is that the PN codes can be detected with a relatively simple receiver structure that does not have to deal with interchip multipath interference. In [Yee93] the authors provide a detailed analysis of this technique, including analysis of performance with the inclusion of diversity combining.

Yet another embellishment of the multicarrier technique is to incorporate the use of error-control coding to combat the effects of frequency-selective fading. The key concept here is to exploit the fact that in frequency-selective fading, there will be some degree of independence of the fading across the subcarriers; and by applying coding across the subcarriers, symbols on faded subcarriers can be corrected. In [Yan94a] this technique was studied for application to indoor wireless channels. Reed–Solomon (n, k) block codes were used to code across n carriers, where k of the carriers transmitted information symbols and the remaining $n - k$ carriers transmitted parity check symbols. The scheme was analyzed for use with BPSK and QPSK modulation on the subcarriers, and the results showed significant improvements in outage probabilities relative to uncoded multicarrier transmission.

8.9.3 Multirate Transmission

Yet another approach to increasing data rate in the presence of multipath fading is to use a multirate modem. A multirate modem provides one or more "fallback" modes of operation for increased reliability of communication under degraded channel conditions. For example, in an area where we have locations of good

TABLE 8.4 Maximum Data Throughput (R_T) Achievable with Single-Rate and Optimum Dual-Rate Modems Operating with Various Orders (M) of Diversity

M	Single-Rate R_T (Mbits / sec)	Dual-Rate R_T (Mbits / sec)	P_{out}	R_1 / R_2
1	0.27	2.6	0.035	16.0
2	2.4	6.1	0.05	3.2
4	7.9	9.6	0.071	1.4
8	11.7	15.5	0.22	1.3

Note: rms multipath spread is 58 nsec for all cases.

signal quality and others of poor quality, we can operate at higher rates in good locations and lower rates in poor locations. This is done in voiceband modems, for example, by using modes which operate with different numbers of points in their signal constellations. Table 8.4 shows the maximum data throughput achievable with single-rate and optimum double-rate modems operating with various orders of diversity [Win85, Zha90]. The results given in the table are based on an assumption of Rayleigh fading and a continuous multipath structure with a fixed rms multipath spread of 58 nsec. We see from the table that we can use a single-rate modem with dual diversity and achieve 2.4 Mbits/sec, a 10-fold increase in data rate over nondiversity operation. However, we also see that by using a dual-rate modem with no diversity we can achieve 2.6 Mbits/sec throughput. That is, we can achieve about the same data rate with either (a) a single-rate modem and two antennas or (b) a dual-rate modem and one antenna. Multirate modems can be incorporated into direct-sequence spread-spectrum systems where the processing gain can be adjusted in accordance with the environment. In this way we can operate with a fixed bandwidth but adjustable data rate. Another approach to adjusting the data rate with a fixed bandwidth is to use the multiamplitude and multiphase modulation techniques. In this method the number of points in the constellation is increased as the channel condition improves. It is also possible to combine the two methods so as to increase the flexibility of the modem.

8.9.4 Multiamplitude and Multiphase Modulation and Coding

Another technique for increasing data rate is to use multiamplitude and multiphase modulation and coding. Table 8.5 shows selected parameters for modems providing several steps of data rate increase over QPSK modulation. Each example

TABLE 8.5 Selected Parameters for Modems Providing Several Steps in Data Rate Increase Over QPSK Modulation

Modulation	Data Rate	Coding
QPSK	R	8-TCM
8-PSK	1.5 R	16-TCM
16-QAM	2 R	32-TCM
64-QAM	3 R	128-TCM

in the table utilizes trellis-coded modulation (TCM) as part of a combined modulation and coding design. In our preceding discussion on increasing data rates, we emphasized two objectives, one to compensate for the power loss caused by fading, the other to increase the data rate in the face of multipath constraints. One might well ask why we cannot use high-order multipoint signal constellations as is done in wireline modems. One problem is power fluctuations in the channel, which make it difficult to reliably demodulate signal sets with large numbers of points. Table 8.5 shows us that using 64-QAM modulation, which is currently a practical limit on modulation alphabet size for use in deep fading, can achieve only a three-fold increase in data rate over QPSK. However, we have seen that with the use of adaptive equalization or sectored antennas, we can increase the data rate by a factor of 10 or more over the simplest systems. For high-speed WLANs, this three-fold increase is not sufficiently attractive, especially when it is compared with other techniques. In the mobile communications industry, where even a factor of 2 or 3 translates into an increase in user channels by that amount, this approach is considered attractive and is being studied for application in new wireless networks.

QUESTIONS

(a)   What are the advantages and disadvantages of fractionally spaced equalizers?

(b)   Why is the DFE technique a popular choice for equalization in radio modems?

(c)   Name two applications in which fast converging algorithms are useful in modem design.

(d)   Name two applications for blind equalization.

(e)   What are typical data rates that a BPSK/DFE modem can provide in LOS indoor areas?

(f)   Explain why a sectored antenna would be used in a wireless LAN (e.g., the ALTAIR product) in preference to a DFE.

(g)   Explain why multiamplitude and multiphase modulation techniques are not very popular in the mobile radio industry.

(h)   Given that a bandwidth constraint does not exist, what is the data rate limitation for a multicarrier system?

(i)   Why do mobile radio systems use equalization rather than time gating and discrete matched filtering?

(j)   Explain the difference between implicit and explicit diversity. Which form of diversity is attainable in flat fading channels?

PROBLEMS

Problem 1. In Chapter 7 we showed that the error rate of a multiamplitude, multiphase modem with coherent detection can be approximated by erfc $\sqrt{d^2/2N_0}$, where d is the minimum distance between the points in the constellation and N_0 is the variance of the additive Gaussian noise. Observe that if we sketch the signal constellation and the decision lines, this equation can be modified to erfc $\sqrt{\delta^2/8N_0}$, where δ is the minimum distance of a point in the constellation from a decision line.

(a) Assume that a channel produces a fixed phase error θ. Give an equation for calculation of the probability of error for a QPSK modem operating over this channel. Use the minimum distance from a decision line, δ, for the calculation.

(b) Assume that the received signal-to-noise is 10 dB, and sketch the probability of error versus the phase error $0 < \theta < \pi/4$.

Problem 2

(a) For a 16-QAM Modem, sketch the probability of symbol error, P_s, versus E_s/N_0, where N_0 is the variance of the noise and E_s is the average energy per transmitted symbol.

(b) Repeat (a) for $10°$ phase error at the receiver, and compare the results with those of part (a).

Problem 3. Assume that a coherent BPSK modem operating at the data rate $R = 1/T$ uses a raised cosine pulse with 50% rolloff ($\alpha = 0.5$).

(a) Sketch the signal-to-intersymbol interference power (the variance of the sum of ISI terms) versus normalized timing error τ/T for $-T/2 < \tau < T/2$.

(b) Assume that the ISI noise forms a Gaussian distributed interference and that the effects of the additive noise are negligible. Sketch the error rate versus normalized timing error τ/T for $-T/2 < \tau < T/2$.

(c) Repeat (b) for a received signal-to-noise ratio of 10 dB.

(d) Repeat (a) by calculating the exact value of the probability of error. To calculate the exact value, every possible bit pattern causing ISI from neighboring symbols has to be considered separately. The error rate is the average of the bit error rates over all bit patterns.

Problem 4

(a) Using Poisson's sum formula

$$\sum_{k=-\infty}^{+\infty} p(t - kT) = \frac{1}{T} \sum_{m=-\infty}^{\infty} P\left(\frac{2\pi}{T}m\right) e^{j(2\pi mt/T)}$$

where $P(\omega)$ is the Fourier transform of $p(t)$, show that

$$\sum_{m=-\infty}^{+\infty} |p(t - mT)|^2 = \frac{1}{T} \sum_{n=-\infty}^{+\infty} Z_n e^{j(2\pi nt/T)}$$

where

$$Z_n = \frac{1}{2\pi} \int_{-\infty}^{+\infty} P(\omega) P^* \left(\omega - n\frac{2\pi}{T}\right) d\omega$$

(b)   Show that $Z_{-n} = Z_n^*$ for all n.

(c)   For $p(t)$ a raised cosine pulse with $0 < \alpha < 1$, show that

$$\sum_{m=-\infty}^{+\infty} |p(t - mT)|^2 = Z_0 + 2\,\mathrm{Re}\big(Z_1 e^{j2\pi t/T}\big)$$

Problem 5. The impulse response of a pair of matched filters which results in a raised cosine spectrum is given by Eq. (8.3.3). A simple way to design these filters is to window the sampled version of the filter impulse response and design an FIR filter with the windowed sampled impulse response. In this problem we examine the time and frequency response of this approach for a particular design specification. Assume that the rolloff factor of the raised cosine pulse is 0.1, the length of each FIR filter are 23 taps, and the sampling rate is $T/4$ with $1/T$ the transmission rate of the pulses.

(a)   If the design uses a rectangular window, sketch the overall back-to-back impulse response of the transmitter and receiver filters. What is the variance of the ISI caused by the filter if the center tap is normalized to 1.

(b)   Sketch the overall frequency response of the channel. What is the minimum attenuation in sidelobes, and the percentage of power in the main lobe.

(c)   Repeat (a) and (b) for a triangular window.

(d)   If the design criteria is to minimize the out-of-band component of the spectrum, which window is your choice? If the design criteria is to minimize the ISI, which window is your choice?

Problem 6. Using Fig. 8.14, determine the equivalent discrete channel model from the information source at the transmitter up to the digital signal processor at the receiver for a 10-Mbit/sec BPSK wireless modem using a raised cosine pulse with rolloff factor of 0.5. Assume that the channel is WSSUS, and its delay power spectrum is given by

$$Q(\tau) = Te^{-\tau/T}$$

where $T = 50$ nsec.

(a)   If we neglect the taps with power 10 dB below the tap with maximum power, how many taps are needed for the model?

(b)   Give the eigenvalues of the covariance matrix of the tap gains.

(c)   Repeat (a) and (b) for a 20-dB threshold.

Problem 7. The equivalent discrete channel model for a fading multipath channel has two taps. The modulation is DBPSK, the transmitted symbols are time-gated to avoid ISI, and the receiver has a two-tap DMF.

(a)   What is the bandwidth efficiency of the system?

(b)   Sketch the probability of outage versus the threshold signal-to-noise ratio γ_{out} if the eigenvalues of the covariance matrix for the equivalent discrete channel model have values of 0.4 and 0.6.

(c) Sketch the average probability of error versus $\overline{\gamma_b}$.

(d) Repeat (a) and (b) for eigenvalues of 0.1 and 0.9 and compare the results found with the two sets of eigenvalues.

Problem 8. A voiceband modem operates over a telephone channel. The equivalent low-pass received signal is given by

$$r(t) = \sum_p a_p h(t - pT) + \eta(t)$$

where $h(t)$ is the impulse response of the channel including the echoes in the line, $\eta(t)$ is the additive white Gaussian noise with variance N_0, and a_p is the transmitted symbol value. For a BPSK system, $a_p = \pm 1$.

To eliminate the echoes for a BPSK modem, the structure shown in Fig. P8.1 is used.

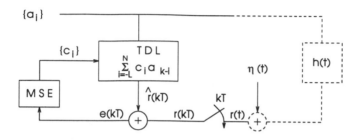

Figure P8.1

(a) By minimizing the MSE, derive the normal equations and solve for the optimum taps of the canceler $\{c_i\}$ in terms of $h(t)$.

(b) Determine the minimum value of the MSE, ξ_{min}, in terms of $\{c_i\}$, $h(t)$ and N_0.

Problem 9. The equivalent baseband impulse response of a channel is given by

$$h(t, \tau) = A\delta(t) - B\delta(t - \tau)$$

A BPSK modem with coherent detection using raised cosine pulses with $\alpha = 0$ and $E_s = 1$ is operating over this channel.

(a) Assume $A = B = 1$ and $E_s/N_0 = 10$ dB and plot the probability of error versus τ_{rms}/T for $0 < \tau_{rms}/T < 5$, where $R = 1/T$ is the bit rate of the channel. Use the Gaussian assumption for the calculation of error rate. With the Gaussian assumption, the ISI caused by channel multipath is treated as an additive Gaussian noise source. The total noise affecting the system is the additive thermal noise with variance N_0 plus the variance of the ISI term.

(b) Repeat (a) for $B = 0.5$, placing both plots in the same graph.

(c) Assume that A and B are independent, slow, time-varying Rayleigh random variables with variance 1 and $E_s/N_0 = 40$ dB and repeat (a). Assume that the variance of B is 0.5 and repeat (b).

Problem 10. Assume that we have a BPSK modem which uses ideal zero rolloff raised cosine pulses for pulse shaping. The modem is operating over a microwave line-of-sight channel with a normalized ($T = 1$) frequency response of

$$H(j\omega) = 1 - 0.1e^{-j.2\omega}$$

(*Hint*: You may use the delay property of the Fourier transform to determine the received overall impulse response).

In this system the transmitted digits are $a_n = \pm 1$ and the received signal-to-noise ratio is $E_b/N_0 = 15$ dB, where E_b is the average received energy per bit and N_0 is the variance of the received noise after prefiltering at the receiver.

(a)   Sketch the mean square error versus τ/T, the normalized sampling time, for $|\tau/T| \leq 1$. Error is defined as the difference between the detected symbols and the sampled signal used for detection.

(b)   Repeat (a) assuming a linear equalizer with three T-spaced taps.

(c)   Repeat (a) assuming a linear equalizer with three $T/2$-spaced taps.

(d)   Repeat (a) assuming a DFE with two T-spaced forward taps and one feedback tap.

(e)   Repeat (a) assuming a DFE with two $T/2$-paced forward taps and one feedback tap.

9

SPREAD SPECTRUM FOR WIN SYSTEMS

9.1 INTRODUCTION

Spread-spectrum communications is now a relatively mature technology with many highly developed subdisciplines, including modulation, coding, and synchronization and acquisition methods. To treat thoroughly all aspects of spread-spectrum communications in one chapter is not feasible, nor is it our purpose here. Instead, we provide an overview of those aspects of spread-spectrum communications that are specific to wireless information networks and principal issues which arise in the application of spread-spectrum techniques in the multiuser wireless environment.

The distinguishing characteristic of spread-spectrum communications is that the signals used for the transmission of information have bandwidth much wider than the underlying information bit rate of the system. There are two basic methods for implementing a spread-spectrum system: direct-sequence spread-spectrum (DSSS) and frequency-hopping spread-spectrum (FHSS). In this chapter we discuss the basic principles of DSSS and FHSS systems, the effects of interference and multipath fading on system performance, and the important topic of code-division multiple access (CDMA), which is built upon spread-spectrum transmission and reception. Detailed treatments of other aspects of the design and analysis of spread-spectrum communication systems, as well as historical accounts of the origins of spread-spectrum technology, can be found in [Coo78], [Dix84], [Sim85], and [Pro89].

In the past several decades, spread-spectrum technology has been used extensively for military communications, where it is attractive because of its resistance to interference and interception, as well as its amenability to high-resolution ranging [Sim85]. More recently, commercial applications of spread-spectrum have attracted considerable attention because of its amenability to CDMA operation, the possibility of spectral overlay, and the availability of unlicensed commercial bands allocated to this technology. Spread spectrum also reduces the harmful effects of multipath, resulting in an increase in signal coverage, mobility, and achievable data rates. After the May 1985 FCC release of the ISM bands for spread-spectrum technology trials [Mar85], various spread-spectrum commercial products, from low-speed fire safety devices to high-speed wireless local area networks (WLANs), appeared in the market. Today, both the voice- and data-oriented wireless information network industries are involved in applications of spread-spectrum technology. However, the two industries have somewhat different motivations for adopting spread-spectrum.

The voice-oriented digital cellular and personal communications services (PCS) industries are considering CDMA spread-spectrum as an alternative to TDMA/FDMA networks in order to increase system capacity, provide a more reliable service, and to provide soft handoff of cellular connections. Much debate in this industry is directed toward the issues of bandwidth efficiency and design complexity of CDMA as compared with TDMA technology. In wireless data communications, spread-spectrum technology is well established in the WLAN industry. WLANs are using this technology primarily because the first bands available for high-speed data communication were the ISM bands, which are specifically designated for spread-spectrum transmission.

Because interfacing with the public switched telephone network (PSTN) is an important aspect of voice-oriented wireless services, manufacturers of digital cellular and personal communication devices have had to wait for the finalization of standards before moving into major marketing and manufacturing efforts. In contrast, to develop and market a WLAN product, a manufacturer needs only a suitable frequency band for high-speed data communication, and the existence of a technical standard is not critical.

One form of spread-spectrum communications receiving much attention in the wireless industry is CDMA. The CDMA technique is a well-proven technology that has been used in a number of military communication systems developed and deployed over the last three decades, and it is now finding application in new digital cellular systems being developed for the commercial mobile communications market. In CDMA spread-spectrum transmission, user channels are created by providing different transmission codes for different users. Privacy of transmitted user information is readily provided by controlling the distribution of user-unique code sequences.

In summary, the principal advantages of spread-spectrum transmission are as follows.

1. Spread-spectrum signals can be overlaid onto bands where other systems are already operating, with minimal performance impact to or from the other systems.

2. The anti-multipath characteristics of spread-spectrum signaling and reception techniques are attractive in applications where multipath is likely to be prevalent. (Achieving good performance in frequency-selective fading may require the use of a Rake receiver, which is in effect a matched filter for a multipath channel.)

3. The anti-interference characteristics of spread-spectrum are important in some applications, such as networks operating on manufacturing floors, where the signal interference environment can be harsh.

4. Cellular systems designed with CDMA spread-spectrum technology offer greater operational flexibility and possibly a greater overall system capacity than do systems built on FDMA or TDMA access methods.

5. The convenience of unlicensed spread-spectrum operation in ISM bands in the United States is attractive to manufacturers and users alike.

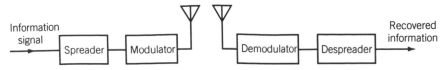

Figure 9.1 Simple block diagram of a direct-sequence spread-spectrum (DSSS) system.

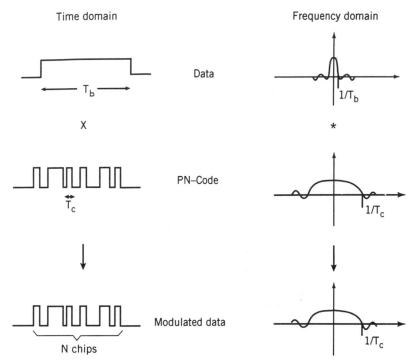

Figure 9.2 Simple example to describe spreading and despreading in DSSS. If a TDMA system is used at the chip rate, we have N times more users. $R_b = 1 / T_b$ (baud rate); $R_c = 1 / T_c$ (chip rate).

9.2 PRINCIPLES OF DIRECT-SEQUENCE SPREAD SPECTRUM

In DSSS transmission we may assume that the information signal is spread at baseband, and the spread signal is then modulated in a second stage. The received signal is first demodulated to recover the spread signal, and the spread signal is despread to recover the original information signal. Figure 9.1 shows the four steps involved in this description. In this approach, the modulation technique is isolated from the spreading-and-despreading operation and we can discuss the baseband spreading and despreading separately. Figure 9.2 provides a simple example to describe spreading and despreading in a DSSS system. A square pulse with duration T_b represents a binary information digit in the time domain, and its Fourier transform is a sinc pulse with zero crossings spaced by $1/T_b$. The information signal is multiplied by a sequence of narrower pulses with time duration T_c and zero crossings spaced by $1/T_c$ to form the spread-spectrum signal. The narrow pulses are referred to as *chips*, and their amplitudes are ± 1. The *bandwidth expansion factor* or *spreading factor* is $N = T_b/T_c$, the baud rate is $R_b = 1/T_b$, and the chip rate of the system is $R_c = 1/T_c$. Because the transmitted power is spread over a bandwidth N times wider than the information symbol rate, the spectral height of the signal is N times lower than it would be in nonspread transmission. The amplitudes of the chips are coded in a periodic random-appearing pattern referred to as the *spreading-code*. Ideally, the spreading code is designed so that the chip amplitudes are statistically independent of one another. Because the chip sequence is coded to appear random, the sequences is referred to as *pseudorandom* or *pseudonoise* (PN) sequences or codes.

Another way of viewing DSSS transmission is to think of it as a rate-$1/N$ coding technique that encodes each bits into N coded chips. The encoding and decoding procedures for this coding algorithm are very simple. The code generation matrix is a simple vector representing one period of the pseudorandom code. At the receiving end, the decoder multiplies the coded sequence by a code generation vector in the reverse direction. The purpose of this coding is not to provide for detection or error correction of errors in additive noise. In fact, as we will see later, the performance of the spread-spectrum signal in additive noise channels is the same as the performance of the uncoded or nonspread signal. It is in the presence of interference and multipath that this coding technique provides a significant performance enhancement. In particular, the spread-spectrum coding provides a systematic method of counteracting the effects of interference. For a fixed information rate, the application of a coding technique generally requires an increase in transmission rate, which in effect spreads the signal spectrum. The spectrum spreading factor is proportional to the inverse of the code rate. Because the code rate of typical DSSS coding, $1/N$, is many times smaller than the rates of typical error-control codes, the spectrum spreading factor is many times larger than that of standard codes. In conventional error-control coding, the expansion of bandwidth is viewed as a penalty which must be paid to provide error-correction power, but the purpose of spread-spectrum coding is to deliberately expand the bandwidth, to provide a number of important advantages relative to non-spread signaling.

One advantage of the spread-spectrum technique is that by reducing the height of the signal spectrum, one may be able to overlay a spread-spectrum system onto other systems already operating in the same band. Another advantage is that by associating each user with a different spreading code, the system provides some degree of privacy for individual users' transmissions. While this scheme does not provide in-depth cryptographic protection for a user's information, it does provide some protection against casual interception. The disadvantage of the spread-spectrum technique is that it requires a wider bandwidth than conventional communication. However, in a multiuser environment, several users can share the same bandwidth using spreading codes which are orthogonal to one another. This approach is referred to as *code-division multiple access* (CDMA). The CDMA design can overcome the cost of bandwidth expansion if the number of CDMA users is larger than the bandwidth expansion factor.

If we assume that the period of the spreading code is T_b, one period of the spreading signal is represented by

$$f(t) = \sum_{i=1}^{N} b_i p(t - iT_c) \qquad (9.2.1)$$

where N is the number of chips per bit, $p(t)$ is the chip pulse shape, and $\{b_i\}$ are the values of the PN chips. The spreading signal is given by

$$s(t) = \sum_{k} f(t - kT_b) \qquad (9.2.2)$$

The transmitted baseband signal is then given by

$$x(t) = \sum_{n} a_n f(t - nT_b) \qquad (9.2.3)$$

where a_n is the information digit.

The cross-correlation function of the transmitted signal and the periodic PN spreading signal is given by

$$R_{xs}(\tau) = \sum_{n} a_n R_{ff}(\tau - nT_b) \qquad (9.2.4)$$

where

$$R_{ff}(\tau) = \int_{-\infty}^{\infty} f(t)f(t + \tau)\, dt \qquad (9.2.5)$$

Figure 9.3 shows the cross-correlation function as given by Eq. (9.2.4). The correlation function is a periodic set of narrow pulses with width twice the chip duration, repeating at intervals of the bit duration T_b. If the transmitted bit is the symbol "1" represented by a positive voltage value, the associated peak at the receiver is positive. If the transmitted symbol is the symbol "0" represented by a negative voltage value, the peaks of the correlation function will be at a negative

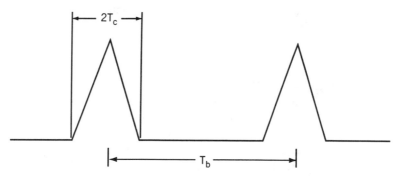

Figure 9.3 Cross-correlation function of the transmitted DSSS signal.

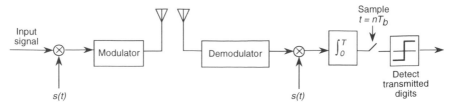

Figure 9.4   Block diagram of a DSSS transmitter and receiver.

value. Therefore, the information bits represented by a constant voltage level with duration T_b at the input to the transmitter are represented by narrow pulses with width twice the chip duration after cross-correlation at the receiver.

To implement a simple receiver, we need only generate the peak of the cross-correlation function; if it is a positive voltage the transmitted information bit is a 1, if it is a negative voltage the transmitted bit is a 0. The peak of the cross-correlation function occurs at $\tau = 0$; therefore, as shown in Fig. 9.4, we need only multiply the received signal by $s(t)$ and integrate it over one symbol interval. The sampled outputs of the integrator form the basis for the decision on the transmitted bit.

The structure of the spreading and despreading operations in Fig. 9.4 is similar to the structure of standard modulation and demodulation, but with a periodic PN spreading code replacing the usual carrier signal. The spread-spectrum operation can then be viewed as a two-layer modulation technique. The spread-spectrum modulation system multiplies the information signal by a random carrier (spreading signal), whereas the standard modulation multiplies the signal by a sinusoidal carrier. The standard demodulator cross-correlates the received signal with a duplicate of the transmitted sinusoid, whereas the spread-spectrum demodulator (despreader) cross-correlates the received signal with a duplicate of the random carrier. The difference between the two layers of modulation is that the generation and synchronization of the random carrier are more complicated processes than are required with the sinusoidal carrier. The modulated sinusoidal signal preserves the shape of the information signal spectrum and only shifts the spectrum to the carrier frequency. The spread-spectrum modulation spreads the spectrum of the information signal but does not shift its center frequency. The two-level modulation technique spreads the signal and shifts the center frequency.

9.2.1   Implementation

Implementation of the spread-spectrum transmitter or receiver is independent of the order in which the two layers of modulation or demodulation are performed. At the transmitter, the information bits can first be modulated onto the random carrier and then the chips modulated onto the sinusoidal carrier; conversely, the bits may be modulated onto a carrier and then multiplied by the spreading signal. The same is true for demodulation at the receiver. The flexibility of being able to perform the modulation steps in different orders allows different implementations that can be tailored to a preferred circuit design.

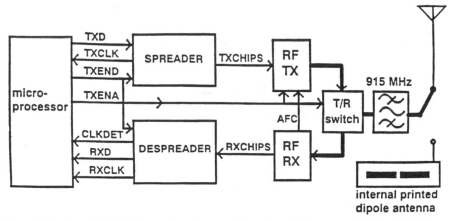

Figure 9.5 A typical DSSS transceiver for operation in the ISM bands at 915 MHz [IEEE91a].

Figure 9.5 shows a block diagram for a typical DSSS transceiver operating in the 915-MHz ISM band. The operation of the transceiver is controlled by a microprocessor. The transmitter enable signal (TXENA) controls the switch that puts the system in transmit or receive mode. Other signals utilized in the transceiver are data (TXD and RXD), clock (TXCLK and RXCLK), end (TXEND), and chips (TXCHIPS and RXCHIPS). Figures 9.6 and 9.7 show the block diagrams of a typical spreader and radio-frequency (RF) transmitter, respectively. The spreading chip rate is 15 MHz, and the transmitter output has selectable power levels of 100 mW or 1 W. The modulation technique is minimum shift keying (MSK), and the transmitter will operate when the TXENA signal allows the voltage-controlled oscillator (VCO) to generate the modulated signal. The TXENA signal also controls the first stage class A amplifier and T/R switch.

One of three different receiver architectures is generally used. Figure 9.8 shows the most commonly used structure, a generic receiver architecture for the *post-demodulation correlator receiver* in which the intermediate frequency (IF) signal is

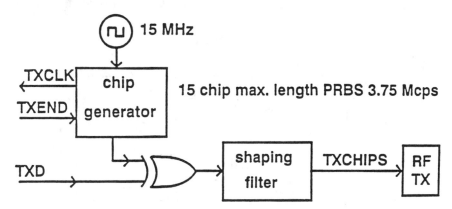

Figure 9.6 Block diagram of a typical spreader for DSSS [IEEE91a].

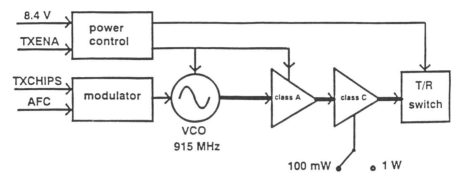

Figure 9.7 Block diagram of a typical RF transmitter for DSSS [IEEE91a].

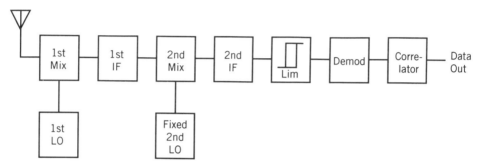

Figure 9.8 Generic receiver architecture for a post-demodulation correlator receiver in which the IF signal is passed through the MSK demodulator [IEEE91a].

passed through the MSK demodulator (limiter and demodulator) to recover the chip sequence. The recovered chips are then passed through the correlator. Figure 9.9 shows a *pre-demodulation despreading receiver* architecture. The code acquisition and tracking subsystem provides the random sequence that performs the despreading before demodulation. Figure 9.10 shows a generic homodyne receiver with a post-detection correlation architecture having two branches that can be used with PSK or QAM modulation. The post-detection correlation

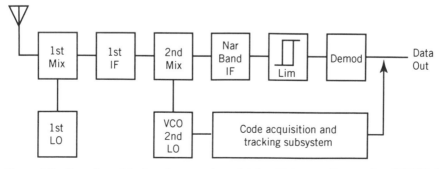

Figure 9.9 Generic architecture for a pre-demodulation IF despreading receiver [IEEE91a].

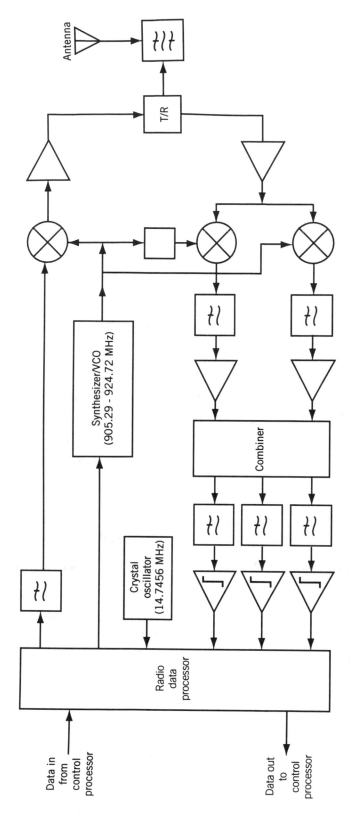

Figure 9.10 A generic receiver post-detection correlation architecture with two branches that can be used for PSK and QAM modulation techniques [IEEE91a].

367

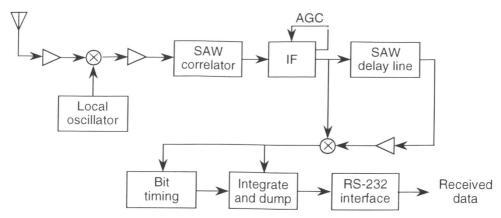

Figure 9.11 A DSSS DBPSK receiver using SAW devices. (From [FER80] © IEEE.)

technique of Fig. 9.8 is the least accurate of the three architectures but provides a less complicated design, which makes it a better choice for inexpensive implementations.

Another way of looking at DSSS modulation is to assume that it is a modulation technique whose pulse-shaping filter $f(t)$ is defined by Eq. (9.2.1). The transmitted baseband signal in this case is given by Eq. (9.2.3), which is same as the transmitted signal for any baseband pulse transmission technique with a pulse-shaping filter given by $f(t)$. The difference between spread-spectrum and other modulation techniques is that the bandwidth of the pulse-shaping filter is approximately N times wider than with a standard modulation with the same information symbol rate. As we saw in Chapter 7, the optimum receiver for the received pulses is a matched filter with impulse response $f(-\tau)$. The output of the matched filter is the cross-correlation given by Eq. (9.2.4) and shown in Fig. 9.3. The output of the cross-correlator represents this function for all values of delay, whereas in other implementations discussed earlier we need generate only the peak of the correlation function.

Surface acoustic wave (SAW) correlators are sometimes used for implementation of the impulse response $f(-\tau)$. The SAW correlator is a tapped delay line with taps set at the pseudonoise (PN) sequence values in the reverse order. In contrast with other implementations, this approach does not require code acquisition circuitry. Figure 9.11 shows a practical implementation of a SAW-based DSSS system using DBPSK modulation. Designs using SAW correlators are more expensive and less stable (due to temperature effects) than other designs, which makes them less attractive for application in commercial products. However, the simplicity of the SAW implementation has made it an ideal candidate for use in experimental systems [Fer80, Kav87]. In commercial implementations, as mentioned before, the received signal is passed through a noise reduction filter followed by a sampler. The sampled digital signal is then processed for purposes of code tracking, timing recovery, phase recovery, and implementation of the correlator.

9.3 PRINCIPLES OF FREQUENCY-HOPPING SPREAD SPECTRUM

The FHSS technique can also be described as a two-layer modulation technique. The first layer can be any standard digital modulation technique, while the second layer is M-ary FSK. The digitally modulated signal makes a PN selection of one of the M frequencies as its carrier frequency. In other words, the carrier frequency of the digitally modulated data is hopped over a wide range of frequencies prescribed by a periodic PN code. The hopping of the carrier produces the desired spreading of the transmitted signal spectrum. The changes in the carrier frequency do not affect the performance in additive noise, and the AWGN performance remains exactly the same as the performance of the digitally modulated system without frequency hopping.

Just as with the DSSS technique, the FHSS technique can allow coexistence of several systems with orthogonal codes in the same frequency band, and can provide a degree of user-signal privacy by association of each user's signal with a randomly selected hopping pattern. One difference between the two spreading methods is that the DSSS technique uses the full system bandwidth throughout the entire transmission time whereas the FHSS uses only a portion of the band at a time. In an FHSS/CDMA system, each user employs a different hopping pattern; in this system, interference occurs when two different users land on the same hop frequency. If the codes are random and independent from one another, the "hits" will occur with some calculable probability. If the codes are synchronized and the hopping patterns selected so that two users never hop to the same frequency at the same time, the multiple-user interference is eliminated. The number of users in this case is limited by the number of frequency slots. This type of FHSS/CDMA scheme is equivalent to TDMA/FDMA with hopping of the center frequency, and it can be referred to as FH/TDMA.

Figure 9.12 shows the block diagram of a typical transmitter and receiver for a FHSS system. At the transmitter, the digital modulation and the modulation over the hop frequencies are implemented in two stages. The hop frequencies are selected randomly using a frequency synthesizer controlled by the PN code generator. A wideband filter is applied to the signal for spectral shaping before the signal is fed to the antenna. The receiver has a wideband front end filter that accommodates the entire system bandwidth. This filter is followed by a PN-code-controlled frequency synthesizer synchronized to the transmitter frequency synthesizer. After desynthesizing, the signal is passed through a noise reduction bandpass filter with the same bandwidth as the transmitted information symbols. The final stage of the receiver is the data demodulator, which demodulates the first-stage digital modulation. Figure 9.13 shows a more detailed receiver structure with timing recovery, and code acquisition and code tracking control.

In an FHSS system the interval of time spent at each hop frequency is referred to as the *chip duration*. In contrast with a DSSS system, the chip duration in an FHSS system is not determined by the inverse of the bandwidth but instead forms an independent design parameter. This is because the system need not use only one hop per information symbol, or let us say per bit. If the chip duration is smaller than the bit duration, we have more than one hop per bit and the system is referred to as a *fast-FHSS system*. If the chip duration is greater than the bit

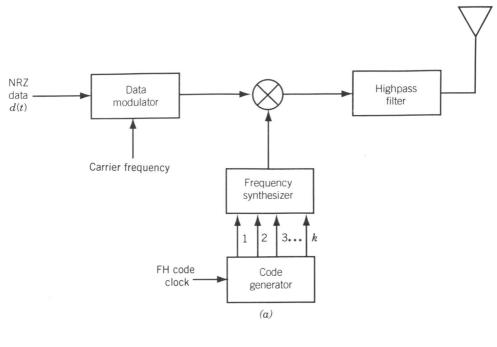

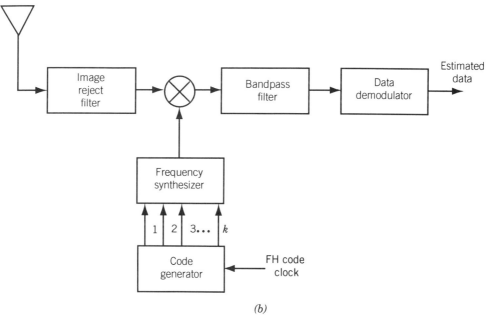

Figure 9.12 Block diagram of a frequency-hopping spread-spectrum modem. (*a*) Transmitter. (*b*) Receiver.

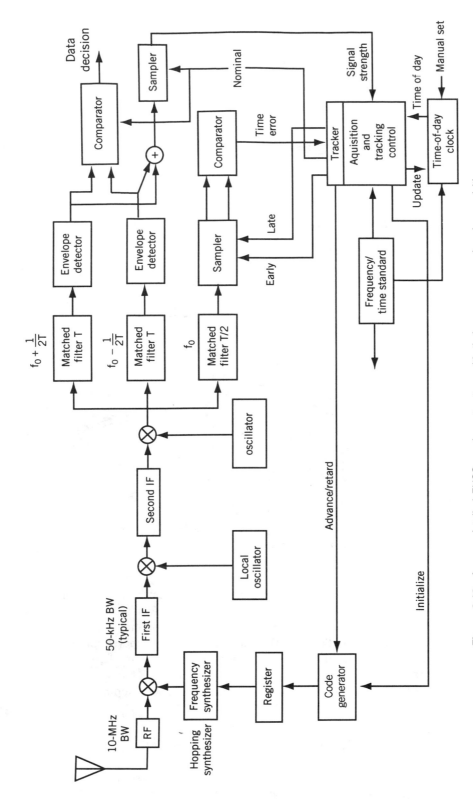

Figure 9.13 A more detailed FHSS receiver structure with timing recovery and code acquisition and tracking control [IEEE91a].

371

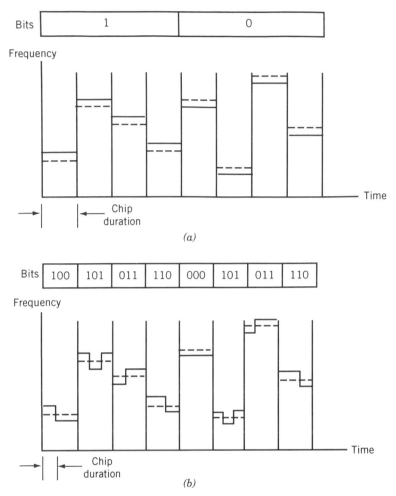

Figure 9.14 Simplified description of fast and slow frequency hopping. (*a*) Fast hopping with 4 hops per bit. (*b*) Show hopping with 3 bits per hop. (From [Skl88] © Prentice-Hall.)

interval, we have more than one bit per hop and the system is called a *slow-FHSS* system. Figure 9.14 illustrates the concepts of slow and fast frequency hopping. Fast frequency hopping is effective in combating narrowband interference and frequency-selective fading. Tracking the hops in slow frequency hopping is relatively simple, which eases the implementation of the modem. Error-correction coding can be very effective in improving the performance of frequency-hopping systems and is more effective for fast than for slow frequency hopping.

9.4 INTERFERENCE IN SPREAD-SPECTRUM SYSTEMS

Analysis of the effects of interference in spread-spectrum systems is similar to analysis of the effects of additive noise. However, interference and additive noise have differences that affect the results of the analysis. The additive noise considered in communication systems is the thermal noise generated by the random motion of electrons in the resistive elements of the receiver circuit. The spectral height of this noise is constant and its bandwidth is unlimited. As we increase the transmission bandwidth of the system, the thermal noise power added to the received signal increases accordingly. However, the source of interference is another transmitter operating in the band of our system. For the sake of this discussion, let us consider the interfering transmitter as having fixed transmission power. In general the bandwidth of the interfering signal is less than or at most equal to the bandwidth of the desired signal, and therefore the interference power is independent of the transmission bandwidth of our system. The spectrum of the interfering signal need not conform to any particular shape. The bandwidth of the interfering signal might be as narrow as the bandwidth of a steady sinusoid generated by an intentional or unintentional source up to a bandwidth as wide as the transmission bandwidth of a similar wideband system coexisting with the intended user in a CDMA environment.

The literature on military communication systems offers many detailed analyses of the performance of spread-spectrum systems in the presence of various intentional interferers or jammers [Sim85]. These jammers are designed to disrupt the operation of a spread-spectrum system, and they can employ relatively sophisticated techniques such as multitone jamming and pulsed jamming. In civilian applications, the interference is neither intentional nor sophisticated. Most often, the interferer is simply another system designed to operate in a portion of, or in the entire band of, operation of our system and the users are generally willing to cooperate so as to minimize the mutual interference. To evaluate the effects of interference, we shall consider narrowband and wideband interference and compare the results with the performance in additive noise. The wideband interferer in commercial applications is another spread-spectrum user sharing the same operating band. Narrowband interference usually occurs in an overlay situation where a wideband spread-spectrum is to coexist with a group of narrowband systems operating in the same band.

The analysis presented here assumes continuous interference. Impulsive interference (jamming) can significantly reduce the effectiveness of the spread-spectrum modulation to provide protection against the interference [Vit79]. Although impulsive jammers are commonly used in military communications, impulsive interference is not a significant concern in the operation of commercial wireless systems.

9.4.1 Interference in FHSS Systems

The spectrum of additive thermal noise is flat and covers all the hop frequencies with the same spectral height. As a result, the received signal-to-noise (SNR) for a FHSS modem on a channel corrupted only by additive noise is the same at every hop. This SNR is given by

$$\gamma_b = \frac{E_b}{N_0} \tag{9.4.1}$$

where N_0 is the height of the two-sided noise power spectral density, and E_b is the

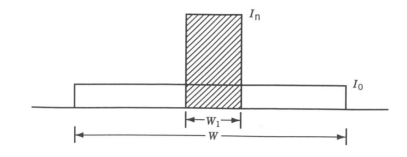

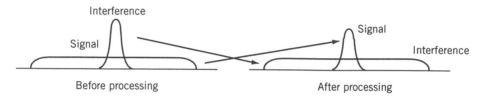

Figure 9.15   Simple mathematical representation of narrowband and wideband interference.

energy per bit:

$$E_b = T_b P_{av} = \frac{P_{av}}{R_b} \tag{9.4.2}$$

with P_{av} the average received power, T_b the bit duration, and R_b the bit rate of the system. The error rate of the FHSS system in additive Gaussian noise remains the same as the error rate of the same system without frequency hopping and is computed from the BER equations for standard modems given in Chapter 7.

While it is true that the frequency hopping does not improve BER performance in additive noise, we shall show that it protects the system, in somewhat different ways, against narrowband and wideband interference. A narrowband interferer interferes with the signal at only one hop, and the signal at all other hops remains unaffected regardless of the power level of the interferer. For the wideband interferer the power of the interference is spread over the entire band and only a portion of this power interferes with the desired signal at each hop.

For a simple mathematical representation of narrowband and wideband interference consider Fig. 9.15. Both narrowband and wideband interference are assumed to have an ideal rectangular power spectrum. The bandwidth of the wideband interference is assumed to be the same as the transmission bandwidth W, and its height is represented by I_0. The bandwidth W_1 of the narrowband interferer is assumed to be narrower than the bandwidth at each hop; that is, $W_1 \leq W/N$, where N is the bandwidth expansion factor of the FHSS system. The power spectral height of the wideband interference is represented by I_n. The total received interference power is given by

$$I_{av} = I_n W_1 = I_0 W \tag{9.4.3}$$

and is assumed to be same for both narrowband and wideband interference cases.

Effects of Wideband Interference. Within the transmission band the power spectrum of wideband interference is the same as the power spectrum of additive noise having spectral height I_0. Therefore, the performance of the system in wideband interference is the same as the performance on an additive noise channel with SNR per bit given by

$$\gamma_{b\text{-WI}} = \frac{E_b}{I_0} = \frac{W}{R_b} \times \frac{P_{\text{av}}}{I_{\text{av}}} \tag{9.4.4}$$

With a fixed ratio of signal to interference power, $P_{\text{av}}/I_{\text{av}}$, the signal-to-noise ratio γ_b used for calculation of the error rate is $N = W/R_b$ times higher than the actual ratio of the signal power to interference power. If the occupied bandwidth at each hop is the same as the bit rate R_b, then N is the number of hops in the transmission band or the bandwidth expansion factor and is referred to as the *processing-gain*. With a fixed interference power, as the transmission bandwidth increases, the spectral height of a uniformly distributed wideband interferer decreases. Reduction of the spectral height of the wideband interference in turn increases the received signal to interference ratio at each hop. Equation (9.4.4) shows that by adjusting the number of hops, we may design a FHSS system to overlay onto an existing wideband system. The existing system is viewed as a wideband interferer to the intended FHSS system, and the FHSS system will be a narrowband interferer to the existing system. The possibility of overlay is one of the attractive features of the spread-spectrum technique for commercial applications.

Example 9.1. Assume that we have a wideband BPSK digital system (the "existing system") that requires $\gamma_b = 10$ dB to operate at an error rate of about 10^{-5}. Further assume that the transmitted signal power is high enough that additive noise can be neglected and only interference need be considered. We can now overlay a BPSK/FHSS system with $N = 100$ hops/bit and a transmitted signal power that is 10 dB below the existing system, and both systems can operate satisfactorily in the same frequency band. For the existing system the γ_b is the same as the signal-to-interference ratio with an FHSS signal as the interferer, which is 10 dB. For the FHSS system the existing system represents wideband interference, and the effective signal-to-interference ratio calculated from Eq. (9.4.4) is also 10 dB. Therefore, both systems operate satisfactorily at the same time in the same band. The data rate of the FHSS system is 100 times lower than the existing system.

This type of coexistence does not require frequency hopping and can be implemented using any narrow-band system. For example, if we overlay a narrow-band signal whose power is 10 dB below the existing system and it occupies only $1/100$ of its bandwidth, in an interference dominated channel, the SNR for both systems is 10 dB. Therefore, in overlaying an FHSS system onto a wideband system no advantage is gained by hopping the carrier frequency.

If we rewrite Eq. (9.4.4) into the form

$$\frac{I_{\text{av}}}{P_{\text{av}}} = \frac{W}{R_b} \times \frac{1}{E_b/I_0} \tag{9.4.5}$$

the value of I_{av}/P_{av} is referred to as the *interference margin*. The interference margin in decibels represents the maximum allowable power difference between the interference and the desired signal in order to maintain proper operation. Proper operation is identified with a minimum required E_b/I_0 needed to deliver data with the maximum allowable error rate.

Effects of Narrowband Interference. For analysis of FHSS systems with narrowband interference, we assume that the interferer interferes with the signal at only one hop and its spectrum is flat over the hop bandwidth. Therefore, the interference bandwidth for a one-bit-per-hertz modulation technique is given by $W_1 = W/N = R_b$. As shown in Fig. 9.15, the spectral height of the narrowband interference is $I_n = I_{av}/W_1$. The energy per bit, given by Eq. (9.4.2), is $E_b = P_{av}/R_b$. As a result, the SNR per bit for the hop that hits the narrowband interferer is given by

$$\gamma_{b\text{-NI}} = \frac{E_b}{I_n} = \frac{W_1}{R_b} \times \frac{P_{av}}{I_{av}} = \frac{P_{av}}{I_{av}} \tag{9.4.6}$$

which is the same as the SNR without frequency hopping. This means we have no protection when the hop hits the interferer. Because there is assumed to be no interference for all other hops, the average signal to interference ratio over all hops is

$$\bar{\gamma}_{b\text{-NI}} = \frac{NE_b}{I_n} = \frac{NW_1E_b}{I_{av}} = \frac{WE_b}{I_{av}} = \frac{E_b}{I_0} = \frac{W}{R} \times \frac{P_{av}}{I_{av}} \tag{9.4.7}$$

which is the same as the SNR in the case of wideband interference, as given by Eq. (9.4.4). Note that this does not mean that if the power of the narrowband and wideband interferences are the same, the average error rate is the same. We can substitute Eq. (9.4.4) into the formula for the probability of error of the modulation technique and determine the error rate of the FHSS system in wideband interference. However, we cannot substitute Eq. (9.4.7) into the same error rate formula to calculate the average error rate in presence of narrowband interference. Instead we should substitute Eq. (9.4.6) into the error rate formula to obtain the error rate in the hop hit by the interferer. Since there is no error in other hops, the average error rate is N times less than this value. The following example will serve to clarify the situation.

Example 9.2. Assume that we have an FH/BPSK modem transmitting 10 hops, operating in an interference environment. Further assume that the effects of additive noise are negligible, the received interference power is at the same level as the received power of the desired signal, and the interference bandwidth is sufficiently narrow that it interferes with only one hop frequency. At this hop frequency the signal-to-interference ratio is $\gamma_b = 1$ and the BPSK error rate, given by 0.5 erfc($\gamma_{b\text{-NI}}$), is approximately 10^{-1}. Other hops are disturbed only by additive noise, with negligible effect. The error rate at these hops is essentially zero. From the user's perspective, the average error rate over the 10 hops is then approximately 10^{-2}. Now consider a case in which the interferer is wideband with the same total power but with its power uniformly distributed over all hops. Here the interference power at each hop is 10 times smaller but all hops are affected by the interference. Therefore, the SNR per bit at each hop is $\gamma_{b\text{-WI}} = 10$ dB, which produces an error rate of 10^{-5} in all hops, three orders of magnitude better than the average error rate over all hops in the presence of the narrowband interferer with the same power. In terms of signal power, the three-orders-of-magnitude improvement in error rate of a BPSK-modulated signal accounts for approximately 5 dB in SNR. Thus the impact on performance with narrowband interference in this example is 5 dB greater than with wideband interference.

In general, the average probability of error over all hops for a fixed narrowband interferer affecting only one hop is given by

$$P_{s\text{-FHNI}} = \frac{P_s(\gamma_{b\text{-NI}})}{N} \tag{9.4.8}$$

where $\gamma_{b\text{-NI}}$ is the signal-to-interference ratio in the hop that hits the narrowband interferer and is given by Eq. (9.4.6), and $P_s(\gamma_b)$ is the error rate equation for the particular modulation technique used for the FHSS system. In Eq. (9.4.8), probability of error is inversely proportional to processing gain N. This equation should be compared with the probability of error for wideband interference covering the entire band, which is given by

$$P_{s-\text{FHWI}} = P_e(N \times \gamma_{b\text{-NI}}) = P_s(\gamma_{b\text{-WI}}) \tag{9.4.9}$$

In this equation the probability of error is an exponential function of the processing gain. Figures 9.16 and 9.17 show the error rate as a function of N and γ_b. The probability of error for narrowband interference is bounded by $P_{s\text{-FHNI}} = 1/(2N)$, which corresponds to the assumption that the information in the hop hitting the interferer is completely lost and $P_s = 0.5$ for that hop. If the interference is another FHSS modem signal with a random hopping pattern, the average fraction of hits remains at $1/N$. Although the FHSS interferer occupies the same band as the modem transmitting the desired signal, when the interference occurs it hits only one hop. Therefore, if the hopping patterns of the signal and interference are independent, the error rate for the desired signal is the same as the error rate in the case of narrowband interference, as given by Eq. (9.4.8). If the hopping pattern in the interfering system is in lock-step with the pattern of the desired signal, the FHSS signal provides no protection and the processing gain is 0 dB. However, this is an unlikely scenario in commercial applications, where the interference is unintentional.

Equation (9.4.9) provides an approximation to the error rate in wideband interference for the case of slow FHSS in which at least one information bit is transmitted per hop. This equation remains the same as the number of bits per hop increases. In contrast, for the case of narrowband interference, given by Eq. (9.4.8), the error rate of the fast-FHSS system improves steadily as the number of hops per bit increases.

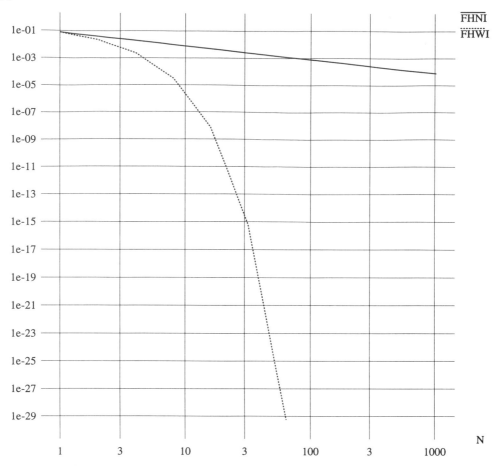

Figure 9.16   The error rate of FHSS-NI and FHSS-WI as a function of N.

Example 9.3. Assume that we have a fast-FHSS system transmitting with three hops per information bit in the presence of a fixed narrowband interference source. If we assume that the fixed interferer hits exactly one of the three hops in each bit, we can use the majority-vote rule to decide the information bit value, and the system is error-free. However, if the interference is another FHSS system hopping randomly over the band used by the desired signal, an information bit decision error occurs whenever two or three errors occur in the three hops, and therefore the probability of symbol error, P_s, is given by

$$P_s = \binom{3}{2}P^2(1-P) + \binom{3}{3}P^3$$

where $P = 1/2N$ and N is the number of frequency slots used by the system. For $N = 1000$ we have $P(\gamma_b) = 10^{-3}$ and $P_s \simeq 3 \times 10^{-6}$. In this case we hop three times in each bit interval, which reduces the data rate by a factor of three. A

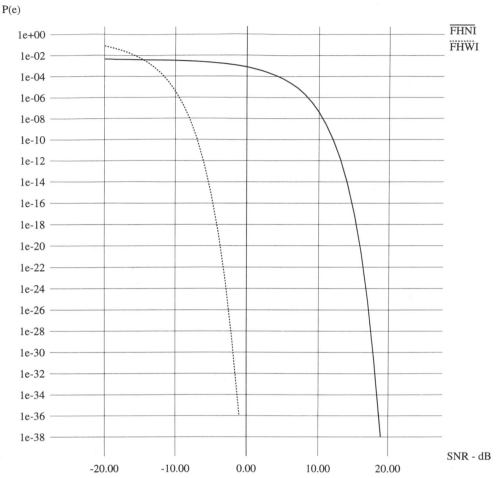

Figure 9.17 The error rate of FHSS-NI and FHSS-WI as a function of γ_b.

threefold reduction in the data rate causes a threefold (5 dB) increase in the processing gain. However, a 5-dB increase in the processing gain does not improve the error rate by five orders of magnitude as in the case of wideband interference. But it does provide three orders of magnitude of error rate improvement, which is much better than the threefold improvement obtained for narrowband interference and slow FHSS.

For M hops per bit we need at least $M/2$ errors per bit to cause a decision error, and the probability of error in given by

$$P_s = \sum_{k=(M+1)/2}^{M} \binom{M}{k} P^k (1-P)^{(M-k)}, \qquad M \text{ odd}$$

$$P_s = \sum_{k=M/2+1}^{M} \binom{M}{k} P^k (1-P)^{(M-k)}, \qquad M \text{ even}$$

where $P = 1/2N$. The use of M hops per bit with a majority decision rule can be viewed as an error-correcting code that encodes each information bit into M coded bits and is capable of correcting up to $M/2$ errors in the set of coded bits. Similar benefits can be obtained by using other error correcting codes. For a block code of length n capable of correcting up to t errors per code block by hard-decision decoding, and given an assumption that coded bits are affected independently by interference hits, the post-decoding symbol error rate is given by

$$P_s \leq \sum_{k=t+1}^{n} \binom{n}{k} P^k (1 - P)^{(n-k)} \qquad (9.4.10)$$

The above equation is an upper bound rather than an exact expression for P_s because in general some error patterns with more than t errors will be detectable, though not correctable. Depending upon the choice of code and the design of the system, the symbols with detectable error patterns may be handled in such a way that they are not regarded as strict-sense errors in the delivered data stream. These points are treated in greater detail in [Mic85] and other coding texts.

As an alternative to block coding, one might use the M hops per information bit to construct a rate-$1/N$ convolutional code having minimum free distance d_{free}. For hard-decision Viterbi decoding and an assumption of statistically independent errors, the probability of an information symbol decision error is bounded [Vit67] by

$$P_s < \sum_{k=d_{\text{free}}}^{\infty} b_k P_k \qquad (9.4.11)$$

where b_k is the number of code trellis paths of weight k times the number of corresponding output (post-decoding) bit errors caused by decoding to an incorrect weight-k path, and P_k is the probability of a weight-k path excursion, which is in turn bounded by

$$P_k < \left[2\sqrt{P(1 - P)} \right]^k$$

9.4.2   Interference in DSSS Systems

Analysis of the effects of interference on the performance of a DSSS system can be done in a more systematic manner. Using the representation of the transmitted spread-spectrum signal given by Eq. (9.2.3), the received signal in a channel disturbed only by interference is given by

$$r(t) = \sum_n a_n f(t - nT_b) + i(t) \qquad (9.4.12)$$

where a_n is the nth information symbol, $f(t)$ defined by Eq. (9.2.1) is the waveform representing one period of the spreading pattern of the signal, and $i(t)$ is the interference signal.

After correlation at the receiver, we have

$$R_{rs}(t) = \sum_n a_n R_{ff}(t - nT_b) + z(t) \qquad (9.4.13)$$

where $R_{ff}(t)$ is the autocorrelation function of $f(t)$, defined by Eq. (9.2.5).

The signal-to-interference ratio per bit at the output of the correlator is given by

$$\gamma_b = \frac{E_b}{I_0} \qquad (9.4.14)$$

where the denominator in Eq. (9.4.14) is given by

$$I_0 = \frac{1}{E_c} \int_{-\infty}^{\infty} |P(f)|^2 \Phi_{zz}(f)\, df \qquad (9.4.15)$$

in which $P(f)$ is the Fourier transform of $p(\tau)$, the impulse response of the pulse-shaping filter used to form the chips, $\Phi_{zz}(f)$ is the power spectral density of the interference signal $z(t)$ at the output of the correlator, and $E_c = E_b/N$ is the signal energy per chip. Finally, we determine the error rate of the DSSS system by substituting γ_b into the appropriate error-rate formula for the given modulation technique.

Effects of Wideband Interference. As in the analysis of FHSS transmission, we assume that the wideband interference has a flat spectrum covering the entire transmission band. If the power spectral density of the interfering signal is I, we have

$$\Phi_{zz}(f) = I \qquad (9.4.16)$$

and we find I_0, the interference component of γ_b, from Eq. (9.4.15) by substituting I for $\Phi_{zz}(f)$, which yields

$$I_0 = \frac{1}{E_c} \int_{-\infty}^{\infty} |P(f)|^2 I\, df = I \qquad (9.4.17)$$

which is independent of the chip pulse shape. The signal-to-interference ratio for the case of wideband interference is

$$\gamma_{b-\mathrm{WI}} = \frac{E_b}{I_0} = \frac{W}{R_b} \times \frac{P_{\mathrm{av}}}{I_{\mathrm{av}}} \qquad (9.4.18)$$

which is the same as the signal-to-interference ratio for the FHSS system with wideband interference given by Eq. (9.4.4).

Note here that the processing gain W/R_b affects the SNR. For a fixed interference power, an increase in the operating bandwidth reduces I_0, resulting in an increase in the SNR. Therefore, as with FHSS transmission in wideband interference, we can control the error rate of the received signal by adjusting the transmission bandwidth. On an additive Gaussian noise channel, $I_0 = N_0$ is constant and the SNR remains fixed as we increase the transmission bandwidth. Of course if the interference consists of several other DSSS systems sharing the same band, the aggregate interference power is simply the sum of the interfering powers.

Effects of Narrowband Interference. To analyze the performance in narrowband interference, we assume that the interferer transmits only a carrier at the center frequency of our DSSS system. The power spectral density of the baseband equivalent of this signal is given by

$$\Phi_{zz}(f) = I_{av}\delta(f) \tag{9.4.19}$$

where I_{av} is the average power of the interfering signal. In this case we have

$$I_0 = \frac{1}{E_c}\int_{-\infty}^{\infty}|P(f)|^2 I_{av}\delta(f)\,df = \frac{I_{av}}{E_c}|P(0)|^2 \tag{9.4.20}$$

where

$$P(0) = \int_{-\infty}^{\infty} p(t)\,dt \tag{9.4.21}$$

with $p(t)$ the chip pulse waveform. The variance of the noise depends on the shape of the chip pulse waveform. The signal-to-interference ratio is given by

$$\gamma_{b\text{-NI}} = \frac{E_c/|P(0)|^2}{R_b} \times \frac{P_{av}}{I_{av}} \tag{9.4.22}$$

For a rectangular pulse, $|P(0)|^2 = T_c E_c = E_c/W$ and

$$\gamma_{b\text{-NI}} = \frac{W}{R_b} \times \frac{P_{av}}{I_{av}} \tag{9.4.23}$$

which is the same as for the case of wideband interference in a DSSS system [as given by Eq. (9.4.18)] or in an FHSS system [as given by Eq. (9.4.4)]. Therefore the performance of a DSSS system with rectangular pulses is the same for narrowband and wideband interference. This is in contrast with our observations in the analysis of FHSS systems. We saw there that the performance of an FHSS system in wideband interference is the same as the performance of a DSSS system and an increase in the processing gain results in an exponential reduction in the error rate. However, we saw that the performance of an FHSS system in narrowband interference is quite different, and the error rate improves as a reciprocal function of the processing gain.

Example 9.4. To observe the effect of choosing a chip pulse shape other than rectangular, let us assume a half sine wave as the pulse waveform. For this waveform, $|P(0)|^2 = 8T_c E_c/\pi^2$. Thus the variance of the noise for the half-sine-wave pulses is 0.9 dB smaller than in the case of rectangular pulses, which results in a 0.9-dB performance improvement in wideband interference.

In summary, if an FHSS system and a DSSS system occupy the same band, the DSSS signal is a source of wideband interference to the FHSS system, while the effect of an interfering FHSS signal on the performance of a DSSS system is the same as the effect of any narrowband interferer. In an FHSS system with narrowband interference we can in principle eliminate the effects of interference by using error-correction coding. The coding benefits achievable with a DSSS system in narrowband or wideband interference are the same as are achievable with an FHSS system operating in wideband interference. In all three cases, coding reduces the required minimum signal-to-interference ratio γ_b by a factor of $R_c d_{\min}$, where R_c is the code rate and $d_{\min}$ is the minimum distance of the code. (We will return to this point in a later paragraph.) However, in the case of narrowband interference against a DSSS system, there is a very effective way of mitigating the effects of the interference, as we describe next.

Narrowband-Interference Suppression. The effects of narrowband interference on the performance of a DSSS system can be greatly reduced or nearly eliminated by the application of adaptive interference suppression techniques. One technique of this type is to apply a *narrowband interference canceller* to the received signal. This device is an adaptive filter, similar to an equalizer, that adjusts its parameters so as to place a deep notch at the location of the narrowband interference. The filter adjusts itself adaptively by using an optimization algorithm operating on the observed spectral or correlation properties of the received signal plus interference. Even if the spectral level of the narrowband interferer is very high relative to the desired DSSS signal, the filter can greatly reduce the interference, though the filtering will cause some change in the correlation properties of underlying desired signal. If the bandwidth of the interferer is very narrow relative to the desired signal, the impact on demodulation of the desired signal is generally found to be very small.

A somewhat more general approach to narrowband interference suppression which has received considerable attention is one based on *transform domain processing* (TDP). Briefly described, a TDP receiver for a DSSS system performs a Fourier transformation of the received signal plus interference and then suppresses interferers by adaptively "excising" or "soft-limiting" appropriate interference spectral components [Gev89]. Narrowband interference suppression in FHSS systems is done in a manner closely related to the TDP technique. That is, techniques are employed to determine whether the signals on some hops are being subjected to interference, and those hop signals are then attenuated or completely excised. In the case of fast-FHSS systems, where there are multiple hops per information symbol, the situation is very much like that of a DSSS system with narrowband interference, and excision of hops hit by interference will greatly reduce the interference effects with relatively little impact on the underlying FHSS signal. This is particularly true where coding can be applied in conjunction with narrowband interference excision.

In slow-FHSS systems, where multiple information bits are contained in each hop, interference excision is less effective, because the excision of a single hop cancels several information bits together with the interference. However, even with slow frequency hopping, interference excision can provide performance benefits when applied in conjunction with coding designs using sufficiently long constraint lengths. Interference suppression techniques have been investigated and applied extensively in military communications systems. Applications to commercial wireless networks are under investigation, with particular attention to the issue of coexistence between new broadband CDMA systems and existing narrowband systems [Mil92].

9.4.3 Effects of Coding

As we saw earlier in this section, the equations used for the calculation of the output SNR of various spread-spectrum systems in the presence of interference is not the same for all systems and all forms of interference. For an FHSS system in wideband interference and a DSSS system in narrowband or wideband interference, the signal-to-interference ratios given by Eqs. (9.4.4), (9.4.18), and (9.4.23) are identical. The same general formula, given in Eq. (9.4.7), is found for *average* signal-to-interference ratio for an FHSS system in narrowband interference, but the effect upon FHSS system performance in narrowband interference is different from the other three cases. In particular, the probability of error for an FHSS system in narrowband interference, as given by Eq. (9.4.8), decreases as a reciprocal of the system processing gain N. However, for the FHSS system in wideband interference, the error probability decreases exponentially with processing gain, as shown by Eq. (9.4.9). With FHSS systems, the same pattern of separation also holds for the effects of coding, with coding being potentially more effective against narrowband interference than against wideband interference.

For an FHSS system operating in narrowband interference, as we explained earlier, an appropriately chosen error-correcting code can be highly effective in eliminating the errors in signal hops affected by the interferer. The general expression for symbol-error probability at the output of a t-error-correcting decoder is given by Eq. (9.4.10). For an FHSS system in narrowband interference, the coding is made most effective by making the block length or constraint length sufficiently long to span many hops. In this way, the hops impacted by the interference are a small fraction of the set of hops spanned by the code, and with a judicious choice of the error-correction limit t, most or all of the errors can be eliminated. In a fast-FHSS system the interference affects only one bit out of N, and therefore a good choice of block-coding parameters is to have $t/n = 1/N$, where n is the code block length. In slow frequency hopping we need an interleaver to disperse the clusters of errors that occur when the signal lands on the hop frequency with interference. With a sufficiently long interleaver the code performance in slow hopping will approach that found in fast hopping. However, interleaving involves buffering operations at both ends of the link, and this creates an added source of time delay in the delivered data stream. For digital voice service, the time delay must be carefully controlled, and this places a practical limit on the amount of interleaving that can be done. For a data service, time delay is not so critical, and longer interleaving buffers can be used.

It is useful to observe that in an FHSS system, the use of coding to combat narrowband interference has much similarity to the use of diversity combining on a fading channel. The redundancy of the error-correction code is similar to the multiplicity of branches in the diversity system. The code effectively combines a small number of erroneous bits with a larger number of error-free bits and uses the mathematical structure of the code in a decoding algorithm to identify and correct the erroneous bits. Similarly, the diversity combiner brings together one or a few low-SNR branches with several higher-SNR branches and applies a combining algorithm that allows the higher-SNR branches to prevail in forming the decision on the information symbol. Here we have one hop with low and many hops with high signal-to-interference ratio. We saw in Section 7.4.2 that the error probability improvements provided by diversity is an exponential function of the diversity order L, and thus with a limited number of diversity branches we can improve the performance by several orders of magnitude. With the use of appropriate coding, the channel error rate given by Eq. (9.4.8) can be brought down to an extremely low post-decoding error rate, corresponding to a substantial lowering of the required signal-to-interference ratio.

In all cases other than that of an FHSS system in narrowband interference, coding will provide more modest reductions in the minimum signal-to-interference power ratio required for the proper operation of the system, at the cost of either reducing the information rate or increasing the system bandwidth. In general, for all these other cases the channel error rate without coding is given by $P_e(\gamma_b)$, where γ_b is the uncoded signal-to-interference ratio and P_e is the appropriate error probability equation for the given modulation technique. Let us say that we use an (n, k) block code to encode the information bits for transmission, and that the Hamming weights of the codewords are $\{W_m\}$, $m = 1, 2, 3, \ldots, M$, where $M = 2^k$ is the total number of codewords in the code. Let $m = 1$ designate the all-zeros codeword and let us assume that this codeword is transmitted. (We could assume transmission of any of the M codewords, but the discussion is made simpler by assuming the all-zeros codeword.) Now consider an ideal decoder at the receiving end, one that decodes the received sequence of n symbols by correlating the sequence with a list of the M possible codewords and selecting the codeword with the highest correlation. This is exactly the same as processing the received sequence with M matched filters, each filter corresponding to one possible codeword. Given this scheme, a decoding error occurs if the received sequence has higher correlation with any one of the $M - 1$ incorrect codewords than it does with the all-zeros codeword actually transmitted in our example. The probability of erroneously deciding in favor of the mth codeword is simply

$$P_m = P_e(\gamma_b R_c W_m)$$

where P_e denotes the probability of a binary decision error, $R_c = k/n$ is the rate of the block code, and W_m is the Hamming weight of the mth codeword. The overall word-error probability for this decoding procedure is difficult to compute precisely, but it can be bounded by taking the union bound of the $M - 1$ possible error events, which we write as

$$P_B \leq \sum_{m=2}^{M} P_e(\gamma_b R_c W_m) \tag{9.4.24}$$

In the limit of high SNR, the decoding error events will be dominated by false decodings to minimum-distance codewords, and if the minimum distance of the code is $d_{\min}$, the word-error probability becomes

$$P_B \simeq A_{d,\min} P_e(\gamma_b R_c d_{\min}), \qquad \gamma_b \gg 1$$

where $A_{d,\min}$ denotes the number of minimum-weight codewords in the code. Similarly, for a convolutional code with minimum free distance d_{free}, and assuming soft-decision Viterbi decoding, we have a union bound on the probability of an information symbol decision error given by

$$P_C \leq \sum_{d=d_{\text{free}}}^{\infty} b_d P_e(\gamma_b R_c d) \qquad (9.4.25)$$

where, similarly to Eq. (9.4.11), b_d is the number of code trellis paths of weight d times the number of corresponding output bit errors caused by decoding to an incorrect weight-d path. These equations are essentially the same as those derived for conventional nonspread modulation over additive noise channels. For a fixed source information rate, coding will improve the interference margin $I_{\text{av}}/P_{\text{av}}$ by a factor of $R_c d_{\min}$ (or $R_c d_{\text{free}}$). For a fixed transmission bandwidth, coding will improve the interference margin by a factor of $d_{\min}$.

9.5 SPREAD-SPECTRUM PERFORMANCE IN FREQUENCY-SELECTIVE FADING

The performance of a spread-spectrum system in flat fading is similar to that of any other modulation. For any modulation method in flat fading, all spectral components of the transmitted signal fade in unison, and these power fluctuations make it necessary to provide additional transmitted power (fade margin) to bring the average BER performance to an acceptable level. As we discussed in Chapter 7, the effects of fading are reduced significantly, and the necessary fade margin reduced accordingly, if we use diversity combining techniques at the receiver. The BER formulas given in Chapter 7 for conventional modulation techniques apply equally well to spread-spectrum modulation systems. In flat fading where the Doppler spread is high (i.e., in rapid flat fading), coding techniques can provide performance improvements similar to those provided by diversity combining. Just as diversity is most effective when the diversity branches fade independently, coding is most effective when the symbols in a codeword fade independently. If the Doppler spread is low, we can use interleaving techniques to break up the clusters of consecutive errors produced by the slow fading, and thereby make the error-correcting codes operate more efficiently. However, in using spread-spectrum techniques on frequency-selective fading channels, new considerations arise, which we discuss next.

As we saw in the previous chapter in discussion of standard non-spread modulation techniques, frequency-selective fading in the frequency domain is manifested as ISI in the time domain. As we showed, the effects of frequency-selective fading can be controlled by several means, including multiple and sectored antennas, adaptive equalization, and multirate modems. However, one of the major features of spread-spectrum is its inherent resistance to frequency-selective fading. In addition, DSSS and FHSS systems both lend themselves to the design of receivers that actually take advantage of the multipath characteristics associated with frequency-selective fading. The approaches used for the design of these receivers are quite different for DSSS and FHSS systems, as we shall see. The effects of frequency-selective fading are, in a sense, similar to the effects of narrowband interference. With a narrowband interferer, a portion of the transmission bandwidth is hit by the interference, and the SNR in that part of the band is degraded relative to other parts of the band. In frequency-selective fading a portion of the band is affected by a deep null in the channel frequency response, and as with narrowband interference, the SNR in that part of the band is degraded. However, despite the close similarity of these situations, the methods used to deal with the two problems are not necessarily the same. In the remainder of this section we analyze the performance of DSSS and FHSS systems on frequency-selective fading multipath channels.

9.5.1 DSSS Transmission in Multipath Fading

Signal demodulation in a DSSS system is similar to adaptive MLSE or DFE in the sense that it takes advantage of the multipath to provide a form of in-band diversity. We begin by considering the case of DSSS transmission on a single-path channel with no interference, for which the output of the receiver correlator was given by Eq (9.2.4). An example of the correlator output when the transmitted pulses are rectangular was shown in Fig. 9.3. The channel in this simple case is represented mathematically by an ideal impulse function. In a multipath environment the channel is represented by a linear summation of L multipath components,

$$h(\tau, t) = \sum_{i=1}^{L} \beta_i \delta(t - \tau_i) e^{j\phi_i} \qquad (9.5.1)$$

where β_i, τ_i, and ϕ_i represent the magnitude, arrival delay, and phase of the ith path. The output of the receiver correlator is then given by

$$R_{rs}(t) = \sum_n a_n \sum_{i=1}^{L} \beta_i R_{ff}(t - iT_b - \tau_i) e^{j\phi_i} \qquad (9.5.2)$$

where a_n is the value of the nth information symbol, T_b is period of the spreading signal, and $R_{ff}(\tau)$ is the autocorrelation function of the spreading signal, given earlier in Eq. (9.2.5).

Figure 9.18 illustrates the output of the correlator for the case of a three-path channel. By comparing this figure with Fig. 9.3, one can observe the effects of multipath on the receive correlator output. The system parameters in both figures are the same, but Fig. 9.3 represents a single-path channel whereas Fig. 9.18 represents a three-path multipath channel. Figure 9.19 shows several correlator

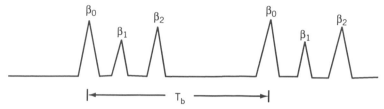

Figure 9.18 The output of the correlator with a three-path channel.

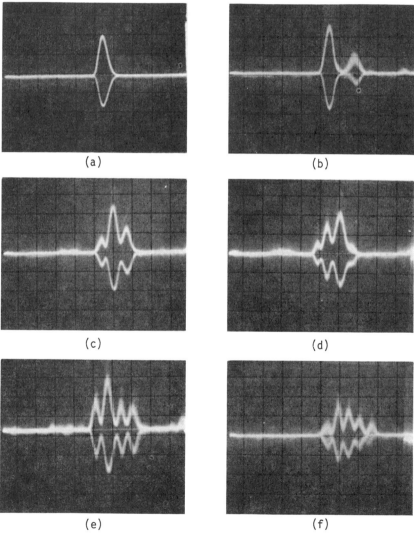

Figure 9.19 A sample correlated signal at the output of a system (shown in Fig. 9.11) that uses a SAW device correlator. (From [Fer80] © IEEE.)

output signals for the system shown previously in Fig. 9.11, in which the receiver correlator was implemented with a SAW device. In Fig. 9.19 one can see that there are up to five isolated paths in the channel, and the base of the correlation function is about 40 nsec wide.

For the example shown in Fig. 9.18, the interarrival delay among the multipath signals is greater than the width of the base of the autocorrelation function, and the delay spread is less than the information bit interval T_b. However, these conditions do not apply in every situation. For instance, if the delay spread of the channel is greater than T_b, this means that the data rate is greater than the coherence bandwidth of the channel, and we have ISI. To avoid interference between detected information symbols, the data rate should be kept well below the coherence bandwidth of the channel. When consecutive signal paths arrive with delay differences greater than the chip duration T_c, the correlator output will exhibit separate peaks, as shown in Fig. 9.18. If the delay between two consecutive paths is significantly less than the chip duration, the two paths will merge and appear as one path equivalent to the phasor sum of the two actual paths. Thus as the signal bandwidth is made narrower, the chip duration becomes correspondingly longer and fewer isolated paths can be resolved at the correlator output. Of course, as paths merge together, the fluctuations in their amplitudes and phases produce an overall fluctuation in the phasor sum, which we observe as fading.

If we operate with a chip duration short enough to resolve individual paths, we can design a receiver to take advantage of the multiple paths to provide diversity and enhance the reliability of the decision on each received information symbol. This sort of diversity is referred to as *implicit*, *internal*, *in-band*, or, simply, *time* diversity. In a DSSS system, a receiver that optimally combines the multipath components as part of the decision process is referred to as a *RAKE receiver*. A typical RAKE receiver structure for a DSSS system is shown in Fig. 9.20. The received signal is passed through a tapped-delay-line, and the signal at each tap is passed through a correlator similar to the one used for standard DSSS receivers. The outputs of the correlators are then brought together in a diversity combiner whose output is the estimate of the transmitted information symbol.

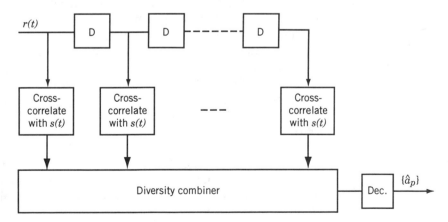

Figure 9.20 A typical RAKE receiver structure for a DSSS receiver.

In the original RAKE receiver [Pri58], the delay between the consecutive taps or "fingers" of the RAKE receiver was fixed at $W/2$ (where W is the chip bandwidth) to provide two samples of the overall correlation function for each chip period. Using this method for a rectangular chip pulse with triangular correlation function, we will have four samples of each triangle in the correlation function. Because the peaks are generally not aligned precisely at multiples of $T_c/2$, it is not possible to capture all the major peaks of the correlation function. But a good approximation of all major peaks will be provided by a RAKE receiver implemented with a sufficiently large number of taps. Modern digitally implemented receivers typically have only a few RAKE taps and the capability to adjust the tap locations. An algorithm is used to search for a few dominant peaks of the correlation function and then position the taps accordingly. If the receiver correlator is designed with a SAW device, there is no need for a tapped delay line. The output of the SAW device provides all values of the correlation function. As noted earlier, Fig. 9.19 shows a sample output of a SAW correlator receiver exhibiting several signal paths. Several preliminary developments of wireless DSSS systems have utilized SAW devices [Fer80]. However, SAW correlators have a stability problem, and it is difficult to integrate them into an ASIC-based modem design. For these reasons, as well as cost considerations, these devices are not widely used in available commercial wireless products.

A RAKE receiver can combine the arriving signal paths using any standard diversity combiner such as a selective, equal-gain, square-law, or maximal ratio combiner (MRC). As we discussed in Chapter 7, the optimum diversity combiner is the MRC, which weighs the received signal from each branch by the signal amplitude at that branch. An MRC RAKE receiver is actually a discrete matched filter for a signal received on a multipath channel and thus provides optimum performance in the available bandwidth.

In summary, in the operation of a DSSS system, multipath does not cause ISI unless the information symbol transmission rate approaches the coherence bandwidth of the channel. Also, it is possible to design a receiver which takes advantage of the isolated arriving paths to improve or even optimize system performance. The wider the transmission bandwidth, the greater the number of the isolated paths that can be resolved, and the greater the order of implicit diversity that can be utilized. The isolated paths will provide a source of implicit diversity to a DSSS receiver, which improves the performance of the system. On the other hand, if not utilized in some diversity combining receiver, the signal arriving from each path is a wideband interference to the signals arriving from other paths, which degrades the performance of the DSSS system.

Performance Evaluation. Now we determine the received SNR for DSSS transmission on a frequency-selective multipath fading channel. The SNRs that we calculated earlier in this chapter apply to the peak of the correlation function and therefore represent the maximum achievable SNR at the output of the correlator. In those calculations the channel was assumed to be a single-path stationary channel. As shown by Eq. (9.5.2), the output of the receiver correlator includes a number of peaks and the SNR is generally different for each peak. Each peak of the correlation function represents a branch of the implicit or time diversity and is associated with the signal arriving from one of the paths of the channel. Here we wish to determine the SNRs for all the diversity branches so that we can use them to analyze the performance of RAKE-type receivers. Because each diversity branch in effect represents a flat fading channel, we begin by treating the SNR at each branch as a random variable.

We begin by determining the impact of multipath components as sources of interference to the signal received on any single path. Assume that we have L resolvable paths and that the amplitudes $\{\beta_i\}$ of the paths, shown earlier in Eq. (9.5.1), are independent random variables. We shall denote the mean-square value $E[\beta_i^2]$ of the magnitude of the ith path in the impulse response as $\overline{\beta_i^2}$. The received energy per bit for the ith branch is then given by $\overline{\beta_i^2}E_b$, where E_b is the transmitted signal energy per bit. The average received power for the ith path is then given by

$$P_{\text{av-}i} = \overline{\beta_i^2}\, E_b R_b$$

where R_b is the information bit rate. The sum of interference power from the other $L-1$ paths, and the additive noise power, is

$$I_{\text{av}} = R_b E_b \sum_{i=1}^{L-1} \overline{\beta_i^2} + N_0 W$$

where N_0 is the one-sided noise spectral density and W is the receiver bandwidth. The average received SNR per bit at ith branch is then given by

$$\overline{\gamma_i} = \frac{W}{R_b} \times \frac{P_{\text{av}}}{I_{\text{av}}} = \frac{\overline{\beta_i^2}\, E_b}{\dfrac{R_b}{W} \displaystyle\sum_{i=1}^{L-1} \overline{\beta_i^2} + N_0} \tag{9.5.3}$$

If we assume the power to be the same on all branches, $\overline{\beta_i^2} = \overline{\beta^2}$, we find the output average SNR per bit to be

$$\overline{\gamma_i} = \frac{\overline{\beta^2}\, E_b}{\dfrac{(L-1)}{N}\overline{\beta^2}\, E_b + N_0}, \qquad \text{all } i \tag{9.5.4}$$

where $N = W/R_b$ is the processing gain of the system. For an interference-dominated environment, the effect of the additive noise is negligible, so that $N_0 \simeq 0$ and finally we have

$$\overline{\gamma_i} \simeq \frac{N}{L-1} \tag{9.5.5}$$

Thus we see that the received average SNR per bit is proportional to the processing gain and inversely proportional to the number of interfering signal paths. Note that for no multipath, $L = 1$, Eq. (9.5.5) gives the unrealistic result $\gamma_i = \infty$, due to the fact that we have neglected the effects of additive noise.

However, for the case $L = 2$, the equation yields $\gamma_i = N$, reflecting the fact that the impact of the equal-strength (0 dB) interferer is reduced by the system processing gain N.

As we discussed earlier, a RAKE receiver can take advantage of the implicit diversity of the received multipath signal to improve performance significantly. Using a RAKE receiver, various methods for diversity combining can be applied to the implicit diversity just as they were applied to the explicit diversity provided by multiple antennas as described in Section 7.4.2. If we assume that each path resolved by a RAKE receiver can be represented by a Rayleigh distributed random variable, then standard equations used for analyzing various diversity combining methods can be applied directly to DSSS systems operating in frequency-selective fading. In carrying out the analysis, the implicit multipath diversity is incorporated into the order of diversity used in the calculation, and the SNR is adjusted to account for the multipath as an additional source of interference. Then, equations given in Section 7.4.2 can be used, with the equations given above for average SNR, to predict the performance of a RAKE receiver.

If we assume that all paths have equal power and that the order of external diversity is one (single antenna), the average error rate of an MRC-RAKE receiver can be obtained by substituting Eq. (9.5.4) or Eq. (9.5.5) into Eq. (7.4.17). If the order of explicit diversity is D, then the total order of diversity to be used in Eq. (7.4.17) is DL, which includes both implicit and explicit diversity. The SNRs, however, remain the same as given in Eq. (9.5.4) because we assume that only the paths arriving at the same antenna interfere with one another.

If the RAKE receiver uses selection combining, it searches among all arriving paths and selects the paths with the highest amplitudes. With D orders of explicit and L orders of implicit diversity, the average probability of error for a selective combiner in Rayleigh fading and assuming equal average power in all branches is given by Eq. (7.4.19) as

$$\bar{P}_s = DL \sum_{k-1}^{DL} \binom{DL - 2}{k - 1} \frac{(-1)^{k-1}}{k} P\left(\frac{\bar{\gamma}_i}{k}\right) \tag{9.5.6}$$

where $P(\bar{\gamma}_i)$ is the expression providing the average error rate over the fading channel for the specific modulation technique used in the system, and $\bar{\gamma}_i$ is given by Eq. (9.5.4) or Eq. (9.5.5).

In the simple design of a spread-spectrum wireless terminal using SAW correlators, shown in Fig. 9.11, the output of the correlator is integrated over all paths before the decision is made on the transmitted symbol. The integrator adds up all the path energies without any weighting factor. Therefore this receiver is an equal gain combiner of the implicit diversity components. Calculation of average probability of error for equal gain combiners is relatively difficult. Calculation of average probability of error for the selective combiner and the MRC as discussed above can be used as lower and upper bounds, respectively, on the performance of the equal gain combiner. The actual performance of the equal combiner is closer to the upper bound,—that is, closer to the performance of the MRC.

The methods discussed in this section for calculation of the average error rate are based on the assumption that the number of paths is fixed, the received power in all paths is the same, and the amplitude of each path is a Rayleigh distributed random variable. As we saw in Chapter 4, the number of paths arriving at different locations is actually a random variable. Furthermore, the average received path powers are typically found to decay exponentially with increasing arrival delay, and received amplitudes are not necessarily Rayleigh. To address these issues properly, somewhat more complicated analysis is required.

If we assume that all paths have the same power, but the number of paths is a random variable, the average probability of error is given by

$$\bar{P}_s = \sum_L \bar{P}_{b/L} P(L) \tag{9.5.7}$$

where $P_{b/L}$ is the probability of error given that we have L equal power paths and $P(L)$ is the probability of having L paths in the profile. For a Poisson model of path arrivals we have

$$P(L) = \frac{\lambda^L}{L!} e^{-\lambda} \tag{9.5.8}$$

where λ is the average number of arriving paths.

If the values of the average received power from different paths are not equal, but the amplitudes are uncorrelated Rayleigh random variables, the average error rate for L paths is given by Eq. (7.4.18) with the average SNR per bit given by Eq. (9.5.3). If we assume that the path arrivals are random and that the received path powers are not equal, Eq. (9.5.7) cannot be used to average the error rate over different number of paths as it was used when the average received power was assumed the same on all paths. For cases in which the path powers are unequal and the path arrivals are random, Eq. (9.5.7) is not applicable and other means must be found to determine the average error rate. A complete solution in such cases requires a Monte Carlo simulation to generate all the path arrival delays. From a simulation of the arrival delays, the number of paths is determined and the power of each path $\overline{\beta_i^2}$ is calculated. The results are then used in Eq. (9.5.3) and Eq. (7.4.18) to determine one value of the error rate. This simulation experiment is then repeated many times, and the resulting error rates are averaged over all the experiments to form an estimate of the overall average error rate.

To gain a better understanding of the above equations, we apply them to a typical example in which we have a set of profiles representing the channel. Saleh's model for indoor radio propagation [Sal87b] is useful for this purpose. As explained in Chapter 6, in this model the path arrivals are described by a Poisson process, the amplitudes are Rayleigh distributed, and the delay power spectrum is an exponential function. The average number of paths is $\bar{L} = \lambda T_m$, where λ is the arrival rate of the paths and T_m is the delay at which the exponential delay power spectrum drops below the SNR of the channel. The RMS multipath delay spread of the channel is also determined easily from the shape and parameters of the delay power spectrum. The normalized average energy per path is $\overline{\beta^2} = 1/\bar{L}$, which is the normalized power divided by the average number of paths. Using these three parameters or using the results of simulations, one can use four different methods to determine the performance of a DSSS system.

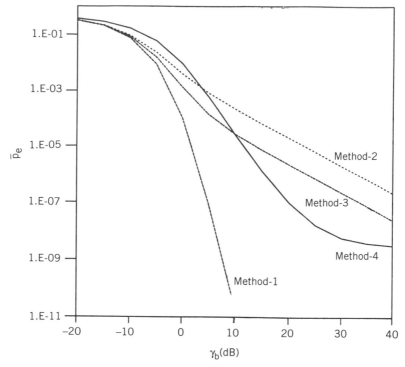

Figure 9.21 A comparison among the four methods for the calculation of the error rate. Single user, threshold $= -10$ dB; bandwidth $= 75$ MHz; spreading code length $= 255$; order of external diversity $= 1$.

Method 1 assumes that all paths have the same power and uses Eqs. (9.5.4) and (7.4.17) with $\overline{L}$ as the number of paths. Method 2 also assumes that the average power is the same for all paths and calculates the number of paths from $\hat{L} = |\overline{\tau}_{rms}/T_c| + 1$, where $|x|$ denotes the integer closest to x and $\overline{\tau}_{rms}$ is the average of τ_{rms} taken over all multipath profiles [Kav87, Pro89]. Method 2 also uses Eq. (7.4.17) for the error rate. Method 3 uses Eqs. (9.5.7) and (9.5.8) and assumes that the number of paths is a random variable but that the average power is the same for all paths. Equation (7.4.17) is used for the calculation of $P_{S/L}$ with Eq. (9.5.4) providing the value of $\overline{\gamma}_b$. Method 4 generates a set of channel profiles and calculates the error rate for each profile using Eq. (7.4.18), with Eq. (9.5.3) providing $\overline{\gamma}_b$. Figure 9.21 compares error rates calculated by these four methods.

All the calculations described in the previous paragraph are based on the assumption that the amplitude of each path is a Rayleigh distributed random variable, and the arrivals of the paths are described by a Poisson process. As we saw in the chapters on channel simulation and modeling, neither of these assumptions is necessarily valid and in most cases the empirical data indicate a lognormal amplitude distribution and modified-Poisson arrival times. The most accurate way to predict performance is to take the results of measurements of the channel impulse response and calculate the error rate directly by generating a set of channel impulse responses. This method is investigated in [Cha93]. By using this method, more details of the receiver structure can be incorporated into the calculations.

9.5.2 FHSS in Frequency-Selective Fading

As is the case with DSSS transmission, FHSS transmission is also resistant to frequency-selective fading. However, performance analysis of FHSS systems in frequency-selective fading and the methods used to take advantage of the in-band diversity are quite different from those described for DSSS systems. Consider the wideband frequency response for a radio channel that exhibits frequency-selective fading,— for example, as shown in Fig. 5.12. The multipath nature of the channel in the time domain accounts for the frequency-selective characteristic in the frequency domain. The depth of fading at certain frequencies can be as much as 30–40 dB below the average received signal power. As we saw in our discussion of statistical-frequency domain modeling, the rate of fluctuations in the frequency domain is proportional to the 3-dB width of the frequency correlation function, which is in turn inversely proportional to the rms multipath spread of the channel. In mathematical terms, fluctuations of the signal strength in the frequency domain follow the same patterns as fluctuations of the strength of a narrowband signal in time domain. The relationship between the rate of fluctuations in the frequency domain and the rms multipath spread is the same as the relationship between the rate of variations in time domain and the rms Doppler spread. In other words, there is a kind of *duality* between the time- and frequency-domain statistics of the channel response. Therefore, the Rice distribution for the number of fades crossing a certain threshold is also applicable to the number of frequency-selective fades crossing certain threshold level, with rms multipath delay spread replacing the rms Doppler spread. Using the Rice fading distribution given by Eq. (4.5.2), with τ_{rms} replacing $B_{D\text{-rms}}$, the average number of fades crossing a threshold A in a bandwidth W is given by

$$K = W\sqrt{2\pi}\,\tau_{\mathrm{rms}}\rho e^{-\rho^2} \qquad (9.5.9)$$

where $\rho = A/A_{\mathrm{rms}}$ is the ratio of the threshold level to the rms value of the amplitudes. Similarly, we may use Eq. (4.5.3) to determine the average width of the fade in the frequency domain. For an FHSS system occupying a band of width W, if τ_{rms} is small enough so that $K < 1$, there is no deep notch in the operating band and the channel is considered a flat fading channel. This situation is analogous to DSSS transmission over a channel having only one resolved path. For larger values of τ_{rms} we have at least one deep fade in the operating band of the system and the FHSS must operate in frequency-selective fading. This situation is analogous to the operation of the DSSS system on a channel with more than one resolved path.

Let us now assume that there is only one deep null in the operating band of an FHSS system and that the width of the fade is such that it affects only one of the hop frequencies. The error rate of the modem is unacceptably high each time the signal hops to the faded frequency, but is extremely low for all the other hops. This situation is similar to that of the same system operating on a channel with a fixed narrowband interferer which disrupts only a single hop frequency. In both cases the SNR drops significantly on certain hops. Therefore, the performance analysis here would lead to the same equations as used to analyze the FHSS system operating in narrowband interference. Also, as in the case of narrowband interference, the FHSS system can be designed with appropriate coding and interleaving to correct the errors occurring on the hops into the frequency-selective fade. System considerations related to the effectiveness of coding in an FHSS system are similar to those related to the effectiveness of the RAKE receiver in a DSSS system. In an FHSS system, coding becomes more effective as the fading on neighboring hop frequencies becomes more independent, and this happens as the transmission bandwidth is increased. Similarly, as the bandwidth increases in a DSSS system, more paths become resolvable, making the RAKE receiver increasingly effective in exploiting the implicit diversity. If the data rate of an FHSS system approaches the coherent bandwidth of the channel, a large number of hops will be affected by the frequency-selective fading; and even with very strong coding and long interleavers, error correction is inadequate and the output error rate becomes unacceptably high. This situation is analogous to operation of a DSSS system with data rates near the coherent bandwidth of the channel, causing ISI, which cannot be exploited by the RAKE receiver.

9.6 PERFORMANCE OF CDMA

With the growing interest in integration of voice, data, and imagery traffic in telecommunication networks, CDMA appears increasingly attractive as the wireless access method of choice. Fundamentally, integration of various types of traffic is readily accomplished in a CDMA environment because coexistence in such an environment does not require any specific coordination among user terminals. In principle, CDMA can accommodate various wireless users with different bandwidth requirements, switching methods and technical characteristics without any need for coordination. Of course, because each user signal contributes to the interference seen by other users, power control techniques are essential in the efficient operation of a CDMA system. In practice today, CDMA is not commonly used in WLAN products because with an efficient channel access technique, such as carrier-sense multiple access with collision detection (CSMA/CD), adequate numbers of users are supported in the currently allocated frequency bands. However, CDMA systems are under development for PCS and cellular telephone applications. For these voice-oriented networks, the volume of traffic is growing rapidly and the available bands have to be used as efficiently as possible in order to support the growing numbers of users. The emerging standards for the CDMA PCS and cellular networks are being written to include the integration of other services as well. The flexibility of CDMA has also made this approach a leading candidate for the evolving experimental multimedia wireless networks. The two common CDMA techniques are direct-sequence CDMA (DS-CDMA) and frequency hopping CDMA (FH-CDMA), which are constructed on the DSSS and FHSS techniques, respectively. The DS-CDMA technique has received by far the greater attention in ongoing developments of new systems, and we address a number of technical issues related to this technique in the remainder of Section 9.6. We shall briefly discuss the FH-CDMA technique in Section 9.7.

9.6.1 Direct-Sequence CDMA

In a DS-CDMA system, as in any DSSS system, the bandwidth of the information signal is expanded by multiplying the signal by a PN waveform having bandwidth much wider than the information signal. In its simplest form, both the information signal and the PN signal are binary waveforms. The PN waveform, commonly called the *chip sequence*, has a transmission rate R_c which is some large integer multiple of the information bit rate R_b. In DS-CDMA, each user's PN sequence is generated by a unique code, and this allows receivers to distinguish among different users' signals. In this way, multiple users can transmit simultaneously using the chip bandwidth R_c and their signals can be separated using their unique PN codes. As in any DSSS system, the receiver essentially reverses the signal process described above: It multiplies the received signal by a replica of the PN chip sequence used at the transmitter, and it integrates or otherwise filters over a bit interval. In the CDMA system, the receiver uses a PN code unique to the desired user signal, and the signals coded with other users' codes simply appear as wideband noise and are thus reduced by the DS processing gain.

CDMA with Ideal Power Control. Here we calculate the signal-to-interference-and-noise ratio for signals received at the base station of a CDMA cellular system, in terms of the number of users simultaneously sharing the channel. (For simplicity, we shall continue to use the abbreviation SNR, even though we are considering combinations of interference and noise.) Having determined the SNR, we can relate it to the error rate requirements and other factors involved in the design of digital cellular systems. Our basic assumption here is one of ideal power control. That is, we assume that the power of each user's transmitter is controlled by the central base station so that precisely the same power is received from every terminal and the power level is constant with time. We further assume a DS-CDMA system in which M users are sending information to the base station simultaneously and the bandwidth of the shared channel is $W = NR_b = N/T_b$, where N is the system processing gain. One of the users is the "target" mobile station, while the other $M - 1$ users are wideband interferers to the target station. At the base station the total interference power from all interferers is $(M - 1)E_b R_b$. The power of the additive noise at the base station is $N_0 W$ and therefore the total interference plus noise at the base station is

$$I_{av} = N_0 W + (M - 1)E_b R_b$$

The power received from the target receiver is

$$P_{av} = E_b R_b$$

and the SNR is therefore given by

$$\gamma = \frac{W}{R_b} \times \frac{P_{av}}{I_{av}} = \frac{E_b/N_0}{1 + [(M - 1)/N](E_b/N_0)} \qquad (9.6.1)$$

If there is no interferer and only one user, $M = 1$ and $\gamma = E_b/N_0$. If the number of users in the system is large and the same power is received from all terminals, the interference noise dominates the background noise. Then the SNR ratio is approximated by

$$\gamma \simeq \frac{N}{M - 1} \qquad (9.6.2)$$

and for some desired level of SNR γ, the number of users is given by

$$M = 1 + \frac{N}{\gamma} \qquad (9.6.3)$$

The bandwidth efficiency η is defined as the aggregate data rate of the users, normalized by the system bandwidth, which we write as

$$\eta = \frac{MR_b}{W} = \frac{M}{N} = \frac{1}{N} + \frac{1}{\gamma} \qquad (9.6.4)$$

The required value of γ depends upon the modulation and coding techniques used in the system, together with the users' requirement on acceptable error rate. In this analysis we have assumed that the same signal power is received from every user terminal. In the radio environment of course, the received power fluctuates due to fading. However, if we extend our definition of ideal power control to mean *ideal instantaneous power control*, the power fluctuations caused by fading are eliminated and performance analysis is identical to that for a nonfading channel.

Example 9.5. For BPSK modulation, $\gamma = 10$ dB will provide an error rate of 10^{-5}, which is a reasonable error rate for many applications using uncoded data. With a processing gain of $N \simeq 100$, the number of simultaneous users and the bandwidth efficiency, calculated from Eqs. (9.6.3) and (9.6.4), are 11 and 0.11, respectively. For $\gamma = 7$ dB the error rate is around 10^{-3}, an acceptable error rate for telephone quality digital voice communications. The number of users and the bandwidth efficiency in this case will increase to 21 and 0.21, respectively.

In general for the evolving digital cellular systems the first terms in Eq. (9.6.3) and Eq. (9.6.4) are much smaller than the second terms and the two equation can be further simplified as

$$M = \frac{N}{\gamma} = \frac{W/R_b}{E_b/I_0} \qquad (9.6.5)$$

and

$$\eta = \frac{1}{\gamma} = \frac{1}{E_b/I_0} \qquad (9.6.6)$$

Example 9.6. Using QPSK modulation and convolutional coding, the existing CDMA digital cellular systems require 3 dB $< \gamma < 9$ dB. Using Eq. (9.6.5) with $N \simeq 100$, the number of users will be in the range $12 < M < 50$ and the bandwidth efficiency will be in the range $0.12 < \eta < 0.5$.

Practical Considerations. In the practical design of digital cellular systems, three other parameters affect the number of users that can be supported by the system as well as the bandwidth efficiency of the system. These are the number of sectors in each base station antenna, the voice activity factor, and the interference increase factor. These parameters are quantified as factors used in the calculation of the number of simultaneous users that the CDMA system can support. The use of sectored antennas is an important factor in maximizing bandwidth efficiency. Antenna sectorization reduces the overall interference, increasing the allowable number of simultaneous users by a *sectorization gain factor* which we denote by G_A. With ideal sectorization the users in one sector of a base station antenna do not interfere with the users operating in other sectors, and $G_A = N_{\text{sect}}$, where N_{sect} is the number of sectors in the antenna pattern. In practice, antenna patterns cannot be designed with ideal sectorization, and due to multipath reflections, users in general communicate in more than one sector. Three-sector base station antennas are commonly used in cellular systems, and a typical value of the sectorization gain factor is $G_A = 2.5$ (4 dB). The voice activity interference reduction factor G_V is the ratio of the total connection time to the active talkspurt time. On the average, in a two-way conversation each user talks 50% of the time. The short pauses in the flow of natural speech reduce the activity factor further to about 40% of the connection time in each direction. As a result, the typical number used for G_V is 2.5 (4 dB). The interference increase factor H_0 accounts for users in other cells in the CDMA system. Because all neighboring cells in a CDMA cellular network operate at the same frequency, they will cause additional interference. The interference is relatively small due to the processing gain of the system; a value of $H_0 = 1.6$ (2 dB) is commonly used in the industry.

Incorporating these three factors as a correction to Eq. (9.6.5), the number of simultaneous calls in a CDMA cell is approximated by

$$M \simeq \frac{N}{\gamma} \times \frac{G_A G_V}{H_0} \tag{9.6.7}$$

If we define the *performance improvement factor* in a digital cellular DS-CDMA system as

$$K = \frac{G_A G_V}{H_0}$$

then assuming the typical parameter values given earlier, the performance improvement factor is $K = 4$ (6 dB).

Example 9.7. If we continue Example 9.6 with the new correction factor included in Eq. (9.6.7), the range for the number of simultaneous users becomes $50 < M < 200$. The range of associated bandwidth efficiencies is $0.5 < \eta < 2.0$. To find the required bandwidth, let us assume that the system is carrying digitized voice encoded at $R_b = 7$ kbits/sec. The occupied bandwidth is $W = 700$ kHz and the bandwidth per active user is between 3.5 kHz and 14 kHz, which should be compared with the 30 kHz per channel used in the existing FM-modulated analog AMPS systems.

9.6.2 Effects of Power Control

The discussion in the previous section presents an ideal situation, which can be approached as a practical matter when power control is used on the link from a mobile station to the base station (usually termed the *uplink*). Power control is an essential function in any DS-CDMA system serving mobile users, because of the near–far problem. On the uplink each signal is received with a different amount of path loss, due to (a) variations in distances of the mobiles from the base site and (b) the statistical variability of the path loss. Therefore if all the mobiles transmit with the same signal power, their signals will arrive at the base station with widely different power levels. As a result, a few mobile stations whose signals arrive with much less path loss than the other stations will overpower the remaining mobiles. This is readily seen from Eq. (9.6.1), which is written for the case where all signals arrive with equal power $E_b R_b$.

Example 9.8. If one user's signal arrives with a power level 10 dB higher than the others, the interference from this user is equivalent to the interference of 10 other users. Therefore, the number of simultaneous users the system is able to support is reduced by nine. The SNR for this user, however, is 10 dB higher than for the others. In principle, it is feasible to design a CDMA system in such a way that user channels with different SNR requirements are overlaid (for example, to provide for simultaneous operation of different types of services having different BER requirements.) This might be done by controlling the power to different levels on different user channels.

The solution to this problem is to control the power transmitted by each mobile station so that the same power level is received at the base station from each mobile. This requires that the base station provide continuous feedback to each mobile station so that the mobile can adjust its power level dynamically. In a practical mobile communication system, it may be necessary to be able to adjust power levels over a range as wide as 80 dB. The speed with which power adjustments must be made depend upon whether the control function follows only the relatively slow lognormal shadowing variations or the more rapid Rayleigh fading. If the power control is to follow the fast multipath fading, the rate of power adjustments should be at least twice the maximum Doppler spread of the channel in order to provide precise power control. Thus for a maximum Doppler shift of 100 Hz, power adjustment would ideally be done every 5 msec. If the power control mechanism controls only the slow lognormal variations due to shadowing, the received signal still exhibits the more rapid Rayleigh fading. If the transmission bandwidth is much narrower than the coherence bandwidth of the channel, the received signal will be characterized by flat fading. If the transmission bandwidth is comparable to the coherence bandwidth, the signal arrives from several isolated paths and with an appropriately designed RAKE receiver it is possible to take advantage of the implicit diversity in the received signal. In the following sections we provide more details of the effects of multipath fading on the number of simultaneous users in a DS-CDMA system using different assumptions on power control.

It should be noted that although fluctuations in the received power have a serious negative impact on the capacity of a DS-CDMA system, the power fluctuations will in fact increase the throughput of contention-based wireless access methods such as ALOHA and CSMA. We explore this point in greater detail in Chapter 11, where we discuss network topologies and network access methods.

9.6.3 DS-CDMA with Average Power Control

In this section we assume that the average received power is controlled so that the effects of path loss and slow lognormal shadow fading are compensated for. However, either there is no dynamic power control or the dynamic power control algorithm is too slow to compensate for the effects of fast multipath Rayleigh fading. When the bandwidth of the spread-spectrum signal is much narrower than the coherence bandwidth of the channel, the channel is not frequency selective and the signal paths are not resolvable. The effects of fading would then appear as fluctuations of the received signal amplitude, with all frequency components of the signal fading in unison. If we have M simultaneous users, the average received power for the target receiver is given by

$$P_{av} = \overline{\beta^2} E_b R_b$$

where $\overline{\beta^2}$ is the average of the squared magnitude of the signal received from the target transmitter, taken with respect to the Rayleigh fading distribution. The average interference plus noise is then given by

$$I_{av} = (M - 1) \overline{\beta^2} E_b R_b + N_0 W$$

where we assume that the same average power is received from all users. The average signal-to-interference ratio is then given by

$$\overline{\gamma} = \frac{W}{R_b} \times \frac{P_{av}}{I_{av}} = \frac{\overline{\beta^2} E_b}{\dfrac{(M - 1) \overline{\beta^2}}{N} + N_0} \tag{9.6.8}$$

where N is the bandwidth expansion factor or processing gain of the system. If we assume that the interference is much higher than the additive noise, then Eq. (9.6.8) becomes

$$\overline{\gamma} = \frac{N}{M - 1} \tag{9.6.9}$$

This equation is essentially equivalent to Eq. (9.6.2) except that here we use the average signal-to-interference ratio $\bar{\gamma}$ and we assume that the instantaneous signal amplitude is Rayleigh distributed. As we did in Eq. (9.6.3), we can relate the number of users to the average signal-to-interference ratio by rearranging Eq. (9.6.9) into

$$M = 1 + \frac{N}{\bar{\gamma}} \qquad (9.6.10)$$

Just as in Eq. (9.6.4), we can express the bandwidth efficiency η as

$$\eta = \frac{MR_b}{W} = \frac{M}{N} = \frac{1}{N} + \frac{1}{\bar{\gamma}} \qquad (9.6.11)$$

Comparing Eq. (9.6.11) with Eq. (9.6.4) we note that in the earlier equation γ is the SNR needed to produce some desired error rate on a steady-signal channel, while in Eq. (9.6.11) $\bar{\gamma}$ is the average SNR needed in Rayleigh fading, and in general $\bar{\gamma} \gg \gamma$ for the same output error rate. As a result, for the steady-signal case the second term $(1/\gamma)$ in Eq. (9.6.4) typically dominates the first term $(1/N)$ while as we see in the following example, for the Rayleigh fading case, the first term is typically dominant and the bandwidth efficiency is therefore approximated by

$$\eta \simeq \frac{1}{N} \qquad (9.6.12)$$

which the reader should compare with Eq. (9.6.6).

Example 9.9. For DPSK modulation in flat Rayleigh fading the average probability of error is given by

$$P(\bar{\gamma}) = \frac{1}{2(1 + \bar{\gamma})} \simeq \frac{1}{2\bar{\gamma}}$$

where $\bar{\gamma}$ is the average SNR. For an error rate of 10^{-3} we need $\bar{\gamma} = 500$, and for an error rate of 10^{-5} we need $\bar{\gamma} = 50{,}000$. With typical value of processing gain $N = 100$, for both cases using Eq. (9.6.11) we have $\eta \simeq 0.01$, which is the same as the result obtained from Eq. (9.6.12). The more important point is that for such a poor bandwidth efficiency, regardless of the error rate requirement, we find from Eq. (9.6.10) that the system cannot support more than one user. This in turn means that the CDMA system is not practical unless we can control the effects of Rayleigh fading in addition to the near–far and shadowing effects.

This result is in sharp contrast with TDMA and FDMA systems, where the increase in average per-user signal power needed to overcome fading has little influence on the number of users supported by the system and therefore no impact on the bandwidth efficiency of the system. In a CDMA spread-spectrum system, as the preceding results show, an increase in the average power transmitted by individual terminals will increase both the desired signal and the total other-user interference by the same proportion and therefore the user's error rate cannot be improved without reducing the interference, which means reducing the number of users on the system. The answer to this problem lies in the use of more rapidly acting power control algorithms. Before discussing practical power control strategies, we continue our discussion of average power control, with attention to the case of frequency-selective fading.

9.6.4 Average Power Control and Frequency-Selective Fading

For standard modems operating on fading multipath channels, an increase in the bandwidth results in destructive intersymbol interference and consequent performance degradation. As we discussed earlier, in spread-spectrum communications, an increase in the bandwidth allows increased resolution of the signal paths, and the resolved paths can then be used for inband diversity combining and a consequent performance improvement. Explicit diversity can also be obtained by using multiple or sectored antennas. In this section we provide a quantitative analysis of the effects of implicit and explicit diversity on the bandwidth efficiency of, and number of users supportable by, a DS-CDMA system. Note that standard communication enhancement techniques such as diversity combining and coding can increase the allowable number of users in a CDMA system by reducing the required per-user signal power. However, in TDMA and FDMA systems, coding and diversity combining can decrease the power required by each user but have no direct effect on bandwidth efficiency of the system.

Figure 9.22 shows a block diagram of a CDMA system designed to operate over a frequency-selective fading channel. The M user signals are assumed to arrive at the central receiver over M different channels, where the general form of the channel impulse response is given by Eq. (9.5.1), and the general form of the signal at the output of the receiver correlator by Eq. (9.5.2). In this model, each of the M channels has its own path amplitude, arrival time, phase, and number of resolved paths. The average power of the desired signal at the ith tap of the RAKE receiver is given by

$$P_{\text{av-}i} = \overline{\beta_i^2} E_b R_b$$

However, the interference has three components: additive noise, multipath interference, and multiuser interference. We designate one receiver as the "target,"

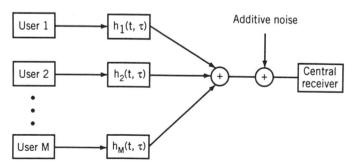

Figure 9.22 Block diagram of a CDMA system operating over a frequency-selective fading channel.

and we write the total average interference power at its ith tap as

$$I_{\text{av}-i} = E_b R_b \sum_{k=2}^{M} \sum_{l=1}^{L_k} \overline{\beta_{lk}^2} + E_b R_b \sum_{j=0,\,j\neq i}^{L_1} \overline{\beta_j^2} + N_0 W$$

where L_k is the number of paths on the kth user channel, and $\overline{\beta_{lk}^2}$ is the average signal power for the lth path on the kth user channel. The first term on the right-hand side of the equation is the multiuser interference while the remaining two terms represent the multipath and additive noise just as was described in Section 9.5.1, where we examined the effects of multipath fading. The multiuser interference term includes the power of all paths for all users except the paths in the target user channel. The average SNR for path i of the target receiver is then given by

$$\overline{\gamma}_i = \frac{\overline{\beta_i^2} E_b}{\dfrac{E_b}{N} \displaystyle\sum_{k=2}^{M} \sum_{l=1}^{L_k} \overline{\beta_{lk}^2} + \dfrac{E_b}{N} \displaystyle\sum_{j=0,\,j\neq i}^{L_1-1} \overline{\beta_j^2} + N_0} \tag{9.6.13}$$

Formulas for the performance of a DS-CDMA system in multipath fading can be obtained by some minor modifications to the equations used earlier for the case of a single user in multipath fading. For example, Eq. (7.4.18) gives the PSK error rate for the case of maximum ratio combining with unequal powers on the diversity branches. If we have D-order external diversity, the average error rate for the kth user can be found by substituting the SNR given by Eq. (9.6.13) into Eq. (7.4.18), with $L_k D$ used as the order of diversity.

To simplify this equation we may further assume (a) the same number L of signal paths on every user channel and (b) a single value of power for all paths on all M user channels. In this case the total multipath and multiuser interference is $(ML - 1)\overline{\beta^2} E_b R_b$ and the average SNR is

$$\overline{\gamma} = \frac{\overline{\beta^2} E_b}{(ML - 1)\overline{\beta^2}\,\dfrac{E_b}{N} + N_0} \tag{9.6.14}$$

Furthermore, if we assume that the additive noise is negligible relative to the interference, the signal to interference ratio is given by

$$\overline{\gamma} = \frac{N}{ML - 1} \tag{9.6.15}$$

The number of users is then given by

$$M = \frac{1}{L} + \frac{N}{L\overline{\gamma}} \tag{9.6.16}$$

and the bandwidth efficiency is given by

$$\eta = \frac{1}{NL} + \frac{1}{L\bar{\gamma}} \tag{9.6.17}$$

The results in this case depend upon the orders of implicit and explicit diversity. As the order D of explicit diversity increases, the value of $\bar{\gamma}$ needed to support a specific error rate decreases, but the relationship among $\bar{\gamma}$, N, M, and L as given by Eq. (9.6.15) remains the same. Therefore the reduction in $\bar{\gamma}$ provides an increase in the allowed number of users and in the bandwidth efficiency, as given by Eqs. (9.6.16) and (9.6.17). As the number L of implicit diversity paths increases, the overall order of diversity increases, which reduces the required average SNR $\bar{\gamma}$. However, examination of Eqs. (9.6.16) and (9.6.17) shows that changes in L and $\bar{\gamma}$ now act in opposition in their effect on M and η. Thus, when implicit diversity increases, there is a gain in performance as measured by required $\bar{\gamma}$ but a simultaneous increase in multipath interference, and the net result must be evaluated in each particular case. In general, as shown in the following example, the reduction in required average SNR is greater than the increase in the number of paths, and therefore $L\bar{\gamma}$ decreases as L increases.

Example 9.10. Figure 9.23 gives performance results for a spread-spectrum CDMA system operating with BPSK modulation and bandwidth expansion factor $N = 255$ over a flat fading channel. The figure shows average probability of error versus the

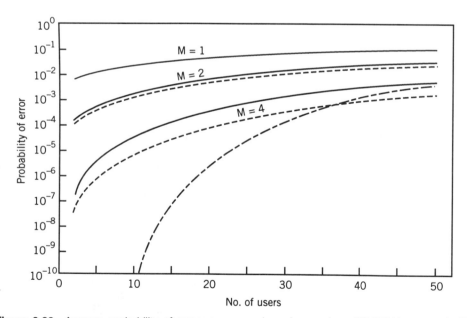

Figure 9.23 Average probability of error versus number of users for a DS-CDMA system in flat fading with different orders of explicit diversity. Diversity components are combined either by maximal ratio combining (*dashed line*) or by selection combining (*solid line*) The *dotted line* represents the reference for a nonfading channel (perfect power control). Bandwidth expansion is $N = 255$, and the modulation technique is BPSK [Pah87].

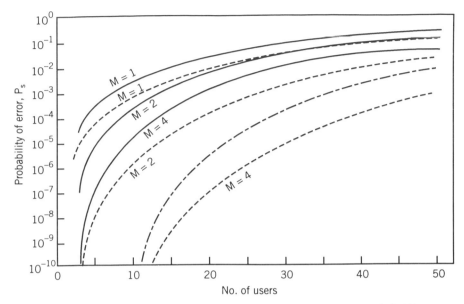

Figure 9.24 Average probability of error versus number of users for a DS-CDMA system in multipath fading with different orders of explicit diversity. The implicit and explicit diversity elements are combined either by maximal ratio combining (*dashed line*) or by selection combining (*solid line*). The *dotted line* represents the reference for a nonfading channel (perfect power control). The number of signal paths is 5, the bandwidth expansion is $N = 255$, and the modulation technique is BPSK [Pah87].

number of active users for single, double, and quadruple explicit diversity. The assumption of flat fading here is equivalent to assuming one signal path per user channel. The solid-line curves correspond to selection combining; the dashed-line curves correspond to maximal ratio combining, and the dotted-line curves correspond to the case of no fading (or ideal power control) of the received user signals. With quadruple diversity and maximal ratio combining, the figure shows that the supportable number of users, and, consequently, the bandwidth efficiency for higher error rates around 10^{-3} are comparable to performance with perfect power control; while for lower error rates of around 10^{-5}, the bandwidth efficiency is considerably better than the performance with perfect power control. Figure 9.24 shows similar curves for fading multipath channels with fifth-order internal diversity ($L = 5$). The bandwidth efficiency of the maximal ratio combiner with fourth-order explicit diversity is better than the case of perfect power control for all meaningful probabilities of error.

The results presented above are based on the assumption that the signal amplitude received from any path is a Rayleigh-distributed random variable, and that we have a fixed number of signal paths on any user channel. As we saw in Chapter 6 where we discussed channel simulation and modeling, these assumptions are not necessarily valid and in most cases the empirical data indicate a lognormal amplitude distribution and modified-Poisson arrival times. The most accurate way of predicting system performance is to use the results of channel

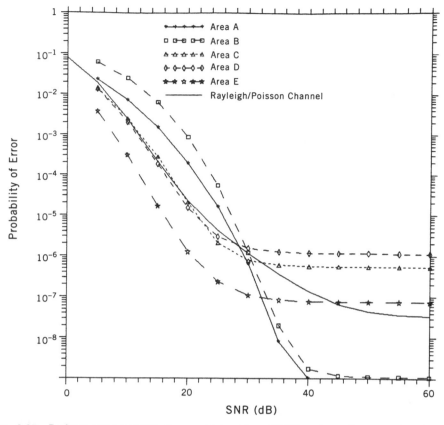

Figure 9.25 Performance prediction curves obtained for a CDMA system with one active transmitter operating over the five manufacturing floors described in Chapter 3. A Rayleigh fading model and a Poisson arrival model are assumed. There are two code sequences per user, with $N = 255$ chips per code sequence [Cha93].

impulse response measurements to calculate the error rate directly. This prediction method was investigated in [Cha93] using impulse responses measured on indoor radio channels; the impulse response measurements are those already described in Chapter 3. Figure 9.25 shows a set of performance prediction curves obtained for a CDMA system operating over channels in five different manufacturing areas.

9.6.5 Practical Power Control

Any practical power control algorithm has a limit on the range of signal variations it can accommodate. Furthermore, the power level adjustments are performed in discrete steps, and in the case of mobile and cellular systems, the speed of adjustment might not be adequate for all vehicular speeds. As a result, the power is not perfectly controlled and we can model the deviation from the ideal control signal as a random variable. The lognormal distribution is a natural model for smaller variations in the received power, measured in decibels. In the digital cellular industry a lognormal distribution has been used for this purpose and the modeling results have shown close agreement with the results of measurements [Vit93, Pad94, Gil91].

To calculate the allowable number of CDMA users by the approach that we have outlined thus far, we need to determine the required signal-to-interference ratio averaged over the lognormal fading distribution. This approach involves integration of the error rate functions (either exponential or erfc) with respect to the lognormal distribution, which is not analytically feasible. An analytical approximation to the solution is provided in [Vit93] and [Pad94], where the allowable number of users is approximated by

$$M = \frac{N}{e^{(\beta\sigma)^2/2+\beta\mu}} \qquad (9.6.18)$$

where $\beta = 0.1 \ln 10$ and μ and σ are the mean and variance of the lognormal power control error, respectively. Details of this derivation are provided in Problem 11 at the end of this chapter. As compared with the case of perfect power control given by Eq. (9.6.5), γ, the minimum required signal-to-interference ratio, is now replaced by an exponential. Generally, γ is smaller than the exponential term in the denominator in Eq. (9.6.18). The ratio of the exponential term to γ represents the reduction in the number of allowable users due to the use of nonideal practical power control.

9.6.6 Orthogonal Codes

Orthogonal codes provide one method of increasing the bandwidth efficiency of a spread-spectrum system [Pah87]. With this approach each user has a set of orthogonal sequences representing a set of symbols for transmission. The set of orthogonal sequences, or *orthogonal-code*, will typically be a Hadamard or Walsh code, discussed in Chapter 7. A stream of information bits is segmented into groups and each group represents a nonbinary information symbol, which is associated with a particular transmitted code sequence. If there are n bits per group, one of a set of $K = 2^n$ sequences is transmitted in each symbol interval. The received signal is correlated with a set of $K = 2^n$ matched filters, each matched to the code sequence of one symbol. The correlator outputs are compared and the symbol associated with the largest output is declared as the transmitted symbol. In the decision process, in addition to multiuser and multipath interference we have an additional form of interference introduced in the decision process when we compare the outputs of the correlators. Assuming a simple one path channel, perfect power control, and negligible additive noise, the signal-to-interference ratio is given by

$$\gamma = \frac{N}{M - 1 + K - 1} \qquad (9.6.19)$$

where $M - 1$ represents the noise from other users and $K - 1$ represents the noise from the outputs of the correlators other than the one corresponding to the correct symbol. The allowable number of users is then given by

$$M = \frac{N}{\gamma} - K + 2 \tag{9.6.20}$$

which indicates that the number of simultaneous users is reduced by the use of orthogonal coding. However, we should note that the data rate of the individual users and consequently the bandwidth efficiency of the system are improved by a factor $\log_2 K$, as shown by

$$\eta = \frac{MR_b}{W} = \frac{M \log_2 KR_s}{W} = \frac{M \log_2 K}{N} \tag{9.6.21}$$

where R_s is the symbol transmission rate.

Example 9.11. Consider a system in which we have $N = 100$, a required minimum SNR of $\gamma = 2$, and an orthogonal code having $K = 2$ symbols. For this case we have $M = 50$ allowable users at a data rate R_b. Now consider a second case in which we have an orthogonal code with $K = 16$ symbols. Here the allowable number of users decreases to $M = 36$, but the data rate for each user increases by a factor $\log_2 16 = 4$. This means that each user in the second case can accommodate four users of the first case—for example, in a TDMA format—and the equivalent number of users with the rates of the first case is $36 \times 4 = 144$, which is almost three times the users supportable (50) in the first case. The disadvantage of the orthogonal signaling scheme is the complexity of the receiver design. In this example we need 16 receiver correlators per user channel instead of only one needed in the simplest system design.

Table 9.1 gives the bandwidth efficiency for different numbers K of code sequences per user, in a system with code length 256 chips, and two orders of explicit diversity. The results are given for 2-, 4-, and 7-path multipath channels. The maximum bandwidth efficiency of 0.3047 is found for the largest code set ($K = 64$) and maximum multipath spread (7 paths). This efficiency will increase if the order of explicit diversity is increased or if the constraint on the probability of error is relaxed from 10^{-4}. For 2-path channels the bandwidth efficiency increases

TABLE 9.1 Bandwidth Efficiency for CDMA Systems Operating Over Fading Multipath Channels with $N = 256$, $K = 2 - 64$, $E_b / N_0 = 30$ dB, Two Orders of Explicit Diversity, Different Number of Symbols, and Different Orders of Implicit Diversity [Pah90a]

K	Two Paths	Four Paths	Seven Paths
2	0.0429	0.0664	0.0742
4	0.0702	0.1171	0.1327
8	0.0937	0.1524	0.1868
16	0.1094	0.1875	0.2343
32	0.1171	0.2148	0.2734
64	0.1409	0.2577	0.3047

TABLE 9.2 Bandwidth Efficiency for CDMA Systems Operating
Over Fading Multipath Channels with $K = 16$, $E_b / N_0 = 30$ dB, Two Orders
of Explicit Diversity, Different Lengths of Spread-Spectrum Code, and
Different Orders of Implicit Diversity [Pah90a]

N	$L = 2$	$L = 4$	$L = 7$
32	0.125	0.25	0.25
64	0.125	0.25	0.25
128	0.125	0.2188	0.25
256	0.1094	0.1875	0.2343
512	0.1016	0.1875	0.2188
1024	0.0977	0.1797	0.2148
2048	0.0938	0.1777	0.2129

by approximately a factor of 3 as the number of code sequences per user increases
from 2 to 64. For 4-path and 7-path channels the corresponding increase is
approximately a factor of 4. Table 9.2 shows the effect of increasing code length N
for $K = 16$ orthogonal sequences and second-order explicit diversity. For a fixed
bandwidth, as the code length increases, the transmission rate and consequently
the bandwidth efficiency decrease. On the other hand, an increase in the code
length reduces the interference from other users, which improves the bandwidth
efficiency of the system. Considering both effects, Table 9.2 shows that for any
number of signal paths, the bandwidth efficiency decreases slightly as the code
length is increased. In practice, the two major drawbacks for implementation of
orthogonal coding systems are (a) the task of finding a large set of orthogonal code
sequences and (b) the complexity of the receivers.

9.7 FREQUENCY-HOPPING CDMA

The FH-CDMA technique is perhaps the easiest to visualize. In each time slot,
users are assigned to distinct frequency bands, but in each successive time slot, the
frequency assignments are permuted. For example, in time slot 1, user 1 occupies
band 1, user 2 occupies band 2, and user 3 occupies band 3. In time slot 2, user 1
hops to band 3, user 2 hops to band 1, and so forth. The modulation scheme used
in an FH-CDMA system will typically be nonbinary FSK.

Let us now assume that we have M FHSS systems with independent random
patterns occupying a bandwidth W, each with a data rate of R_b and a bandwidth
expansion factor $N = W/R_b$. The error rate performance of the FH-CDMA
system is the same as the performance of the FHSS system in random narrowband
interference. The symbol error probability without coding is given by Eq. (9.4.8)
and with coding is determined from Eq. (9.4.10). As we have done earlier, let us
designate one of the M users as the target user and consider the effect of the M-1
interfering user signals on the target receiver. The probability that exactly k hops
of the $M - 1$ simultaneous users hit the hop used by the target receiver is given by

$$P_{k\text{-hit}} = \binom{M - 1}{k} P^k (1 - P)^{M-1-k} \qquad (9.7.1)$$

where $P = 1/N$ is the probability of having a hop at a specific hop frequency. The target receiver will have no error on a hop if no other user hits that hop frequency at the same time. This is obtained by evaluating Eq. (9.7.1) at k = 0, which yields

$$P_{0\text{-hit}} = \begin{pmatrix} M - 1 \\ 0 \end{pmatrix} P^0 (1 - P)^{M-1} = (1 - P)^{M-1} \qquad (9.7.2)$$

The probability of a hit is $1 - P_{0\text{-hit}}$ and the probability of error given a hit is $1/2$, which yields an overall error probability

$$P_{s-\text{FHCDMA}} = \frac{1}{2}(1 - P_{0\text{-hit}}) = \frac{1}{2}\left[1 - \left(1 - \frac{1}{N}\right)^{M-1}\right] \qquad (9.7.3)$$

For a single user, $M = 1$, and the probability of error is zero because we neglect additive noise in this analysis. For two users $M = 2$ and we have $P_s = 1/2N$, which is consistent with the derivations for FHSS operation in random narrowband interference.

As discussed earlier, system performance can be improved significantly by the use of coding. Continuing the line of analysis begun above, we would find that the error rate achievable with some coding technique is independent of the chip modulation technique or the SNR. The assumption here is that the SNR is high enough that the error rate in a single-user environment is negligible. This is in contrast with our derivation of the error rate for the DS-CDMA, where the error rate was shown to be dependent on the chip modulation technique. The simple derivation of the error rate presented here can be modified to include the effects of the choice of modulation method [Yeg93], but the effects are not as important as the effects of modulation choice on the performance of a DS-CDMA system.

If power control is not used, or if the power control technique is unable to keep the power from all the receivers at precisely the same level, the effects are not as serious as in a DS-CDMA system. If the users' signals arrive with different power levels, there will be instances in which we have two hops at the same frequency but the difference between the received powers is large enough that one of the signals will survive. Analysis of this phenomenon is similar to the analysis of the effects of capture in contention-based access methods operating on fading channels, a topic we discuss in Chapter 11. Theoretically, the performance of an FS-CDMA system with received power fluctuations is expected to be slightly better than the predictions given above. However, we should note that with large fluctuations in the power we need to increase the transmitted power to achieve a negligible error rate in the absence of a hit. In wireless networks where battery-powered terminals are to be used, an increase in the transmitted power is an unattractive design choice, and for that reason power control techniques should be applied. In frequency-selective fading, when hops of different users' signals coincide or when a hop lands in a deep fade, errors will occur. In this situation the performance analysis must include the effects of faded hops as well as hits from other users.

In some applications such as GSM, the FHSS technique is combined with TDMA. In these systems the hops are coordinated so that there are no hits among different users' signals. A system of this type is referred to as *FHSS/TDMA* and can be considered as an FD-CDMA system with nonrandom orthogonal codes. The capacity in this case is the same as the capacity of a TDMA system with the number of user slots equal to the number of hop frequencies. With this approach, several packets are transmitted at each hop and the advantage is that only a portion of the packets for each user will suffer from the frequency-selective fading characteristics of the channel. For voice users the loss a fraction of the packets generally does not cause a significant degradation in the quality of the reconstructed voice. For data users, the packets lost in a hop exposed to frequency selective fading will be retransmitted, very likely at hop frequencies where the channel is not faded. Given these considerations, both the voice and data oriented wireless information network industries are evaluating frequency hopping systems for possible future application.

QUESTIONS

(a) Name five major advantages of spread-spectrum technology for WIN systems.

(b) What is the major reason for using spread spectrum for digital cellular?

(c) Why is spread spectrum used in wireless LANs?

(d) What is the simplest method of implementing the correlator for a spread-spectrum receiver?

(e) Is coding effective in DSSS systems operating over slow flat fading channels? Explain.

(f) Is coding effective in FHSS systems operating over slow flat fading channels? Explain.

(g) Is coding effective in DSSS systems operating over frequency-selective fading channels? Explain.

(h) Is coding effective in FHSS systems operating over frequency-selective fading channels? Explain.

(i) What are the advantages of a RAKE receiver?

(j) Name the two major advantages of using power control in multiuser environments.

(k) Can a DSSS wireless LAN operate efficiently without power control? Explain.

(l) Can an FHSS/CDMA system operate efficiently without power control?

(m) What is the advantage of using a sectored antenna in a CDMA system?

(n) What is the advantage of using a sectored antenna for high-speed data communications?

PROBLEMS

Problem 1. The transmission bandwidth of a TV station is 5.5 MHz, and the minimum acceptable received signal level is 30 dB above the background noise. We want to transmit a BPSK/FHSS digital broadcast signal in the same band using the same transmitter antenna.

(a) What is the maximum transmission power for the digital broadcasting service that allows the performance of the TV station to remain acceptable.

(b) What is the minimum processing gain of the FHSS system if the maximum error rate of the digital broadcasting system is to stay at 10^{-5}.

Problem 2. Repeat Problem 1 assuming BPSK/DSSS is used. Neglect the effects of multipath.

Problem 3. Use the bound in Eq. (9.4.10) to find P_s in Example 9.3. Compare the results of the bound with the exact calculation provided in the example.

Problem 4. Using the MatLAB or MathCAD, sketch the Fourier transform of the half-sinewave pulse discussed in Example 9.4. Compare the spectrum with that of the rectangular pulse. How does the power in the main lobe compare in the two cases? What is the difference, in decibels, between the maximum sidelobe attenuations in the two cases?

Problem 5. The pulse shape

$$g(t) = A \sin^2 \frac{\pi t}{T}, \qquad 0 < t < T$$

is used for shaping the chips of a DSSS system.

(a) Show that

$$A = \sqrt{\frac{16E_c}{3T_c}}$$

where E_c is the energy in the pulse shape.

(b) Determine the dc $G(0)$ value of the waveform.

(c) Compare the signal-to-narrowband interference ratio of a DSSS system that uses these pulses for pulse shaping with a DSSS system that uses a rectangular waveform for pulse shaping.

Problem 6. Repeat Problem 3 for the waveform described in Problem 4.

Problem 7. We have two 1-W, BPSK/DSSS WLANs from two different manufacturers (using different codes) transmitting in the 902-928MHz ISM band. The two WLANs communicate with separate receivers and operate in the same area. At the target receiver, one of the transmitters provides the information source and the other transmitter acts as a source of wideband interference. The received power from the information source and the interference source are independent from one another, and they vary according to relative location of the two transmitters with respect to the target receiver.

(a) If the processing gain of both systems is 10 dB and the transmitters are at the same location (i.e. the received power from each transmitter is the same), what is the error rate at the receiver for the target transmitter? If the received power from each transmitter forms a zero-mean lognormal random variable with variance of 10 dB, what is the outage rate for the target transmitter? Assume that the threshold error rate for calculation of outage probability is 10^{-5}.

(b) Repeat (a) if the processing gain was 0 dB (no spreading and only BPSK modulation).

Problem 8. Repeat Problem 7 for BPSK/FHSS WLANs with the same power and processing gains. Assume that the frequency patterns of the two WLANs form random variables independent of one another.

Problem 9

(a) Determine the number of users in an uncoded QPSK spread spectrum CDMA system that has perfect central power control. Assume that the bandwidth expansion factor is 128, the acceptable error rate is 10^{-3}, and the effects of additive noise are negligible.

(b) If trellis-coded modulation with 8-PSK modulation is applied and the coding gain is 3 dB, what is the maximum number of simultaneous users?

(c) Repeat part (a) if average power control is applied to cancel the lognormal shadow fading but the Rayleigh multipath fading still disturbs the system. Assume that the receiver uses one antenna and that the fading in the channel is flat.

(d) Repeat (c) if the receiver uses two antenna with uncorrelated received signals.

Problem 10. Derive an equation that gives the probability of outage versus threshold error rate of a binary DPSK-modulated CDMA system in flat fading. Assume that the effects of additive noise are negligible.

Problem 11. In this chapter we introduced a method to calculate the number of simultaneous users in a DS-CDMA system with perfect power control. In reality the power cannot be controlled perfectly and the signal to interference ratio for user i

$$\gamma_i = \frac{E_{bi}}{I_o} = \frac{W}{R}\frac{P_i}{N_0W + \sum\limits_{j=1}^{M-1} P_j} \approx \frac{W}{R}\frac{P_i}{N_0W + \sum\limits_{j=1}^{M} P_j}$$

forms a lognormal distribution function with mean and variance of μ dB and σ dB, respectively. In this equation, $P_i = E_{bi}R$ is the received power for user i and M is the number of simultaneous users in the system.

(a) If we define the total signal to interference ratio as

$$x = \frac{R}{W}\sum\limits_{i=1}^{M}\gamma_{bi}$$

show that

$$E(x) = \frac{R}{W}Me^{(\beta\sigma)^2/2+\beta\mu}$$

and

$$\text{Var}(x) = \left(\frac{R}{W}\right)^2 Me^{(\beta\sigma)^2+2\beta\mu}\left[e^{(\beta\sigma)^2} - 2\right]$$

where $\beta = \ln(10)/10$.

(b) Show that the received total power to noise ratio is given by

$$\gamma = 1 + \sum_{i=1}^{M} \frac{P_i}{N_0W} = \frac{1}{1-x}$$

(c) If we define the received signal above noise level in dB by $Z = 10\log_{10}\gamma$, show that the commulative distribution function of Z is given by

$$CDF: f_Z(z) = \int_{-\infty}^{\frac{1-e^{-\beta z}-E(x)}{\sqrt{\text{Var}(x)}}} e^{-y^2/2}\,dy$$

Let m be the median of Z and show that

$$E(x) = 1 - 10^{-m/10}$$

(d) Using results of **(b)** and **(c)** show that the following relation exists between the median of the received signal above the noise background and the simultaneous number of users:

$$M = \left(1 - 10^{-m/10}\right)\frac{W/R}{e^{(\beta\sigma)^2/2+\beta\mu}}$$

(e) Compare the results of **(d)** assuming that the background noise is negligible $(m \to \infty)$ with the result of derivations with perfect power control analyzed in this chapter. Assume $\mu = 7.9$ dB and $\sigma = 2.4$ dB [Pad94].

10

WIRELESS OPTICAL NETWORKS

10.1 INTRODUCTION

Optical communication technology has several features that are well-suited to wireless indoor applications. Infrared (IR) transmitters and receivers can be built at relatively low cost, and with small size and low power consumption suitable for battery-supported operation. Optical modems can be built less expensively than radio-frequency (RF) equipment, at costs comparable to wired connections. IR transmissions will not interfere with existing RF systems, and IR systems do not fall under any regulations of the FCC. Because IR signals do not penetrate walls, these systems provide a considerable degree of privacy simply by confinement of the transmissions within an office or other work area. The only way for IR signals to be detected outside the installation area is through windows, which can easily be covered with curtains or shades. The confinement of IR signals by walls also allows concurrent usage of similar systems in neighboring offices without mutual interference. Therefore, in a cellular network architecture all units can be identical—in contrast with RF systems, in which the operating frequencies of neighboring cells must be different. In an IR network, terminals within a cell communicate with a node or "satellite" installed on the ceiling, and these nodes are interconnected to the rest of the network with wires, cables, radio, or fiber optics lines [Gfe82, Yen85]. The cell might be a small office or a section of a large open office, depending on the architecture of the building. Figure 10.1 shows two typical setups for optical networks. The first setup is a manufacturing floor divided into sub-areas covered by satellites. The second setup comprises three areas covered by separate satellites. The satellites in either example are connected by a wired backbone network to a central control station. Wireless optical networks are sensitive to shadowing, and therefore coverage will usually suffer to some extent from blind corners or areas with outage. In these instances, coverage can be extended by installing passive or active reflectors.

IR communication technology is dominant in the low-speed remote control market [Cia82] and is also finding some application in cordless phones, wireless keyboards for personal computers, and wireless local area networks (WLANs). The major drawbacks in the use of IR technology in wireless networks are its data rate limitations, extensive power fluctuation, and susceptibility to interference from ambient light. In particular, there is a lack of a simple and reliable multiple-access

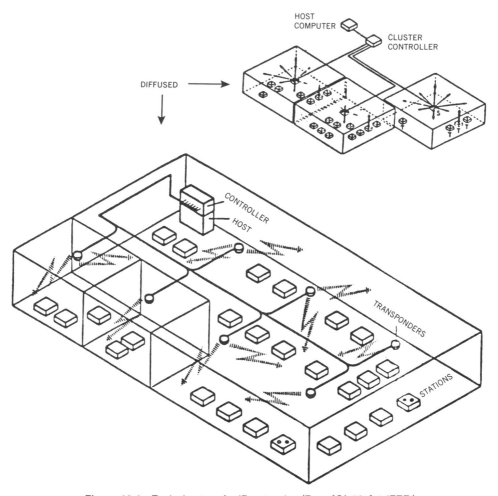

Figure 10.1 Typical setups for IR networks. (From [Gfe79a] © IEEE.)

technique for IR systems capable of counteracting the effects of ambient light and objects moving near the transmitter or receiver.

10.2 IMPLEMENTATION ISSUES

In the past decade, most of the developments in wireless optical communications have concentrated on diffused IR radiation (DFIR) technology. Figure 10.2 shows a typical DFIR network configuration. The primary advantage of this mode of transmission is that it does not require a direct line of sight between the transmitter and the receiver [Gfe79b]. Instead the receiver can collect a transmitted signal by way of reflections from walls, ceiling, and other objects in the work area.

Therefore, the installation of the network does not require precise alignment to establish a communication link, and this facilitates portability of the user terminals. As a result, DFIR networks are well-suited for applications requiring portability, such as cordless phones or networking of laptop or pen-pad computers.

The disadvantages of DFIR transmission are as follows:

1. It consumes relatively high power to cover the entire installation area.

2. The data rate is limited due to the effects of multipath.

3. Diffusion of the IR signals elevates the risk of eye exposure.

4. In simultaneous two-way communications, each receiver collects its own transmitted reflections which are stronger than the transmitted signal from the other end of the connection.

An alternative to diffused transmission is directed-beam IR (DBIR) communication, which has been investigated for application to wireless information networking [Yen85]. In the WLAN industry, this technology has not been as widely adopted as the DFIR technology, but some products using this technology have emerged in the market. Figure 10.3 shows a typical DBIR network. The transmitted radiation pattern is aimed in the direction of the receiver. The advantage of DBIR transmission is that it requires less optical power for communication, it does

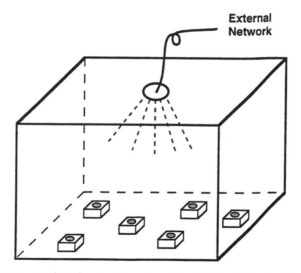

Figure 10.2 Diffuse infrared (DFIR) network configuration. Light is diffusely reflected and scattered in the room from a wide-angle source or a diffusing spot formed on a reflector. It is received by wide-acceptance-angle photodetector portable units located anywhere in the room. The advantages of this configuration are that there are no alignment requirements and that there is freedom of movement.

not suffer from extensive multipath propagation, and it can handle bidirectional communications better than diffused radiation. As a result, higher data rates and larger areas can be covered using this method of transmission. The disadvantages are (a) the need for alignment of transmitters and receivers and (b) signal interruption caused by shadowing. This transmission method is used in applications where the terminals are in fixed locations during the transmission interval. In typical installations, the transmitters are placed on high posts to avoid shadowing. In such installations, care must be taken to minimize direct eye exposure to the directed beams.

Semiconductor laser diodes and light-emitting diodes (LEDs) are used as the radiating elements in wireless optical networks. A semiconductor laser emits a narrow optical beam, and for diffused transmission applications a diffusing lens or some other optical device must be used to increase the coverage area. The more

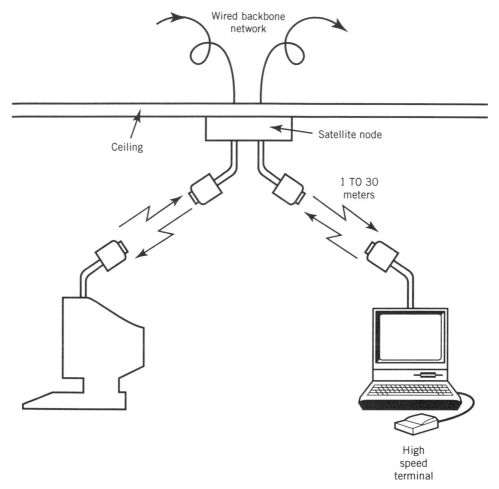

Figure 10.3 Directed beam infrared (DBIR) network configuration. Two collimated IR beams connect a terminal to the network, one for the uplink and one for the downlink.

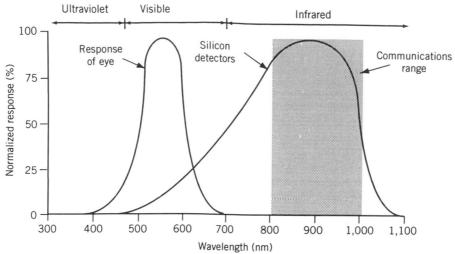

Figure 10.4 Responses of the human eye and of silicon detectors. The shaded portion of the figure shows the range of wavelengths for IR communication systems.

expensive semiconductor laser diodes have a more linear electrical-to-optical power conversion characteristic, and they provide higher radiated optical power. The radiating element plays a role similar to that of the last-stage power amplifier in a radio transmitter. Linearity of this element provides a more flexible environment for employing a variety of modulation techniques. For example, multiamplitude or multifrequency modulation requires use of a linear transmitter. At the receiver, an avalanche photodiode (APD) or a *p*-intrinsic-*n* (PIN) photodiode is used to convert the received optical power to an electrical signal. These diodes are analogous to the front-end RF circuits in radio modems. The more expensive APD has a higher internal gain, which supports higher signal-to-noise ratios needed for wideband communication links. Figure 10.4 shows the response of the human eye, and of silicon detectors and LEDs.

The choice of optical wavelength is another implementation issue. In addition to extremely inexpensive LEDs and PIN diodes, inexpensive gallium arsenide (GaAs) laser diodes and low-noise silicon APDs at 850 nm are available in the market. The decreased safety hazards of optical radiation at longer wavelengths allow higher transmitted power, which facilitates higher transmission rates. Disadvantages are that devices at these frequencies are more expensive and noisier than today's implementations. Today's IR technology operates primarily at about 900 nm, but the evolving technology includes 1.5-μm devices.

10.3 EYE SAFETY

The human eye acts similarly to a camera, focusing the energy density of the incident light onto the retina by a factor of 100,000 or more. As a result, the

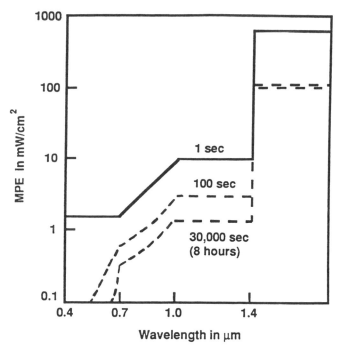

Figure 10.5 Levels of maximum permissible exposure (MPE). The sharp increase in MPE for $\lambda > 1.4$ μm is due to the filtering property of the eye's cornea. (From [Chu87] © IEEE.)

maximum permissible exposure (MPE) levels are quite small. The outer layer of the eye, the cornea, acts as an optical bandpass filter passing wavelengths ranging roughly from 0.4 μm to 1.4 μm. Light energy at wavelengths outside this range is absorbed by the cornea and does not pass through to retina. Because of this, wavelengths of 1.5 μm and higher are relatively safe for the human eye. The low cost of GaAs laser diodes and silicon APDs, which operate at 0.85 μm, make them popular devices, but care must be taken in systems using them not to exceed acceptable MPE levels. The acceptable levels of MPE are specified by standards organizations. Figure 10.5 shows a typical curve representing acceptable MPE levels for various wavelengths.

10.4 IR CHANNEL CHARACTERIZATION AND DATA RATE LIMITATIONS

There are three principal limitations for IR communications, caused by (1) interference from ambient light, (2) multipath channel characteristics in the case of diffused transmission, and (3) transient response of the IR devices. The rise and fall times of inexpensive LEDs limit the data rate to about 1 Mbit/sec. Explanation of the other two limitations requires more detail discussion, which follows.

10.4.1 Optical Propagation and Multipath Effects

As implied in the preceding discussion, IR communication in an indoor environment makes use of signals arriving at a receiver by a multiplicity of paths, and therefore the channel is a multipath channel.

The basis for calculation of optical signal propagation is the Lambertian law. The Lambert equation

$$P(\theta) = \frac{n + 1}{2\pi} P_t \cos^n \theta, \qquad -90° < \theta < 90°$$

gives the power per unit solid angle $P(\theta)$ received from an IR source or a reflector with total transmitted or reflected power P_t, where n is the *mode order* representing the directionality of the source. Figure 10.6 shows the radiation pattern for different values of n. In simulation of optical propagation, the ceiling and the walls in an indoor area are modeled as diffuse Lambertian reflectors with $n = 1$. To determine the reflected power, in a manner similar to ray tracing in radio propagation modeling, a reflection coefficient ρ is included in the calculation to represent the ratio of the total power reflected into the hemisphere to the power incident upon the surface. The reflection coefficient ρ is a positive number less than 1. A typical value of ρ for plaster walls and most ceiling materials is in the

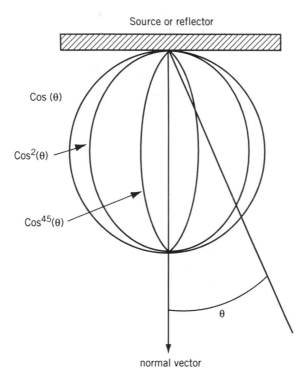

Figure 10.6 Lambertian radiation intensity pattern. The equation for the generalized Lambertian pattern is $P(\theta) = [(n + 1) / 2\pi]P_T \cos^n \theta$, where $-90° < \theta < 90°$.

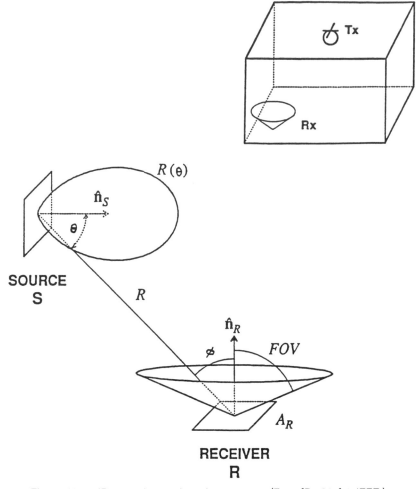

Figure 10.7   IR transmitter and receiver patterns. (From [Bar91a] © IEEE.)

range 0.7 to 0.85, while for painted, wooden, and textile surfaces it varies from 0.4 to 0.9.

Figure 10.7 shows a simple room with a transmitter and a receiver, a model used for simulation of optical radiation in [Bar91a]. The receiver has a photosensitive area A_R with a field of view (FOV) determined by the packaging of the photosensitive diode. The power received from an area radiating or reflecting power W is given by

$$P_R = WA_R \sin^2(\text{FOV}), \qquad 0° < \text{FOV} < 90°$$

As indicated by this equation, the received power P_R is independent of the position and angular orientation of the photodetector relative to the radiating element, if the radiating element covers the entire FOV. In practice, the FOV is typically reduced by device packaging to some angle less than 90°, and this

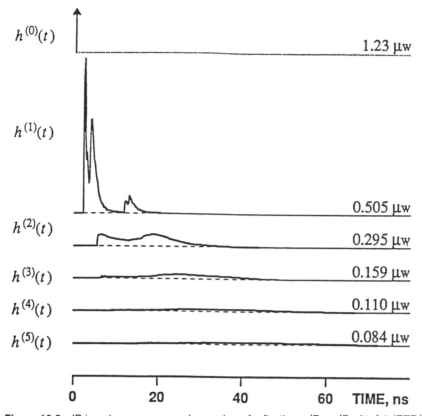

Figure 10.8   IR impulse response and a number of reflections. (From [Bar91a] © IEEE.)

constrains the application of diffused IR systems in open environments. However, in indoor areas the photosensitive diode can absorb power reflected from the walls, ceiling, and other objects located within the FOV.

To simulate the channel impulse response, the signal received in the FOV forms the line-of-sight (LOS) power. The power from the first-order reflection is obtained by calculating the incident and reflected power over all the areas of the walls and ceiling. For higher-order reflections, reflections from the floor are also included. Figure 10.8 shows the result of simulation of the channel impulse responses for different orders of reflections in a 5-m × 5-m × 3-m room [Bar91a]. The reflection coefficient is assumed to be 0.85 for the walls and the ceiling and 0.3 for the floor. The only active radiating element is a Lambertian point at the center of the ceiling pointing downward. The receiver is assumed to have a FOV of 85° and is located at coordinates (0.5 m, 1 m, 0 m) in the three-dimensional model. Figure 10.9 shows the overall impulse response and the associated frequency response of the channel. The existence of a strong LOS signal is represented by an impulse. The four subsequent peaks represent the reflections from the four walls. The overall power fluctuation in the entire 8-MHz bandwidth is less than 8 dB. The deep fades on the order of 30 dB typically observed on radio channels are not evident here.

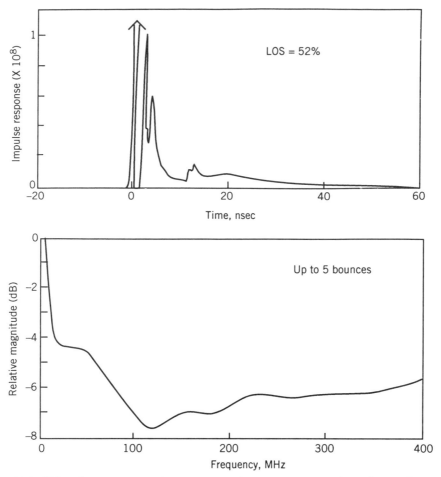

Figure 10.9 IR impulse response and corresponding frequency response. (From [Bar91a] © IEEE.)

There are some important differences between the simulations of radio and optical channels. In simulation of a radio channel, some of the signal energy incident upon a wall is assumed to pass through the wall and it has to be traced, whereas for optical transmission there is no penetration through walls or other structures. In simulation of the optical channel the entire surface of a wall or ceiling is used for the calculation of the received power, whereas in ray tracing of radio propagation only the paths whose reflections passes through the receiver are included in the calculation. Therefore, ray tracing is less computationally complex. The signals arriving from different paths in optical propagation simply add in their power levels, whereas in radio propagation modeling, the amplitudes and phases of the arriving signal paths are added together vectorially. Characteristics of the radio channel may change drastically with movement of the antenna by a fraction of a wavelength. However, the power received by a photodetector remains almost unchanged as we move the antenna. Figure 10.10 shows the received power in an IR simulation. This figure should be compared with Fig. 3.8 for the received power

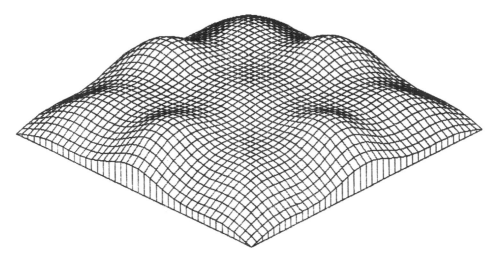

POWER RECEIVED AS FUNCTION OF TERMINALS

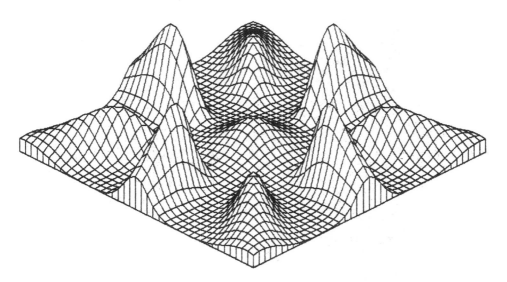

RADIANT EMITTANCE PRODUCED BY 9 LED'S

Figure 10.10 Received power in an IR channel simulation.

in a radio channel. Signal shadowing is much more evident on optical wireless channels. The optical signal with a FOV is analogous to a radio signal observed in one sector of a multisector antenna.

The multipath on an optical channel causes time dispersion of the transmitted symbol, and the resulting intersymbol interference limits the maximum digital transmission rate, just as it does on a radio channel. As in radio propagation, as room dimensions become larger, the multipath spread increases and the maximum

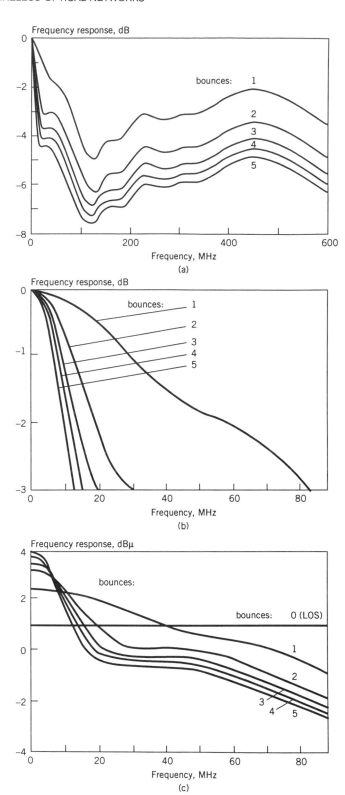

(a)

(b)

(c)

supportable bit rate decreases. The theoretical limit on the data rate is proportional to the bandwidth of the channel frequency response, which is in turn related to the number of reflections included in the model. Figure 10.11 shows the 3-dB width of the frequency response for different numbers of reflections. As the number of reflections increases, the bandwidth reduces, lowering the supportable transmission rate. The theoretical limit for the transmission rate with only first-order reflections is 260 Mbits-m/sec [Gfe79b]. Inclusion of up to fifth-order reflections will reduce this number to 85 Mbits/sec [Bar91a]. Therefore, for a room with a length of 10 m, one expects an achievable transmission rate of 26 Mbits/sec if only first-order reflections are considered, and 8.5 Mbits/sec if reflections up to fifth order are included in the simulation.

10.4.2 Effects of Ambient Light

Ambient light is the main source of noise in wireless optical communications. The infrared content of ambient light interferes with the reception of IR radiation and, if extensive, can overload the receiver photodiode and drive it beyond its operating point. The current in the photodetector due to the ambient light is modeled as Gaussian shot noise plus a strong direct current (DC) component. As long as the DC component does not saturate the photodiode, its effect is negligible and therefore it is the shot noise that determines the quality of the received signal. Three sources of ambient light are daylight, incandescent illumination, and fluorescent lamps, all of which potentially interfere with IR communications. Figure 10.12 shows the power spectral density of the shot noise component of these three ambient light sources over a range of wavelengths from 0.4 μm to 1.4 μm. Incandescent light, being rich in long-wavelength (red) light, has the worst effect because its spectral peak (around 0.7 μm) overlaps that of GaAs diodes.

Daylight contains less IR radiation than incandescent light; but if sunlight falls directly onto the receiver lens, whether it is indoors or outdoors, the DC component can saturate the photosensitive diode, preventing proper operation of the receiver. Fluorescent light, the predominant type of illumination in office and professional buildings, contains a relatively small amount of IR radiation. However, unlike sunlight and incandescent light, the optical interference generated by a fluorescent light follows the pattern of the line voltage, causing a strong 120-Hz interference baseband signal rich in harmonics reaching as high as 50 kHz [Ank80]. Figure 10.13 shows the spectrum of harmonics of the interfering signal from a fluorescent bulb.

One way of avoiding this dominant source of interference is to modulate the transmitted signal onto a carrier frequency which is at least several hundred kilohertz. Figure 10.14 shows the maximum theoretical transmission speed of pulse code modulation (PCM) systems for different room sizes in the presence of

Figure 10.11 Frequency responses for different numbers of reflections. In (a) the range is 600 MHz and in (b) the range is 90 MHz, with both curves normalized to unity DC gain to highlight the −3-dB bandwidths. In (c) the range is 90 MHz but the curves are not normalized, and one can see the higher DC gain provided by multiple reflections. The curves are labeled with the highest number of considered reflections.

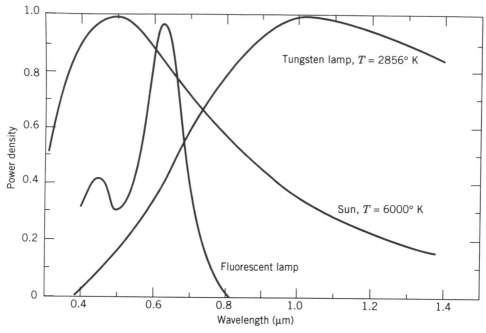

Figure 10.12   Power spectral densities of sources of ambient light, and the spectral center of the GaAs diodes used for diffused infrared communications. (From [Gfe79b] © IEEE.)

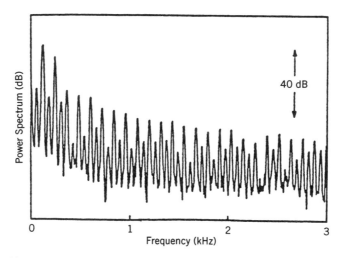

Figure 10.13   Harmonics contained in fluorescent light. The spectral components appear at multiples of the line frequency, extending to over 100 kHz. (From [Bar91b] © IEEE.)

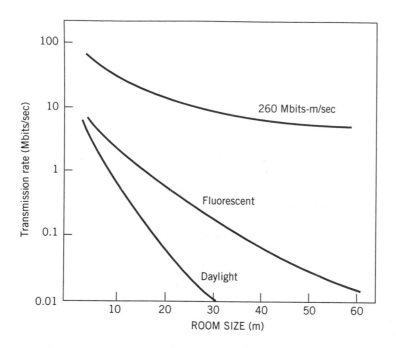

Figure 10.14 Maximum theoretical transmission rates in megabits per second versus room size in meters. Optical power, 1 W; receiver area, 1 cm^2; λ = 950 nm. (From [Gfe79b] © IEEE.)

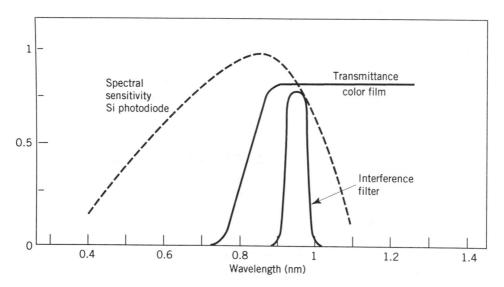

Figure 10.15 Noise reduction using an optical filter. Visible light can be blocked using developed unexposed color film. (From [Gfe79b] © IEEE.)

fluorescent light, daylight, and multipath distortion, for a probability of error equal to 10^{-9}. The noise from the ambient light can be suppressed by an optical filter in front of the photodetector. Figure 10.15 shows a typical response for such a filter. These filters are implemented with a stack of dielectric slabs. One drawback of mounting a filter of this sort in front of the photodiode is that it will reduce the FOV of the receiver.

10.4.3 Photodiode Capacitance and Noise Enhancement

The light incident upon the surface of a photodetector causes electrical charges on the surface, which are discharged in the subsequent preamplifier. The photodetectors appear to the preamplifier as an input capacitance or first-order low-pass filter. Let us assume that the incident light is intensity-modulated by a sinusoid of frequency f and that we are measuring the SNR at the output of the preamplifier. If the only source of noise is the thermal noise of the preamplifier, the measured SNR will decrease as the modulation frequency increases. In other words, the capacitance of the photodetector attenuates the high-frequency components of the received signal, but has no effect on the noise subsequently introduced in the preamplifier. However, referred to the input signal, the effective noise component increases as the modulation frequency increases. When a first-order low-pass filter is followed by white Gaussian noise, the equivalent noise referred back to the input of the low-pass filter has a power spectrum that is quadratic in frequency and therefore is referred to as the f^2 noise [Bar91b]. In order to collect sufficient signal power, it would appear that we should use a photodetector with large area. However, an increase in the area of the photodetector increases the value of the capacitance, which in turn reduces the bandwidth of the low-pass filter and thus worsens the noise penalty.

To counter the effect of f^2 noise, we divide the required photodetector surface area into n smaller areas, each smaller photodetector having its own preamplifier. In this way we reduce the effect of f^2 noise at the expense of increasing the effect of thermal noise. The optimal number of photodetectors is determined by striking a balance between the f^2 noise and the thermal noise. More detailed discussions of f^2 noise and the methods for counteracting its effects are available in [Gfe79] and [Bar91b].

10.5 MODULATION TECHNIQUES FOR OPTICAL COMMUNICATIONS

The most common applications of IR communications in the past decade include remote control, hearing aids, wireless audio systems, cordless phones, wireless keyboards, and WLANs. Various modulation and multiple access techniques have been examined by various manufacturers for use in these systems. In this section we provide a brief summary of these techniques.

In optical communications, modulation is ordinarily performed in two stages, as indicated in Fig. 10.16. The message signal first modulates a carrier frequency, and the resulting signal then modulates the emitted optical light. The second or optical stage of modulation can be performed using intensity (amplitude), frequency (wavelength), or phase modulation. With *intensity modulation* the amplitude of the

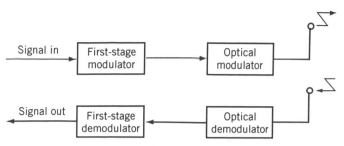

Figure 10.16 Block diagram of a two-stage modulator.

emitted light is varied in accordance with the voltage of the information signal, and the implementation is straightforward. In *wavelength modulation* the information is impressed on the optical signal by modulating the amplitude of the signal at different wavelengths. The implementation of wavelength modulation requires a bank of expensive narrowband optical filters, and this generally makes the technique impractical. *Coherent optical modulation*, in which the phase of the optical signal is modulated directly, has not been considered for wireless optical communications because it is an extremely expensive approach.

The most typical form of second-stage or optical modulation is intensity modulation, in which the voltage of a diode is changed in accordance with the signal amplitude. At the receiver the detector is a photodiode generating a voltage proportional to the light incident on the photodetector plate. A variety of modulation techniques are used for the first stage of modulation in different applications, and in the remainder of this section we describe several of the commonly used techniques.

10.5.1 Analog Modulation Techniques

Analog modulation techniques are typically used for the first stage of modulation in wireless optical communications for audio applications. One motivation for modulating the audio signal is to move the transmitted signal spectrum away from the lower end of the band where there is much interference from the harmonics of fluorescent light. Another motivation for using modulation in wireless optical audio systems is to provide a convenient means of supporting multiuser applications using frequency division multiplexing (FDM). Though FDM is commonly used in radio systems, it is not widely used in wireless optical systems.

Amplitude modulation is generally not suitable for wireless optical communications because of its relatively poor resistance to noise. FM is the standard choice for the first stage of modulation in an analog audio communications system. The frequency-modulated signal is used in the second stage to modulate the intensity of the optical signal, as shown in Fig. 10.17. A carrier frequency of 95 kHz with a frequency deviation of 50 kHz is commonly used in single-channel wireless optical audio systems [Ank80]. The carrier frequency can be readily increased up to 500 kHz. In a multichannel environment, a nine-channel FDM system has been developed for simultaneous interpretation in multilingual conferences and exhibitions [Ank80].

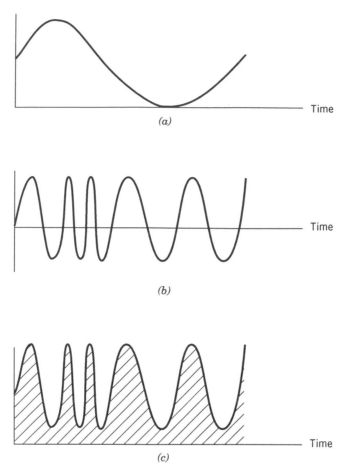

Figure 10.17 Example of analog AM / FM modulation. The FM-modulated audio signal is used to modulate the intensity of the IR light. (*a*) Message amplitude. (*b*) FM-modulated signal. (*c*) Intensity of the light.

10.5.2 Pulse Modulation Techniques

Pulse modulation techniques include pulse amplitude modulation (PAM), pulse duration modulation (PDM), and pulse position modulation (PPM). PAM is a form of amplitude modulation and therefore is not suitable for optical wireless media because of relatively poor performance in additive noise. PPM is the most popular modulation technique for wireless optical communications. For analog signal applications such as cordless phones the audio signal to be transmitted is sampled and the position of the transmitted pulse is adjusted in accordance with the sample amplitude [Bra80]. Figure 10.18 illustrates the principle of operation of PPM for a sampled analog signal. For transmission of digital messages, the pulse position is adjusted in accordance with the value of the transmitted digit.

For low-speed data, the remote control and wireless keyboard applications are very similar. In wireless keyboards, the transmitted alphabets are larger and the transmission rates are higher than are needed for remote control functions. For

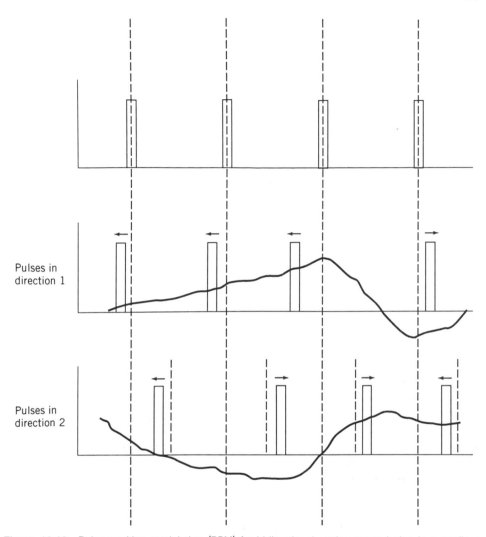

Figure 10.18   Pulse position modulation (PPM) for bidirectional analog transmission in a cordless phone system.

remote control, various modulation techniques have been examined by different manufacturers. The most commonly used modulation technique is simple on–off keying (OOK), which we discuss later. For wireless keyboards, a form of digital PPM is used in which the IR pulse is transmitted at the beginning or at the middle of the bit interval to represent transmission of a 1 or a 0. A 40-MHz carrier frequency is virtually a standard for both applications [LaR84], and in either application the data rate is below 2400 bits/sec. In typical practice several consecutive narrow pulses are used to represent one modulation symbol. This technique saves transmitted signal power, thus increasing battery life.

Figure 10.19 shows a biphase data stream for two PPM systems used in wireless keyboard applications. In one of the PPM systems, three short pulses are used to represent each modulation symbol; in the other, nine are used. A disadvantage of

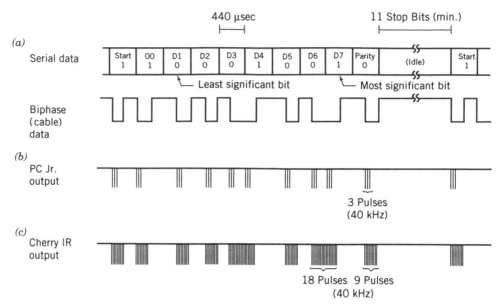

Figure 10.19 PPM for a wireless keyboard application.

transmitting multiple pulses per modulation symbol is that the transmission bandwidth is much wider than the information bit rate. In WLAN applications, where very high data rates are desirable, this is not practically feasible. In addition, the wide bandwidth requirement introduces significant f^2 noise into the system, which in turn restricts the area of coverage, leaving this method suitable only for short-range applications. However, the PPM approach has the advantage of low power consumption, which translates into long battery life.

10.5.3 Digital Modulation Techniques

Binary phase shift keying (PSK) and frequency shift keying (FSK) have been examined for use in high-speed wireless optical data communication networks. Figure 10.20 shows a typical binary PSK system for optical wireless communications. The phase-modulated signal is used to intensity-modulate the emitted optical radiation. An experimental 64-kbit/sec binary PSK system is reported in [Gfe79a]. This system uses a carrier frequency of 256 kHz. The receiver has two phase-locked loops for carrier and bit synchronization. The high carrier frequency eliminates the effects of ambient light from fluorescent lamps. The receiver operates with a low error rate of 10^{-7} in a 380-lux ambient light environment [Gfe79b]; but during the 100-msec switching time of a lamp, the error rate increases to 10^{-3}.

An FSK experimental system using 200-kHz and 400-kHz carrier frequencies for downlink and uplink, respectively, is reported in [Gfe81]. Carrier synchronization in a multipath environment would add to the complexity of the design, and therefore differentially encoded PSK and noncoherent FSK are preferable. To

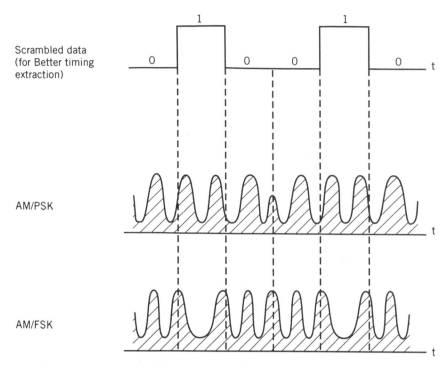

Figure 10.20   AM / BPSK modulation in optical wireless communications.

achieve higher data rates, multiphase PSK and multifrequency FSK are the natural design choices. However, these systems require higher SNRs for proper operation.

Another approach to increasing the data rate is to use multicarrier modulation. Optical multicarrier systems are popular in video distribution applications [Bar91b]. In a multicarrier system, the incoming data bits are divided into several interleaved data streams each modulated onto a separate carrier frequency. The number of carriers should be chosen so that the data rate in each subcarrier is low enough that the intersymbol interference (ISI) caused by multipath is negligible for individual carrier signals. Multicarrier communication over a frequency-selective channel offers advantages similar to those described for frequency hopping spread-spectrum systems (Chapter 9). Coding can be applied across bits on different carriers so that the errors on a carrier falling into a deep fade can be recovered from the bits received correctly on the other carriers. Another approach to improving the performance is to adjust the power or the data rate of the carrier impacted by the frequency-selective fading. Multicarrier communication requires linearity in the transmitter final stage and the receiver front end. Otherwise, harmonic distortion will cause interference among the carriers. Another difficulty with the multicarrier systems is that they require high peak transmission power.

10.5.4   Baseband PCM

The simplest and yet most effective modulation technique for high-speed wireless optical communications is baseband pulse code modulation (PCM). The PCM-

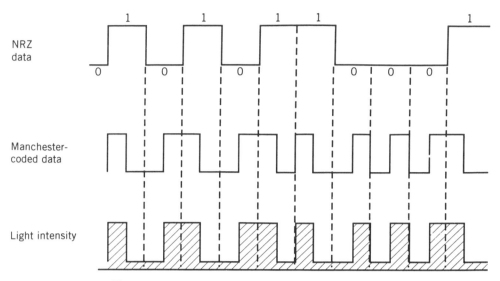

Figure 10.21 Manchester-coded pulse code modulation.

coded bits are modulated directly onto the emitted optical signal. To transmit one of the binary symbols, the light is emitted with the highest power and for the other symbol nothing is transmitted. This is equivalent to OOK for the second stage of modulation. This method is suitable for high-speed data communications. To avoid interference from fluorescent lights, the incoming data stream is Manchester-coded, which reduces the low-frequency portion of the transmitted spectrum. Figure 10.21 represents a typical Manchester-coded baseband PCM system.

The spectrum of the Manchester-coded data is shown in Fig. 10.22. While the Manchester-coded signal has the advantage of reduced frequency components at lower frequencies, it has the disadvantage that it doubles the transmission rate and consequently the required transmission bandwidth. As the data rate is increased, the main lobe of the spectrum shifts further away from the low frequencies impacted by ambient light. With the increase in data rate the ISI caused by

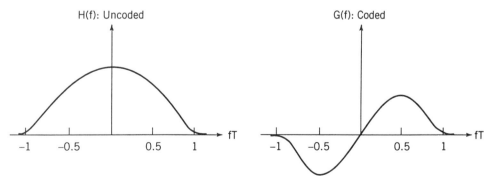

Figure 10.22 Spectrum of the Manchester-coded signal.

TABLE 10.1 Published Accounts of Wireless Infrared LANs

Date	Organization	Bit Rate	Directionally	Duplex Mux	Multiple Access	Subcarrier Frequency	Modulation	Wavelength	Power	Area
1979, 1981	IBM Zurich	64 kbits / sec, 125 kbits / sec	Diffuse	Subcarrier	CSMA / CD	200 / 400-kHz baseband	Subcarrier BPSK Manchester	950 nm	100 mW	0.67 cm^2
1983	Fujitsu	19.2 kbits / sec	LOS, diffuse Narrow	Subcarrier	—	1 / 1.5 MHz	Subcarrier FSK	880 nm	15 mW	1 cm^2
1985	Fujitsu	48 kbits / sec	up, wide down Narrow	Subcarrier	—	1 / 1.5 MHz	Subcarrier BPSK	880 nm	—	—
1984	Hitachi	250 kbits / sec, 1 Mbits / sec	up, wide down	Subcarrier	CSMA / CD polling	> 10 MHz	Subcarrier FSK	—	300 mW	—
1985	HP Labs	1 Mbits / sec	Narrow LOS	Wavelength	CSMA / CD	Baseband	Manchester	660 / 880 nm	165 mW	1 cm^2
1986	Motorola	50 kbits / sec	Widebeam LOS Narrow	—	TDMA	Baseband	RZ OOK	950 nm	16 mW	7.6 mm^2
1987	Bell Labs	45 Mbits / sec	beam LOS Narrow	—	—	Baseband	OOK	800 nm	1 mW	—
1988	Matsushita	19.2 kbits / sec	up, wide down	Subcarrier	Spatial	650 / 950 kHz	Subcarrier FSK	880 nm	—	—

Note: Directionality refers to a system's dependence on accurate alignment between transmitter and receiver; diffuse systems have no directionality, while line-of-sight (LOS) narrowbeam systems are highly directional. Duplex Mux specifies how interference between uplink and downlink is avoided. Power refers to the total optical power of the base station, and Area refers to the total area of the receiver photodetectors. (From [Bar91b] © IEEE.)

multipath will affect the performance of the system. An experimental PCM system using diffused optical propagation and having a data rate of 125 kbits/sec was first reported in [Gfe79b]. Using the latest fast LEDs and PIN diodes, data rates of the order of 1–2 Mbits/sec are feasible with baseband PCM modulation. As with applications to radio channels, the DFE has been studied for mitigating the effects of multipath on high-speed wireless optical communication channels [Bar91b]. As shown in Fig. 10.9, the impulse response of the optical wireless channel has a strong LOS component followed by multipath arising from reflections. The DFE in this case does not need a forward equalizer, making the implementation simpler and more cost-efficient. Using DFE technology and direct optical propagation, very-high-speed optical links can be designed.

Table 10.1 [Bar91b] shows the details of several experimental systems that have been reported in the literature. Commercial products with data rates of 1–2 Mbits/sec providing coverage of an indoor area for a portable computer are available in the market. Indoor coverage is aided by active (receive-transmit unit) or passive (mirror) reflectors. For lower data rates on the order of 19.2 kbits/sec, the received SNR is adequate to cover a large room without the need for a reflector.

10.6 MULTIPLE ACCESS AND DATA RATE

WLANs or wireless PBX systems are intended to support many users in a multiple access environment, and there are requirements to provide much higher information transmission rates than those discussed above. Therefore, more work has to be done to address these applications. Experimental data systems have been implemented with DFIR, using PCM at a data rate of 125 kbits/sec [Gfe79b]. That system showed poor performance in the vicinity of ambient light, due to interference at low optical frequencies. To avoid this problem, a PSK system with a data rate of 64 kbits/sec was examined in a continuation of the same development. In a multiple-access environment, an experimental system was developed that operated with FSK modulation and a carrier-sense multiple-access (CSMA) protocol [Gfe79a]. (We discuss multiuser access protocols in Chapter 11.) In that experiment the uplink and downlink were separated in frequency, operating at 200-kHz and 400-kHz center frequencies, respectively, and data rates up to 100 kbits/sec were supported in this system. In this experiment, transponders were installed in the ceiling, and communication between terminals and transponders used the IR links while transponders were connected by fixed wiring to the controller. The wired portion of the network in such a configuration does not have to be changed when terminals are relocated, and therefore terminals can be relocated at minimal cost. None of these systems has been successfully marketed. As can be seen in Table 10.1, subsequent developments of IR WLANs have typically incorporated forms of CSMA as the multiple access protocol, though time-division multiple access (TDMA) and spatial separation have also been used. The current direction of research in IR WLANs is toward higher-order data rates (tens of Mbits/sec), using directed beam optical transmission.

QUESTIONS

(a) Name three major advantages and three major disadvantages of IR communication for WIN systems.

(b) What is the dominant mechanism in optical wave propagation?

(c) Describe the field of view of an IR photosensitive diode.

(d) What are the major sources of ambient light affecting IR transmissions?

(e) Why does the spectrum of fluorescent light contain a 120-Hz interference component and its harmonics?

(f) What is f^2 noise and how we can counter its effects?

(g) Why are analog signals modulated onto a carrier before modulation onto IR light?

(h) Explain why FM modulation is used for analog modulation of an audio signal onto an IR signal. What carrier frequency and frequency deviations are typically used in practice?

(i) In analog modulation onto IR signals, how high can the carrier frequency be made?

(j) What modulation techniques are typically used in digital IR systems?

(k) In lower-speed IR data communication systems, why are several narrow pulses used to represent an IR pulse? Is this a bandwidth-efficient solution? Explain. Why is this method not applied to higher-speed IR communications systems?

(l) Explain how the baseband modulation techniques counter the effects of ambient light at low frequencies.

(m) What are the advantages and disadvantages of DBIR transmission as compared with DFIR transmission?

(n) What type of multiple-access methods are used in IR wireless LANs?

(o) What access method is used in multichannel analog IR systems?

(p) What is the maximum data rate that can be supported with IR transmission?

11

NETWORKS AND ACCESS METHODS

11.1 INTRODUCTION

In Chapters 3 through 10 we concentrated on the characteristics of individual point-to-point wireless links and the technologies that are used for digital communication over these links. In this chapter we turn our attention from links to networks, and we examine both the topologies of networks and the wireless access methods which are used to enable a number of users to communicate with one another across the network.

Just as the nodes in a wired computer network can be connected in a variety of ways, the terminals in a wireless network can interconnected using a variety of topologies. The reader familiar with the configurations of wired communication networks such as the public telephone network, long-haul computer networks, and wired local area networks (LANs) will find a few similarities, but several important differences, between wireless and wired networks. The differences are due to the fact that wireless communications is fundamentally a broadcast medium. One consequence of this characteristic is that, since terminals are not connected together by wired circuits, transmitted message can in principle be received by an arbitrary (and perhaps unknown) number of other users. At the same time, given the absence of wired connections and the uncertainties of signal propagation, one cannot always be guaranteed a link from every transmitting terminal to every intended receiving terminal. Another consequence of the broadcast nature of wireless communications is that any user, in transmitting a message to any other user, is utilizing a scarce resource (some portion of the system bandwidth, perhaps the entire system bandwidth) for the duration of the message. Therefore, means have to be provided for the fair and efficient utilization of available bandwidth. Yet another consequence of the characteristics of wireless communications is that transmitted signal power becomes an important parameter in the operation of the system. Enough signal power must be used to achieve reliable communication from terminal to terminal, but care must also be taken to avoid excessive interference to other networks operating in the same frequency band. These various considerations have led to the use of two types of topologies for wireless networks: (1) centralized or "hub-and-spoke" networks and (2) peer-to-peer networks. Representative configurations of each type are discussed in the next subsection. As is the case in any area of system design, there are advantages and disadvantages with each configuration, and these will be outlined and discussed.

An ultimate requirement in designing a wireless network is to allow users to communicate at rates comparable to the rates achievable in wired networks for the same category of application. This implies rates close to high-performance data modems for mobile applications and rates close to wired-LAN rates in the case of wireless LANs (WLANs). To approach or reach these goals, efficient channel access methods must be provided, and we find that different access methods are appropriate for use with different network topologies. In Section 11.2 we discuss the basic topologies used in portable and mobile communications networks for voice and data services, and in Section 11.3 we give particular attention to the design of cellular topologies. Sections 11.4 to 11.6 provide detailed descriptions of various access methods used in wireless networks. Section 11.4 is devoted to fixed-access methods typically used in voice-oriented networks. Section 11.5 provides insight into random-access methods used in wireless data communications and analyzes the effect of fading on the performance of these access methods. We discuss controlled random-access techniques in Section 11.6, and various methods for integration of voice and data in Section 11.7.

11.2   NETWORK TOPOLOGIES

One can describe a *network* simply as a collection of communication links interconnecting a collection of devices equipped to exchange information. The literature on communication networks often uses the name "station" to refer to any terminal, computer, printer, or other data-handling device which is connected to the network. We use the name "terminal" in the same general way, and shall use the names station and terminal interchangeably.

There are two fundamental types of topologies used in wireless networks. They are (1) the *centralized* or "hub-and spoke" network and (2) the *peer-to-peer* network, both shown in Fig. 11.1. In the centralized configuration, Fig. 11.1a, station number 1 serves as the hub of the network and the user stations are located at the ends of the spokes. Any communication from one user station to another—that is, between peers—goes through the hub. The hub station controls

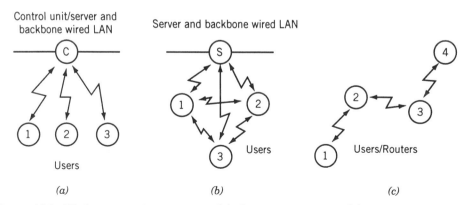

Figure 11.1   Wireless network topologies. (*a*) Centralized network. (*b*) Distributed network. (*c*) Multihop network. *Circles* denote wireless terminals, and *lines* denote communication paths.

the user stations and monitors what each station is transmitting. Thus the hub station is involved in managing the access by user stations to the network's allocated bandwidth. There is no provision in the centralized topology for direct peer-to-peer communication. The centralized topology is a common one in telephone networks, where all the subscribers in a local exchange area are connected through the central office switch. The star topology for WLANs is another example of a centralized configuration. Cellular mobile telephone systems use a centralized network configuration to serve the mobile terminals currently operating within the coverage area of any one cell base station. In the WLAN industry, the Windata and ALTAIR products use a centralized configuration.

Figure 11.1*b* and *c* includes two variations of the peer-to-peer network topology. Figure 11.1*b* shows a *fully- connected network* in which, as the name implies, every user terminal has the functional capability to communicate directly with any other user terminal. In some systems, where users may be distributed over a wide area, any user terminal may be able to reach only a portion of the other users in the network, due to signal blockage or transmitter power limitations. In this situation, user terminals will have to cooperate in carrying messages across the network between widely separated stations. Networks designed to function this way are called *multihop networks*, shown in Fig. 11.1*c*. A fully connected network is practical in an application where connectivity can be assured between every pair of terminals in the network, regardless of the site in which the network is installed. This is often feasible in WLAN applications and in fact one WLAN system, manufactured by NCR, uses the fully connected network configuration. The multihop network configuration is used in military tactical networks, where reliable communication must be provided under unpredictable propagation conditions and over widely varying geographic areas.

11.2.1 Comparison of Network Topologies

There are a number of issues related to the connectivity among user terminals in a network, which bear upon the choice of a topology for the network. We briefly discuss several key issues in relation to the two types of network topology.

Centralized Topology. One important advantage of the centralized topology is that the network can be designed to operate with relatively efficient use of signal transmission power. For example, if contrasted with a fully connected peer-to-peer network, user stations in the centralized network can reach stations at twice the distance with the same signal power. A related advantage is that the hub station can be placed in a location that is optimized with respect to the set of user stations. For example, in a WLAN application, the central hub might be placed in an appropriate location to facilitate unobstructed propagation between user stations and the hub. For infrared (IR) networks, the central unit is often installed on a ceiling, while for 18-GHz indoor radio LANs, it is usually recommended that the central unit be located high on a wall. In the centralized configuration, the central hub also provides a natural point at which connection is made to a backbone network. For this latter reason, many WLANs have this centralized configuration. Another advantage of this topology is that the user terminals can be made functionally simple, while the more sophisticated control functions are concentrated in the single hub station.

Power control is an effective means of minimizing the radiated power of individual user terminals, and this in turn conserves battery power and controls interference. Power control is essential in code-division multiple-access (CDMA) networks, where it is a key ingredient in achieving high levels of bandwidth efficiency. Power control also increases the capacity of time-division multiple-access (TDMA) cellular networks by minimizing co-channel interference from other cells. In applications where battery-operated portable units are to be used, power control serves the important requirement of maximizing battery life. If power control is to be employed in a radio network, a centralized configuration is necessary. The central unit in a network also provides the home for other centralized control functions such as provision of a common timing reference among the terminals.

One disadvantage of the centralized topology is the presence of a single failure point. If the central control module fails, the entire network is disabled. Another disadvantage is the delay characteristic of the network. Because all user-to-user communications go through the hub, the store-and-forward delay is twice that of a fully connected peer-to-peer network. Another measure of communication efficiency is channel occupancy measured in hertz-seconds (bandwidth × time). This parameter is gaining favor as a measure of channel occupancy, quantifying the fair time sharing of a common bandwidth among multiple users. This measure represents the volume of frequency that is occupied for the duration of the transmission. If we use this measure, we see that the occupancy with a centralized network is twice that of the fully connected peer-to-peer network, because each message is broadcast twice.

A further disadvantage of the centralized topology is that it does not offer the functional flexibility needed to deal with unpredictable propagation environments. Nor can it cover wide areas where user-to-user connections can exceed the range of a single link in the network. The centralized topology is not suitable for ad-hoc networks where users might want to use laptops for file transfer or multitask sharing, without any previous networking arrangements having been made. The more complicated central unit is not likely to be available in such situations. However, the centralized network topology is favored for existing alternating-current (AC)-powered WLANs designed for use with portable terminals operating over a reasonably restricted area. The next-generation battery operated WLANs will most likely incorporate other topologies as well, to provide flexibility in networking personal communications service (PCS)-data terminals.

Fully Connected Peer-to-Peer Topology. An important advantage of the fully connected topology is that it has no single point of failure. Another advantage is that peer-to-peer messages do not suffer the store-and-forward delay of the centralized topology, and thus time delay and channel occupancy as measured by hertz-seconds are both halved. Given that no routing functions need be implemented in any of the stations, the complexity of equipment in this design can be minimized. Connection to a backbone network is provided by an additional terminal acting as a server. The server also acts as a bridge or gateway to convert communication protocols from the WLAN to the backbone network. If many of the network nodes are equipped with this connection capability, implementation complexity and cost become issues. A major disadvantage of the fully connected peer-to-peer network is that performance will degrade, or enhanced levels of transmitter power will have to be provided, in operation across large networks. A related issue is the presence of a *near–far problem*, owing to the fact that transmitters needing to operate at high-power level can interfere with unintended receivers in close proximity to the transmitting station. However, the fully connected topology offers an attractive alternative for small networks where reliable connectivity can be ensured between all pairs of user stations. The peer-to-peer network topology should prove attractive for high-speed PCS-data applications where limited number of low-power personal computing devices such as laptops can be networked without prior arrangements. The nodes need not be complex, and they can be networked in the absence of a central unit or a server.

Multihop Peer-to-Peer Topology. A major advantage of the multihop topology is power efficiency, which derives from the fact that message transmission between widely separated users is accomplished with multiple shorter hops. In some applications, such as military tactical communication over a wide operational area, multihop networks provide the only practical approach to achieving reliable connectivity among mobile users. A prominent example of a multihop network design for military application is the U.S. Army Mobile Subscriber Equipment (MSE) network [Sch84, Li87]. Multihop networks using repeater stations are also used in the land mobile radio industry, where, for example, networks are required to serve municipal and state public safety organizations over wide geographic areas.

As with a fully connected network, connection to a backbone or other networks is provided by equipping one or more nodes with the appropriate connection capability. One disadvantage of the multihop topology is the added complexity needed in user terminals to implement efficient message routing and control algorithms. A further disadvantage is the accumulated store-and- forward delay incurred by multiple hops connecting widely separated users. Associated with the store-and-forward capability is a considerable amount of transmission overhead carried with transmitted messages. While multihop topologies have found important applications in military radio and public safety communications networks, they have not been widely adopted in the wireless information network industry.

11.2.2 Cellular Networks and Frequency Reuse

Another important network topology is that of cellular networks employing frequency reuse. This is the network architecture employed in cellular mobile telephone networks, personal communication networks, mobile data networks, and some WLANs. This network configuration is particularly well suited to serving large numbers of mobile users operating over wide geographic areas. In cellular systems, a large service area is divided into smaller areas each of which is served by a fixed cell site. The cell sites are distributed in an approximately regular geometric pattern to cover the entire service area with the level of signal strength needed for acceptable service quality. The spacing between cell sites is about 5–10 miles for typical cellular systems. Each cell in effect is a centralized network, with all

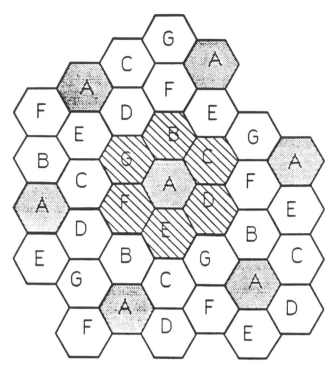

Figure 11.2 Typical cellular frequency reuse pattern, with seven-cell clusters. Cells with the same-letter designation use the same set of frequencies simultaneously.

communications to and from a mobile user in the cell area passing through and controlled by the cell base station. Typically, all of the cell sites in the service area are connected by landlines (e.g., T1 lines) to a mobile telephone switching office (MTSO). The MTSO controls call establishment and manages the cell-to-cell call handoffs as mobile users move about the system. The MTSO also provides the connection to the public switched telephone network. Thus an MTSO in the cellular topology is similar to the control module in the centralized topology

Cellular networks are designed with frequency reuse, so as to maximize the overall capacity attainable with given set of frequency channel allocations. Cellular telephone systems in the United States operate with 30-kHz channels in the 800- to 900-MHz band. A typical cellular frequency reuse scheme is shown in Fig. 11.2. In the figure, the cells are organized into seven-cell clusters, with the cells in a cluster designated as A through G. Seven different sets of frequency channels are used in each cluster, one set in each cell. At a sufficient distance from any cell, the same set of frequencies is used simultaneously. Using this design scheme, the overall system capacity can in principle be made as large as desired by steadily reducing the area of each cell, while controlling power levels to avoid *co-channel interference*—that is, interference to other users operating in another cell on the same frequency channel. System designs are now under development for use in cities, where very small cells called *microcells* will each cover an area about the size of a city block and will serve users carrying low-power pocket-size phones.

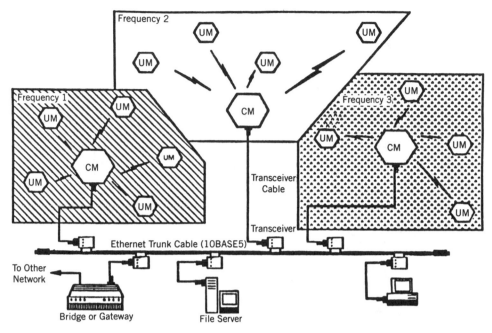

Figure 11.3 Wireless LAN in a cellular environment. (From [Buc91] © IEEE.)

Microcell configurations are also used in WLANs; we say more about this in the next section.

While cellular networks provide an excellent system design to support wide-area mobile and portable communications services, they represent a level of complexity which is not needed in today's WLAN systems, where user terminals are relatively immobile and generally constrained to a well-defined area of use, such as a factory or office building. In fact, a WLAN network with centralized topology can be viewed as a single-cell network, and connection to another "WLAN cell" is typically provided by installation of a bridge or router. Furthermore, the propagation characteristics of WLAN links are such that co-channel interference between separated networks (e.g., networks installed in separate buildings or even separate floors of the same building) can be avoided. Thus cellular configurations with frequency reuse have found only minimal adoption to date in the WLAN industry. Figure 11.3 shows a typical cellular WLAN setup with microcells interconnected through an Ethernet cable.

The growing demand for wireless data services implemented with battery operated portable computers leads to requirements for wide-area coverage supported by a cellular architecture. The backbone for such cellular networks could be a privately owned cellular architecture such as those used by ARDIS or Mobitex. It could also be a data service such as CDPD overlaid onto the cellular phone infrastructure, or a data service integral to a cellular network such as IS-95 or GSM.

Infrared LANs and CDMA spread-spectrum cellular systems reuse the same frequencies in adjacent cells. The base stations in a CDMA cellular system or the satellite nodes in an IR LAN cover designated areas and are connected together

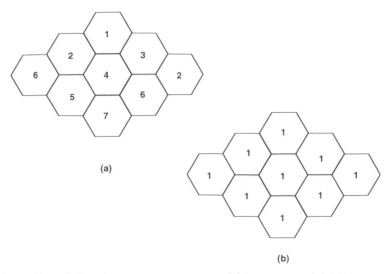

(a)

(b)

Figure 11.4   Cellular frequency reuse patterns. (*a*) FDMA reuse. (*b*) CDMA reuse.

with a wired backbone network. These setups can be considered as cellular with a frequency reuse factor of 1. Figure 11.4 shows the frequency reuse concept for frequency-division multiple access (FDMA) and CDMA cellular systems. With FDMA with frequency reuse factor 7, the available frequency band is divided into seven sub-bands used in different cells, as we described earlier. With CDMA, all cells use the entire band and the frequency reuse factor is 1. Figure 11.5 shows three cells of an IR network with overlapped or redundant coverage. Figure 11.6 shows two topologies for IR networks. It can be seen from Figs. 11.6*a* and 11.6*b* that by adjusting the distance between satellite nodes, the redundancy in coverage is varied.

The existing cellular systems are evolving into a hierarchical cellular architecture in which cells have radius on the order of several kilometers to cover major travel routes and wide urban and suburban areas, *microcells* have coverage of the order of a few hundred meters to cover the streets in densely populated downtown areas, and *picocells* cover indoor areas.

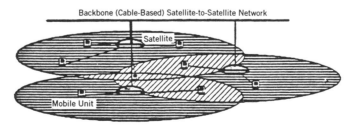

Figure 11.5   Infrared signal coverage with overlapping cells. (From [Les88] © IEEE.)

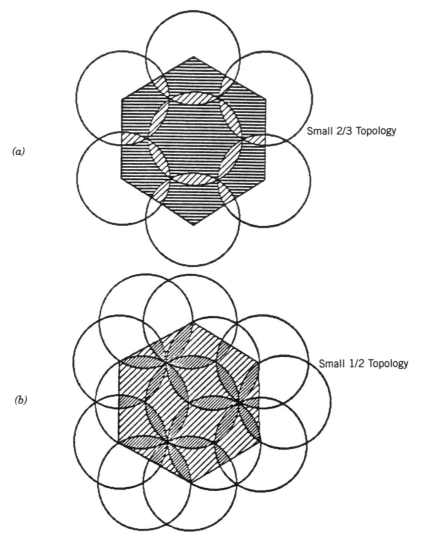

Small 2/3 Topology

(a)

Small 1/2 Topology

(b)

Figure 11.6 Coverage adjustment of IR networks with alternative topologies. (a) Small 2/3 topology. (b) Small 1/2 topology. (From [Les88] © IEEE.)

11.3 RADIO PERFORMANCE

Next we examine some of the basic factors determining communication performance in radio networks. Here we are specifically addressing cellular configurations, where we treat a WLAN with backbone interconnection as a special case of a cellular network. The fundamental consideration is that communication performance in cellular networks is interference-limited rather than noise-limited. That is, cellular networks are generally designed in such a way that additive noise is low enough that the performance of any receiver is limited by the level of interference to the desired signal. In general the interference is a combination of co-channel

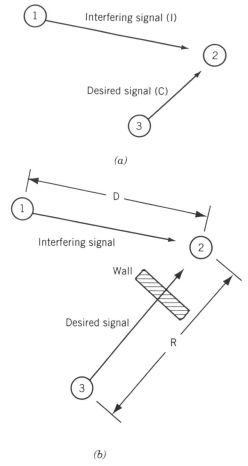

(a)

(b)

Figure 11.7 Signal reception with co-channel interference. (*a*) Desired and interfering signals. (*b*) The desired signal is attenuated by an obstruction, while the interfering signal is unobstructed.

interference and multipath or intersymbol interference. If adaptive equalizers are used, the interference is primarily co-channel interference. The sketches in Fig. 11.7 depict three nodes in a network with the desired signal propagating from node 3 to node 2, and an interfering co-channel signal propagating from node 1 to node 2. In Fig. 11.7*a*, both signals arrive at the receiver without obstruction. In Fig. 11.7*b*, the intended signal is attenuated by an obstruction, whereas the interfering signal arrives without obstruction. The intended and interfering transmitters are located at distances R and D, respectively, from the receiver. Now, the carrier-to-interference ratio C/I is found using the power–distance relationship discussed in Chapter 4; that is,

$$\frac{C}{I} = \frac{\text{Power in the desired signal}}{\text{Power in the interfering signal}}$$
$$= \frac{A_d R^{-\alpha_d}}{A_i D^{-\alpha_i}}$$

where α_d and α_i are the power–distance-law gradients on the desired and interfering paths, respectively, and A_d and A_i are the received powers at 1 m from the desired and interfering terminals, respectively. If the attenuation factors and the received powers at 1 m are the same along the two paths and equal to α, the carrier-to-interference ratio is simply

$$\frac{C}{I} = \left(\frac{D}{R}\right)^\alpha$$

The level of C/I that the intended receiver requires for acceptable performance is a function of the system design. The required C/I level will typically be in the range 15–20 dB, and its exact value will depend upon the adopted modulation and coding techniques and the bit-error-rate (BER) quality needed for the intended application. In designing a cellular layout, we take the distance R to be the radius of a cell—that is, the maximum range for the mobile user relative to the cellular base station. We then take D to be the spacing between cell sites operating with the same the frequency, which would represent the average distance from an interfering transmitter in that cell. It can be seen from the preceding equation that because C/I depends on the ratio of the two distances, D and R can be scaled down proportionally, allowing greater reuse of the available frequencies and correspondingly greater overall system capacity.

In some situations, illustrated by Fig. 11.7b, the desired signal is attenuated by obstacles whereas an interfering signal arrives unobstructed, and there is a consequent degradation in C/I. This situation may prevent the design of a logical layout for the cells in an indoor area. However, if the signal is contained in one room and does not penetrate the walls of the room, the walls can be used to define the boundaries of the cells. This situation exists for infrared and microwave (above 20 GHz) WLANs, wherein each room constitutes one cell of the network.

Cellular arrangements of WLANs are sometimes used in large office buildings or factories, and the relatively small area covered by each network justifies use of the name *microcell*. Figure 11.8 gives an example of a cellular layout in an area of an office building. There are several advantages in using smaller rather than larger microcells. With smaller microcells, signal coverage is better due to fewer obstructions, and higher data rates can be achieved due to reduced multipath dispersion. Typically, the microcells will be spaced so as to provide a prespecified C/I level at the cell boundaries.

If a microcell WLAN configuration is used to cover a large work area, the microcells will ordinarily be connected by a wired backbone, as shown in Fig. 11.9. Though wired interconnections are typical, wireless bridges and routers are coming into increasing use. The backbone can be used to interconnect microcells implemented in different configurations. The figure shows the interconnection of a centralized microcell with a fully connected peer-to-peer microcell. In this case the fully connected network must have one station implemented with a backbone connection. Where WLAN products from different manufacturers are to be interconnected, compatible software must be provided. Implementing combinations of WLAN products from different manufacturers raises the question of interference between systems. Some combinations of manufacturers' products will

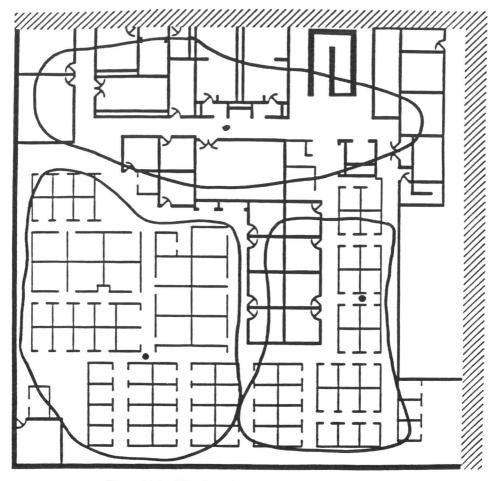

Figure 11.8 Office floor plan with a microcell WLAN layout.

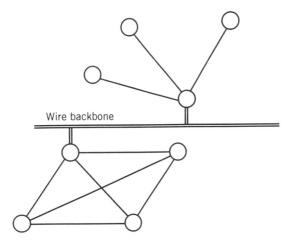

Figure 11.9 Microcell interconnection.

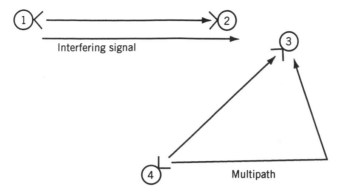

Figure 11.10 Directional antennas.

mutually interfere, whereas others will not. The IEEE 802.11, WINForum, and HIPERLAN Committees on Wireless LAN Standards address the issue of interference among WLAN products.

While a frequency reuse plan would seem to be necessary as part of a WLAN microcell layout, the reuse plan can be simplified and the reuse distances reduced by the use of directional antennas, as shown in Fig. 11.10. With directional antennas, one can minimize the instances of any transmitting station interfering with unintended receivers. Directional antennas can also reduce multipath interference by reducing the number of indirect signal arriving at any receiving terminal.

11.4 FIXED-ASSIGNMENT CHANNEL-ACCESS METHODS

From the preceding discussion of network topologies and the different ways in which they facilitate communication between user stations, it should be apparent to the reader that channel-access methods and methods for sharing a channel among multiple users are essential ingredients in achieving efficient operation and good performance in a wireless network. Users in a wireless network seldom have need to access a channel for a long period of time. Thus schemes are needed for providing multiuser access, usually referred to simply as *multiple-access*, to the frequency-and-time resources of the network in an orderly manner and in a way that minimizes transmission overhead while maximizing overall network capacity. In this section and the following two sections we outline the principal categories of channel access methods and describe the main characteristics of each method. There are three major categories of channel-access methods for the wireless communications environment: fixed assignment, random access, and controlled random access. Access techniques in all three categories are commonly used for wired communications networks as well, but here we discuss access techniques from the perspective of their use in wireless networks.

In fixed-assignment access method, a fixed allocation of channel resources (frequency or time, or both) is made on a predetermined basis to a single user. We will briefly describe the three basic access methods—FDMA, TDMA and CDMA

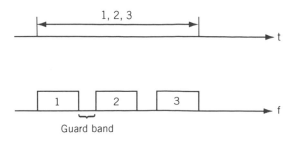

Figure 11.11 Fixed-assignment FDMA format.

—and will draw some comparisons among them. We describe them here in their simplest forms. However, these schemes can and often are implemented with various multiuser access algorithms to form random-access schemes. We discuss random-access schemes separately in Section 11.5.

11.4.1 Frequency-Division Multiple Access

The *frequency-division-multiple-access* (FDMA) technique is built upon the well-known *frequency-division multiplexing* (FDM) scheme for combining nonoverlapping user channels for transmission as a wider-bandwidth signal. FDMA, shown in Fig. 11.11, is the simplest and oldest form of multiplexing and has long been used in the telephone industry and in the commercial radio and television broadcasting industries. In fixed-assignment FDMA a fixed subchannel is assigned to a user terminal and is retained until released by the user. At the receiving end, the user terminal filters the designated channel out of the composite signal. FDMA is used in the current generation of cellular mobile telephone systems and in VHF and UHF land-mobile radio systems. It is also the most common form of multiple access in satellite networks [Bha81]. This access method is efficient if the user has a steady flow of information to be sent, as in digitized voice traffic, but can be very inefficient if the user data are sporadic in nature, as is the case with bursty computer data or short-message traffic.

11.4.2 Time-Division Multiple Access

In time-division multiplexing (TDM), a digital stream is apportioned among multiple users by making a deterministic allocation of time intervals, called *time-slots*, to each user. The time slots, one or more for each user plus associated overhead bits, are typically organized into *frames*, as illustrated in Fig. 11.12. A prominent example of the TDM scheme is the transmission format called *T1 carrier*, used throughout the public telephone network [Jam72]. In the hierarchy of digital transmission rates standardized throughout North America, the basic building block is the 1.544-Mbits/sec rate known as T1. A T1 transmission frame is formed by multiplexing 24 PCM-encoded voice channels. Each voice channel is sampled at an 8-kHz rate and each sample is encoded into 8 bits, producing a 64-kbit/sec data rate for each channel. Framing bits are multiplexed with the 24

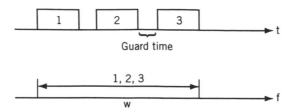

Figure 11.12 Fixed-assignment TDMA format.

digitized voice channels to produce the aggregate 1.544-Mbit/sec data stream.

The *time-division multiple-access* (TDMA) scheme is built upon a TDM trans-mission format. In fixed-assignment TDMA operation a transmit controller serves to assign users to time slots, and an assigned time slot is held by a user until the user releases it. At the receiving end, a user station synchronizes to the TDMA signal frame, and extracts the time slot designated for that user. As is the case with FDMA, this access method makes efficient use of channel resources when a user has a steady flow of data to be transmitted, but can be very inefficient if the data source is bursty or sporadic. The TDMA concept was developed in the late 1960s for use in digital satellite communication systems [Skl88] and first became opera-tional commercially in the mid-1970s [Kwa75]. The fixed-assignment form of TDMA is seldom used in satellite systems, and more flexible multiple access methods are used instead. However, TDMA is the access method used in new digital cellular systems currently under development in North America (IS-54), Europe (GSM), and Japan (JDC). As currently defined, these systems, which we describe further in Chapter 12, operate in a fixed-assignment manner for digital voice service and for circuit-mode data and facsimile services.

A form of TDMA being used in some wireless systems is *time-division duplex* (TDD), in which alternating time slots on the same carrier signal are assigned to the forward and reverse directions of communication. The European DECT system for low-power local-area voice applications, described further in Chapter 12, uses the TDD/TDMA format.

11.4.3 Comparison of FDMA and TDMA

Figure 11.13 shows FDMA, TDD/FDMA, TDMA/FDMA, and TDMA/ TDD/ FDMA access methods in time, frequency, and amplitude dimensions. The advan-tage of TDD is that the reciprocity of the channel allows for exact open-loop power control and simultaneous synchronization of the forward and reverse channels, because only one frequency is needed for duplex operation. The TDD technique is used in systems intended for low-power local area communications where interference must be carefully controlled and low complexity and low electronic power consumption in the portable devices are very important. Compar-isons between the FDMA and TDMA access methods involve a number of performance and implementation issues, and some of the issues have greater or lesser importance depending upon the type of system in which the access method

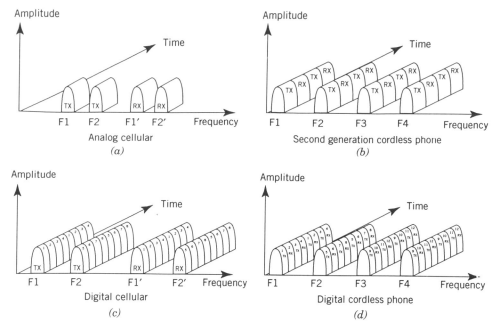

Figure 11.13 Multiple-access techniques. (*a*) FDMA. (*b*) TDD / FDMA. (*c*) TDMA / FDMA. (*d*) TDMA / TDD / FDMA.

is to be employed. Here we briefly describe some of the major points of comparison and comment on the choices that might be made in certain applications.

Format Flexibility. A major advantage of the TDMA format, as compared with FDMA, is its flexibility. Because of the fully digital format and the flexibility of buffering and multiplexing functions, time-slot assignments among the multiple users are readily adjusted to provide different access rates for different users. This feature can be particularly useful in a network where a combination of different services at different rates is to be supported. As an example, the digital backbone of the PSTN is the T-carrier system which as we noted earlier is a TDMA-based architecture merging lower data rate carriers with higher data rate carriers. This architecture is therefore well-suited to supporting multimedia applications built upon a hierarchy of data rates. The TDMA format is also advantageous where a system design is to evolve over time from one multiplexing format to another. This is the case for example in the planned two-phase development of the IS-54 digital cellular system. In the first phase of implementation of the system, three user channels are supported in each six-slot TDMA frame, with each user channel comprising two slots. In the next phase of IS-54 evolution, when reduced-bit-rate vocoders are available, up to six users (one time slot per user) will be supported in each TDMA frame. Multislot assignment will also be available to provide a wider choice of data rates per user [EIA94a, EIA94b]. A further advantage of the TDMA format is that it lends itself more readily to integration of digital voice and data services. It is significantly more difficult to provide these various forms of flexibility, system evolution, and services integration in an FDMA system, particularly if channel splitting must be implemented.

Fading, Diversity, and Related Issues. The difference in bandwidth between an individual FDMA user channel and the composite multiuser TDMA channel is very important in multipath fading environments. In a typical FDMA system channel bandwidth is usually smaller than the coherence bandwidth of the transmission channel and there is no need to use an adaptive equalizer at the receiver. At the same time, this situation removes the opportunity for the implicit frequency diversity gains which are achievable when signal bandwidth approaches the coherence bandwidth. With the TDMA format, in many instances the composite TDMA signal bandwidth is close to the coherence bandwidth of the channel and an adaptive equalizer is needed, that as a benefit provides a form of implicit frequency diversity, mitigating the effects of frequency-selective fading. Equalization of the channel requires transmission of a training sequence as a part of each transmitted packet. The length of the packet should be much longer than the length of the training sequence so that only a small fraction of transmission time is devoted to training of the equalizer. On the other hand, the length of the packet should be short enough so that the between two consequitive transmissions of training sequences the channel remains almost stationary.

Bit-Rate Capability and Message Delay. From a broad perspective, it might appear that FDMA and TDMA should provide the same capability for carrying information from a set of users over a network. In fact this is true with respect to bit-rate capability, if for simplicity we neglect all overhead elements such as guard bands in FDMA and guard times in TDMA. That is, assuming M users each with the same source data rate, and assuming that the overall data rate capability for the system is the same value R for either system, then a data rate of R/M is available to each user in either system. However, if the measure of performance is the average delay encountered by a message packet, the results are different for the two access methods. For example, it is a simple matter to show that for M users generating data at a constant uniform rate of R/M bits/sec, and given FDMA and TDMA systems that each transmit a packet of N bits every T seconds, the average packet delay (packet waiting time before transmission plus packet transmission time) is

$$D_{\text{FDMA}} = T$$

for the FDMA system and

$$D_{\text{TDMA}} = \frac{T}{2}\left[1 - \frac{1}{M}\right] + \frac{T}{M} = D_{\text{FDMA}} - \frac{T}{2}\left[1 - \frac{1}{M}\right]$$

for the TDMA system [Skl88]. Therefore for two or more users, TDMA is superior to FDMA with respect to average packet delay. Note that for large numbers of users, the difference in packet delay is approximately $T/2$, where T is the time interval between transmitted packets. While this result applies to the case of constant deterministic rates for the M sources, the smaller average delays for TDMA are exhibited for any independent message arrival process [Rub79, Ham86].

Amplifier Backoff. The FDMA system requires the multiple modulated carriers to be added together to form the transmitted signal. This addition of carriers produces an amplitude modulation in the composite signal feeding the transmit power amplifier. In some systems this makes it necessary to reduce the input to the amplifier from its maximum drive level in order to control intermodulation distortion. In satellite repeaters, this procedure is referred to as *input backoff* and is an important factor in maximizing the power efficiency of the repeater. This issue, as it relates to overall capacity, was a major factor in the adoption of TDMA in satellite systems. However, this is not necessarily as important in terrestrial communication systems, because a mobile terminal will only be transmitting one channel at a time. The situation for the base station in a centralized network is similar to that for the satellite repeater, but the power consumption in the base station electronics is not a major design issue.

Spurious Interference. In transmission environments where spurious narrowband interference is a problem, the FDMA format, with a single user channel per carrier, has an advantage in that a narrowband interferer can impair the performance of only one user channel. With the wider-bandwidth TDMA format, a single narrowband interfering signal can affect the performance of all user channels in the TDMA stream.

11.4.4 Code-Division Multiple Access

It is useful to think of the FDMA and TDMA access methods as distinct methods for apportioning the time-frequency signal domain of the system among multiple users. In FDMA, the system signal domain is sliced into frequency bands, perhaps with guard bands left between the user bands. In fixed-assignment access, each user has use of an assigned band for an unlimited amount of time. In TDMA, the system signal domain is sliced into time slots, perhaps with guard time intervals left between slots or frames, and each user occupies the entire signal bandwidth, but only during each of his assigned slot times. Thus the user channels are cleanly separated in either the frequency or time domain. A third access method, called *code-division multiple access* (CDMA), can be viewed as a hybrid combination of FDMA and TDMA. With CDMA, multiple users operate simultaneously over the entire bandwidth of the time-frequency signal domain, and the signals are kept separate by their distinct user-signal codes. As we discussed in Chapter 9, the CDMA method is built upon spread-spectrum signaling. It should be noted, however, that the use of spread-spectrum signals does not necessarily imply the use of CDMA. Spread spectrum is used, however, to provide sufficient degrees of freedom to be able to separate the user signals in the time-frequency domain. As discussed in Chapter 9, the two common CDMA techniques are (1) direct sequence CDMA (DS-CDMA) and (2) frequency hopping CDMA (FH-CDMA), which are constructed on the DS and FH spread-spectrum techniques, respectively.

CDMA in Portable and Mobile Radio Networks. Much attention is being given to the adoption of CDMA methods in emerging new wireless networks. The personal and mobile communication industries in the United States are in the process of developing standards based on DS-CDMA (PCS at 2 GHz, and IS-95), while the Pan European RACE project and Japanese digital cellular organization are evaluating CDMA for the next generation of portable and mobile devices, and some countries such as Korea have adopted QUALCOMM's spread-spectrum system for digital cellular service. Some companies are considering FH-CDMA as an alternative for the DS-CDMA [Bor93]. The advantages of the CDMA approach in these systems are as follows:

1. *Timing Flexibility.* The chief advantage of CDMA relative to TDMA, as mentioned above, is that it can operate without timing coordination among the various simultaneously transmitted user signals. The separation of user signals is ensured by the design of the signal codes and is unaffected by transmission-time variations.

2. *Performance in Frequency-Selective Fading.* In frequency-selective fading, one portion of the signal spectrum can be attenuated while other portions are received without attenuation. In an FDMA system, a user signal transmitted in the attenuated portion of the spectrum will be degraded as long as the fade persists. However, in an FH-CDMA system, a user's signal will be attenuated only during the time interval when the signal hops into that part of the spectrum. Therefore the FH-CDMA format will tend to distribute the effects of frequency-selective fading over all the users' signals. As we mentioned in Chapter 9, error-correction coding with interleaving can significantly improve performance by correcting the errors caused by signal fades.

3. *Interference Resistance.* Because both forms of CDMA are constructed on spread-spectrum signals, they provide an inherent resistance to both intentional and unintentional interference. The performance of spread-spectrum communication techniques in interference and jamming is discussed in greater detail in Chapter 9.

4. *Communication Privacy.* In CDMA systems, communication between a particular transmitter/receiver pair can be made private by use of a signal code that is known only to that pair of stations. This technique is the basis of many secure communications systems used in military applications.

5. *System Capacity.* CDMA provides more users per cell, as we discussed in Chapter 9.

6. *Soft Handoff.* Because adjacent cells in a spread-spectrum cellular network use the same frequency, when a mobile moves from one cell to another the handoff can be made "seamless" by the use of signal combining. When the mobile station approaches the boundary between cells, it communicates with both cells and combines the signals with a RAKE receiver. When a reliable link has been established with the new base station, the mobile stops communicating with the previous base station and communication is fully established with the new base station. This technique is referred to as *soft handoff.*

7. *Soft Capacity Limit.* With CDMA there is no hard limit on the number of users that the system can support, as there is in an FDMA or TDMA system.

Rather, each user is a source of noise for all other users as the users occupy the same time and frequency space with distinct user signal codes. Thus theoretically we may arbitrarily increase the number of users at the expense of degradation in the performance seen by every user. As a practical matter, systems are designed to deny access by new users as the BER of the individual user signals approaches a prescribed threshold, around 10^{-3}. The number of users at which the threshold is reached will depend upon the particular geographic distribution of mobiles being served by the system.

8. *Overlay.* The new digital cellular systems will be deployed in the same bands currently occupied by analog AMPS systems. The transition to digital service will be accomplished gradually as digital mobile terminals and base stations replace analog equipment. Spread-spectrum transmissions can overlay the existing analog systems and allow the two systems to coexist during the transition phase.

9. *Interference Control with Antenna Sectorization.* Another advantage of spread-spectrum technology is that the sectored antennas used for interference control will increase network capacity. The reduction of interference allows more users to operate simultaneously in the network.

10. *Time Diversity.* As we discussed in Chapter 9, spread spectrum provides a means of combatting multipath through the use of RAKE receivers. As a result, a spread-spectrum system is expected to provide better signal coverage than standard radio systems by exploitation of multipath as a form of implicit time diversity.

The main disadvantages of the CDMA technique are:

1. *Implementation Complexity.* Spread spectrum is a two-layer modulation technique requiring greater circuit complexity than conventional modulation schemes. This in turn will lead to higher electronic power consumption and higher weight and cost for the mobile terminals. Gradual improvements in battery technology, however, will eventually make the issue of electronic power consumption less critical. Also, the issue of battery life is not as important for mobile radio users as it is for PCS users. Users will also see steady improvements in the weight and cost of mobile terminals as integrated circuit technology progresses.

2. *Power Control.* As we discussed in Chapter 9, the capacity of a spread-spectrum CDMA system operating without power control is extremely limited. As a result, power control is essential for the practical operation of CDMA systems. In TDMA–FDMA networks, power control also reduces interference and improves the quality of received signals. With CDMA, power control is the key ingredient in maximizing the number of users that can operate simultaneously in the system. Another motivation for using power control is to conserve transmitted signal power, thereby increasing the battery recharging cycle. Power control is performed by using either an open-loop or closed-loop method. The open-loop method is based on the similarity of power loss in the forward and the reverse directions. The mobile terminal keeps the sum of the transmitted and received power at a constant level (typically -73 dBm). Thus by constant monitoring of the received power, the transmitted power is, in turn, adjusted. In closed-loop power control the base station monitors the received power from the mobile and commands the mobile to

adjust its power up or down one step at a time. In the Qualcomm CDMA system, adjustments are made 800 times per second (every 1.25 msec) and the step size is 1 dB.

Example 11.1. An example of a DS-CDMA system is the Qualcomm system used by the TIA Subcommittee TR45.5 as the basis from which it developed the IS-95 standard. This system operates in the existing cellular frequency bands, 824–849/869–894 MHz. The transmit power for mobile stations ranges from 2.2 mW to 6 W. The modulation method for base station transmission is QPSK, whereas OQPSK, which has a more nearly constant envelope characteristic, is used for transmission from the mobile. The mobile-to-base signal structure includes a rate-1/3 $K = 9$ convolutional code, while the base-to-mobile channel uses a rate-1/2 $K = 9$ code. Variable-rate speech coding is used, with the speech coded at 8000 bits/sec during talk spurts and at 1000 bits/sec during listening intervals and pauses between talk spurts. This variable-rate scheme serves to minimize radio interference, which enhances the capacity of the CDMA system while minimizing battery drain. The chip rate in the Qualcomm system is 1.2288 Mbits/sec, and the receiver uses a three-tap RAKE receiver. The number of antenna sectors per base station is three.

While the CDMA scheme is being used in a few systems under development for digital cellular service, it is not yet being used in the WLAN industry, although spread-spectrum signaling techniques are used in all ISM band products. Even without the use of CDMA, spread spectrum has a number of advantages for use in WLANs, including the following:

1. Spread-spectrum signals can be overlaid onto bands where other systems are already operating, with minimal performance impact to or from the other systems.
2. The anti-multipath characteristics of spread-spectrum signaling and reception techniques are advantageous in locations where multipath is likely to be prevalent.
3. The anti-interference characteristics of spread spectrum are important in some locations, such as on manufacturing floors, where the signal interference environment can be harsh.
4. The use of spread spectrum provides the convenience of unlicensed operation (in the United States) in ISM bands.

Example 11.2. An example of a WLAN using spread-spectrum technology is NCR's WaveLAN [Tuc91], which operates in the first ISM band (915 MHz). The symbol transmission rate is 1 Msymbol/sec, and the processing gain is 11, yielding a system bandwidth of 11 MHz. The modulation is DQPSK with two bits per symbol, which provides a 2-Mbit/sec effective bit rate. The transmitter power is 250 mW (24 dBm) and the antenna gain is 2 dB.

Digital cellular standards groups are at work defining data services, structured on the IS-95 DS-CDMA standard, to operate at rates up to about 9.2 kbits/sec.

Within the existing ISM bands used for spread-spectrum WLANs, it is possible to develop a baseband CDMA system that uses a number of orthogonal codes for simultaneous data transmission over an existing ISM band. In this way a wide range of data rates can be accommodated with a modulation scheme obeying the constraints of operation in the ISM bands. The IEEE802.11 offers a set of standards for physical and media access control using DSSS and FHSS technologies operating in 2.4 GHz ISM bands with data rates of 1 and 2 Mbps.

11.5   RANDOM-ACCESS METHODS

In our discussion of fixed-assignment access methods, we noted that such methods make relatively efficient use of communication resources when each user has a steady flow of information to be transmitted. This would be the case, for example, with digitized voiced traffic, data file transfer, or facsimile transmission. However, if the information to be transmitted is intermittent or bursty in nature, fixed-assignment access methods can result in communication resources being wasted much of the time. Furthermore, in a system such as cellular mobile telephone, where subscribers pay for service as a function of channel connection time, fixed-assignment access can be an expensive means of transmitting short messages. Thus *random-access* methods provide a more flexible and efficient way of managing channel access for communication of short messages. Here we discuss the most commonly used random-access methods: (1) the basic ALOHA scheme; (2) two enhanced versions of the basic scheme, slotted-ALOHA and reservation-ALOHA; and (3) carrier-sense multiple access with collision detection. In contrast with fixed-assignment access schemes, random-access schemes provide each user station varying degrees of freedom in gaining access to the network whenever information is to be sent. A natural consequence of randomness of user access is that there is contention among the users of the network for access to a channel, and this is manifested in collisions of contending transmissions. Therefore these access schemes are sometimes called contention-based schemes or simply *contention schemes*. We begin by describing the simplest and least disciplined of these access methods, the ALOHA protocol.

11.5.1   Pure ALOHA

The original ALOHA protocol is sometimes called *pure ALOHA* to distinguish it from subsequent enhancements of the original protocol. This protocol derives its name from the ALOHA system, a communications network developed by Abramson and his colleagues at the University of Hawaii and first put into operation in 1971 [Abr70, Abr73, Kob77, Ray84, Leu86]. The initial system used ground-based UHF radios to connect computers on several of the island campuses with the university's main computer center on Oahu, by use of a random-access protocol which has since been known as the *ALOHA protocol*.

The concept of pure ALOHA, shown in Figure 11.14, is very simple: Users transmit whenever they have information to send. A user sends information in packets, each packet encoded with an error-detection code. Of course, because users transmit packets at arbitrary times, there will be collisions between packets

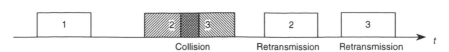

Figure 11.14 Collision mechanism in ALOHA.

whenever packet transmissions overlap by any amount of time, as indicated in Fig. 11.15. Thus after sending a packet, the user waits a length of time equal to the round-trip delay for an acknowledgment (ACK) from the receiver. If no acknowledgment is received, the packet is assumed lost in a collision and is transmitted again with a randomly selected delay to avoid repeated collisions.

Let us assume for simplicity that all packets have a standard length and each packet requires the same amount of time T_p for transmission. Now, referring to Fig. 11.15, consider the transmission of packet A beginning at time t_0, and let us determine the interval of time during which packet A is vulnerable to collision. If another user starts the transmission of packet B between times $t_0 - T_p$ and t_0, the end of packet B will collide with the beginning of packet A. This can occur if long propagation times make it impossible for the sender of packet A to know that the transmission of packet B had already begun. Similarly, if another user begins transmitting packet C between times t_0 and $t_0 + T_p$, the beginning of packet C will collide with the end of packet A. From this, we can see that the vulnerable interval for packet A is $2T_p$, twice the packet transmission time. If two packets overlap by even the smallest amount of time, each packet will suffer one or more errors, which will be recognized at the receiver by failure of the error-detection parity check bits on each packet. The receiver will not be able to acknowledge receipt of either packet, and both will have to be retransmitted.

Let us now determine the channel *throughput S*, which we define as the average number of successful packet transmissions per time interval T_p. Also, let us assume an infinite population of users and let G be the *traffic intensity* or total traffic "offered" to the channel, defined as the number of packet transmissions attempted per packet time T_p, including new packets as well as retransmissions of old packets. The standard unit of traffic flow is the *erlang*, named for the Danish mathematician A. K. Erlang. For our purposes here, we can define an erlang by thinking of the channel time being segmented into intervals of T_p seconds each; then a traffic flow of one packet per T_p seconds has a value of one erlang. By our

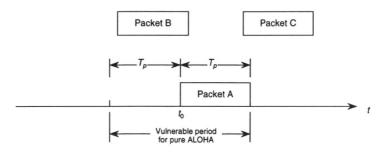

Figure 11.15 Packet collision in pure ALOHA. The vulnerable period is two times the packet interval.

definition of the throughput S, we see that S cannot exceed 1 erlang without collisions, and thus we expect the throughput to be bounded as $0 < S < 1$. We note further that if the offered traffic load is very low, $G \simeq 0$, there will be very few collisions and in turn very few retransmissions, so that we expect $S \simeq G$ at low traffic load. At very high traffic loads, we expect a large number of collisions and consequent retransmissions, so that we will have $S \ll G$, and S will eventually decrease toward 0.

To calculate throughput S as a function of the traffic load G, we make the standard assumption a traffic model in which the probability that k packets are generated during a given packet time obeys a Poisson distribution with a mean of G packets per packet time; that is,

$$P(k) = \frac{G^k e^{-G}}{k!} \qquad (11.5.1)$$

The throughput S is simply the traffic load G times the probability that a packet transmission is successful, which we write as

$$S = GP_0$$

where we define P_0 as the probability of no collision, which from our previous discussion is the probability that no other packet is generated during the vulnerable interval of two packet times. Using Eq. (11.5.1), the average rate of packet arrivals in two packet slots is $2G$, and therefore the probability that 0 packets are generated in an interval that is two packet times long is

$$P_0 = e^{-2G}$$

and thus the throughput is

$$S = Ge^{-2G}$$

which is plotted in Fig. 11.16. The maximum throughput occurs at traffic load $G = 0.5$, where $S = 1/2e$, which is about 0.184. This means that the best channel utilization achievable with the pure ALOHA protocol is only about 18%. The Poisson arrival model, which assumes an infinite population of users, generally provides a good approximation for a network serving large numbers of terminals. For a finite population a more accurate model is provided by the binomial distribution [Gan87].

As we noted above, the offered traffic G is the aggregate of all traffic on the network, including newly generated packets as well as retransmitted old packets.

Example 11.3. As a practical example, assume that we have a centralized network that supports a maximum data rate of 10 Mbits/sec and serves a large set of user terminals with the pure ALOHA protocol. The terminals contending for access to the central module can altogether succeed in getting at most 1.84 Mbits/sec of information through the network. At that peak the total traffic from the terminals is 5 Mbits/sec, which is composed of 1.84 Mbits/sec of successfully delivered

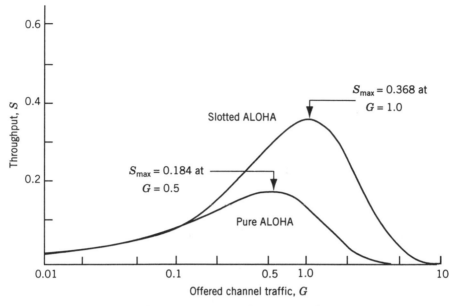

Figure 11.16 Throughput versus traffic load for ALOHA protocols.

packets (some mixture of new and old packets) and 3.16 Mbits/sec of packets doomed to collide with one another.

11.5.2 Slotted ALOHA

To increase the efficiency of the ALOHA protocol, the *slotted ALOHA* scheme was proposed [Rob75]. In this scheme, shown in Fig. 11.17, the transmission time is divided into time slots, with each slot exactly equal to a packet transmission time. All the users are then synchronized to these time slots, so that when a user terminal generates a packet of data, the packet is held and transmitted in the next time slot. (Synchronization can be accomplished by transmitting a periodic synchronization pulse from one designated station in the network.) With this scheme of synchronized time slots, the interval of vulnerability to collision for any packet is reduced to one packet time from two packet times in pure ALOHA.

Because the interval of vulnerability is reduced by a factor of two, the probability of no collision, using Eq. (11.5.1), is now

$$P_0 = e^{-G}$$

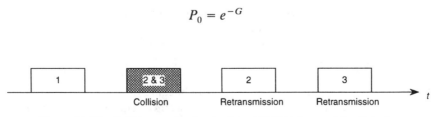

Figure 11.17 Collision mechanism in slotted ALOHA (no partial collision).

and therefore the throughput for slotted ALOHA is

$$S = Ge^{-G}$$

which is plotted in Fig. 11.16 along with the throughput for pure ALOHA. The throughput for slotted ALOHA peaks at $G = 1.0$, where $S = 1/e$ or about 0.368, double that of pure ALOHA. Note that at the point of peak throughput, 37% of the slots are carrying successfully transmitted packets, while the same fraction of slots (37%) are empty (this is the probability of having no packets in a time slot, with $G = 1$), and therefore about 26% are in collision. If we try to operate at higher traffic loads, the numbers of empty and successful slots both decrease and the number of collisions increases rapidly. By any reasonable measure, this cannot be considered a very efficient scheme. Somewhat higher levels of throughput are attainable with ALOHA schemes in the presence of signal capture effects; we shall examine capture effects in Section 11.5.6. But first we describe other access methods which provide higher levels of throughput efficiency by imposing greater discipline on access to the channel.

Slotted ALOHA is used in wireless data communication applications where long transmission delays are encountered. The protocol is used in satellite communications networks as well as in the Mobitex system and some other mobile data communication applications. More is said on this point in Chapter 12.

11.5.3 Stability Considerations

In using any contention-based multiple-access technique, there are important issues as to the stability of operation of the network. The fundamental question here is: What user data rates can be supported by the network in stable operation over an acceptable finite period of time? As it happens, this is usually a difficult question to answer for any networks except for those using the simplest of the random access protocols. It is beyond the purpose of this book to delve into this issue in mathematical detail, but we shall provide a brief overview of the topic.

In the preceding derivations of throughput functions S versus G for the pure and slotted ALOHA protocols, an important assumption of statistical equilibrium was made in modeling the offered traffic as a Poisson process with fixed average arrival rate. This model assumes a steady-state condition. That is, the offered traffic, which is a combination of new data packets and retransmissions of earlier packets suffering collisions, remains constant at the value G. However, an examination of Fig. 11.16 will reveal that this assumption may not always be valid.

For example, let us examine the slotted ALOHA throughput curve in Fig. 11.16 and consider some traffic load value $G_1 > 1.0$, that is, an operating point somewhere to the right of the peak of the curve. Let us assume that the network is initially in stable operation at the selected point and then G increases somewhat, due to a statistical fluctuation in the offered load. In response to this increase in G, the throughput S will decrease, as indicated by the shape of the S versus G curve. This reduction in throughput means that there are fewer successful packet transmissions and more collisions. This in turn means that the number of retransmissions increases, further increasing both the backlog of messages to be transmitted and the traffic load G, in turn decreasing the throughput S. The end result of this situation is that the operating point keeps moving to the right and the throughput eventually goes to zero. This is referred to as *channel saturation* and is the unavoidable result of operating at values of G greater than the optimum value $G = 1.0$.

The curves plotted in Fig. 11.16 were derived with an assumption of an infinite number of users, a condition that greatly simplifies the derivation of the throughput expressions. The reader might well ask if the unstable operation described above is simply a mathematical artifact arising from the infinite-population assumption. The answer is that with an infinite number of users, all contention-based protocols are inherently unstable, while with finite numbers of users, a network may be either stable or unstable, subject to design choices. We will discuss these two points in order.

Let us once more examine the slotted ALOHA curve in Fig. 11.16, and now consider an operating point somewhere to the left of the peak of the curve. Again we assume initial operation at some steady load value, say G_2, and assume a sudden small increase in G, one which keeps the value of G well within the range $0 < G < 1.0$. Because of the shape of the curve, this increase in G results in an increase in throughput S, which decreases the backlog of messages to be retransmitted and consequently reduces G toward its initial steady-state value G_2. Thus the system can operate stably around operating points in the region $0 < G < 1.0$ for some length of time. However, if a short-term increase in the traffic load is enough to move the operating point to a point to the right of the peak where the throughput is lower than the initial operating point, unstable operation will ensue, and the throughput will eventually go to zero. A key element of this scenario is that with an infinite number of users, even a very large increase in the backlog of messages waiting to be transmitted does not diminish the number of new data packets offered for transmission. This is a heuristic explanation of the statement that contention-based protocols are inherently unstable with infinite user populations. However, the situation is different with a finite number of users, as we shall describe next.

With a finite number of users, a short-term increase in the number of collisions causes some number of the user stations to go into the *blocked state*, thereby reducing the flow of new data packets while the previously unsuccessful packets are cleared through the network by retransmission. This does not mean that having a finite number of user terminals ensures stability of the network. However, with a finite number of terminals, the retransmission strategy can be designed to ensure stable operation. The key idea here is that stability can be ensured by providing a *backoff algorithm* that spreads retransmissions out over an interval of time sufficient to ensure that short-term increases in traffic load do not trigger a decrease in throughput.

A number of authors have carried out detailed analyses of stability characteristics of contention-based multiple access protocols. For any given protocol, the analysis begins with derivation of the delay-versus-throughput characteristic, which incorporates the assumed backoff algorithm for management of retransmissions. For some of the simpler multiple-access protocols, this relationship can be found mathematically. In other cases, simplifying assumptions must be invoked or computer simulations used to develop the delay-throughput relationship. For a given access scheme and retransmission strategy, one then finds that the mean transmission time for a data packet increases as a monotone function of throughput S to a point S' such that, for levels of throughput equal to S' or greater, the average packet-transmission delay is infinite. To delve further into the delay and stability analyses for specific protocols is beyond the objective of this book. We shall simply conclude the discussion by noting that for networks with finite numbers of users, which of course is the situation of practical interest, stable network operation can always be ensured by proper choice of the buffer size and the retransmission backoff parameters.

Detailed analyses of the performance of ALOHA schemes, including stability considerations, can be found in the papers of Lam and Kleinrock [Lam74, Lam75, Kle75b] and in the text by Hammond and O'Reilly [Ham86], as well as in other references cited therein.

11.5.4 Carrier-Sense Multiple Access

The relative inefficiency of the ALOHA schemes lies in the fact that users take no account of what other users are doing when they attempt to transmit data packets, and this leads to a high rate of packet collisions. The collision problem can be dealt with only by adjusting the offered traffic load for maximum throughput, which in turn has the effect of leaving a significant portion of the channel transmission time unused. Much better use can be made of channel resources if a user station listens to the channel before attempting to transmit a packet. This technique, shown in Fig. 11.18, is the basis of several protocols termed *carrier-sense multiple access* (CSMA), sometimes called the *listen-before-talk* protocol. Here we describe several common protocols of this type, which differ primarily in the rules that are followed by a station with data ready to send, after sensing the state of the channel. Detailed descriptions and analyses of CSMA systems can be found in papers by Kleinrock, Tobagi, and others [Kle74, Kle75c, Tob75, Tak85, Tak87, Soh87]. Variations of CSMA are used in applications with low transmission delays. The CSMA technique has been widely used in both wired and wireless LANs. A version of CSMA called *data-sense multiple access* (DSMA) has been adopted for use in many of the evolving wireless packet data networks such as CDPD and TETRA.

1-Persistent CSMA. The simplest form of CSMA is one in which each user terminal with data to transmit first listens to the channel to determine if other users are transmitting. In its basic form, which we describe here, the protocol is unslotted. If the channel is busy, the user terminal listens continuously, waiting until the channel becomes free, and then sends a data packet immediately. The name of this protocol, *1-persistent CSMA*, signifies the transmission strategy, which is to transmit with probability 1 as soon as the channel is available. After sending

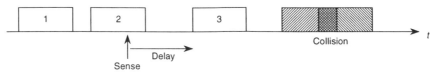

Figure 11.18 Collision mechanism in CSMA.

the packet, the user station waits for an ACK, and if none is received in a specified amount of time, the user will wait a random amount of time and then resume listening to the channel. When the channel is again sensed idle, the packet is retransmitted immediately. Clearly, the objective of this strategy is to avoid collisions with other user packets. However, propagation delays can have a significant effect on CSMA protocols; and with delays, collisions occur despite the discipline of waiting for the channel to be clear. For example, let us say that stations A and B are widely separated, and Station A begins to transmit. Until station A's transmission reaches station B, the latter will sense a quiet channel and will be free to start its transmission, which will result in collision. Even without propagation delay, there can be collisions. Suppose, for example, that both stations A and B have data ready to send and are sensing the channel, which is occupied by a transmission from Station C. As soon as Station C stops transmitting, the first two stations will begin transmitting simultaneously, again resulting in a collision. Although this protocol does not completely eliminate collisions, and its performance depends upon propagation delay, its throughput characteristic for realistic ranges of propagation delay is appreciably better than that of either pure or slotted ALOHA. We shall summarize the throughput calculations later, after describing other versions of the CSMA scheme.

The 1-persistent CSMA scheme also has a slotted form, in which user stations are synchronized and all transmissions, whether initial transmissions or retransmissions, are synchronized to the time slots [Kle75c, Tob75]. The throughput for this scheme is given later.

Nonpersistent CSMA. With *nonpersistent CSMA*, a user station does not sense the channel continuously while it is busy. Instead, after sensing the busy condition, it waits a randomly selected interval of time before sensing again. As with 1-persistent CSMA, a user with data to send begins transmitting immediately when the channel is sensed to be idle. But here the randomized waiting times between channel sensings eliminate most of the collisions that would result from multiple users transmitting simultaneously upon sensing the transition from busy to idle condition. This leads to throughput values much higher than 1-persistent CSMA at high traffic loads and maximum throughput values of 80% or higher given realistic propagation delays. However, at low traffic loads, the throughput of nonpersistent CSMA is somewhat poorer than that of 1-persistent CSMA, because the waiting strategy is of no benefit when few users are trying to transmit. As with 1-persistent CSMA, there is also a slotted version of nonpersistent CSMA, which is especially beneficial at higher traffic loads [Kle75c].

p-Persistent CSMA. The *p-persistent CSMA* protocol is a generalization of the 1-persistent CSMA scheme, applicable to slotted channels. The slot length is typically chosen to be the maximum propagation delay. When a station has data to send, it senses the channel; and if the channel is sensed idle, it transmits with probability p. With probability $q = 1 - p$ the station defers action to the next slot, where it senses the channel again. If that slot is idle, the station transmits with probability p or defers again with probability q. This procedure is repeated until either the frame has been transmitted or the channel is sensed to be busy. When the channel is detected busy, the station then senses the channel continuously; and when it becomes free, it starts the above procedure again. If the station initially senses the channel to be busy, it simply waits one slot and applies the above procedure.

The analysis of throughput for this scheme is rather tedious, and it is not presented here. We simply note that for low-to-intermediate values of propagation delay and with the parameter p optimized, the throughput of p-persistent CSMA lies between that of slotted and unslotted nonpersistent CSMA, whereas for long propagation delays, its throughput somewhat exceeds that of either of the other two. Plots of throughput for a few selected cases are found in [Tan88], and detailed analyses are found in [Kle75c], [Tob75], [Tak85], and [Tak87].

11.5.5   Throughput Performance Comparisons

Table 11.1 summarizes the throughput expressions for the ALOHA and the 1-persistent and nonpersistent CSMA protocols, including the slotted and unslotted versions of each. The expressions for p-persistent protocols are very involved and are not included here. The interested reader should refer to [Kle75c], [Tob75], and [Tak85], where the derivations of the other CSMA expressions can also be found. The expressions in the table are also derived in [Ham86] and [Kei89].

In Table 11.1, throughput S is the number of successfully delivered packets per packet transmission time T_p, and G is the offered traffic load in packets per packet time. The four CSMA throughput expressions include the parameter a, defined as $a = \tau/T_p$, where τ is the propagation delay. The parameter a corresponds to the time interval, normalized to packet duration, during which a transmitted packet can suffer a collision in the CSMA schemes. (Recall that when one station's transmitted packet reaches another station and is sensed there, the second station is inhibited from transmitting.) The propagation delay (3.33 μsec/km) is generally much smaller than the packet transmission time, and thus values of a on the order of 0.01 are usually of interest.

Figure 11.19 shows plots of throughput S versus offered traffic load G for the six protocols listed in Table 11.1, with normalized propagation delay $a = 0.01$. (For $a < 0.01$, the throughput curves for the slotted and unslotted versions of

TABLE 11.1   Throughput Expressions for Two ALOHA and Three CSMA Protocols

Protocol	Throughput
Pure ALOHA	$S = Ge^{-2G}$
Slotted ALOHA	$S = Ge^{-G}$
Unslotted 1-persistent CSMA	$S = \dfrac{G[1 + G + aG(1 + G + aG/2)]e^{-G(1+2a)}}{G(1+2a) - (1 - e^{-aG}) + (1 + aG)e^{-G(1+a)}}$
Slotted 1-persistent CSMA	$S = \dfrac{G[1 + a - e^{-aG}]e^{-G(1+a)}}{(1+a)(1 - e^{-aG}) + ae^{-G(1+a)}}$
Unslotted nonpersistent CSMA	$S = \dfrac{Ge^{-aG}}{G(1+2a) + e^{-aG}}$
Slotted nonpersistent CSMA	$S = \dfrac{aGe^{-aG}}{1 - e^{-aG} + a}$

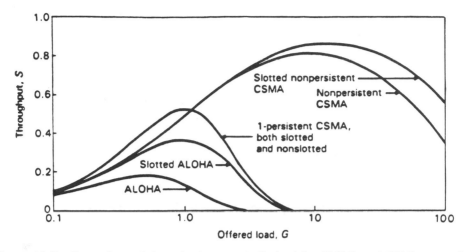

Figure 11.19 Comparison of throughput versus traffic load for ALOHA and CSMA protocols. Normalized propagation delay is $a = 0.01$. (From [Ham86] © 1986 Addison–Wesley, reprinted with permission.)

1-persistent CSMA are essentially indistinguishable.) It can be seen from the figure that for low levels of offered traffic, the persistent protocols provide the best throughput, but at higher load levels the nonpersistent protocols are by far the best. It can also be seen that the slotted nonpersistent CSMA protocol has a peak throughput almost twice that of persistent CSMA schemes.

The equations in Table 11.1 can also be used to calculate capacity, which is defined as the peak value S_{max} of throughput over the entire range of offered traffic load G [Ham86]. Curves of capacity versus normalized propagation delay are plotted in Fig. 11.20 for the same set of ALOHA and CSMA schemes. The curves show that for each type of protocol the capacity has a distinctive behavior as a function of normalized propagation delay a. For the ALOHA protocols capacity is independent of a and is the largest of all the protocols compared, when a is large. The reason for this is that with long propagation delays relative to packet transmission time, the channel state information arrives too late to be used effectively in reducing collisions. In other words, there is a relatively large time interval in which a sender's packet is vulnerable to collision because the presence of a potentially colliding packet has not yet been sensed.

The curves in Fig. 11.20 also show that the capacity of 1-persistent CSMA is less sensitive to normalized propagation delay for small a than is nonpersistent CSMA. However, for small a, nonpersistent CSMA yields a larger capacity than does 1-persistent CSMA, though the situation reverses as a approaches the range of 0.3–0.5 [Ham86].

Example 11.4. The nonpersistent CSMA protocol has been a popular choice for LANs. For an efficient system design the normalized propagation delay a should be smaller than about 0.01. The maximum distance usually considered for a wireless LAN installation is 100 m, which yields a propagation delay of 0.33 μsec. Therefore the packet length should be no smaller than about 33 μsec. If we

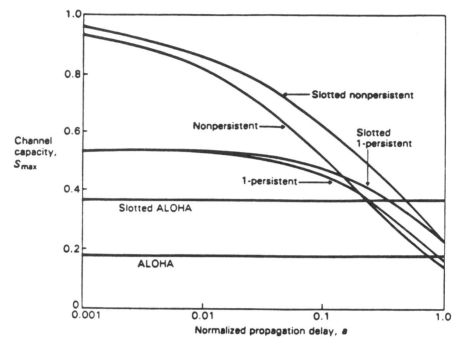

Figure 11.20   Capacity or peak throughput vs. normalized propagation delay for ALOHA and CSMA protocols. (From [Hor86] © 1986 Addison – Wesley, reprinted with permission.)

assume a WLAN operating with a data rate of 1 Mbits/sec, the associated minimum packet length will be about 33 bits. If the data rate is 10 Mbits/sec, the minimum packet length increases to 333 bits.

11.5.6   Wireless Access and Radio Channel Characteristics

As we discussed earlier, the received power in a radio communication system is subject to extensive power fluctuations caused by the near–far effect as well as by the locations and movements of terminals and the fading characteristics of the medium. In point-to-point communications the extensive power fluctuation caused by the movements of terminals may cause temporary degradation of a communication link or may even completely hide a terminal from direct communication with another terminal. The effects of power fluctuation in a multiuser environment depend largely on the multiple-access method employed in the system. In FDMA/TDMA cellular systems, power fluctuations will increase adjacent channel and co-channel interference. The additional interference results in degradation of the quality of the communication link, but the throughput or capacity of the channel remains unaffected. In CDMA systems, as we discussed in Chapter 9, the capacity of the channel is limited by the interference from other terminals sharing the same time and frequency space. Therefore if power fluctuations in a CDMA system are not controlled, the high power levels received from some terminals will sharply limit the number of users than can be on the system, resulting in drastic reduction in the overall capacity of the system. The quantitative analysis of capacity in the presence of multipath fading was discussed in Chapter 9. Power control is the traditional method for counteracting the effects of power fluctuations in multiuser radio systems. Power control will serve to minimize the interference in systems using fixed assignment multiple-access methods, resulting in improvement of the quality of the received signal in FDMA/TDMA and improvement of the capacity in the CDMA systems. The superior performance offered by CDMA relative to other access methods cannot be realized unless effective power control is implemented. For FDMA/TDMA systems, power control is an added feature that improves communication quality. Another attractive feature of power control for any portable or mobile network is the saving in electronic power consumption, which is important for battery-powered applications.

In networks using contention access methods, the difference in power levels received from different stations is sometimes helpful. If there is a large difference in power levels received from two terminals and data packets from the two terminals collide, one of the packets will survive the collision and only one packet will be destroyed, resulting in an increase in the throughput of the network. In reality the throughput from the terminal with higher power will increase while the throughput from the other terminals will remain the same. This phenomenon, which we discuss in the following subsection, is referred to as *capture*.

Capture Effects in Contention-Based Protocols. In the analyses of throughput given in the previous sections, it was assumed that if two packets appear in the channel with any amount of overlap in their transmission intervals, the resulting collision renders both packets unusable. The terminals sending the colliding packets then each wait a randomly selected amount of time and retransmit. The throughput expressions for ALOHA and CSMA protocols summarized in the previous paragraphs were all derived with the assumption that no packets survive their collisions. In radio channels, sometimes collision of two packets may not destroy both packets. Because of signal fading, packets from different transmitting stations can arrive with very different power levels, and the strongest packet may survive a collision. In some situations, the differences in received signal power levels may be due simply to large differences in the lengths of transmission paths in the network, which is referred to as the *near–far effect*. This situation, in which one of the packets involved in a collision is successfully received, is termed *capture*, and the surviving packet is said to have *captured* the packets with which it collided. The possibility of a received packet surviving a collision depends upon channel characteristics as well as the design of the modulation and coding scheme, the average received SNR, and the length of the packets. This is usually quantified by determining, for a given receiver, the minimum power ratio which one arriving packet must have relative to the other colliding packets in order that it can be received successfully. This power ratio is referred to as the *capture parameter* or *capture ratio* of the receiver.

Capture effects in a random-access protocol lead to increased levels of throughput relative to results derived without accounting for capture. Many studies of capture effects on contention-based protocols have been reported, many dealing with ALOHA systems [Rob75, Met76a, Abr77, Kah78, Fra80, Kup82, Kle84, Nel84, Nam84a, Nam84b, Goo87, Gan87, Arn87, Pra88, Hab89, Zha89, Zha92] and others with CSMA systems [Zdu89, Zha91]. In [Goo87] and [Gan87], the analysis focuses on the near–far effects. The ratio of powers received from two terminals is formed and compared with a capture parameter, which takes into account the path loss as a function of locations of transmitting terminals relative to the receiver. If the power ratio exceeds the capture ratio, the packet with higher received power survives the collision. The difference in the received power levels is due solely to the different distances of user terminals from the central station, the terminals assumed to be uniformly distributed over the coverage area. The studies reported in [Arn87] and [Pra88] deal with Rayleigh-fading and lognormal-fading channels, respectively and also consider a general spacial distribution for the radio terminals. Other investigations have in addition considered the effects of signal design in the analysis of capture. An analysis for a system with uniformly distributed terminals using ALOHA on indoor radio channels is given in [Hab89], where the performance of modulation and coding in fast Rayleigh fading are taken into account. The capture discussions in the remainder of this section closely follow Zhang and Pahlavan [Zha91, Zha92], where slotted ALOHA and CSMA are treated with general distributions of terminals and an assumptions of fast and slow Rayleigh fading.

Performance of Slotted ALOHA System in Capture. Let us consider a system configured as an ideal slotted ALOHA network with a base station located in the center and terminals distributed around it with a given distribution. We shall assume that the system has negligible propagation delay, perfect acknowledgments from the receiver, and an infinite number of terminals.

For a slot length T_p and an average packet generation rate from all the terminals λ, the average number of packets arriving in a slot is $G = \lambda T_p$. If the arrival process is Poisson, the probability $P(k)$ that k packets arrive in a slot is calculated from Eq. (11.5.1).

At the beginning of every slot, we assume that the terminals generate a total of $k + 1$ packets. For this analysis, one of these packets is randomly chosen to be the *test packet*, which is phase-locked to the receiver, while the other k packets are considered to be interference to the test packet. Defining $P_C(k)$ as the probability that the test packet captures the k interfering packets, the average throughput of the system associated with this probability of capture is the average number of packets received successfully per time slot, given by

$$S = \sum_{k=0}^{\infty} P(k + 1) P_C(k) \tag{11.5.2}$$

In the conventional analysis of slotted ALOHA, it is assumed that in each collision all colliding packets are destroyed, and a packet survives only if it is received without collision. This implies $P_C(0) = 1$ and $P_C(k) = 0$, for $k \geq 1$, which yields

$$S = Ge^{-G} \tag{11.5.3}$$

for conventional ALOHA. Equation (11.5.3) provides an absolute lower bound for the performance, if transmission errors are neglected, and is usually referred to as the *case with no capture*. In reality there always are transmission errors and $P_C(0)$ is not exactly equal to 1. However, with packet error rates in the range of 10^{-3} to 10^{-5}, $P_C(0)$ can be assumed approximately equal to 1.

In the presence of capture, some of the packets involved in a collision will survive. In an ideal situation, one packet survives all collisions with k interfering packets. This case is referred to as *perfect capture* for which $P_C(k) = 1$ for all values of k. Substituting $P_C(k) = 1$ into Eq. (11.5.2), perfect capture provides an upper bound on the throughput of slotted ALOHA with capture, given by

$$S = 1 - e^{-G} \qquad (11.5.4)$$

For large values of G, the throughput approaches 1 and the channel is fully utilized in this ideal case of perfect capture in every time slot. The results presented in the remainder of this section are for practical cases, and the performance results in each instance will fall between the two bounds provided in Eqs. (11.5.3) and (11.5.4).

In general, the probability of capture is a function of modulation and coding, distribution of user terminals, signal-to-noise ratio (SNR), and packet length. Calculations of this sort are beyond the scope of this book. Zhang and Pahlavan [Zha92] provide an accurate method of analysis and two bounds on a generalized probability of capture on slow Rayleigh fading channels. The results in [Zha91] are derived for ring- and bell-shaped distributions of user terminals. The ring distribution is equivalent to an assumption that the same average power is received from each terminal, which models a system having *average-power control* but no means of tracking the instantaneous power fluctuations caused by multipath fading. The bell-shaped distribution assumes that the distance power gradient is 4 and the normalized distance r between transmitter and receiver obeys a distribution of the form

$$\rho(r) = 2re^{-(\pi/4)r^4}$$

Using the capture model developed earlier, $P_C(k)$ is obtained by taking the average value of $P_C(k|a_0, \mathbf{G})$ which is defined as the probability of capturing a packet given a_0 and $\mathbf{G}_k = [g_1, \ldots, g_k]$, where a_0 is the amplitude of bits in the test packet and the $\{g_i\}$ define the interference from other packets. Given the probability density functions of a_0 and g_i, the average probability of capture for the test packet $P_C(k)$ can be obtained from the following k-dimensional integral:

$$P_C(k) = \int_0^\infty da_0 \int_{-\infty}^{+\infty} dg_1 \cdots \int_{-\infty}^{+\infty} dg_k \, f_{A_0}(a_0) f_{G_1}(g_1) \cdots f_{G_k}(g_k) P_C(k|a_0, G_k)$$

$$(11.5.5)$$

Let us assume that each packet of L bits is protected by a BCH block code with block length L capable of correcting up to t bit errors per block. (For an uncoded system, $t = 0$.). Then $P_C(k|a_0, \mathbf{G}_k)$ is given by

$$P_C(k|a_0, \mathbf{G}_k) = \sum_{i=0}^{t} \binom{L}{i} (1 - \bar{P}_b)^{L-i} \bar{P}_b^i$$

where $\bar{P}_b$ is given by

$$\bar{P}_b = \frac{1}{2^k} \sum_{\alpha_1 = \pm 1} \cdots \sum_{\alpha_k = \pm 1} \frac{1}{2} \operatorname{erfc} \left[\frac{a_0 + \sum_{i=1}^{k} \alpha_i g_i}{\sqrt{N_0}} \right]$$

Equation (11.5.5) provides an exact solution for calculation of the probability of capture for a given modulation and coding technique. The probability distribution function of the test bit amplitude, $f_{A_0}(a_0)$, and of the interfering bit amplitudes, $f_{G_i}(g_i)$, are functions of the distance-power gradient and the distribution of the terminals. These distribution functions for ring- and bell-distributed terminals are available in [Zha92]. Calculation of the probability of capture from Eq. (11.5.5) is rather tedious, and simpler upper and lower bounds are introduced in [Zha92]. Figure 11.21 shows upper and lower bounds on probability of capture for slotted ALOHA, together with an exact calculation and the results of Monte Carlo computer simulations, all plotted versus the number k of interference packets [Zha92]. The exact calculation and simulations were done for binary phase shift keying (BPSK) modulation, SNR = 20 dB, packet size $L = 16$ bits, and the bell-shaped terminal distribution given earlier. The SNR is defined here for a

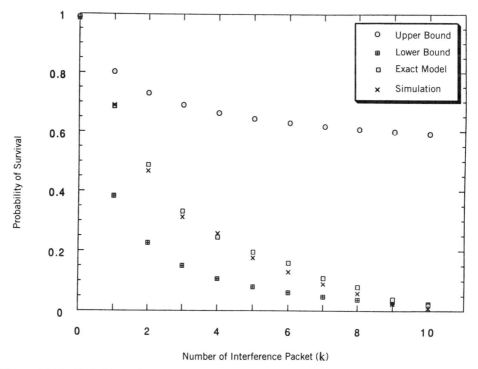

Figure 11.21 Probability of capture using calculation and Monte Carlo simulation. $L = 16$ bits, SNR = 20 dB, the modulation is BPSK, and the distribution is bell-shaped.

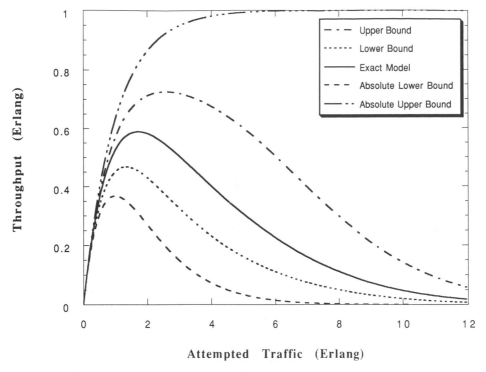

Figure 11.22 Throughput versus attempted traffic for a slotted ALOHA system operating at SNR = 20 dB in slow Rayleigh fading: Absolute upper and lower bounds, and upper and lower bounds and an exact calculation derived for packet length $L = 16$ bits and a bell-shaped distribution of the terminals. The modulation is BPSK.

terminal located at the median distance from the central station. The simulation results show close agreement with the exact calculation.

To determine the throughput for an average traffic intensity of G, we use Eq. (11.5.2) with $P_C(k)$ and $P(k)$ calculated from Eq. (11.5.5) and Eq. (11.5.1), respectively. For the calculation of the upper and lower bounds we shall use the appropriate equations for the $P_C(k)$ provided in [Zha92] rather than Eq. (11.5.5).

Figure 11.22 shows throughput versus traffic intensity for a slotted ALOHA system operating under the same conditions as in the previous figure: BPSK modulation at SNR = 20 dB, slow Rayleigh fading, packet length $L = 16$ bits, and the bell-shaped distribution of terminals. The figure gives five curves, consisting of two sets of bounds and an exact calculation. The "absolute" upper and lower bounds are the cases of no capture and perfect capture, as given by Eqs. (11.5.3) and (11.5.4), respectively. The upper bound on throughput is calculated by assuming that the interfering bit patterns have the same pattern as that of the test packet. The lower bound is calculated by assuming that the signal fading affects each bit in a packet independently from any other bit in the packet. This is essentially equivalent to assuming extremely fast fading, and provides a lower bound on the throughput for slotted ALOHA with capture, because the probability of capture is lower than that of slow fading, where all the bits in a packet fade in unison. As we

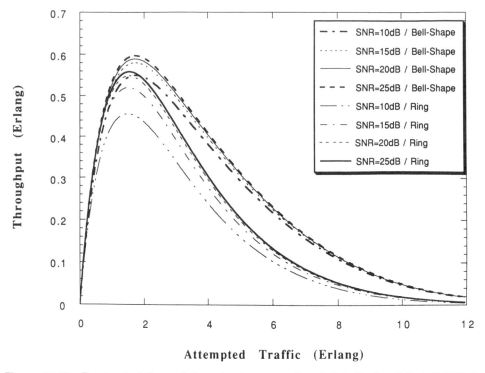

Figure 11.23   Exact calculations of throughput versus attempted traffic for slotted ALOHA in Rayleigh fading at different levels of SNR, assuming two distributions of terminals. For both bell- and ring-shaped distributions, the modulation is BPSK and the packet length is $L = 16$ bits.

discussed earlier, the details for calculation of these bounds is given in [Zha92]. The curve for the exact calculation shows throughput approaching 60% in Rayleigh fading, a significant improvement over the 36% throughput achievable without capture.

It is also of interest to consider how slotted ALOHA throughput is affected by variations in other system parameters. Figure 11.23 shows exact calculations of the relationship between throughput S and the attempted traffic G for BPSK modulation and packet length $L = 16$, given different levels of SNR in Rayleigh fading. Results are shown both for bell-shaped and ring-shaped distributions of terminals. With the ring distribution, user terminals are located at a fixed distance from a central station, as we stated earlier. The greater effect of variations in SNR is found for the ring distribution, which shows an approximate 18% drop in peak throughput when the SNR is reduced from 25 to 10 dB. Because of the wider variations in power received from terminals in the bell-shaped distribution, that distribution is less sensitive to SNR changes. The performance with the bell-shaped distribution, where no power control of any sort is assumed, is better than with the ring distribution with its assumed average-power control. The technique of instantaneous power control provides the same received power from each terminal,

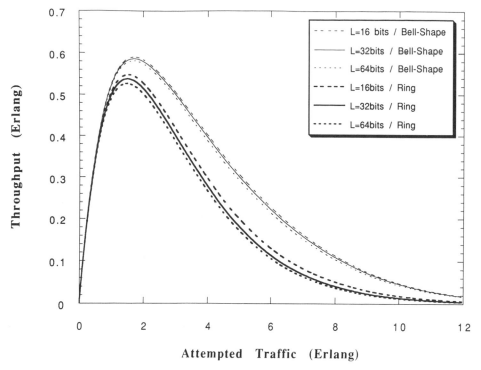

Figure 11.24 Relationship between *S* and *G* for different packet lengths and with both the bell- and ring-shaped distributions of terminals. The greater variation of throughput with packet length is observed for the ring-shaped distribution, where the throughput with 64-bit packets is about 8% lower than with 16-bit packets. SNR = 20 dB, and the modulation is BPSK.

and the case of zero capture gives the closest approximation to this case. As one might expect, a greater degree of control on received power leads to lower probability of capture and in turn lower achievable throughput.

Figure 11.24 shows the relationship between *S* and *G* for slotted ALOHA with different packet lengths and with both the bell- and ring-shaped distributions of terminals. The greater variation of throughput with packet length is observed for the ring-shaped distribution, where the throughput with 64-bit packets is about 8% lower than with 16-bit packets.

Figure 11.25 shows the effect of different choices of modulation method on throughput versus traffic intensity. Again, slotted ALOHA in Rayleigh fading is assumed, the packet length is *L* = 16 and SNR = 20 dB. The modulation schemes are PSK, FSK, and NCFSK, and both the bell-shaped and ring-shaped distributions are considered. The greater effect of modulation choice on throughput is observed for the ring distribution, where the maximum throughput with PSK modulation is about 5% higher than the maximum throughput with NCFSK.

Figure 11.26 shows the effect on throughput resulting from the use of error-correction coding with slotted ALOHA in slow fading. Results are shown for two cases, in which a 64-bit information packet is encoded with a 71-bit or 127-bit BCH

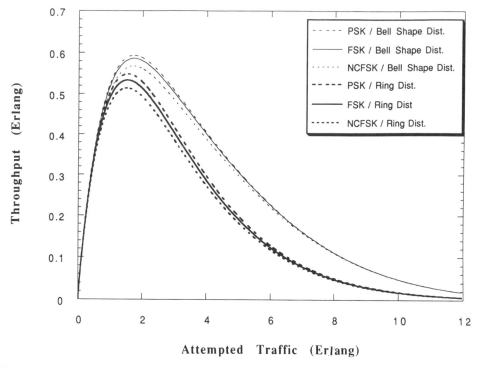

Figure 11.25 Effect of different choices of modulation method on throughput versus offered traffic. Slotted ALOHA in Rayleigh fading is assumed, the packet length is $L = 16$, and the SNR $= 20$ dB. The modulation schemes are PSK, FSK, and NCFSK, and both the bell-shaped and ring-shaped distributions are shown.

block code. Transmission of uncoded 64-bit packets is included for comparison. The modulation is PSK, the channel SNR is 20 dB, and the bell-shaped distribution of terminals is assumed. The (71,64) code can correct one error in any code block, while the (127,64) code can correct up to 10 errors per block. It can be seen in the figure that the use of error-correction coding has very little effect on the achievable throughput in slow Rayleigh fading. The reason for this is that when the fade durations are long relative to the code block length, errors within a code block are highly correlated and thus bursty, rendering the error-correction coding ineffective [Lev76, Eav77]. Figure 11.27 is similar to Fig. 11.26, but now with 16-bit packets coded into length-31 or -63 BCH code blocks. Again, error-correction coding provides minimum benefit in slow Rayleigh fading. We can conclude from Figs. 11.22 to 11.27 that the major influence on throughput for slotted ALOHA in slow Rayleigh fading comes from the spatial distribution of user terminals in the network or, equivalently, from the choice of power-control scheme.

Throughput of a CSMA Channel and the Capture Model. With $P_C(k)$ the probability that the test packet survives with k interfering packets, the throughput of a nonpersistent CSMA system with perfect acknowledgments on channels with

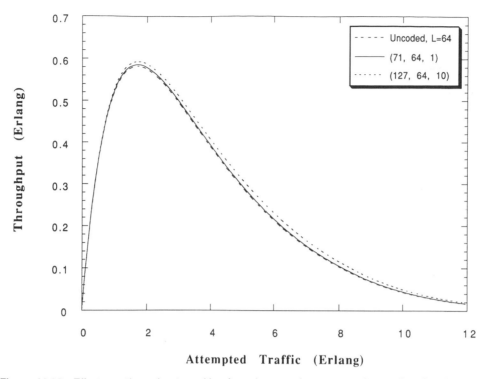

Figure 11.26 Effects on throughput resulting from the use of error-correction coding. Results are shown for two cases, in which a 64-bit information packet is encoded with a 71-bit or 127-bit BCH block code. Transmission of uncoded 64-bit packets is included for comparison. The modulation is BPSK, the channel SNR is 20 dB, and the bell-shaped distribution of terminals is assumed.

capture is given by Zdunek et al. [Zdu89] as

$$S = \frac{\sum_{k=0}^{\infty} \frac{1}{k!} P_C(k) e^{-aG} (aG)^k}{1 + 2a + \frac{e^{-aG}}{G}}$$

where G is the average traffic intensity during the duration of a packet and a is the ratio of the maximum transmission delay to the duration of a packet. This equation is valid for versions of CSMA such as BTMA [Tob75], which deal with the *hidden terminal problem*. The capture probability $P_C(k)$ can be calculated as described in the previous subsection.

The average probability of capture $P_C(k)$ can be determined by Monte Carlo integration of Eq. (11.5.5) using the probability density functions $f_{A_0}(a_0)$ and $f_{G_i}(g_i)$ calculated for a bell-shaped distribution function of the terminals. Figure 11.28 shows the relationship between the average throughput S and the traffic intensity G for BPSK modulation with SNR = 20 dB and packet lengths of

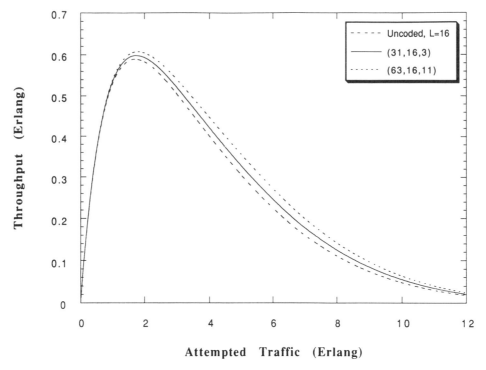

Figure 11.27 Effects on throughput with 16-bit packets coded into length-31 or -63 BCH code blocks. The modulation is BPSK, the channel SNR is 20 dB, and the bell-shaped distribution of terminals is assumed.

16, 64, and 640 bits for both nonpersistent CSMA and slotted ALOHA. The curves show the effects of packet length on achievable throughput and assume $a = 0.01$ for all three lengths. Also shown for comparison are the curves for conventional nonpersistent CSMA and slotted ALOHA without capture. With capture, the maximum throughput of CSMA with packet length 16 bits is 0.88 erlang, which is 0.065 erlang more than the case without capture. The maximum throughput for slotted ALOHA with the same packet length is 0.591 erlang, which is 0.231 erlang higher than the case without capture. In slow fading channels, if the terminal generating the test packet is in a "good" location, the interference from other packets is small and all the bits of the test packet survive the collision. In contrast, for a test packet originating from a terminal in a "bad" location, all the bits are subject to high probability of error and the packet does not survive the collision. As a result, the system shows minimal sensitivity to the choice of packet length, which is consistent with our assumption of slow fading.

Figure 11.29 shows the effect of error-correction coding on the throughput of CSMA in slow fading for 64-bit information packets BCH encoded into 71-bit and 127-bit packets, respectively. The case of uncoded 64-bit packets is included in the figure for comparison. The results show that error-correction coding is of negligible value in improving CSMA throughput in slow fading. The reason for this is essentially the same as given in the previous paragraph in explaining the lack of

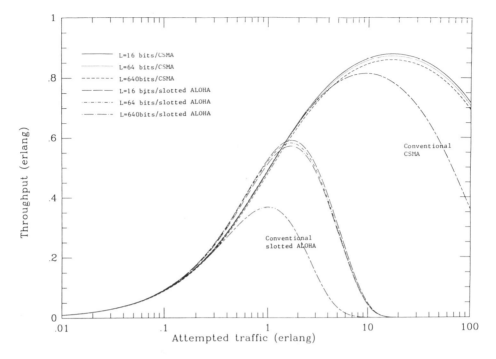

Figure 11.28 Effects of packet length on throughput for CSMA and slotted ALOHA with capture. The modulation is BPSK, and SNR = 20 dB.

sensitivity of throughput to packet length. In simple terms, the presence of slow fading causes the bits in the test packet to be received as "all good" or "all bad" relative to interfering packets, and coding has little effect in such circumstances. Coding can be effective only when bit errors in a block are essentially independent, as occurs in additive Gaussian noise or perhaps in extremely rapid fading.

In summary, the performance results given in Figs. 11.28 and 11.29 have shown that the capture effect results in only about 0.06-erlang improvement in the maximum throughput of a CSMA packet radio network operating in slow fading. This improvement is considerably less than the improvement of approximately 0.2 erlang found for slotted ALOHA. The throughput has also been shown to have minimal sensitivity to the length of the packet and the complexity of the coding technique.

Delay versus Throughput. The delay-versus-throughput characteristic for CSMA can be analyzed using an approximation suggested in [Kle75c]. This approximation assumes that the acknowledgment of a received packet is always correct and that all channel delays other than those defined below are ignored. The expression for the average delay of the CSMA [Zha91] is

$$D = \left(\frac{G}{S} - 1\right)(1 + 2\alpha + \delta) + G - \frac{H}{S}\delta + 1 + \alpha$$

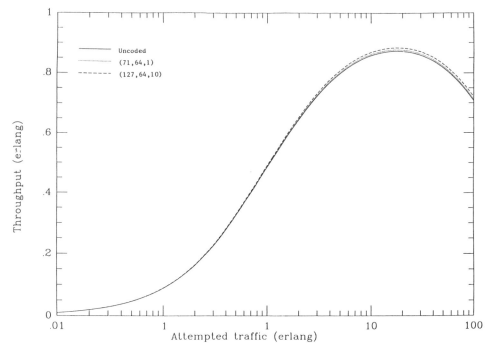

Figure 11.29 Effects of coding on throughput of CSMA. The modulation is BPSK and SNR = 20 dB.

where α is the normalized propagation delay, $H = (1 + \alpha G)/(1 + 2\alpha + e^{-\alpha G}/G)$, which is the number of actual transmissions, and δ is the average retransmission delay.

Figure 11.30 shows delay versus throughput for the CSMA system with and without capture for a 640-bit packet. The average retransmission delay is six packet lengths. As expected, the capture effect improves the throughput and reduces the delay at high traffic load.

CSMA with Busy-Tone Signaling. In wireless networks, it cannot always be assumed that all user terminals are within range and line-of-sight of each other. Typically in a radio network, two terminals can each be within range of some intended third terminal but out of range of each other, separated by excessive distance or by some physical obstacle that makes direct communication between the two terminals impossible. This situation is referred to as the *hidden terminal problem*. Consider as an example a packet radio network with centralized topology, in which the central station is positioned so as to have line-of-sight communication with all the other stations, but some of the stations cannot communicate directly with certain other stations. This is a likely situation in cases of radio networks covering wide geographic areas in which hilly terrain blocks some groups of user terminals from other groups. If a CSMA protocol is being used in the network, it will successfully prevent collisions among the users of one group but will fail to prevent collisions among users in groups hidden from one another. In this case,

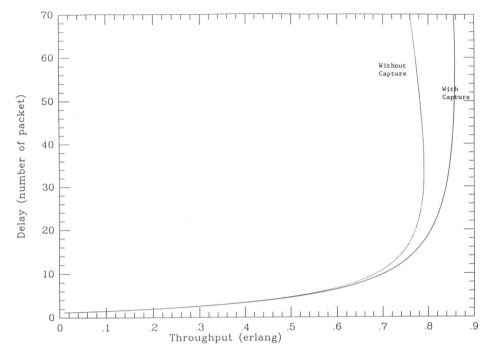

Figure 11.30 Delay versus throughput of CSMA. The modulation is BPSK, and SNR = 20 dB.

given that there is a centralized station in range of all the users, the hidden terminal problem can be solved using a technique called *busy-tone multiple access* (BTMA).

In the BTMA scheme, the system bandwidth is divided into two channels: a *message channel* and a *busy-tone channel*. Whenever the central station senses signal energy on the message channel, it transmits a simple busy-tone signal (e.g., a sinusoid) on the busy-tone channel, and this is in principle detectable by all the user stations. With this method, a user station with a packet ready to send first senses the busy-tone channel to determine if the network is occupied. The procedure the user station then follows in transmission of the packet depends upon the particular version of CSMA being used in the network, and any of the various CSMA schemes might be chosen. The BTMA scheme has been used in a number of packet radio networks having multihop topology. In this topology when a terminal senses a signal on the channel, it turns on its busy tone. In other words, as a terminal detects that some user is on the message channel, it sounds the alarm on the busy tone channel in an attempt to inform other users. In a military tactical situation—for example, where mobile units must stay in communication with each other at all times—the BTMA scheme is the solution of choice.

Exact analysis of the BTMA scheme, particularly for a multihop network, involves several issues which do not arise in evaluations of conventional CSMA techniques. These issues revolve around the requirement for reliable detection of the busy-tone signal in a narrowband busy-tone channel. The parameters involved are the busy-tone detection time window, the false-alarm probability, and the fraction of system bandwidth devoted to the busy-tone signaling. (The situation is rather less complicated for a centralized network, where the throughput can be approximated by that of CSMA [Zdu89].) What is interesting to note is that even if there are no hidden terminals, BTMA, with parameters properly chosen, can provide better throughput than other CSMA schemes. A detailed analysis of the nonpersistent version of the BTMA scheme is given by Tobagi and Kleinrock [Tob75]. A brief summary of BTMA is given in [Tob80a], where a number of packet communication protocols are discussed and compared.

Data-Sense Multiple Access. Digital or data-sense multiple access (DSMA) is very popular in full-duplex wireless data communication networks such as CDPD, ARDIS, and TETRA. In these networks, communication from the mobile to base station and from base to mobile station are performed in separate frequency channels using different access methods. The link from the base to the mobile is referred to as the *forward channel* or *downlink* and uses a version of TDMA in which the messages for different mobile stations occupy non-overlapping time slots. In the downlink, power control, synchronization, and other centralized control signals are also time-division multiplexed with the information packets. The communication channel from the mobile to the base station, which is referred to as the *reverse channel* or *uplink*, uses a version of CSMA called DSMA. Interleaved among other signals broadcast on the forward channel, the base station transmits a *busy-idle* bit in each time frame to report the status of the reverse channel. A mobile station will check this *flag* bit before transmission. If it indicates that the channel is idle the mobile station will proceed to send its packet in the following time slot. As soon as the transmission is started the base station will switch the flag bit to the busy state until the transmission from the mobile terminal is completed.

With DSMA we cannot completely eliminate the collisions and thus we have to devise a retransmission strategy similar to CSMA. The announcement of the occupancy of the reverse channel from the base station is very similar to the busy-tone multiple access (BTMA) protocol used in a centralized topology with the difference that with BTMA a separate channel is dedicated to the announcements on the status of the reverse channel, whereas with DSMA only one bit in each frame of the TDMA forward channel is devoted to the reverse channel status report.

CSMA with Collision Detection. The persistent and nonpersistent CSMA protocols improve upon the ALOHA schemes by ensuring that no user station begins to transmit when it senses that the channel is occupied. One might well ask if a sending station can make further use of channel state information after its own transmission is underway. We shall briefly describe a technique which does precisely this. The technique is called *carrier-sense multiple access with collision detection* (CSMA/CD) and is sometimes referred to as the *listen-while-talk proto-col*. The CSMA/CD technique can be used in nonpersistent, 1-persistent, or *p*-persistent variations of CSMA, each with a slotted or unslotted version. The defining feature of CSMA/CD is that it provides for detection of a collision shortly after its onset, and each transmitter involved in a collision stops transmission as soon as it senses a collision. In this way, colliding packets can be aborted promptly, minimizing the channel transmission time occupied by transmissions destined to be unsuccessful.

In the operation of CSMA/CD, if the channel is sensed to be idle or busy, a user station with a packet to send first follows the procedure dictated by the particular protocol variation in use. However, each station uses hardware that not only monitors the channel before transmission, but also monitors while transmitting. Unlike CSMA, which requires an acknowledgment (or lack of an acknowledgment) to learn of the status of a packet collision, CSMA/CD requires no such feedback information, because the collision detection mechanism is built into the transmitter. If a collision is detected, the transmission is immediately aborted, a jamming signal is transmitted, and a retransmission backoff procedure is initiated, just as in CSMA. The jamming signal is not an essential feature of CSMA/CD, but it is used in many implementations of this access method. The jamming signal serves to force consensus among users as to the state of the network, in that it ensures that all other stations know of the collision and go into backoff condition. As is the case with any random-access scheme, proper design of the backoff algorithm is an important element in ensuring stable operation of the network.

The delay and throughput characteristics of CSMA/CD have been analyzed by Lam [Lam80] and others [Tob80b, Cho85, Hey86, Coy83, Coy85, Tas86, Med83]. Throughput and stability analyses of CSMA/CD can also be found in the texts by Hammond and O'Reilly [Ham86] and Keiser [Kei89].

The initial development of CSMA/CD was done at Xerox in the early 1970s for application in local area networks [Met76b]. Further development work was done in a joint effort by Digital Equipment Corporation, Intel, and Xerox, leading to the detailed specification for Ethernet, one of the first commercially available LAN products [Dig82]. While the name Ethernet is often used, somewhat inaccurately, as a protocol designation, it in fact refers to a specification encompassing a set of products produced by several manufacturers and endorsed by preliminary IEEE standards. The IEEE 802.3 CSMA/CD standard for LANs is based on the Ethernet specification and is nearly identical to it [ANS85, ANS88]. The Ethernet specification and the IEEE 802.3 standard specify use of the 1-persistent version of CSMA/CD.

While collision detection is easily performed on a wired network simply by sensing voltage levels, such a simple scheme is not readily applied to wireless channels. One thing that can be done is to have the transmitting station demodulate the channel signal and compare the resulting information with its own transmitted information. Disagreements can be taken as an indicator of collisions, and the packet can be immediately aborted. However, on a wireless channel the transmitting terminal's own signal dominates all other signals received in its vicinity, and thus the receiver may fail to recognize the collision and simply retrieve its own signal. To avoid this situation the station's transmitting antenna pattern should be different from its receiving pattern. Arranging this situation is not convenient in radio terminals because it requires directional antennas and expensive front-end amplifiers for both transmitter and receiver. Therefore, detecting collisions with an acknowledgment scheme is the approach typically taken in radio data networks. However, the remainder of the protocol structure is usually chosen to conform to IEEE 802.3 in order to be compatible with backbone wired LANs.

The approach called *CSMA with collision avoidance* (CSMA/CA) is actually preferred for use in many WLANs. The specific strategy for collision avoidance differs from one manufacturer to another. Some WLAN manufacturers implement CSMA with an exponential backoff strategy and an acknowledgment scheme, and they refer to the exponential backoff strategy as a collision avoidance mechanism. Others employ a DSMA or even a packet reservation scheme, discussed later, and refer to that as the collision avoidance strategy.

The CSMA/CD scheme is used in many infrared LANs, where both transmission and reception are inherently directional and the design of the receiver front end is inexpensive. In DBIR LANs the transmitted optical signal is narrowly focused by design. In DFIR LANs the radiation is made directional by the packaging of the LEDs. The field of view of the receiver photodiode provides the directionality for the received signal in both DFIR and DBIR LANs. In such an environment, detection of a station's own transmitted signal is readily feasible. In general, the CSMA/CD scheme is not as well-suited to wireless networks as it is to wired networks. However, as we discussed in Chapter 1, compatibility is very important for WLANs, and therefore designers of these networks have had to consider CSMA/CD in order to make the networks compatible with the Ethernet backbone LANs that dominate the wired-LAN industry.

11.6   CONTROLLED RANDOM ACCESS METHODS

The fundamental characteristic of contention-based protocols which limits their achievable throughput is the high incidence of packet collisions under heavy loads of offered traffic. The refined versions of these access methods help to mitigate the collision problem, with combinations of slotted formats and backoff strategies; but even with these refinements, the peak throughput of ALOHA is only about 37% and that of CSMA is about 50%. (We omit consideration of capture effects in this discussion.) A key factor limiting throughput in these access methods is that when a transmission is initiated, the full time and frequency resources of the channel are being used even though the sender cannot be certain that the transmitted packet will not encounter a collision. Thus for those packets destined to collide with each other, the channel resources are in effect being wasted while the transmissions are in progress. With a view toward this problem, other protocols have been devised which exercise more control over the access method and as a result avoid some of the inefficiencies of the uncontrolled random-access schemes.

In the remainder of Section 11.6 we briefly describe three such protocols: reservation ALOHA, polling, and token-passing. These protocols represent somewhat different approaches to the allocation of the control functions in the network.

11.6.1   Reservation ALOHA

The *reservation ALOHA* (R-ALOHA) schemes were devised for use on multiuser satellite systems, but they are equally applicable to terrestrial radio systems and have in fact been adopted in some WLAN products. There are several versions of the R-ALOHA scheme, each of which can be viewed as a combination of slotted-ALOHA and TDM protocols, with the apportioning of transmission time

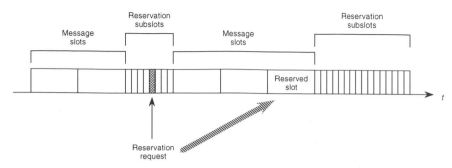

Figure 11.31   Reservation ALOHA.

to each protocol being varied in response to the traffic demand. Here we briefly describe one of the schemes, due to Roberts [Rob73]. Let us visualize the time axis as divided into fixed-length frames, with the frame length chosen to be longer than the longest propagation delay in the network. Each frame is then divided into equal-length slots, some of which are designated as *message slots* while some are further subdivided into short *reservation subslots*. The mixture of message slots and reservation subslots is different for each of two modes of operation, as we now describe.

Consider the diagram in Fig. 11.31. In the *unreserved mode*, there are no message slots, and every slot is composed of reservation subslots. User stations with data to send transmit short reservation requests in the reservation subslots using the slotted ALOHA protocol. After transmitting a reservation request, a user waits for the intended recipient to return a positive acknowledgment and a slot assignment. If the reservation request packet does not suffer a collision, the returned reservation acknowledgment advises the sending station when to send its first packet. (The reservation request may have asked for a single slot or multiple slots, up to some design limit.) The system then switches to the *reserved mode*.

In the reserved mode, one assigned slot in each frame is composed of reservation subslots, and all the remaining slots in the frame are available for use as message slots, on a reservation basis. A sending station that has been granted a reservation sends its packets in successive message slots, skipping over the designated reservation slots when they are encountered. Because all reservation exchanges are heard by all stations in the network, each sending station knows which message slots to skip over before starting its own transmission. This access scheme can be viewed as a kind of flexible TDMA, in which the contention-based reservation exchanges are confined to the relatively short reservation subslots, while the message slots are shared among stations with data to send, in an orderly noninterfering manner. The choice of the number of reservation subslots relative to the number of message slots is a design tradeoff issue. The number of subslots should be small enough to keep the transmission overhead low, but large enough to accommodate the expected number of reservation requests. Given the throughput characteristic of slotted ALOHA, a reasonable design choice is to provide about three (i.e., $1/0.36$) reservation subslots per message slot.

It is interesting to note that there is no centralized control function in the R-ALOHA scheme, but instead the control of the system is distributed among all

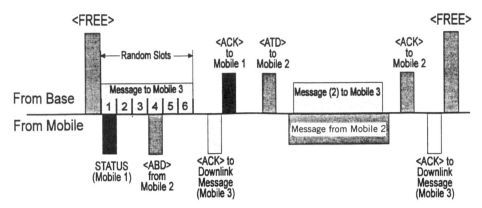

Figure 11.32 An example for operation of dynamic slotted ALOHA used in Mobitex.

the stations in the network. In order to make this procedure work properly, each station maintains information on the queue of outstanding reservations for all the other stations, as well as for the slots at which its own reservations begin. When the queue length drops to zero, the system returns to the unreserved mode, in which all the slots revert to reservation subslots.

Other versions of the R-ALOHA scheme have been developed by Crowther et. al, [Cro73] and by Binder [Bin75]. A description of several of the R-ALOHA schemes can be found in [Tan88] and [Tob80a]. Another multiple access scheme similar to R-ALOHA, called *packet reservation multiple access* (PRMA), is discussed later in this chapter.

Example 11.5. In Mobitex mobile data networks, a version of reservation ALOHA is used. This version is referred to as dynamic slotted ALOHA and operates in the full-duplex scenario of the Mobitex system, in which forward and reverse links operate in separate channels. Figure 11.32 shows a typical example of the operation of this access method by illustrating all packets communicated between a base station and three mobile users. In this example Mobile 1 transmits a status report that consists of a short packet that can be accommodated in one random slot. Mobile 2 has a longer message that requires a reservation. First Mobile 2 sends a request for transmission and then transmits the actual message. The base station controls the entire process and at the same time in this example, needs to send some messages to Mobile 3. Operation is initiated from the base station by transmitting a ⟨FREE⟩ signal. This signal announces to all mobile users that six short slots are available for contention, the number of contention slots being determined by the size of the message intended for Mobile 3. Mobile 1 and Mobile 2 each select one of the six slots randomly with Mobile 1 picking slot 1 and Mobile 2 picking slot 4. If the same slot is used by the two mobiles, we will have a collision and the process will be repeated in the next free slot. In this example, however, we have no collision and base station will receive the request for transmission from Mobile 2 and the status report from Mobile 1 while it is sending its message to Mobile 3. After completion of reception by Mobile 3, this mobile station sends an acknowledgment to the base station and the base station sends an acknowledgment to Mobile 1 confirming reception of the status report and an acknowledgment to Mobile 2 to grant the following time slot for its message to be sent to the base station. While Mobile 2 is sending its message to the base station, the base station will send another message to Mobile 3. At the end of this period, the base station sends an acknowledgment to Mobile 2 and Mobile 3 sends its acknowledgment to the base station. At this stage the transaction cycle among all terminals is completed and another transaction cycle is initiated by the base using another free-slot announcement.

11.6.2 Polling Techniques

Another way of imposing discipline on a network of independent users is to equip one station in the network as a controller that periodically polls all the other stations to determine if they have data to transmit. In marked contrast with R-ALOHA, where control is distributed among all user terminals, this *polling* technique utilizes very centralized control. Our brief discussion here follows [Tob80a]. The controller station may have a polling list giving the order in which the terminals are polled. If the polled station has something to transmit, it starts sending. If not, a negative reply (or lack of any reply) is detected by the controller, which then polls the next terminal in the sequence. Polling requires the constant exchange of control message between the controller and the terminals, and it is efficient only if (1) the round-trip propagation delay is small, (2) the overhead due to polling messages is low, and (3) the user population is not large and bursty. Polling has been analyzed by Konheim and Meister [Kon74], and their analysis has been applied to the radio environment [Tob76], [Aca87]. The results show that as the population of users increases, thus containing more and more bursty users, the performance of polling degrades significantly. Polling is widely used in dedicated telephone networks for data communications, such as networks serving ATM machines and airline reservation systems. However, it has generally not been adopted in existing mobile data networks or WLANs.

11.6.3 Token Passing

Another method of exerting discipline in the network is a technique which has been used in wired LANs connecting computers. Here a *ring* (or *loop*) topology is used. With this topology, messages are not broadcast but instead are passed from station to station along unidirectional links, until they return to the originating station. A simple scheme suitable for this topology, called *token-passing*, consists of passing the access privilege sequentially from station to station around the ring. Any station with data to send may, upon receiving the token, remove the token from the ring, send its message, and then pass on the control token. While this scheme is well-suited to wired LANs, it has not gained wide adoption in wireless networks. However, it has found an application in some DBIR networks with a dual ring structure.

11.7 INTEGRATION OF VOICE AND DATA TRANSMISSION

As the wireless communications industry moves ahead in developing increasingly sophisticated systems, one of the important objectives is the use of a single wireless system to support a variety of communications services, including voice, data, and imagery in various forms and combinations. A key technical problem to be dealt with in such integrated systems is that of multiuser access. An access method that efficiently supports one category of service may be unsuitable for another category of service. Here we specifically address the integration of digital voice and data services in a single multi-user wireless network.

Contention-based packet communications protocols such as ALOHA and CSMA are well-suited to many forms of wireless data communications. They are especially well-suited to networks comprising many user stations each with low average data rate and potentially high peak rates. Those protocols can operate with little or no centralized control, and can generally accommodate variable numbers of users in the network. However, contention-based schemes can become very inefficient in sharing the communications resources when the traffic load is heavy, as the system throughput degrades and the transmission delays increase. The unpredictable and possibly excessive time delays make these access methods less attractive for voice communication service, where minimum delay is essential for user acceptance of the service. Fixed access methods such as FDMA and TDMA are more appropriate for voice communication, but they lack the flexibility needed for efficient integration of multiple services.

In a packet communication environment, voice and data have different requirements. Packets of voice can tolerate errors and even packet loss (a loss of 1–2% of voice packets has insignificant effect on the perceived quality of reconstructed voice [Kum74]), whereas data packets are sensitive to loss and errors but can generally tolerate delays. As a result, as we saw earlier, voice- and data-oriented networks use different multiple-access methods. In wireless access, the simplest approach is to assign different bands to isochronous (voice) and asynchronous (data) packets. The problems encountered with this approach are as follows:

1. The fixed-assignment access methods used in voice-oriented networks are designed to support a certain number of simultaneous users. When the number of active users falls below that number, some portion of the transmission resources is wasted.

2. A typical two-way conversation does not make full use of the call connection time, because only one user talks at a time. Furthermore, the flow of natural speech is actually composed of *talkspurts* with intervening short pauses. It is generally estimated that in a two-way voice connection, the average *voice activity factor* for each talker is in the vicinity of 40% and thus about 60% of available transmission time goes unused.

If a data service can be efficiently integrated with a voice service the wasted transmission resources can be used for data, which typically do not have a stringent delay requirement. In TDMA and FDMA systems the data users may use free time slots and free frequency bands, respectively, as they become available. In a CDMA system the situation is somewhat different. The structure of CDMA is such that the entire bandwidth-time space is used simultaneously by all active users, and the resource to be managed is signal power. With application of efficient power control algorithms, the signal levels transmitted by mobile stations and the base station are continually adjusted in response to the changing locations of mobiles and the number of users on the system at any given time. In a CDMA network, the integration of data calls with voice calls is straightforward in principle, because various numbers of both categories of calls are readily mixed together, with each call accessing the channel with its unique user signal code. Therefore in a CDMA system no modification need be made to the channel-access scheme to accommodate integration of voice and data "channels," and the information rate for voice or data traffic in any one channel can in principle be varied by a variable-rate scheme such as is used for voice service in the IS-95 standard. The integration of voice and data services in a single user channel is not necessarily straightforward.

The CDPD packet data system described in Chapter 12 uses available frequency channels in the existing analog FDMA cellular telephone network (AMPS) to provide an overlaid packet data service supporting data rates similar to voiceband modem rates (up to 19.2 kbits/sec). In its present form, this system does not exploit the pauses between talkspurts but simply takes advantage of the frequency bands temporarily unused by mobile telephone users in each cell area. The multiple-access method is simple; it is briefly described in Chapter 12. The integration of data into the IS-95 CDMA digital cellular system is also described in

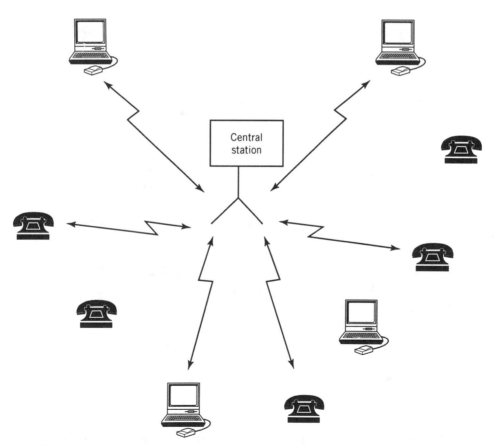

Figure 11.33   Channel-access diagram for a stand-alone integrated voice / data system.

Chapter 12. The integration of voice and data services in TDMA-based networks is more involved, and several implementations are described in the literature. The remainder of this section is devoted to this topic.

11.7.1 Efficient Integration in TDMA

An efficient method of integrating voice and data packets is *movable boundary TDMA with silence detection*. This method has been applied in the time-assignment speech interpolation (TASI) system used in T1-carrier telephone networks [Fis80] to maximize the number of voice users carried and to integrate data into the channel. Using this basic idea, it is possible to design a *TDMA/framed-polling* protocol to integrate voice and data packets in a WLAN. This system, shown in Fig. 11.33, consists of a number of voice and data terminals and a central station which coordinates all the transmissions. The system can operate as a stand-alone system or as a subsystem in a cellular structure as shown in Fig. 11.34. An example

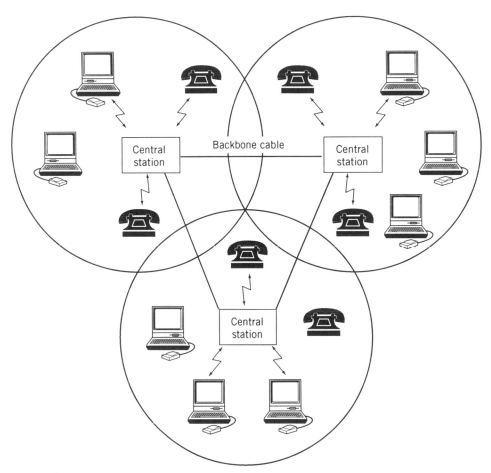

Figure 11.34 Cellular architecture for an indoor wireless integrated voice / data network using the system in the previous figure in each cell.

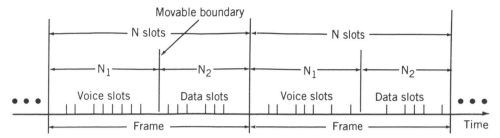

Figure 11.35 Frame structure in a movable-boundary frame-polling system.

of such a system is described in Section 11.7.5. Closely following Zhang and Pahlavan [Zha90], we first describe a simulation method and four analytical methods for studying an integrated voice/data network using the movable boundary approach. Then we provide the results of performance evaluation using computer simulations. The simulation model and the analytical frame work are applicable to other integration methods with similar structures.

The protocol for integration of voice and data packets in a movable boundary TDMA scheme is shown in Fig. 11.35. A frame is divided into two regions with a boundary between them. The first region is used for both voice and data traffic where the voice traffic has priority. If not all the slots in this region are occupied by voice packets, the remaining slots are used for data traffic. The second region is reserved exclusively for data traffic. The boundary between the voice and data regions moves in accordance with the number of active voice packets in each frame. The maximum number of voice packets per frame is N_1 which is assigned an appropriate value to ensure some minimum data traffic capacity and to keep the blockage of voice packets below a selected value (2% in [Zha90]). As will be shown later, this probability can be determined from a well-known queuing theory formula (the Erlang-B equation). Next we describe an approach to simulation of this integrated voice and data system.

11.7.2 The Simulation Model

Figure 11.36 shows a block diagram of the system used for computer simulation [Zha90]. The simulation program generates the number of slots N, the voice packets for the first region of the frame, and the data packets for the remainder of the frame. It then determines the throughput and delay characteristics of the data stream for a given number of voice users.

The simulation model for voice packets is based on the *dynamic talker model* first suggested by Weinstein [Wei78]. This model uses a simple *birth–death Markov approximation* for the generation of talkspurts in the system [Ham86, Tan88]. When the speaker population is sufficiently large ($n \geq 25$), this approximation leads to a computationally efficient model which can replace the more complicated Brady model [Bra69]. In the model suggested by Weinstein, voice traffic is modeled by a Poisson talkspurt arrival process of rate λ_v, and exponentially distributed call holding time with mean $1/\mu_v$. The number of talkspurts in progress at a given time n_v is generated by a *Markovian loss-system* model [Kle75a].

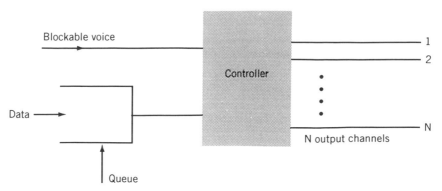

Figure 11.36   Queing model used in analysis of a movable-boundary TDMA system for integration of voice and data traffic.

Let us assume that at a particular time, $n_v = k$, and for a time interval τ it is held at k. Because the arrivals are Poisson, τ is drawn from an exponential distribution with average value given by

$$\bar{\tau} = \begin{cases} 1/(\lambda_v + k\mu_v), & k < N_1 \\ 1/(N_1\mu_v), & k = N_1 \end{cases}$$

At the end of each interval τ, the number of active voice users n_v is updated to $k + 1$ with probability

$$P_k = \begin{cases} \lambda_v/(\lambda_v + k\mu_v), & k < N_1 \\ 0, & k = N_1 \end{cases}$$

or updated to $k - 1$ with probability $1 - P_k$.

The data traffic is modeled as Poisson arrivals with an aggregate arrival rate of λ_d packets per frame. Because none of the data packets is retransmitted, the throughput is the same as the arrival rate of the data packets, λ_d. The number of data packets arriving in a frame time n_d is generated by a Poisson process with average arrival rate λ_d. The number of packets being transmitted during the frame time $n_{d,t}$ is equal to the number of slots left out by the voice packet; that is,

$$n_{d,t} = N - n_v$$

The queue size for the data traffic is given by

$$Q_d = \max\left[\left(Q_d^{\text{old}} + n_d - n_{d,t}\right), 0\right] \tag{11.7.1}$$

which represents the instantaneous delay for the data packets.

11.7.3 Analytical Methods

Analysis of movable boundary integrated voice/data TDMA systems [Mag82, Gav82, Kum74] can be applied to both the uplink and downlink of the system described above. Using the same model as in the simulation, this section presents four different analytical approaches for determining the throughput and delay of the proposed system.

Analysis Method I. One approach to analyzing a movable-boundary TDMA voice/data system is to approximate the system by a fixed-boundary TDMA system. In this analysis, it is assumed that the voice packets always occupy the N_1 available slots. Therefore, the results provide lower bound on the throughput of data traffic in the actual movable boundary system. The fixed-boundary TDMA system is analyzed by assuming a one-dimensional M/D/1 (Markov arrival, definite service time and single server) queuing model [Kle75b].

The queuing delay for $M/D/1$ is given by [Kle75a]

$$Q_d = \frac{\lambda_d}{2\mu_d^2(N - N_1)^2(1 - \rho_d)} \tag{11.7.2}$$

where

$$\rho_d = \frac{\lambda_d}{\mu_d(N - N_1)}$$

and N is the number of slots in a frame.

Analysis Method II. The one-dimensional M/G/1 (Markov arrival, general service time, and single server) queuing model provides another approach to analyzing the system. This method assumes a variable service rate and depends upon the probability of voice occupancy of the slot. The analysis includes the mechanism used for moving the boundary by including the variance of the service rate in the calculation. As will be shown later, this method provides an upper bound on throughput in the overload region.

The queuing delay for M/G/1 is given by [Kle75a]

$$Q_d = \lambda_d \frac{\overline{x^2}}{2(1 - \lambda_d \bar{x})}$$

where

$$\overline{x^2} = \sum_{i=0}^{N_1} P(i) x_i^2$$

$$\bar{x} = \sum_{i=0}^{N_1} P(i) x_i$$

Here $x_i = 1/(\mu_d(N - i))$ is the service time for $N - i$ slots of data, and $P(i)$ is the probability that the voice traffic would require i slots for transmission. This is identical to the probability that $N - i$ slots are occupied by data. The probability $P(i)$ is defined by the Erlang-B distribution [Kle75a],

$$P(i) = \sum_{k=0}^{N_1} \frac{\rho_v^i / i!}{\rho_v^k / k!}$$

where $\rho_v = \lambda_v / \mu_v$. The probability that a voice packet is blocked is $P_B = P(N_1)$.

Analysis Method III. The delay versus throughput characteristic for the movable boundary integrated voice and data TDMA system can be derived using a recursive method analyzed by Williams and Leon-Garcia [Wil84]. This model uses a two-dimensional Markov chain to model the voice and data traffic. Results obtained by this analytical method lie closest to the results of simulations.

The queuing delay of the data traffic, using the recursive method introduced in [Wil84], is given by

$$Q_d = \text{Tr}\left[R(I - R)^{-1} \pi \right]$$

where $\text{Tr}[\ ]$ is the trace operator, which sums all elements of a vector, R is an $(N_1 + 1) \times (N_1 + 1)$ matrix, I is the identity matrix, and π is a vector of length $N + 1$.

The elements of π are given by

$$\pi_i = \Pr[V = i] = \frac{\rho_v^i / i!}{\displaystyle\sum_{j=0}^{N_1} \rho_v^j / j!}$$

and elements of R are obtained from the following recursions. For the diagonal elements,

$$r_{ii}^{\text{new}} = \frac{1}{2}\left\{ a_{ii} - \left[a_{ii}^2 - 4\left(\sum_{k=0, k \neq i}^{N} \left(r_{ik}^{\text{old}} - d_{ik} \right) r_{ki}^{\text{old}} - b_{ii} \right) \right]^{1/2} \right\}$$

and for the off-diagonal elements,

$$r_{ij}^{\text{new}} = \frac{\displaystyle\sum_{k=0, k \neq i}^{N} a_{ik} r_{kj}^{\text{old}} + b_{ij} - \sum_{k=0, k \neq i, j} r_{ik}^{\text{old}} r_{kj}^{\text{old}}}{r_{ii}^{\text{old}} + r_{jj}^{\text{old}} - a_{ii}}$$

where a_{ij} and b_{ij} are calculated from

$$\begin{aligned}
a_{i\,i+1} &= -(i + 1)\mu_v / ((N - i)\mu_d), & 0 \leq i < N_1 \\
a_{ii} &= (\lambda_v + i\mu_v + \lambda_d + (N - i)\mu_d)/((N - i)\mu_d), & i \leq N_1 \\
&= (N_1 \mu_v + \lambda_d + (N - N_1)\mu_d)/((N - N_1)\mu_d), & i = N_1 \\
a_{i\,i-1} &= -\lambda_v / ((N - i)\mu_d), & 0 < i \leq N_1
\end{aligned}$$

and

$$b_{ii} = -\lambda_d/((N-i)\mu_d)$$

The $\{a_{ij}\}$ and $\{b_{ij}\}$ not defined in the above equations are all zero.

Analysis Method IV. This method was proposed by Occhiogrosso et al. [Occ77] for an approximate analysis of the underload region in which the data throughput is smaller than the number of slots left in the frame after voice transmission. Here the parameter ρ_d is given by

$$\rho_d \leq N - E\{i\} = N - \rho_v(1 - P_B)$$

where E{ } denotes the statistical expectation.

Next we define $K = N - i$ as the number of slots in a frame available for data traffic. The probability of having n packets waiting for service is given by the Erlang-C distribution [Kle75a],

$$p_n = \begin{cases} \dfrac{(K\rho_d)^n}{n!} p_0 & n \leq K \\ \dfrac{K^K \rho_d^n}{K!} p_0 & n \geq K \end{cases} \quad \text{where} \quad \rho_d = \frac{\lambda_d}{K\mu_d}$$

and

$$p_0 = \left[\sum_{n=0}^{K-1} \frac{(K\rho_d)^n}{n!} + \frac{1}{1 - \rho_d} \frac{(K\rho_d)^K}{K!} \right]^{-1}$$

The probability that a data packet is delayed, given K slots available, is

$$P_{d,K}(\rho_d) = \sum_{n=K}^{\infty} p_n = (K\rho_d)^K \frac{p_0}{(1 - \rho_d)K!}$$

The average number of packets not being served, during a frame with K slots available for data, is

$$Q_{d,K} = \sum_{n=K+1}^{\infty} (n - K)p_n = \frac{\rho_d}{K - \rho_d} P_{d,K}(\rho_d)$$

Therefore, the average delay is given by [Occ77] as

$$Q_d = \sum_{i=0}^{N_1} P(i)Q_{d,(\bar{N}-i)} \simeq \frac{\rho_d \sum\limits_{i=0}^{N_1} P(i)P_{d,N-i}(\rho_d)}{a}$$

The approximation is made by assuming $a = N - \rho_d = N - [\rho_v(1 - P_B) + \rho_d]$.

11.7.4 Comparison of Performance Assessment Methods

We now compare performance results obtained by computer simulation with analytical results obtained by the four methods described above. Figure 11.37 shows the delay versus throughput characteristic for the data traffic given 59 voice users and a data rate of 2.6 Mbits/sec. Monte Carlo simulation results and analytical results for methods I–IV are included in the figure. For high throughput, the results obtained with method III agree closely with the simulation results. Method IV and the approximate calculations using methods I and II do not match simulation results as well as method III. Method I provides a lower bound on performance, whereas method II gives an upper bound on performance in the overload region. Method IV provides a close approximation to simulation results in the underload region.

Figure 11.38 shows the throughput versus delay characteristic of the data traffic for data rates of 2.4, 7.9, and 11.7 Mbits/sec, corresponding to a receiver with antenna diversity of orders 2, 4, and 8, respectively, determined using method III in a typical indoor radio channel with rms multipath spread of 58 nsec [Aca87]. For antenna diversity of order 2 we have 20 voice users, while for antenna diversity of orders 4 and 8, we have 59 voice users.

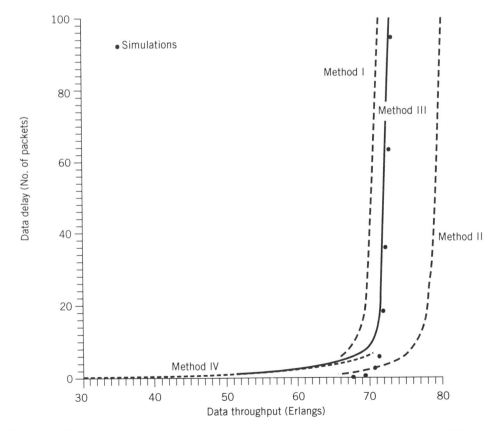

Figure 11.37 Delay versus throughput for a dual-rate system with dual diversity. Diversity (*M*) = 2, voice users = 59.

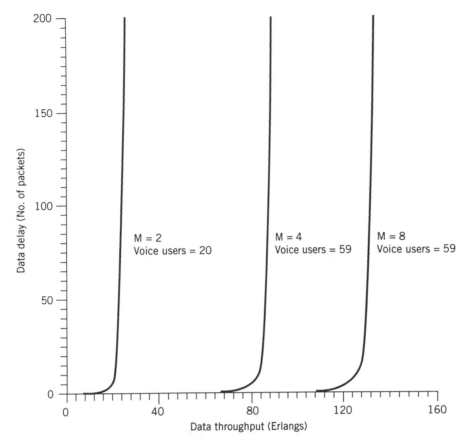

Figure 11.38 Delay versus throughput for a single-rate movable-boundary TDMA system with $M = 2$, 4, and 8 orders of diversity, determined using analysis method III.

Figure 11.39 shows data throughput versus the number of active voice users given a delay of no more than 10 msec and three different data rates (or orders of diversity). These results were obtained using analytical method III. The curves in these figures can be represented by best-fit linear functions to give a relationship between (a) the allowed data throughput for a bounded delay and (b) the number of voice users supported by the system. The linear fit is given by

$$D = -AV + (R_T - \delta)$$

where D is the data throughput, V is the number of voice users, R_T is the average transmission rate of the system, and δ and A are the correction factors for the D–V relationship and the slope of the linear function, respectively.

As shown in [Zha90], A and δ can be approximated by 0.032 and 0.29, respectively; and therefore the relationship between the supported data rate and the number of active voice channels is given by $D = 0.032V + (R_T - 0.29)$. For example, if $R_T = 6.9$ Mbits/sec and the number of active voice users is 30, the data throughput with maximum delay of 10 msec is 5.44 Mbits/sec. A system with

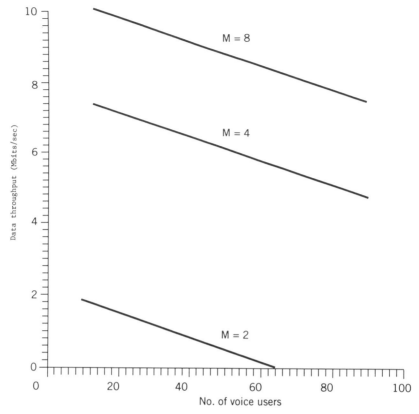

Figure 11.39 Throughput for a single-rate system versus number of voice users for data delay less than 10 msec with diversity $M = 2$, 4, and 8.

transmission rate of $R_T = 2.39$ Mbits/sec can provide a data rate of only 1.14 Mbits/sec, if 30 voice users are active in the system.

11.7.5 A Radio Network with Movable Boundary

Here we give an example of a radio network using the movable-boundary TDMA scheme analyzed in the previous subsections [Zha90]. Figure 11.33, shown earlier, gives the general overview of a system in which several voice and data terminals communicate with a central station. Depending on the location and time, each user terminal and the central station communicate at one of two selectable data rates. In operation of the system, the choice between the two data rates will depend upon the order of diversity and the outage threshold of the system. We assume that this decision is made using the SNR of the signal received from the central station when the terminal is polled.

The channel time is partitioned into frames of fixed length. The length of the frame is chosen to be $T = 10$ msec. This ensures that the delay time between two consecutive voice packets from the same voice terminal does not affect the quality of reconstructed speech. Because of different transmission times for a packet when two rates are used, the number of slots N in the frame is a variable which depends on the time and the location of the terminals. In contrast, for a single rate system, N is always fixed. The average number of slots per frame is $\overline{N} = \text{Int}[R_T T/L]$, where T is the frame time and L is the packet length in bits, and the function $\text{Int}[x]$ gives the integer closest to the variable x.

For two-way communication, the channel is divided into two frequency bands. One band is assigned to the downlink (control station to user terminal), on which TDMA is used, while the other band is assigned to the uplink, on which framed polling is the channel-access protocol. Transmission on the downlink channel is straightforward, because the central station can send its packets to all voice and data users consecutively in a TDMA format. The uplink transmission is not as simple, because there is no direct communication between terminals to coordinate their transmissions. Particularly with two transmission rates, the information packets from different terminals have different time durations depending on the data rate of the terminal. Therefore, even if the terminals are synchronized, we cannot arrange slots in a regular TDMA structure unless each terminal knows the data rates of all other terminals and computes the start of each of its packet transmission times. In order to overcome this difficulty, the uplink transmission is coordinated by the central station using a framed polling access protocol. In this method, the central station grants the terminals permission to transmit when the channel is free.

During call setup, the central station assigns sequential identification (ID) numbers to the users. The downlink operates similarly to the uplink, except that all the packets are coming from the central station and there is no need to exchange any control signal between the terminals and the central station. For the uplink channel, at the beginning of a frame, the voice user with the lowest ID number is given permission to start transmission with its appropriate data rate. If there is no packet in the buffer, a signal is sent to notify the central station. After completion of the first message, the central station informs the user having the next sequential ID number to start transmission. This process continues until the boundary is reached or the last voice ID number has arrived.

The transmission of data packets begins immediately following the voice packets and stops at the end of the frame time. The data traffic is controlled by conventional polling. Once a data terminal is polled and receives permission to transmit, it can use all the remaining slots until its buffer is emptied. Then, the central station polls another data terminal, and the same process repeats.

11.7.6 Packet Reservation Multiple Access

Another example of a system for integrating voice and data services is the work done by David Goodman and his colleagues in developing the concept of *packet reservation multiple access* (PRMA) [Goo89]. PRMA is a method for transmitting, in a wireless environment, a variable mixture of voice packets and data packets. The PRMA system is closely related to reservation-ALOHA, in that it merges characteristics of slotted ALOHA and TDMA protocols. PRMA has been developed for use in centralized networks operating over short-range radio channels. Short propagation times are an important ingredient in providing acceptable delay characteristics for the voice service. Our description here closely follows [Goo89] and [Goo91b].

The transmission format in PRMA is organized into frames, each containing a fixed number of time slots. The frame rate is identical to the arrival rate of speech packets. The terminals identify each slot as either "reserved" or "available," in accordance with a feedback message received from the base station at the end of the slot. In the next frame, a reserved slot can be used only by the user terminal that reserved the slot. An available slot can be used by any terminal, not holding a reservation, that has information to transmit.

Terminals can send two types of information, referred to as *periodic* and *random*. Speech packets are always periodic. Data packets can be random if they are isolated, or they can be periodic if they are contained in a long unbroken stream of information. One bit in the packet header specifies the type of information in the packet. A terminal having periodic information to send starts transmitting in contention for the next available time slot. Upon successfully detecting the first packet in the information burst, the base station grants the sending terminal a reservation for exclusive use of the same time slot in the next frame. The terminal in effect "owns" that time slot in all succeeding frames as long as it has an unbroken stream of packets to send. After the end of the information burst, the terminal sends nothing in its reserved slot. This in turn causes the base station to transmit a negative acknowledgment (NACK) feedback message indicating that the slot is once again available.

To transmit a packet, a terminal must verify two conditions. The current time slot must be available, and the terminal must have permission to transmit. Permission is granted according to the state of a pseudorandom number generator, with permissions at different terminals being statistically independent. The terminal attempts to transmit the initial packet of a burst until the base station acknowledges successful reception of the packet or until the packet is discarded by the terminal because it has been held too long. The maximum holding time, D_{max} seconds, is determined by delay constraints on speech communication and is a design parameter of the PRMA system. If the terminal drops the first packet of a burst, it continues to contend for a reservation to send subsequent packets. It drops additional packets as their holding times exceed the limit D_{max}. Terminals with periodic data (as opposed to voice) packets to send store packets indefinitely while they contend for slot reservations (equivalent to setting $D_{max} = \infty$). Thus as a PRMA system becomes congested, both the speech packet dropping rate and the data packet delay increase.

In [Goo91b], Goodman and Wei analyze PRMA efficiency, which they quantify as the maximum number of conversations per channel that the system can support within a chosen constraint on packet-dropping probability. In their work they adopted a constraint of $P_{drop} < 0.01$. They used a speech source rate of 32 kbits/sec and a header length of 64 bits in each packet. Using computer simulation methods, they investigated the effects of six system variables on PRMA efficiency: (1) channel rate, (2) frame duration, (3) speech activity detector, (4) maximum delay, (5) permission probability, and (6) number of conversations. Over the range of conditions examined, they found many PRMA configurations capable of supporting about 1.6 conversations per channel and found that this level of efficiency could be maintained over a wide range of conditions. Their overall conclusion was that PRMA shows encouraging potential as a statistical multiplexer of speech packets. However, they judged that there are still many questions to be answered in order to verify that PRMA will perform properly in short range radio systems. Issues requiring further investigation include (a) the effects of mixing random information packets (data) and periodic information packets (speech) in PRMA and (b) the effects of packet transmission errors on PRMA efficiency.

QUESTIONS

(a)   Why is the use of error correction coding more prevalent in voice-oriented networks than in data-oriented networks?

(b)   Why is error detection important in a packet data system?

(c)   What are the most popular access methods for packet voice communications? Why are they not used for data communications?

(d)   What is the best topology for ad-hoc laptop networking?

(e)   What is the main obstacle to utilizing CSMA/CD in radio channels?

(f)   Explain the difference between the reservation ALOHA and TDMA access methods.

(g)   Compare the throughput of reservation ALOHA with that of pure ALOHA. Explain the difference.

(h)   Name an advantage for integration of voice and data on wirelines that does not hold for wireless media.

(i)   Explain how CDPD integrates data service with cellular voice service

(j)   Explain the use of Brady's model.

(k)   Why do real-time packet voice systems not use acknowledgment packets?

(l)   Why is CDMA not used in wireless LANs?

(m)   Why does the use of sectored antennas at the base station increase the capacity of a CDMA digital cellular network? What is the effect of antenna sectorization on the performance of a TDMA cellular system?

(n)   Name two methods of overlaying data service onto the cellular telephone networks. Give an example for each method.

PROBLEMS

Problem 1. Figure 8.2 shows a typical cellular frequency reuse scheme with hexagonal cells. In this topology the frequency reuse factor is $K = 7$ (seven frequency sets, A to G, are assigned to the cells).

(a)   Assign the frequency sets for hexagonal cells with frequency reuse factors of $K = 4$, 12, and 19.

(b) If the minimum distance between the centers of two cells using the same frequency sets is represented by D and the maximum distance from the center of a cell to any of its corners is represented by R, give the D/R ratios for $K = 4$, 7, 12, and 19.

(c) Show that the relationship $D = \sqrt{3K} R$ holds in general.

Problem 2

(a) Show that the minimum received signal-to-noise ratio of a cellular topology with frequency reuse factor of $K = 7$ can be approximated by

$$\frac{C}{I} = \frac{1}{6}\left(\frac{R}{D}\right)^{-\alpha}$$

where α is the gradient of the distance–power relationship.

(b) Repeat (a) for $K = 4$, 12, and 19.

(c) If the gradient of the distance-power relation is approximated by $\alpha = 4$, give the minimum signal to noise ratio in the hexagonal cellular topology with frequency reuse factors of $K = 4$, 7, 12, and 19.

Problem 3. Using the throughput formulas given in Sections 11.5.1 and 11.5.2, show that the maximum values S are $1/2e$ for pure ALOHA and $1/e$ for slotted ALOHA.

Problem 4. A number of terminals use the pure ALOHA protocol to transmit to a central control station over a shared 8-kbit/sec channel. Each terminal transmits a 100-bit packet, on the average, every minute.

(a) What is the maximum number of terminals that the channel will support?

(b) Repeat for the slotted ALOHA protocol.

(c) Repeat for both protocols for packet lengths of 200 and 400 bits.

(d) Describe how the results change if the transmission rate of the channel increases to 16 kbits/sec.

Problem 5. A small slotted ALOHA network has k user stations, each of which transmits with probability $1/k$ (original transmissions and retransmissions combined) during any slot. Calculate the channel throughput as a function of k. Evaluate the throughput for $k = 2$, 4, 8, and 16, and for the limiting case of $k \to \infty$.

Problem 6. A large network (assume an infinite population of user terminals) operates with the slotted ALOHA protocol. Each station waits an average of W slots before retransmission after a collision. Calculate and plot the delay-versus-throughput curves for this network for $W = 2$, 4, 8, and 16.

Problem 7. In a slotted CDMA packet data network we assign an orthogonal code of length 16 to each individual user and the modulation technique is DPSK.

(a) Give the received signal-to-noise power ratio as a function of M, the number of simultaneously arriving packets. Assume that perfect power control is applied to the system and that the received signal-to-noise power ratio in the absence of interference is 10 dB.

(b)   Determine the signal-to-noise ratio per bit for M simultaneously arriving packets and give the probability of error per bit.

(c)   Give an expression for the calculation of the probability of capture $P_C(M)$ of a packet with length of L bits as a function of M. Give the numerical value of $P_C(1)$, $P_C(2)$, and $P_C(10)$ for $L = 10$.

(d)   Give the expression for the calculation of the throughput as a function of offered traffic G if the packets arrive with a Poisson distribution.

Problem 8

(a)   Use the Erlang B equation to calculate the required number of slots for a TDMA system with probability of blockage of 2% and $\rho = \lambda/\mu = 0.5$.

(b)   Repeat (a) for probability of blockage of 20% and 0.2%.

(c)   Repeat (a) for $\rho = 0.9$.

Problem 9. In the US AMPS system the total available band for reverse channel (uplink) is 12.5 MHz and the frequency reuse factor of $K = 7$ is used with three sector antennas at the base stations.

(a)   What are the number of channels per sector? What is the number of users (Erlang capacity) per sector for a 2% probability of blockage?

(b)   Repeat (a) for a digital cellular system in which three TDMA users operate simultaneously in each analog channel.

Problem 10. If the maximum number of calls per hour in a sector of a cell is 5000 and the average calling time is 1.76 min, how many radio channels are needed, if the acceptable blocking probability is 2%?

Problem 11. A TDMA/TDD PCS system accommodates 50 users (100 time slots) with a speech coding rate of 32 kbits/sec. If a movable boundary scheme is implemented on this system to integrate the voice and data for the same number of voice users, what is the data throughput given that the delay remains below 10 msec? Use the approximate equation in Section 11.7.4.

Problem 12. In a slotted ALOHA wireless network the received signal amplitude from each terminal forms a Rayleigh distributed random variable with average received power of P. In this problem we examine the throughput of this network using a simple power-based capture model. In the power-based capture model, if there is a collision of $n + 1$ packets, the target packet will be captured if its instantaneous power, p_s, is at least z times more than the total power of the n interfering packets, p_n. In words, the target packet is destroyed if

$$\gamma = \frac{p_s}{p_n} < z$$

where z is the capture threshold and γ is the signal-to-interference ratio at the collision.

(a)   Show that the probability density functions of the received power from the target user and the interferece power from other terminals are given by

$$f_{P_s}(p_s) = \frac{1}{P}e^{-p_s/P}$$

and

$$f_{P_n}(p_n) = \frac{1}{P} \frac{(p_n/P)^{n-1}}{(n-1)!} e^{-p_n/P}$$

(b) Show that the probability density function and probability distribution function of the signal-to-interference ratio are given by

$$f_\Gamma(\gamma) = n(\gamma + 1)^{-n-1}$$

and

$$F_\Gamma(\gamma) = 1 - \left(\frac{1}{\gamma + 1} \right)^n$$

(c) Show that if the arrivals of the packets obey a Poisson distribution, the throughput of the system is given by

$$S = G \left[1 - \sum_{n=1}^{\infty} \frac{G^n}{n!} e^{-G} F_\Gamma(z) \right] = Ge^{-Gz/(z+1)}$$

where G is the offered traffic.

(d) Use MathCAD or MatLAB to sketch S versus G for the capture parameter values of $z = 0, 3, 6, 20$, and ∞ (no-capture) dB.

12

STANDARDS AND PRODUCTS

12.1 INTRODUCTION

In this chapter we describe the major digital wireless systems and products in use or under development in various parts of the world. As we did in Chapter 1, we categorize the systems as either voice-oriented or data-oriented systems. Many of these systems—in particular, most of the voice-oriented systems—have been defined by standards bodies, whereas many of the data-oriented systems are proprietary designs and do not conform to standards. As pointed out in Chapter 1, standards are of greatest importance for systems which are intended to serve large numbers of users who require portability and mobility over wide areas, with cellular telephone and emerging personal communications services (PCS) being the best examples. For products such as wireless local area networks (WLANs), where users require wireless access over a well-defined area, standards have not been as important, and proprietary designs have been more common. For more details on the applicable standards bodies the reader may wish to refer to Chapter 2 and references cited therein.

12.2 VOICE-ORIENTED SYSTEMS

As we did in Chapter 1, we divide voice-oriented services into two subcategories: (1) high-power, wide-area systems and (2) low-power, local-area systems. A summary of wireless technologies employed in several worldwide standards for the wireless voice industry is given in Table 12.1.

12.2.1 Wide-Area, High-Power Systems (Cellular and Mobile Radio)

GSM. The Pan-European standard for digital cellular telephony, called Global System for Mobile Communication (GSM), was motivated by the need for a common mobile standard throughout Europe, where six incompatible analog cellular systems have been in use in different geographic areas. At the start of work on GSM in the early 1980s, spectrum was reallocated throughout much of Europe and surrounding regions so that a completely new system, based on digital technology, could be developed. The bands allocated to GSM are 890–915 MHz

TABLE 12.1 Summary of Wireless Technologies Employed in Several Worldwide Standards for the Wireless Voice Industry

	Digital Cellular				Low-Power Systems				
System	IS-54	GSM	JDC	IS-95	DECT	PHP	CT-2	CT-3	Bellcore UDPC
Multiple access	TDMA / FDMA	TDMA / FDMA	TDMA / FDMA	FDMA / CDMA-SS	TDMA / FDMA	TDMA / FDMA	FDMA	TDMA / FDMA	TDM / TDMA / FDMA
Frequency band Base TX (MHz):	869 – 894	935 – 960	810 – 826 1477 – 1489	869 – 894	1800 – 1900 (Europe)	1895 – 1907 (Japan)	864 – 868 (Europe & Asia)	862 – 866 (Sweden)	Emerging technology (U.S.)
Mobile TX (MHz):	824 – 849 (U.S.)	890 – 915	940 – 956 1429 – 1441 1501 – 1513 1453 – 1465	824 – 849 (U.S.)					
Duplexing:	FDD	FDD	FDD	FDD	TDD	TDD	TDD	TDD	FDD
Channel spacing (kHz):	30	200	25	1250	1728	300	100	1000	350
Modulation:	π / 4- QDPSK	GMSK	π / 4- QDPSK	BPSK / QPSK	GFSK	π / 4- QDPSK	GFSK	GFSK	π / 4- QDPSK
Portable transmit power, max / avg:	600 mW / 200 mW	1 W / 125 mW		200 mW	250 mW / 10 mW	80 mW / 10 mW	10 mW / 5 mW	80 mW / 5 mW	200 mW / 20 mW
Frequency assignment:	Fixed	Dynamic	Fixed		Dynamic	Dynamic	Dynamic	Dynamic	Autonomous
Power control Handset:	Yes	Yes	Yes	Yes	No		No	No	Yes
Base:	Yes	Yes	Yes	Yes	No		No	No	No
Speech coding and rate (kbits / sec):	VSELP 8	RPE-LTP 13	VSELP 8	QCELP 1 – 8 (var.)	ADPCM 32	ADPCM 32	ADPCM 32	ADPC 32	ADPCM 32
Speech channels per RF channel:	3	8	3	—	12	4	1	8	10
Channel rate (kbits / sec): Uplink (kbits / sec):	48.6	270.833	42	1,228.8 (Chip rate)	1152	96	72	640	500
Downlink (kbits / sec):	48.6	270.833	42						500
Channel coding:	Rate-1 / 2 conv.	Rate-1 / 2 conv.	Rate-1 / 2 conv.	R-1 / 2 forward R-1 / 3 reverse CRC	CRC	CRC	None	CRC	CRC
Frame duration (msec):	40	4.615	20	20	10	5	2	16	2

for downlink (mobile to base) and 935–960 MHz for uplink (base to mobile) communications. The band in either direction is divided into 124 channels, with carriers spaced 200 kHz apart. Each cell site in a GSM system has a fixed number of channels (two-way) ranging from only one to usually not more than 15. The cells range in size from 1 to several kilometers [Rah93]. The GSM standard is primarily intended to serve vehicular terminalsand provides a considerably enhanced roaming capability, relative to the first-generation analog cellular systems, through automatic network location detection and registration. The GSM standard went into operation in 1992 and is rapidly gaining adoption in Europe and elsewhere in the world, particularly the Asia-Pacific region. As of April 1993, 32 operators in 22 countries had committed to implementing GSM, with a number of other countries considering its adoption [Rah93]. As currently implemented, GSM is providing only digital voice service, but data services are being planned for later introduction. A good overview of the considerations leading to the GSM standard is given in [Hau94]. A frequency-shifted version of GSM, DCS-1800, operates at 1.8 GHz.

The GSM network architecture is shown in Fig. 12.1. The three basic network elements, essential for cellular wireless access to the wired public network, are the mobile terminals, base stations, and switching centers. Each base station provides the wireless communication to and from mobile terminals in its cell coverage area, and the switching centers provide connections from the base stations to the wired public network, including both the PSTN and the ISDN. In addition to providing the wireless-to-wired network connections, the switches control the assignment of radio channels to mobile terminals and manage the handoff of calls from one cell site to another as mobile users move through the service area.

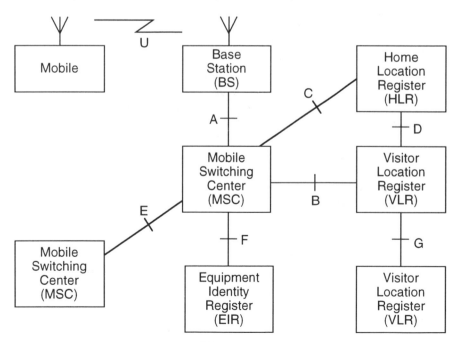

Figure 12.1 GSM network architecture.

The GSM also provides support to roamers, users who move from one service system to another. The management of intersystem roaming is implemented using two databases, shown in Fig. 12.1, called the Home Location Register and the Visitor Location Register. GSM provides personal mobility by use of a subscriber identity module (SIM), which carries the personal number assigned to the user. The mobile user identifies himself to the network by insertion of the SIM into a mobile terminal. In this way, users may access a GSM network from arbitrary locations, including public GSM terminals. The various functions concerned with handoff of mobile calls from cell to cell or from one service system to another, or access from remote locations, are all grouped under the name *mobility management*. The mobility management functions in GSM are described in some detail in [Rah93]. Figure 12.1 also includes another database, the Equipment Identity Register. This database holds subscribers' equipment identities, used for identification of unauthorized subscriber equipment, which is dealt with by denial of service.

All of the network elements and databases shown in Fig. 12.1 are contained in all of the second-generation cellular systems currently under development. However, in GSM, all of the labeled interfaces A-G and U are defined by GSM standards. This gives the cellular operators flexibility in configuring their individual systems and in procuring equipment from a variety of manufacturers. In the North American digital cellular systems, only a few of the interfaces are being standardized.

In GSM, each frequency channel supports multiple user channels with a TDMA signal format. This FDMA/TDMA radio channel structure is used in the North America (IS-54) TDMA standard and the Japanese Digital Cellular (JDC) standard as well, though the details of the channel structure are somewhat different from one system to another. In GSM, each 200-kHz FDM channel uses an aggregate bit rate of 270.833 kbits/sec, carried over the radio interface using GMSK modulation with bandwidth-time product 0.3. The use of a transmitted bit rate as high as 270 kbits/sec requires the implementation of adaptive equalization techniques to deal with channel multipath, and GSM specifications require that equipment be built to accommodate time dispersion up to 16 μsec. The GSM standard provides for the use of slow frequency-hopping as a means of reducing other-user interference, though no GSM systems currently use this technique.

The 270-kbit/sec data stream in each FDM channel is divided into eight fixed-assignment TDMA channels termed *logical channels*. These logical channels are organized into a hierarchical frame structure that provides each mobile terminal with a two-way traffic channel and a separate two-way control channel. The numbering of time slots is offset between the two directions on the radio link to prevent a mobile terminal from transmitting and receiving at the same time.

The GSM radio-interface standard provides a variety of traffic channels and control channels defined in a hierarchy built upon the basic 8-slot TDMA transmission format. The frame hierarchy, depicted in Fig. 12.2, ranges from a bit interval of 3.69 μsec to a hyperframe (used for initializing an encryption algorithm) of length 3 hr 28 min 53.76 sec. The basic building block of the frame hierarchy is a 4.615-μsec frame, shown in Fig. 12.3. Each frame comprises eight 577-μsec time slots, and the composition of a time slot is shown in the expanded portion of the figure. The time-slot interval is equivalent to the transmission time for about

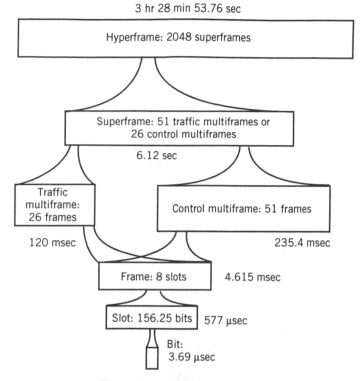

Figure 12.2   GSM frame hierarchy.

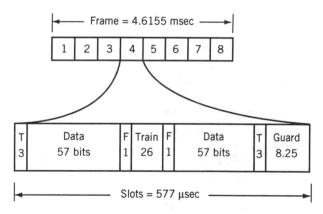

T: Tail bits
F: Flag
Train: Equalizer training sequence
Guard: Guard time interval

Figure 12.3   GSM frame and time-slot structure.

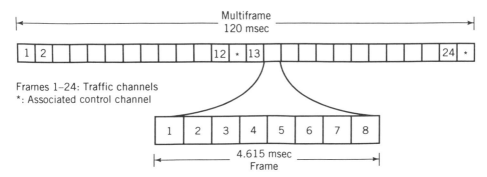

Figure 12.4 GSM multiframe and frame structure.

156.25 bits, though only 148 bits are actually transmitted in each slot interval. The remaining time, 8.25 bits time duration or about 30.5 μsec, is guard time in which no signal is transmitted, to prevent overlapping of signal bursts arriving at a base station from different mobile terminals. Each slot carries two data bursts, each burst consisting of 57 bits of user information and a flag bit to identify the 57 bits as either digitized speech or other information. The remaining bits are a 26-bit equalizer training sequence in the middle of the slot and two sets of "tail bits," all logical zeros. The tail bits are used in the convolutional decoding of the 57-bit segments of user data.

The next level in the hierarchy is a GSM multiframe, shown in Fig. 12.4. Each 120-msec multiframe is composed of 26 frames, each containing eight time slots. In each multiframe, 24 frames carry user information, while two frames carry system control information in *associated control channels*. In the first phase of GSM system implementation, each "full rate" traffic channel utilizes all 24 user-information frames in each multiframe. From Figs. 12.3 and 12.4, we see that one full-rate traffic channel utilizing one slot in each of the 24 frames in a 120-msec multiframe has a data rate of $24 \times 2 \times 57/0.120 = 22{,}800$ bits/sec. The speech coder specified in the first phase of GSM is described as linear predictive coding with regular pulse excitation [Var88]. This speech coder has a data rate of 13 kbits/sec, and the addition of error-detection and error-correction coding brings the transmission rate up to 22.8 kbits/sec. A later phase of GSM implementation, termed "half-rate," will improve spectrum utilization by allocating only 12 frames per multiframe to user channels, and the error-protected speech transmission rate will be reduced to 11.4 kbits/sec.

As noted above, two frames in each 120-msec multiframe are allocated to control channels. In the initial or full-rate implementation of GSM, only one of the frames, the one near the center of the multiframe, is used. It provides eight *slow-associated control-channels* (SACCH), one for each logical traffic channel, to carry link control information between mobile terminals and base stations. In the half-rate GSM implementation, the last frame in each multiframe will be used to provide the additional eight SACCH channels. The transmission format also provides for establishing a *fast associated control channel* (FACCH) on demand within any traffic channel. This is done by usurping one or both 57-bit data blocks within a slot and by indicating this using the flag bits in each slot time. The

FACCH is used by either the mobile terminal or the base station for conveying signal quality information and other information used in managing call handoffs.

Figure 12.2 shows that the 8-slot frames may be organized into control multiframes rather than traffic multiframes. Control multiframes, which we shall not discuss in detail, are used to establish several types of signaling and control channels used for system access, call setup, synchronization, and other system control functions. Either traffic or control multiframes are grouped into superframes, which are in turn grouped into hyperframes. A detailed description of the GSM frame structure, including the specified allocations of bits in the frame hierarchy, can be found in the GSM Recommendations [Rah93].

North American TDMA Digital Cellular (IS-54). The North American Digital Cellular (NADC) TDMA standard has been developed by the EIA/TIA TR45.3 digital cellular standards subcommittee as replacement for the existing analog AMPS system. As with GSM, the NADC radio channel structure is a combination of FDM and TDMA, with user traffic and control channels built upon the logical channels provided by TDMA time slots. The designated frequency channels for this digital standard are the same as those in AMPS, with carriers spaced 30 kHz apart. This channel plan was selected by EIA/TIA so as to allow conversion of individual analog channels to digital operation, as needed by a cellular operator to gradually expand system capacity. In the initial phase of TDMA implementation, the mobile phones are dual-mode devices, capable of operating on either AMPS analog channels or TDMA digital channels, and cellular operators are required to continue supporting AMPS users as digital service is introduced. The overall network architecture of the NADC system is much like that of GSM, with a cellular design supported by base stations and switching centers and also with roaming capability supported by home and visitor location registers. However, unlike GSM, not all the NADC network interfaces are standardized. At this writing, only the U_m air interface and the E interface between mobile switching centers have been standardized in Interim Standards IS-54 and IS-41, respectively. The mobile/base station "air interface" specification is defined in Interim Standard 54, and thus is often referred to as the IS-54 standard [EIA92]. As in GSM, the IS-54 standard was developed to efficiently support digital voice service, though digital data services will be provided in later implementations.

In contrast with the constant-envelope modulation schemes used in analog AMPS and in most mobile radio dispatch networks, the IS-54 standard specifies the radio channel modulation as $\pi/4$-shift DQPSK, to be implemented with square-root RC filtering and rolloff parameter 0.35. The principal advantage of this modulation scheme is bandwidth efficiency. The channel transmission rate is 48.6 kbits/sec; and with a channel spacing of 30 kHz, this yields channel utilization of 1.62 bits/sec/Hz, a 20% improvement over GSM. The principal disadvantage of a linear modulation scheme is a poorer power efficiency relative to constant-envelope modulation, which is reflected in the size and weight of hand-portable terminals.

In IS-54, each 30-kHz digital channel has a channel transmission rate of 48.6 kbits/sec. The 48.6-kbit/sec stream is divided into six TDMA channels of 8.1 kbits/sec each. The IS-54 slot and frame format, shown in Fig. 12.5, is much simpler than that of the GSM standard. The 40-msec frame is composed of six

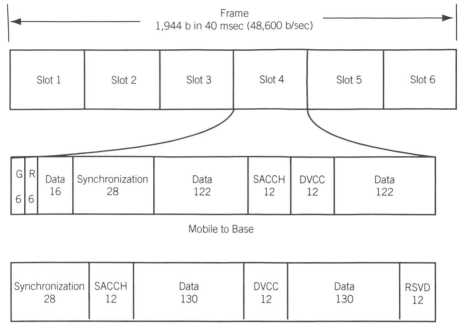

Figure 12.5 IS-54 slot and frame structure. G, guard time; R, ramp time; DVCC, digital verification color code; RSVD, reserved for future use.

6.67-msec time slots. Each slot contains 324 bits, including 260 bits of user data, and 12 bits of system control information in a slow associated control channel (SACCH). There is also a 28-bit synchronization sequence, along with a 12-bit digital verification color code (DVCC) used to identify the frequency channel to which the mobile terminal is tuned. In the mobile-to-base direction, the slot also contains (1) a guard time interval of 6-bit duration when no signal is transmitted and (2) a 6-bit ramp interval to allow the transmitter to reach its full output power level. As in GSM, the IS-54 standard includes provision for a fast associated control channel (FACCH), by usurpation of the 260 bits of user data in a user-channel time slot to convey a FACCH message.

As in the GSM standard, IS-54 defines full-rate and half-rate logical traffic channels. The first phase IS-54 implementations, currently being deployed, provide only full-rate channels, which utilize two slots per 40-msec frame (slots 1 and 4, 2 and 5, or 3 and 6) for a single user channel. This provides 520 bits of user data per 40 msec, for a transmitted data rate of 13 kbits/sec. The full-rate IS-54 speech coder is a form of codebook-excited linear predictive coding called Vector Sum Excited Linear Prediction (VSELP) [Ger90]. The VSELP coder operates at 7.95 kbits/sec, and the addition of error-detection and error-correction coding brings the transmitted bit rate to 13 kbits/sec. In the half-rate IS-54 system, only one slot per 40-msec frame will be allocated to a user channel, and a speech coder (yet to be selected) will be specified that will operate with an error-protected channel bit rate of about 6.5 kbits/sec.

Japanese Digital Cellular. The Japanese Digital Cellular (JDC) system is a digital cellular system that operates at 800/900 MHz and 1.5 GHz. While the spectrum allocations for GSM and IS-54 are 50 MHz for each system, the allocation for JDC is 110 MHz [Nak90, Kin91]. As with GSM and IS-54, the radio channel structure for JDC is a combination of FDM and TDMA. For the 800/900-MHz band the duplex separation is 130 MHz, whereas in the 1.5-GHz band the separation is 48 MHz. The carrier spacing is 25 kHz, which is compatible with the existing analog system in Japan.

The air-interface standard for JDC, established in early 1991, is very similar to IS-54. The modulation method is $\pi/4$-shift QPSK, with a rolloff factor of 0.5, and the channel bit rate is 42 kbits/sec. In the first phase of JDC implementation, there are three TDMA user channels per frame; and in a future implementation, there will be six per frame. In the full-rate system, there are 224 user data bits per 20-msec frame, for a transmitted bit rate of 11.2 kbits/sec, comprising coded voice bits and error-protection coding. In the half-rate phase of implementation, the coded bit rate will be reduced to 5.6 kbits/sec. Details of the JDC frame structure are given in [Nak90] and [Kin91].

Digital Land-Mobile Radio. The Telecommunications Industry Association (TIA) and the Association of Public Safety Communications Officers (APCO) have been collaborating in the development of a new digital standard for next-generation land-mobile radios. This standards effort is known as APCO Project 25. When a standard has been adopted by Project 25, it will then be considered for adoption as a new U.S. Federal Standard, to be known as FedStd 1102. At this writing, a proposed "Common Air Interface Standard" is undergoing review and revision [TIA93a].

North American CDMA Digital Cellular (IS-95). Subsequent to the development of the IS-54 TDMA standard, a proposal was brought forward for a new digital cellular system based upon CDMA technology [CTI89]. This initiative was based on a cellular system design developed by Qualcomm, Inc. [Gil91]. Proponents of the CDMA system projected that the system would provide a significantly larger capacity increase (over analog AMPS) than would be achieved with the IS-54 TDMA standard. Subsequently, the EIA/TIA established subcommittee TR45.5, which began developing a standard based on the Qualcomm CDMA design. The resulting air-interface specification IS-95 was released in July 1993.

In contrast with the IS-54 standard, which uses the same set of carrier frequencies, spaced 30 kHz apart, as are used by the analog AMPS system, the CDMA system uses a spread-spectrum signal with 1.2288-MHz spreading bandwidth, a frequency span equivalent to 41 AMPS channels. (The forward and reverse links are actually using two separate carrier frequencies, spaced 45 MHz apart.) Clearly, this design does not lend itself to selective channel replacement of analog, unlike the TDMA system. Instead, large blocks of channels will be replaced at one time by the CDMA system.

Because in a CDMA system every user is a source of interference to every other user on the system, control of the mobile station power is a critical element of the system design. The Qualcomm CDMA system is a relatively sophisticated design, using several power-setting algorithms to optimize system operation [Vit93, Sal91]. In addition, the system uses (a) powerful error-correction coding with interleaving, (b) speech activity detection and variable-rate speech encoding, and (c) RAKE receiver techniques to maximize system capacity.

A simplified description of the IS-95 CDMA system is given by Fig. 12.6, which shows the signal processing functions for transmission from base station to mobile unit. The underlying data rate for the system is 9600 bits/sec, which represents the speech coder rate of 8550 bits/sec augmented by error-correction coding which is tailored to the speech-coding technique. (The speech coder actually detects speech activity and changes the data rate to lower values during quiet intervals, but 9600 bits/sec is the maximum error-protected data rate.) The 9600-bit/sec stream is segmented into 20-msec blocks and then further encoded with a $K = 9$, rate-1/2

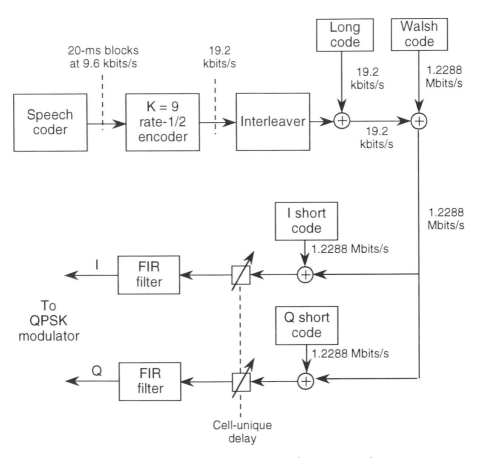

Figure 12.6 IS-95 CDMA cellular system (base to mobile).

convolutional code, bringing the data rate to 19.2 kbits/sec. (In the reverse direction, mobile to base, rate-1/3 convolutional coding is used, which brings the coded data rate up to 28.8 kbits/sec.) The convolutional encoding is followed by interleaving over each 20-msec interval for burst-error protection in the radio channel. The 19.2-kbit/sec stream is then modified by use of a so-called "long code," which serves as a privacy mask. The modified stream is then encoded for spread-spectrum transmission using Walsh codes [Pro89] of dimension 64. This produces a 64-fold spreading of the data stream, resulting in a transmitted bit rate of 1.2288 Mbits/sec. The structure of the Walsh code provides 64 orthogonal sequences, and one of the 64 sequences is assigned to a mobile station during call setup. In this way, 64 orthogonal "channels" are established by CDMA encoding on the same carrier frequency. After Walsh encoding, the spread data stream is separated into I and Q streams, each of which is modified using a unique "short code." The spread-spectrum stream is carried over the air interface with filtered QPSK modulation. In other cells the same carrier frequency is used again, with the same set of 64 Walsh codes, but each cell applies a unique time offset to the two short codes, which allows the CDMA channels in each cell to be uniquely identified. The same two short codes are used by all cell sites and all mobile stations.

When the signal is received, The short codes are removed, using the cell-specific time offset, and the channel-specific Walsh code is decoded by correlation decoding. The rate-1/2 convolutional code is then decoded, producing the received 9600-bit/sec data stream. A more detailed description of the CDMA cellular system can be found in [Gil91].

12.2.2 Low-Power Systems

Here we briefly summarize the characteristics of the principal second-generation cordless systems. Thus far, systems in this category have been deployed primarily in Europe, including the United Kingdom. We address CT2, CT2 Plus, CT-3, and DECT. Key parameters of these systems are shown in Table 12.2 [Mic91].

CT2 Telepoint and CT2 Plus. CT2 is a standard for second-generation cordless telephones. CT2, also known as Telepoint, began as a British initiative, part of the government's drive to engender greater competition in its information technology industry. Telepoint service was initiated commercially in the United Kingdom in September 1989, and has since been introduced in several other European countries as well. Telepoint is the digital successor to the current generation of simple cordless telephones (termed CT-0 and CT-1), allowing use of a single handset in conjunction with home, office, and public base stations. With this system, the user subscribes to public Telepoint service and uses a CT2 phone to gain wireless connection to Telepoint base stations which are in turn connected to the switched telephone network. Charges are billed to the caller's home or office. Public base stations are located in airports, railway stations, and other heavily trafficked areas. The CT2 phone is able to make calls within a range of about 200 m of the base station. The CT2 phone user is able to initiate but not receive calls, and the system cannot provide call handoff to another base station.

TABLE 12.2 Key Parameters of CT and DECT Standards

Parameter	CT2	CT2 plus	CT3	DECT
Bandwidth	4 MHz	4 / 8 MHz	4 / 8 MHz	20 MHz
Band (MHz)	864 – 868	800 – 1000	800 – 1000	1880 – 1900
Access method	FDMA	FDMA	FDMA / TDMA	FDMA / TDMA
Carrier bandwidth (kHz)	100	100	1000	1728
Number of carriers	40	40 / 80	4 / 8	11
Channels per carrier	40	40 / 80	4 / 8	11
Number of channels	40	38 / 76[a]	32 / 64	132
Handoff	No	Yes	Yes	Yes
Two-way calling	[b]	Yes	Yes	Yes
Speech rate (kbits / sec)	32	32	32	32
Maximum cell radius (m)	30 – 100	30 – 100	30 – 100	30 – 100

[a] 2 / 4 Carriers reserved for signaling channels.
[b] Two-way calling possible for public base stations; planned for home and office.

The CT2 standard uses the 864- to 869-MHz band and provides 40 channels, with carrier frequencies spaced 100 kHz apart. The access method is frequency-division multiple access (FDMA); each carrier supports one call, with time-division duplex (TDD) for two-way conversation. The channel bit rate of CT2 is 72 kbits/sec; and its timing structure, shown in Fig. 12.7, is relatively simple. A 2-msec frame contains two time slots, one for portable station to fixed station (mobile to base) and the other for fixed station to portable station. Each slot contains 64 bits of user information and 4 bits of control information. (The standard also allows an alternative multiplex arrangement with only 2 bits of control information per slot and correspondingly longer guard times between slots.) This totals 136 bits per frame, to which CT2 adds two guard times of 8 bits total duration for a grand total of 144 bits or bit intervals per 2-msec frame.

The CT2 modulation technique is binary frequency shift keying (BFSK). With a channel spacing of 100 kHz, the bandwidth efficiency of CT2 is 0.72 bits/sec/Hz. The speech-coding technique is standard adaptive differential pulse code modulation (ADPCM) operating at 32 kbits/sec [CCI84b].

CT2 Telepoint service has been slow to build in the United Kingdom. Initially the reason for this was the insufficient numbers of public base stations. The earliest CT2 Telepoint service providers had set up proprietary (i.e., noncompatible) systems. Thus the subscribers were not able to use the CT2 networks of other service providers, and often could not determine where Telepoint bases were located for their service provider. There was also customer dissatisfaction with the size and complexity of some handsets, and with access ranges.

The service and consumer acceptance are expected to improve as CT2 service providers begin to implement a common air interface (CAI), now approved by ETSI. To achieve a mass market for CT2, it was essential that all operators and manufacturers involved should adhere to a common set of standards. The British Department of Trade and Industry was instrumental in bringing together British manufacturers to produce a common standard for CT2 Telepoint; the standard became known as CAI. Thirteen operators in the United Kingdom and the

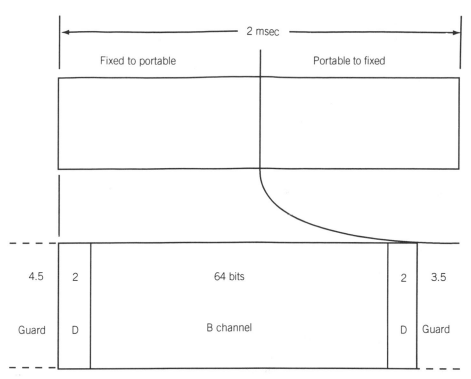

Figure 12.7 CT2 frame and slot structure.

European Common Market signed a Memorandum of Understanding committing themselves to establishing CT2 Telepoint systems using the CAI.

The CAI is highly flexible, and it allows a cordless handset to connect to the PSTN or to an integrated services digital network (ISDN). The same handset can be used in a residential, business, or public access mode. However, in the public access mode, the handset can only originate calls. However, this is a function of the license requirements placed upon the Telepoint operators, rather than a limitation of the technology. CAI is designed to support two-way call establishment and does so in office and residential situations. A major feature of the CAI is its ability to vary transmit power levels in response to particular situations. Under command of the base station, a handset is able to reduce its power output level to maintain good service, without causing interference to other users. The base station determines the power level in the received signal; and if this power is greater than necessary to ensure good service, the handset is instructed to make a power reduction. This capability is most important in an environment of high signal density.

ETSI undertook the task of developing standards for CT2, and this led to disagreements within ETSI as to whether or not CT2 should be recognized by the European Community (EC) as an interim standard for wireless office communications. The issue was resolved in late 1990 in favor of CT2, which has been adopted as an interim European standard for digital cordless telephones pending introduction of the Digital European Cordless Telephone (DECT).

An enhancement of CT2, termed CT2 Plus, offers substantially more capabilities than does CT2. This system, advocated by Northern Telecom, primarily in Canada, is compatible with existing CAI-compliant CT2 handsets and base stations, while introducing common-channel signaling via one or more designated channels for control information. A significant feature of this system is that a subscriber can register with a public base station and receive calls at that location. Thus CT2 Plus is an important step toward the full-service PCN. The British government is planning to upgrade Telepoint to CT2 Plus, while the French Pointel service introduced in 1991 began as a CT2 Plus service.

DECT. The Digital European Cordless Telephone (DECT) standard is a direct contender to CT2, differing significantly from CT2 in its level of sophistication and its range of services offered. The DECT signal structure comprises 10 carriers, at 1.728-MHz spacing, each carrying 12 channels in a time-division multiple-access (TDMA) format. DECT, as does CT2, uses time-division duplex to support a two-way conversation on the same carrier. However, the TDMA signal structure of DECT, in contrast with FDMA in CT2, conveniently provides power savings by being able to turn the signal off intermittently during a call when there is no information to send.

The slot-and-frame format for DECT is shown in Fig. 12.8. The frame duration is 10 msec, with 5 msec for portable-to-fixed station and 5 msec for fixed-to-portable. The transmitter transfers information in signal bursts which it transmits in slots of duration $10/24 = 0.417$ msec. With 480 bits per slot (including a 64-bit guard time), the total bit rate is 1.152 Mbits/sec. Each slot contains 64 bits for system control (C, P, Q, and M channels) and 320 bits for user information (I channel).

The modulation technique specified for DECT is Gaussian minimum shift keying (GMSK), in common with the Pan-European digital cellular standard GSM. The normalized bandwidth of the Gaussian filter is $BT = 0.5$, where B is the 3 dB bandwidth (in Hertz) and T is the inverse of the bit rate. The bandwidth efficiency, 1.152 Mbits/sec in 1.728 MHz, is comparable to that of CT2. In common with CT2, the speech coder in DECT is ADPCM at 32 kbits/sec.

DECT began deployment in late 1992, as the standards were completed. A key difference between DECT and CT2 systems is that DECT, at least initially, is being sold as a complete system to a closed group of users, rather than simply as a handset and service. Unlike CT2, DECT will allow users to receive as well as initiate calls. Furthermore, "seamless" handoff of calls between base stations is an integral part of the DECT standard.

DECT was initiated by Ericsson, sponsored by the governments of Scandinavia, and is strongly supported by Siemens and a number of other European manufacturers. DECT is particularly suited for high-density, small-cell applications such as cordless Private Branch Exchanges (PBXs) that connect the telephones in a private organization to the outgoing lines. DECT has now been formally adopted by the EC as its standard for future advanced cordless and wireless PBX systems for business use. ETSI has been developing standards for DECT, with the goal of implementing it in a relatively modest 20 MHz in the 1.88- to 1.9-GHz frequency band across the EC's 12 member countries; this began in 1992. The EC sees DECT as an important step in the harmonization of telephone standards across Europe, similar in this respect to the GSM standard for digital cellular services [Goo91a].

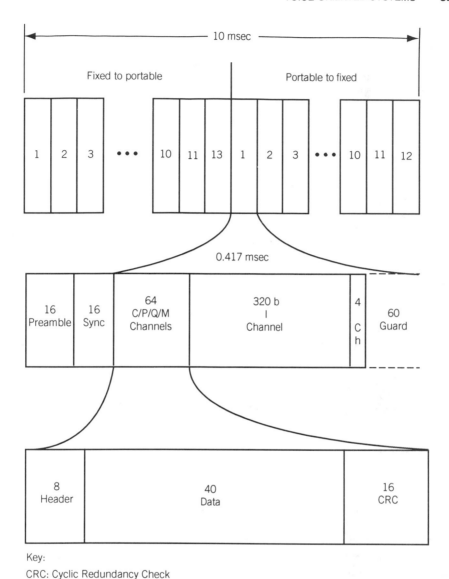

Key:
CRC: Cyclic Redundancy Check
Ch: Parity check bit
Sync: Synchronization

Figure 12.8   DECT slot and frame structure. CRC,

Some observers see DECT as suited primarily for use in office environments (cordless PBX), due to its relatively high data rate and ability to perform seamless handoffs, while CT2 is seen as mainly a telepoint voice service [EDG93]. This is partly due to the fact that with DECT, closely spaced base stations must be synchronized to avoid time slot collisions arising from two or more callers on the same frequency. This is readily achieved in an office environment where there is only one system in operation. In the case of telepoint, there may be two or three different systems with base stations in close proximity where synchronization would be difficult, and there the FDMA structure of CT2 has an advantage.

It is worthwhile noting that DECT represents a standard which approaches the characteristics that users would like to have in a wireless LAN. The data rates needed in a LAN can be provided by combining connections into groups, the maximum data rate achievable being one 614.4-kbit/sec stream suitable for conveying computer data from one access point to another. It typically takes 30 msec to set up and clear such a connection; this limits the practical throughput in a LAN application to about 300 kbits/sec [Hay91].

CT3. The Cordless Telephone CT3 is a hybrid of CT2 and DECT technologies. It is a proprietary technology developed by Ericsson and available as an Ericsson product called DCT900 (Digital Cordless Telephone operating at 900 MHz). CT3 systems became available commercially beginning in 1990.

CT3 has many similarities to DECT. As DECT does, it uses TDMA with time-division multiplexing, but operates at 800–1000 MHz instead of 1.88 GHz (see Table 12.2). CT3 uses dynamic channel allocation with TDMA, which allows all possible time slots to be used in every cell. The DCT900 system provides 16 time slots per 1-MHz spectrum segment, one slot for base- to-mobile and a second slot for mobile-to-base, resulting in 8 channels per carrier frequency. Ericsson comments that one of the major applications of DCT900 is wireless PBX, similar to DECT.

Televerket, the Swedish PTT, and Ericsson have advocated that Ericsson's CT3 system be adopted by ETSI as an interim standard. Many European countries have decided to implement or test CT3 as a predecessor to DECT.

12.3 DATA-ORIENTED SYSTEMS

12.3.1 Mobile Data Systems

In this section we describe existing and future mobile data systems as well as the applications they are intended to serve. In the United States, packet data services currently available for mobile applications include (a) ARDIS, formed by IBM and Motorola, and (b) the RAM Mobile Data Network, which uses Ericsson Mobitex data technology. Future mobile data services, currently the subject of standardization efforts, include (a) Cellular Digital Packet Data (CDPD), designed to transport data as a supplementary service overlaid onto the existing North American analog cellular telephone network, and (b) the EIA/TIA digital cellular standards, IS-54 and IS-95, which will define an array of data services, including both circuit-mode and packet-mode services, for the next-generation North American digital cellular systems. In Europe, the European Technical Standards Institute (ETSI) has begun developing a public standard for trunked radio and mobile data systems, designated as TETRA.

ARDIS. ARDIS (Advanced Radio Data Information Service) is a two-way radio service which was developed as a joint venture between IBM and Motorola and which was first implemented in 1983. The ARDIS network consists of four network control centers with 32 network controllers distributed through 1250 base station

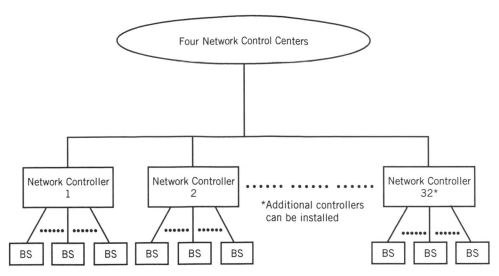

Figure 12.9 ARDIS network architecture.

in 400 cities in the United States. The service is suitable for two-way transfers of data files of size less than 10 kbytes, and much of its use is in support of computer-aided dispatching, such as is used by field service personnel, often while they are on customers' premises. Remote users access the system from laptop radio terminals, which communicate with the base stations. Each of the ARDIS base stations is tied to one of the 32 radio network controllers, as shown in Fig. 12.9. The backbone of the network is implemented with leased telephone lines. The four ARDIS hosts, located in Chicago, New York, Los Angeles, and Lexington, Kentucky, serve as access points for a customer's mainframe computer, which can be linked to an ARDIS host using async, bisync, SNA or X.25 dedicated circuits.

The operating frequency band is 800 MHz, and the radio-frequency (RF) links use separate transmit and receive frequencies separated by 45 kHz. The system was initially implemented with an RF channel data rate of 4800 bits/sec per 25-kHz channel; but this has been upgraded to 19.2 kbits/sec, with a user data rate of about 8000 bits/sec. The system architecture is cellular, with cells overlapped to increase the probability that the signal transmission from a portable transmitter will reach at least one base station. The base station power is 40 W, which provides line-of-sight coverage up to a radius of 10–15 miles. The portable units operate with 4 W of radiated power. The overlapping coverage, combined with (a) designed power levels and (b) error-correction coding in the transmission format, ensures that ARDIS can support portable communications from inside buildings, as well as on the street. This capability for in-building coverage is an important characteristic of the ARDIS service. The modulation technique is 4-ary FSK, the access method is FDMA, and the transmission packet length is 256 bytes.

While the use of overlapping coverage, almost always on the same frequency, does provide reliable radio connectivity, it poses the problem of interference when signals are transmitted simultaneously from two adjacent base stations. The ARDIS network deals with this by turning off neighboring transmitters, for 0.5–1 sec, when an outbound transmission occurs. This scheme has the effect of constraining overall network capacity.

The laptop portable terminals access the network using *data-sense multiple access* (DSMA), discussed in Section 11.5.6. A remote terminal listens to the base station transmitter to determine if a "busy bit" is on or off. When the busy bit is off, the remote terminal is allowed to transmit. However, if two remote terminals begin to transmit at the same time, the signal packets may collide, and retransmission will be attempted, as in other contention-based multiple access protocols. The busy bit lets a remote user know when other terminals are transmitting, and thus reduces the probability of packet collision.

Mobitex. The Mobitex system is a nationwide, interconnected trunked radio network developed by Ericsson and Swedish Telecom. The first Mobitex network went into operation in Sweden in 1986, and other networks have been implemented in Norway, Finland, Canada, the United Kingdom, and the United States. In the United States, Mobitex service was introduced by RAM Mobile Data in 1991 and now covers almost 100 metropolitan areas. While the Mobitex system was designed to carry both voice and data service, the U.S. and Canadian networks are used to provide data service only. Mobitex is an intelligent network with an open architecture which allows establishing virtual networks. This feature facilitates the mobility and expandability of the network [Kil92, Par92].

The Mobitex network architecture is hierarchical, as shown in Fig. 12.10. At the top of the hierarchy is the Network Control Center (NCC), from which the entire network is managed. The top level of switching is a national switch (MHX1) that routes traffic between service regions. The next level comprises regional switches (MHX2s), and below that are local switches (MOXs), each of which handles traffic within a given service area. At the lowest level in the network, multichannel

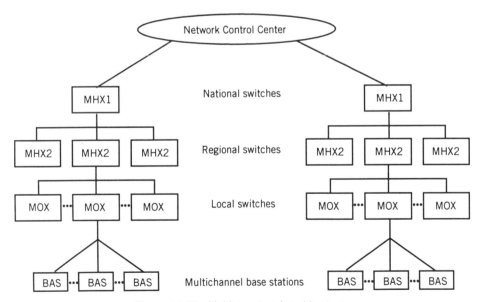

Figure 12.10 Mobitex network architecture.

trunked-radio base stations communicate with the mobile and portable data sets. Mobitex uses packet-switching techniques to allow multiple users to access the same channel at the same time. Message packets are switched at the lowest possible network level. If two mobile users in the same service area need to communicate with each other, their messages are relayed through the local switch, and only billing information is sent up to the network control center.

The base stations are laid out in a grid pattern using the same system engineering rules as are used for cellular telephone systems. In fact, the Mobitex system operates in much the same way as a cellular telephone system, except that handoffs are not managed by the network. That is, when a radio connection is to be changed from one base station to another, the decision is made by the mobile terminal, not by the network computer as in cellular telephone systems.

To access the network, a mobile terminal finds the base station with the strongest signal and then registers with that base station. When the mobile terminal enters an adjacent service area, it automatically re-registers with a new base station, and the user's whereabouts are relayed to the higher-level network nodes. This provides automatic routing of messages bound for the mobile user, a capability known as *roaming*. The Mobitex network also has a store-and-forward capability.

The mobile units transmit at 896–901 MHz, and the base stations transmit at 935–940 MHz (SMR band). The system uses dynamic power setting, in the range of 100 mW to 10 W for mobile units and 100 mW to 4 W for portable units. The GMSK modulation technique is used, with noncoherent demodulation. The transmission rate is 8000 bits/sec half duplex in 12.5-kHz channels, and the service is suitable for file transfers up to 20 kbytes. The packet size is 512 bytes with 1 to 3-sec delay. Forward-error correction, as well as retransmissions, are used to ensure the bit-error-rate quality of delivered data packets. The system uses the dynamic slotted-ALOHA random-access method, discussed in Chapter 11.

RAM Mobile is planning to have installed approximately 800 base stations in 100 metropolitan areas throughout the United States by the end of 1993, providing coverage to about 90% of the United States. By locating its base stations close to major business centers, the RAM Mobile system provides a degree of in-building signal coverage.

CDPD. The Cellular Digital Packet Data (CDPD) system is designed to provide packet data services as an overlay to the existing analog cellular AMPS network. CDPD was developed by IBM in collaboration with nine cellular carriers: McCaw, GTE, Contel Cellular, Ameritech, Bell Atlantic, NYNEX, Pacific Telesis, Southwestern Bell, and US West. These companies will cover 95% of the United States, including all major urban areas. A basic goal of the CDPD system is to provide data services on a noninterfering basis with the existing cellular telephone services using the same 30-kHz channels. To do this, CDPD is designed to make use of AMPS channels that are not being used for voice traffic, as well as to move to another channel when the current channel is allocated to voice service. The compatibility of CDPD with the existing cellular telephone system allows it to be installed in any analog cellular system in North America, providing data services that are not dependent upon support of a digital cellular standard in the service area. A preliminary field demonstration system was operated during the second half of 1992, and the consortium issued a specification for the CDPD system [CDPD93]. Commercial service began in 1994. Intended applications for CDPD service include: electronic mail, package delivery tracking, inventory control, credit card verification, security reporting, vehicle theft recovery, traffic and weather advisory services, and a potentially wide range of information retrieval services.

Although CDPD cannot increase the number of channels usable in a cell, it can provide an overall increase in user capacity if data users use CDPD instead of voice channels. This capacity increase would result from the inherently greater efficiency of a connectionless packet data service relative to a connection-oriented service, given bursty data traffic. That is, a packet data service does not require the overhead associated with setup of a voice traffic channel in order to send one or a few data packets. In the following paragraphs we briefly describe the CDPD network architecture and the principles of operation of the system. Our discussion follows [Qui93] closely.

The basic structure of a CDPD network (Fig. 12.11) is similar to that of the cellular network with which it shares transmission channels. Each mobile end system (M-ES) communicates with a mobile database station (MDBS) using the protocols defined by the air-interface specification, to be described below. The MDBSs are expected to be collocated with the cell equipment providing cellular telephone service, to facilitate the channel-sharing procedures. All the MDBSs in a service area will be linked to a mobile data intermediate system (MD-IS) by microwave or wireline links. The MD-IS provides a function analogous to that of the Mobile Switching Center (MSC) in a cellular telephone system. The MD-IS

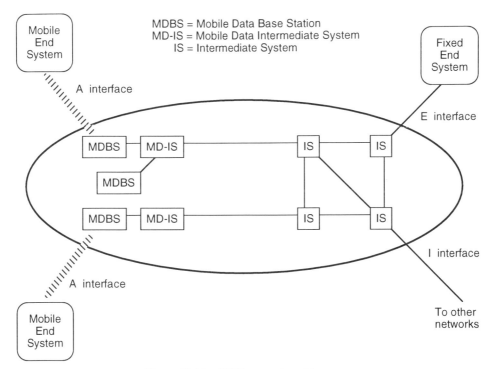

Figure 12.11   CDPD network architecture.

may be linked to other MD-ISs and to various services provided by end-systems outside the CDPD network. The MD-IS also provides a connection to a network management system, and it supports protocols for network management access to the MDBSs and M-ESs in the network.

Service endpoints can be local to the MD-IS or remote, connected through external networks. An MD-IS can be connected to any external network supporting standard routing and data exchange protocols. An MD-IS can also provide connections to standard modems in the PSTN by way of appropriate modem interworking functions (modem emulators). Connections between MD-ISs allow routing of data to and from M-ESs that are roaming—that is, operating in areas outside their home service areas. These connections also allow MD-ISs to exchange information required for mobile terminal authentication, service authorization, and billing.

CDPD employs the same 30-kHz channelization as is used in existing analog cellular systems throughout North America. Each 30-kHz CDPD channel will support channel transmission rates up to 19.2 kbits/sec. However, packet collisions and typical radio channel conditions will limit the actual information payload throughput rate to lower levels and will introduce additional time delay due to the error-detection and retransmission protocols.

The CDPD radio link physical layer uses GMSK modulation at the standard cellular carrier frequencies, on both forward and reverse links. The Gaussian pulse-shaping filter is specified to have bandwidth-time product $B_b T = 0.5$. The specified B_T product ensures a transmitted waveform with bandwidth narrow enough to meet adjacent-channel interference requirements, while keeping the intersymbol interference small enough to allow simple demodulation techniques. The choice of 19.2 kbits/sec as the channel bit rate yields an average power spectrum that satisfies the emission requirements for analog cellular systems and for dual-mode digital cellular systems.

The forward channel carries data packets transmitted by the MDBS, whereas the reverse channel carries packets transmitted by the M-ESs. In the forward channel the MDBS forms data frames by adding standard HDLC terminating flags and inserted zero bits, and then it segments each frame into blocks of 274 bits. These 274 bits, together with an 8-bit *color-code* for MDBS and MD-IS identification, are encoded into a 378-bit coded block using a (63,47) Reed–Solomon code over a 64-ary alphabet. A 6-bit synchronization and flag word is inserted after every 9 code symbols. The flag words are used for reverse link access control. The forward link block structure is shown in Fig. 12.12.

In the reverse channel, when an M-ES has data frames to send, it formats the data with flags and inserted zeros in the same manner as in the forward link. That is, the reverse link frames are segmented and encoded into 378-bit blocks using the same Reed–Solomon code as in the forward channel. The M-ES may form up to 64 encoded blocks for transmission in a single reverse channel transmission burst. During the transmission, a 7-bit transmit continuity indicator is interleaved into each coded block; it is set to all ones to indicate that more blocks follow, or all zeros to indicate that this is the last block of the burst. The reverse channel block structure is shown in Fig. 12.13.

The media access control (MAC) layer in the forward channel is relatively simple. The receiving M-ES removes the inserted zeros and HDLC flags and

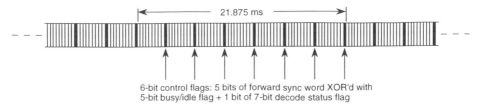

Figure 12.12 CDPD forward link block structure. The base station transmission is broadcast to all mobiles.

reassembles data frames that were segmented into multiple blocks. Frames are discarded if any of their constituent blocks are received with uncorrectable errors.

On the reverse channel (M-ES to MDBS), access control is more complex, because several M-ESs must share the channel. CDPD uses the multiple-access technique called digital-sense multiple access (DSMA), which is closely related to CSMA/CD, as discussed in Chapter 11.

The network layer and higher layers of the CDPD protocol stack are based on standard ISO and Internet protocols. It is expected that the earliest CDPD products will use the Internet protocols.

The selection of a channel for CDPD service is accomplished by the radio resource management entity in the MDBS. Through the network management system, the MDBS is informed of the channels in its cell or sector that are available as potential CDPD channels when they are not being used for analog voice service. There are two ways in which the MDBS can determine whether the channels are in use. If a communication link is provided between the AMPS system and the CDPD system, the AMPS system can inform the CDPD system directly about channel usage. If such a link is not available, the CDPD system can use a forward power monitor ("sniffer" antenna) to detect channel usage on the AMPS system. Circuitry to implement this function can be built into the cell sector interface.

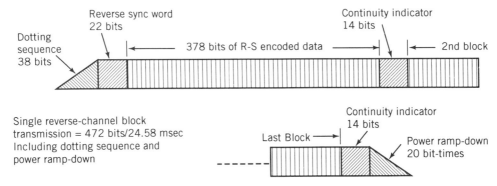

Figure 12.13 CDPD reverse link block structure. Mobiles transmit to the base station with a contention-based multiple-access protocol.

Digital Cellular Data Services. As discussed in Section 12.2.1, the Cellular Telecommunications Industry Association (CTIA) and the Telecommunications Industry Association (TIA) have been developing standards for new digital cellular systems to replace the existing analog AMPS systems. Two air-interface standards have now been published. The IS-54 standard specifies a three-slot TDMA system, and the I-95 standard specifies a CDMA spread-spectrum system. In both systems, a variety of data services are being planned.

The general approach taken in the definition of IS-95 data services has been to base the services on standard data protocols, to the greatest extent possible [Tie93]. The previously specified physical layer of the IS-95 protocol stack was adopted for the physical layer of the circuit-switched data services, with an appropriate radio link protocol (RLP) overlaid. The current standardization effort is directed to defining three primary services: (1) asynchronous data, (2) Group 3 facsimile, and (3) packet data access carried over a circuit-mode data call connection. Later, attention will be given to other services, including synchronous data and connectionless packet data services.

IS-95 asynchronous data will be structured as a circuit-switched service. For circuit-switched connections, a dedicated path is established between the data devices for the duration of the call. This type of service will provide point-to-point connectivity with standard PCs or fax devices in the PSTN. There are several applications which fall into this category. For file transfer involving PC-to-PC communications the Asynchronous Data Service is the desired cellular service mode. The service will employ an RLP to protect data from transmission errors caused by radio channel degradations at the air interface. The RLP employs automatic repeat request (ARQ), forward error correction (FEC), and flow control. Flow control and retransmission of data blocks with errors are used to provide an improved error performance in the mobile segment of the data connection at the expense of variations in throughput and delay. Typical raw channel rates for digital cellular transmission are measured at approximately a 10^{-2} bit-error rate. However, acceptable data transmission usually requires a bit-error rate of approximately 10^{-6}, and achieving this will require the design of efficient ARQ and error-correction codes to deal with error characteristics in the mobile environment.

At this writing, the TIA CDMA data services Task Group has drafted interface standards for asynchronous data and for Group 3 fax services. Work on defining packet data service standards has recently begun. In parallel with the CDMA data services effort, another TIA task group, TR45.3.2.5, has been defining standards for digital data services for the TDMA digital cellular standard IS-54 [Sac92, Wei93]. As with the IS-95 data services effort, initial priority is being given to standardizing circuit-mode asynchronous data and Group 3 facsimile services. A standard for asynchronous data service was released in mid-November of 1994 [TIA94].

TETRA. As is the case in the United States and Canada, there is interest in Europe in establishing fixed wide-area standards for mobile data communications. While the GSM system will provide an array of data services, data will be handled as a circuit-switched service, consistent with the primary purpose of GSM as a voice service system. Therefore the European Telecommunications Standards Institute (ETSI) has begun developing a public standard for trunked radio and

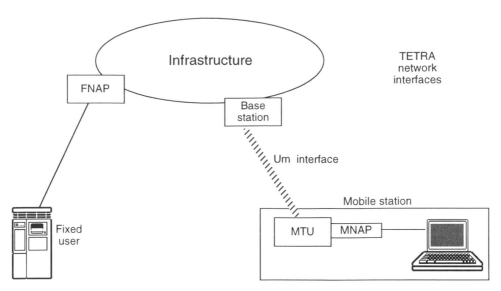

Figure 12.14   TETRA network.

mobile data systems. The standards, which are known generically as Trans-European Trunked Radio (TETRA), are the responsibility of the ETSI RES 6 Sub Technical Committee [Hai92].

TETRA is being developed as a family of standards. One branch of the family is a set of radio and network interface standards for trunked voice (and data) services. The other branch is an air-interface standard which is optimized for wide-area packet data services for both fixed and mobile subscribers and which supports standard network access protocols. Both versions of the standard will use a common physical layer, based on $\pi/4$-DQPSK modulation operating at a channel rate of 36 kbits/sec in each 25-kHz channel.

Figure 12.14 is a simplified model of the TETRA network, showing the three interfaces at which services will be defined. The U_m interface is the radio link between the base station (BS) and the mobile station (MS). At the other side of the network is the fixed network access point (FNAP) through which mobile users gain access to fixed users. Fixed host computers and fixed data networks typically use standardized interfaces and protocols, and it is intended that the mobile segments of the TETRA network will utilize these same standards. Finally, there is the interface to the mobile data user, the mobile network access point (MNAP). This is shown in the figure as a physical interface between a mobile terminating unit (MTU) and a data terminal. However, the MNAP may be only a logical interface and not a physical interface, if the MS is an integrated device with no external data port. It is envisioned that all three interfaces—FNAP, MNAP, and the air interface U_m—will support data services with packet-mode protocols.

It is planned that TETRA will provide both connection-oriented and connectionless data services [Hai92]. While work is still in progress on defining protocols for the three interfaces in the TETRA network model, some broad decisions have been made. It has been decided that the physical data port at the

TABLE 12.3 Characteristics and Parameters of Five Mobile Data Services

System:	ARDIS	Mobitex	CDPD	IS-95[a]	TETRA[a]
Frequency Band					
Base to mobile (MHz):	(800 band,	935 – 940[b]	869 – 894	869 – 894	(400 and 900 bands)
Mobile to base (MHz):	45-kHz sep.)	896 – 901	824 – 849	824 – 849	
RF channel spacing:	25 kHz (U.S.)	12.5 kHz	30 kHz	1.25 MHz	25 kHz
Channel access / multiuser access:	FDMA / DSMA	FDMA / dynamic-S-ALOHA	FDMA / DSMA	FDMA / CDMA-SS	FDMA / DSMA and SAPR
Modulation method:	FSK, 4-FSK	GMSK	GMSK	4-PSK / DSSS	π / 4-QDPSK
Channel bit rate (kbits / sec):	19.2	8.0	19.2	9.6	36
Packet length:	Up to 256 bytes (HDLC)	Up to 512 bytes	24 – 928 bits	(Packet service-TBD)	192 Bits (short) 384 Bits (long)
Open architecture:	No	Yes	Yes	Yes	Yes
Private or public carrier:	Private	Private	Public	Public	Public
Service coverage:	Major metropolitan areas in United States	Major metropolitan areas in United States	All AMPS areas	All CDMA cellular areas	European trunked radio
Type of coverage:	In-building and mobile	In-building and mobile	Mobile	Mobile	Mobile

[a]IS-95 and TETRA data services standardization in progress.
[b]Frequency allocations in the United States (In the United Kingdom, 380 to 450-MHz band is used.)

Note: DSMA, data-sense multiple access; DSSS, direct sequence spread spectrum; S-ALOHA, slotted ALOHA; SAPR, slotted-ALOHA packet reservation.

mobile station will be a true X.25 interface, which means that any attached data terminal must implement the X.25 protocol. This will provide true peer-to-peer communication between the mobile and fixed ends of the call connection. For connectionless data services, the protocol definition is still incomplete, but it is envisioned that the protocols will be based on ISO standards for connectionless-mode network service [ISO87].

The protocol for the radio link has not yet been defined, but it will certainly employ a combination of forward-error correction (FEC) coding, CRC error-detection coding, and an ARQ scheme. It has been reported that the TETRA standard is being designed to accommodate two popular forms of multiuser access, namely, slotted ALOHA (with and without packet reservation), and data-sense multiple access.

Table 12.3 compares the chief characteristics and parameters of the five wireless data services described above.

12.3.2 Wireless LANs

Three distinct technologies are currently in use for wireless LAN (WLAN) systems: infrared light systems, licensed cellular systems operating at 18–19 GHz [Buc91], and unlicensed spread-spectrum systems operating in the ISM bands (902–908 MHz, 2.4–2.5 GHz, and 5.8–5.9 GHz) [Mar91]. We summarize the chief characteristics of current WLAN systems in Table 12.4, where we distinguish between diffused (DFIR) and directed-beam infrared (DBIR) systems. A variety of systems are available in the marketplace, and there are no industry standards currently in place.

An important factor in selection of a WLAN system is the ease of installation and flexibility of the software to accommodate various protocols for interconnecting with the backbone-wired LANs. The evolving next generation of WLANs is designed to be incorporated into the laptop, notebook and pen-pad computers now coming into the market, where significant reductions in size and power consumption are needed. It will be ideal if these devices also provide low-speed wireless access with wide-area coverage. In the following paragraphs we discuss the different categories of WLAN systems and cite a few examples of products currently available in the market.

Infrared. The transmission technology used in IR LANs is the same as that found in familiar consumer electronic devices used for remote control of television sets,

TABLE 12.4 A Comparison of Available Radio and IR Wireless LANs

Technique:	DF / IR	DB / IR	RF	RF / SS
Data rate:	1 Mbit / sec	10 Mbits / sec	15 Mbits / sec	2 – 6 Mbits / sec
Mobility:	Good	None	Better	Best
Detectability:	Negligible	Negligible	Some	Little
Range:	70 – 200 ft	80 ft	40 – 130 ft	100 – 800 ft
Frequency / wavelength:	$\lambda = 800 - 900$ nm	$\lambda = 800 - 900$ nm	$f = 18$ GHz	$f = 0.9, 2.4, 5.7$ GHz
Radiated power:	—	—	25 mW	< 1 W

VCRs, and other entertainment products, as well as in wireless keyboards and wireless printer connections for laptop computers. Transmitters are simple and inexpensive LEDs, and receivers are similarly simple avalanche photodiodes (APDs) or p-intrinsic-n (PIN) photodiodes. In addition to relatively simple design, IR LANs offer considerable flexibility of installation, because they do not come under FCC regulation and they neither interfere with, nor are impacted by, radio systems. Furthermore, since IR systems operate at very low power levels, separate IR LANs are not likely to interfere with each other, unless placed very close together. Unlike radio transmissions, IR signals will not penetrate walls, and this is advantageous in applications where signal confinement within a walled office is desirable. IR LANs can suffer interference from the sun or other light sources, but this can be controlled by careful placement of system components and by the use of filters, as discussed in Chapter 10. Disadvantages for IR LANs in some applications include limited range relative to radio LANs and a sensitivity to shadowing. Diffused and directed-beam IR LANs offer somewhat different characteristics [Kah91], which can be traded off for different applications.

DFIR LANs offer simplicity of installation, because there is no need for a direct line-of-sight path between transmitter and receiver. Receivers can collect signals reflected from walls or a ceiling, or from a passive reflector mounted in a central location on a ceiling or high on a wall, and this type of system is generally not susceptible to signal blockage. DFIR LANs provide moderate data rates and coverage, because transmitted signal power is scattered over a considerable area and because the consequent multipath limits achievable data rates. As indicated in Table 12.4, data rates for currently available DFIR LANs are limited to about 1 Mbit/sec. DFIR LANs are suitable for moderate-size offices and short-distance battery-oriented applications, such as portable-to-printer links. Most IR LANs developed over the last 15 years have utilized DFIR transmission, and IEEE 802.11 is developing standards for DFIR LANs. A current example is Photonics Corporation's PhotoLink system, which provides data rates up to 1 Mbit/sec.

Within the last few years, much attention has been given to DBIR technology for wireless networks, and DBIR LANs have begun to appear on the market. DBIR LANs use focused beams, and transmitter and receiver must be carefully aligned. With directed-beam operation, this type of IR LAN provides higher received signal-to-noise ratio and less multipath dispersion than is found with DFIR LANs, and thus longer ranges and higher data rates can be achieved. Because of the need for alignment, DBIR LANs are best suited for applications employing fixed terminals. DBIR LANs are suitable for large-file transfers between main frames and servers as well as large open offices with many fixed terminals. An example is the InfraLAN Token-Ring System, manufactured by InfraLAN Technologies, Inc. As its name implies, the system uses the token-ring multiple-access protocol and is in fact fully compliant with the IEEE 802.5 token ring standard. Each pair of optical nodes in the ring layout is aligned with the aid of LED indicator lights. At this writing, InfraLAN products are available at data rates of 4 and 16 Mbits/sec.

Microwave LANs. WLANs operating at tens of gigahertz offer higher data rates than are currently achievable with most IR LAN products. In particular, WLAN products have been developed for operation at 18–19 GHz, where a number of frequencies have been allocated by the FCC for such low-power applications. These frequencies are licensed and coordinated within geographic areas to prevent interference between systems. The signal power levels for these systems are fractions of a Watt, and the signals are confined within a general area in which the network is to be used. The nature of propagation at these frequencies is such that signals are significantly constrained by dense structures such as concrete floors or many building walls. Therefore signal transmission can be very satisfactory throughout an open office area but confined enough by floors and walls to permit reuse of the same frequencies in other portions of the same building or office complex, without interference between systems. A disadvantage of the 18-GHz technology is that the radio transmitters and receivers for this band are rather specialized and expensive.

At this writing, in the United States only Motorola is manufacturing WLAN products for the 18-GHz band. Their main WLAN products are the Altair and Altair Plus systems, which are deployed in microcells, much like cellular telephone systems [Fre91a, b]. Because of the attenuation of the 18-GHz transmissions by floors and walls, multiple microcells reuse the same frequencies in different parts of a building. The Altair products provide compatibility with the IEEE 802.3 Ethernet protocol, and they also provide transmission rates as high as 15 Mbits/sec. These products are suitable for high speed communications in large partitioned areas such as large offices buildings or complexes, or similar open areas such as some libraries. In Europe, the HIPERLAN committee is considering proposals for WLAN allocations at microwave frequencies (5.2 and 17.1 GHz), and thus microwave LAN product developments can be expected to follow.

Spread-Spectrum LANs. Spread-spectrum WLANs designed for use in the 900-MHz, 2-GHz, and 5-GHz ISM bands can be operated without the need for FCC licensing. These systems are limited to power levels less than 1 W and are typically intended to provide signal coverage up to about 500–800 ft. This is generally larger coverage than is provided by either IR or microwave LANs. The use of spread-spectrum transmission, combined with effective multi-user access protocols such as CSMA, make it feasible to deploy multiple systems in the same general area, even though signal coverage is overlapping. Also, these systems can be deployed with special directional antennas, allowing longer distances to be spanned, such as links between buildings on a campus or in an office park. The freedom from licensing requirements makes deployment of ISM-band LANs very convenient. However, a drawback of unlicensed operation is that one must be able to operate with interference from other users of the same bands. The IEEE 802.11 committee is developing WLAN standards for operation at transmission rates of 1 Mbits/sec and 2 Mbits/sec in the 2.4–2.485 GHz band. The standards will specify the two data rates for both frequency-hopped and direct-sequence spread spectrum (FHSS and DSSS) radio transmission (as well as 1 Mbit/sec for DFIR transmission). The standards will deal with the mutual-inteference problem by providing five overlapping sub-bands for DSSS operation and three hopping patterns for FHSS operation, all selectable by the user.

The propagation characteristics in the ISM bands make these products suitable for applications where penetration through building floors is desired. Thus spread-spectrum WLANs are well-suited for small-business applications where a few terminals are distributed over several floors of a building and can be served by a single system. At this time, a number of spread-spectrum LANs are available in the market. As examples, we will briefly mention two systems: NCR's WaveLAN and Windata's FreePort, representing two different network configurations.

The NCR WaveLAN product is a spread-spectrum wireless LAN system that uses the 902- to 928–MHz band and operates at data rates up to 2 Mbits/sec. The WaveLAN hardware is a network adapter card that plugs into each PC in the network. The card accommodates a module housing multiple antennas. The network has a peer-to-peer topology, with one special terminal functioning as server or access point connecting the other terminals to the backbone network. The access method employed is CSMA/CA with exponential backoff. Depending upon the physical constraints of the environment, WaveLAN can operate over distances from 100 to 800 ft.

Windata's FreePort spread-spectrum LANs operate in the 2.400- to 2.485–GHz and 5.725- to 5.850–GHz bands. FreePort systems are composed of a central hub and wireless transceiver units that each link one or more computers. The transceivers transmit data at 2.440 GHz, and the hub transmits data at 5.780 GHz. The FreePort System is fully IEEE 802.3 Ethernet compatible and can be integrated into existing wired Ethernet backbone networks. The FreePort system is able to resolve and adaptively combine multipath signals to optimize signal reception. FreePort systems provide data rates up to 16 Mbits/sec and cover ranges up to about 260 ft in an office environment. Results of performance evaluation of spread-spectrum wireless LANs is available in [Bro93].

QUESTIONS

(a)   How does the bandwidth efficiency of GSM compare with that of IS-54?

(b)   Although many aspects of IS-54 are similar to those of GSM, the transmission bandwidth per TDMA channel is 30 KHz, more than six time smaller than that of GSM. Explain why this difference came about.

(c)   How does GSM interface to ISDN and the PSTN?

(d)   What are the basic blocks of a mobile radio network?

(e)   What is the advantage of defining standardized interfaces within a wireless network?

(f)   How many interfaces are defined by the GSM standard, and what are they?

(g)   Which system, GSM or IS-54, is more vulnerable to multipath fading effects? Explain.

(h)   What is the frame size and the number of TDMA slots per frame in GSM, and how does it compare with the IS-54 format?

(i)   What are the differences among the speech coding rates in GSM, IS-54, and JDC?

(j)   How many layers of coding are used in the Qualcomm CDMA system? What is the type and the purpose of each of these coding layers?

(k)   What is the spreading factor of the codes used in the Qualcomm CDMA?

(l)   What is the maximum data rate supportable by each channel in the Qualcomm CDMA system?

(m)   Compare the speech coding rate, access method and transmission power of the CT2 Telepoint service with the GSM digital cellular system.

(n)   From the user's point of view, what are the differences between a CT2 service and a GSM service?

(o)   Compare the speech coding rate, access method, transmission power, and the frequency of operation of the CT2 and the DECT systems.

(p)   From the user's point of view, what are the differences between a CT2 service and a DECT service?

(q)   Discuss the differences between the DECT and the DCS-900 systems.

(r)   Compare the data rate per user, frequency of operation, transmission power, and the access method of ARDIS and IS-54.

(s)   What are the similarities and the differences between the ARDIS and the Mobitex systems?

(t)   What are the basic blocks of the CDPD system?

(u)   Compare the access methods of CDPD, ARDIS, and Mobitex.

(v)   Name two major applications for circuit-switched data services.

(w)   What is the most popular transmission technology for wireless LANs?

(x)   What is the major drawback in using spread spectrum for wireless LANs?

(y)   What transmission technologies are expected to evolve in unlicensed data PCS bands?

(z)   Name the two major technical specifications for wireless LANs.

REFERENCES

[Abr68] M. Abramowitz and I. A. Stegun, *Handbook of Mathematical Functions*, 5th ed., Dover, New York (1968).

[Abr70] N. Abramson, "The ALOHA System—Another Alternative for Computer Communications," *AFIPS Conf. Proc., Fall Joint Comput. Conf.*, 37, 281–285 (1970).

[Abr73] N. Abramson, "The ALOHA System," in *Computer-Communication Networks*, N. Abramson and F. Kuo (Eds.), Prentice–Hall, Englewood Cliffs, NJ (1973).

[Abr77] N. Abramson, "The Throughput of Packet Broadcasting Channels," *IEEE Trans. Commun.*, **COM-25**, 117–128 (1977).

[Aca87] A. S. Acampora and J. H. Winters, "A Wireless Network for Wide-Band Indoor Communications," *IEEE J. Selected Areas Commun.*, **SAC-5**, 796–805 (1987).

[Ada92] E. Adams and C. R. Frank, "WARC Embraces PCN," *IEEE Commun. Mag.*, **30**, No. 6, 44–46 (1992).

[Ahm74] N. Ahmed, T. Natarajan, and K. Rao, "Discrete Cosine Transform," *IEEE Trans. Comput.*, **C-23**, 90–93, (1974).

[Aka87] Y. Akaiwa and Y. Nagata, "Highly Efficient Digital Mobile Communications with a Linear Modulation Method," *IEEE J. Selected Areas Commun.*, **SAC-5**, 890–895 (1987).

[Ake88] D. Akerberg, "Properties of a TDMA Pico-Cellular Office Communication System," *Proc. IEEE GLOBECOM*, Hollywood, FL, 1343–1349 (Dec. 1988).

[Ale82] S. E. Alexander, "Radio Propagation Within Buildings at 900 MHz," *IEE Electron. Lett.*, **18**, 913–914 (1982).

[Ale83a] S. E. Alexander, "Characterizing Buildings for Propagation at 900 MHz," *IEE Electron. Lett.*, **19**, 860 (1983).

[Ali94] M. H. Ali, A. S. Parker, and K. Pahlavan, "Time and Frequency Domain Modeling of Wideband Radio Propagation for Personal, Mobile and Indoor Applications," *Proc. IEEE PIMRC'94*, The Hague, Netherlands (1994).

[Amo80] F. Amoroso, "The Bandwidth of Digital Data Signals," *IEEE Commun. Mag.*, **18**, No. 6, 13–24 (1980).

[And93] H. R. Anderson, "A Ray-Tracing Propagation Model for Digital Broadcast Systems in Urban Areas," *IEEE Trans. Broadcasting*, **AB-39**, 309–317 (1993).

[Ank80] H. A. Ankerman, "Transmission of Audio Signals by Infrared Light Carrier," *SMPTE J.*, **89**, 834–837 (1980).

[ANS85] ANSI/IEEE Std. 802.3-1985, "Carrier Sense Multiple Access with Collision Detection (CSMA/CD)," 1985.

[ANS88] ANSI/IEEE STd. 802.3a,b,c,e-1988, "Supplement to Carrier Sense Multiple Access with Collision Detection (CSMA/CD)," 1988.

[Arn87] J. Arnbak and W. Blitterswijk, "Capacity of Slotted ALOHA in Rayleigh-Fading Channels," *IEEE. J. Selected Areas Commun.*, **SAC-5**, 261–269 (1987).

[Arn89] H. W. Arnold, R. L. Murray, and D. C. Cox, "815 MHz Radio Attenuation Measured Within Two Commercial Buildings," *IEEE Trans. Antennas Propag.*, **37**, 1335–1339 (1989).

[Arn93] J. C. Arnbak, "The European (R)evolution of Wireless Digital Networks," *IEEE Commun. Mag.*, **31**, No. 9, 74–82 (1993).

[Arn73] G. A. Arredondo, W. H. Chriss, and E. H. Walker, "A Multipath Fading Simulator for Mobile Radio," *IEEE Trans. Commun.*, **COM-21**, 1325–1328 (1973).

[Art62] E. Arthurs and H. Dym, "On the Optimum Detection of Digital Signals in the Presence of White Gaussian Noise," *IEEE Trans. Commun. Syst.*, **COM-10**, 336–372 (1962).

[Ata79] B. S. Atal, and M. R. Schroeder, "Predictive Coding of Speech Signals and Subjective Error Criteria," *IEEE Trans. ASSP*, **ASSP-27**, 247–254 (1979).

[Ata82] B. S. Atal and J. R. Remde, "A New Model of LPC Excitation for Producing Natural-Sounding Speech at Low Bit Rates," *Proc. ICASSP '82*, 614–617 (1982).

[Ata84] B. S. Atal and M. R. Schroeder, "Stochastic Coding of Speech Signals at Very Low Bit Rates," *Proc. IEEE ICC '84*, 1610–1613 (May 1984).

[Bar86] P. J. Barry and A. G. Williamson, "Radiowave Propagation Into and Within a Building at 927 MHz," *IEE Electron. Lett.*, **22**, 248–249 (1986).

[Bar91a] J. R. Barry et al., "Simulation of Multipath Impulse Response for Indoor Diffuse Optical Channel," *Proc. IEEE Workshop on Wireless LANs*, Worcester, MA, 81–87 (1991).

[Bar91b] J. R. Barry et al., "High-speed Nondirective Optical Communication for Wireless Networks," *IEEE Network Mag.* **Nov.**, 44–54 (1991).

[Bel63a] P. A. Bello, "Characterization of Randomly Time-Variant Linear Channels," *IEEE Trans. Commun. Syst.*, **CS-11**, 360–393 (1963).

[Bel63b] P. A. Bello and B. D. Nelin, "The Effect of Frequency Selective Fading on the Binary Error Probabilities of Incoherent and Differentially Coherent Matched Filter Receivers," *IEEE Trans. Commun. Syst.*, **CS-11**, 170–186 (1963).

[Bel65] P. A. Bello, "Selective Fading Limitations of the Kathryn Modem and Some System Design Considerations," *IEEE Trans. Commun. Technol.*, **COM-13**, 320–333 (1965).

[Bel69] P. A. Bello, "A Troposcatter Channel Model," *IEEE Trans. Commun. Tech.*, **COM-17**, 130–137 (1969).

[Bel79a] Bell Laboratories, "Special Issue on Advanced Mobile Phone Service," *Bell Syst. Tech. J.*, **58**, 1–269 (1979).

[Bel79b] C. A. Belfiore and J. H. Park, Jr., "Decision-Feedback Equalization," *Proc. IEEE*, **67**, 1143–1156 (Aug. 1979).

[Bel83] J. Bellamy, *Digital Telephony*, Wiley, New York (1983).

[Bel84] P. A. Bello and K. Pahlavan, "Adaptive Equalization for SQPSK and SQPR Over Frequency Selective Microwave LOS Channels," *IEEE Trans. Commun.*, **COM-32**, 609–615 (1984). (Also presented at *IEEE ICC '82*.)

[Bel88] P. A. Bello, "Performance of Some RAKE Modems Over the Non-disturbed Wide Band HF Channel," *IEEE MILCOM '88 Conf. Rec.*, 89–95 (Oct. 1988).

[Bha81] V. K. Bhargava et al., *Digital Communications by Satellite—Modulation, Multiple Access and Coding*, Wiley, New York (1981).

[Big91] E. Biglieri et al., *Introduction to Trellis-Coded Modulation with Applications*, Macmillan, New York (1991).

[Bin75] R. Binder, "A Dynamic Packet Switching System for Satellite Broadcast Channels," *Proc. IEEE ICC '75*, San Francisco (June 1975).

[Bin88] J. Bingham, *The Theory and Practice of Modem Design*, Wiley, New York (1988).

[Bin90] J. A. C. Bingham, "Multicarrier Modulation for Data Transmission: An Idea Whose Time Has Come," *IEEE Commun. Mag.*, **28**, No. 5, 5–14 (1990).

[Bla91] U. Black, *The V Series Recommendations*, McGraw–Hill, New York (1991).

[Ble80] F. H. Blecher, "Advanced Mobile Phone Service," *IEEE Trans. Veh. Technol.*, **VT-29**, 238–244 (1980).

[Bod84] D. Bodson, G. F. McClure, and S. R. McConoughey (Eds.), *Land-Mobile Communications Engineering*, Selected Reprint Series, IEEE Press, New York (1984).

[Bor93] D. Borth, "Slow Frequency CDMA for High-Mobility Personal Lan Systems," *Proc. Allerton Conf.*, Allerton House, Monticello, IL, 325–334 (1993).

[Bra69] P. T. Brady, "A Model for Generating On–Off Speech Patterns in Two-Way Conversation," *Bell Syst. Technol. J.*, **48**, 2445–2472 (1969).

[Bra75] K. Brayer (Ed.), *Data Communications Via Fading Channels*, Selected Reprint Series, IEEE Press, New York (1975).

[Bra80] E. Braun and S. Schon, "A Cordless Infrared Telephone," *Telecom Rep.*, **3**, 83–86 (1980).

[Bra90] M. S. Brandenstein et al., "A Real-Time Implementation of the Improved MBE Speech Coder," *Proc. ICASSP '90*, Alburquerque, NM, 5–8 (April 1990).

[Bro91] I. Brodsky, "Wireless MANs/WANs Offer Data to Go," *Business Commun. Rev.*, **Feb.**, 45–51 (1991).

[Bro93] G. Bronson, K. Pahlavan, and H. Rotithor, "Performance Prediction of Wireless LANs Based on Ray Tracing Algorithm," *Proc. PIMRC '93*, Yokohama, Japan, 151–156 (Sept. 1993).

[Buc88] C. Buckingham, G. K. Walterink, and D. Akerberg, "A Business Cordless PABX Telephone System on 800 MHz Based on DECT Technology," *IEEE Commun. Mag*, **29**, No. 1, 105–110 (1991).

[Buc91] D. Buchholz, P. Odlyzko, M. Taylor, and R. White, "Wireless In-Building Network Architecture and Protocols," *IEEE Network Mag.*, **5**, No. 6, 31–38 (1991).

[Bul47] K. Bullington, "Radio Propagation Above 30 Megacycles," *Proc. IRE*, **35**, 1122–1136 (1947).

[Bul77] K. Bullington, "Radio Propagation for Vehicular Communications," *IEEE Trans. Veh. Technol.*, **VT-26**, 295–308 (1977).

[Bul87] R. J. C. Bultitude, "Measurement, Characterization and Modeling of Indoor 800/900 MHz Radio Channels for Digital Communications," *IEEE Commun. Mag.*, **25**, No. 6, 5–12 (1987).

[Bul89] R. J. C. Bultitude and S. A. Mahmoud, and W. A. Sullivan, "A Comparison of Indoor Radio Propagation Characteristics at 910 MHz and 1.75 GHz," *IEEE J. Selected Areas Commun.*, **SAC-7**, No. 1, 20–30 (1989).

[Bur83] W. D. Burside and K. W. Burgener, "High Frequency Scattering by Thin Lossless Dielectric Slab," *IEEE Trans. Antennas Propag.*, **AP-31**, 104–110 (1983).

[Cal88] G. Calhoun, *Digital Cellular Radio*, Artech House, Norwood, MA (1988).

[Cal89] M. Callendar, "International Standards for Personal Communications," *Proc. 39th IEEE Veh. Technol. Conf.*, San Francisco, 722–728 (May 1989).

[Cam91] J. P. Campbell, T. E. Tremain, and V. C. Welch, "The Federal Standard 1016 4800 bps CELP Voice Coder," *Dig. Signal Proc.*, **1**, 145–155 (1991).

[Cas90] E. Casas and C. Leung, "A Simple Digital Fading Simulator for Mobile Radio," *IEEE Trans. Veh. Technol.*, **39**, 205–212 (1990).

[Cas93] G. Castagnoli, S. Brauer, and M. Hermann, "Optimization of Cyclic Redundancy Check Codes with 24 and 32 Parity Bits," *IEEE Trans. Commun.*, **COM-41**, 883–892 (1993).

[CCI84a] CCITT Recommendation V.32, "A Family of 2-Wire, Duplex Modems Operating at Data Signalling Ranges of Up to 9600 Bit/s for Use on the General Switched Telephone Network and on Leased Telephone-Type Circuits," Malaga-Torremolinos (1984).

[CCI84b] CCITT, "32-kb/s Adaptive Differential Pulse Code Modulation (ADPCM)," Recommendation G.721, fascicle III.3, Malaga-Torremolinos (1984).

[CCI90] International Consultative Committee on Radio (CCIR), "Future Public Land Mobile Telecommunication Systems," Report M/8, XVII Plenary Assembly, Dusseldorf, (1990).

[CCI91] CCITT Recommendation H.261, "Video Codec for Audio-Visual Services at $P \times 64$ kbits/s," 1991.

[CDPD93] Cellular Digital Packet Data Specification, Release 1.0, Version 1.0, July 1993, CDPD Industry Coordinator, P. O. Box 97060, Kirkland, WA, 98083-9760, USA.

[Cha72] D. Chase, "A Class of Algorithms for Decoding Block Codes with Channel Measurement Information," *IEEE Trans. Inf. Theory*, **IT-18**, 170–182 (1972).

[Cha75] D. Chase and P. A. Bello, "A Combined Coding and Modulation Approach for High Speed Data Transmission Over Troposcatter Channel," *Proc. NTC '75*, 28–32 (Dec. 1975).

[Cha82] K. A. Chamberlain and R. J. Leubbers, "An Evaluation of Longley–Rice and GTD Propagation Models," *IEEE Trans. Antennas Propag.*, **AP-30**, 1093–1098 (1982).

[Cha91] L. F. Chang, "Throughput Estimation of ARQ Protocols for a Rayleigh Fading Channel Using Fade- and Interfade-Duration Statistics," *IEEE Trans. Veh. Technol.*, **40**, 223–229 (1991).

[Cha93] M. Chase and K. Pahlavan, "Performance of DS-CDMA Over Measured Indoor Radio Channels Using Random Orthogonal Codes," *IEEE Trans. Veh. Technol.*, **42**, 617–624 (1993).

[Che82] P. Chevillat and G. Ungerboeck, "Optimum FIR Transmitter and Receiver Filters for Data Transmission Over Band-Limited Channels," *IEEE Trans. Commun.*, **COM-30**, 1909–1915 (1982).

[Cho85] G. Choudhury and S. S. Rappapoport, "Priority Access Schemes Using CSMA-CD," *IEEE Trans. Commun.*, **COM-33**, 620–626 (1985).

[Cho90] A. Chouly and H. Sari, "Coding for Multicarrier Digital Radio Systems," *Proc. IEEE ICC '90*, 1755–1759 (Apr. 1990).

[Chu87] T. S. Chu and M. J. Gans, "High Speed Infrared Local Wireless Communication," *IEEE Commun. Mag.*, **25**, No. 8, 4–10 (1987).

[Cia82] S. Ciarcia, "Use Infrared Communication for Remote Control," *Byte*, **Apr.** 40–49 (1982).

[Cla68] R. H. Clarke, "A Statistical Theory of Mobile-Radio Reception," *Bell Syst. Tech. J.*, **47**, 957–1000 (1968).

[Cla81] G. C. Clark, Jr., and J. B. Cain, *Error-Correction Coding for Digital Communications*, Plenum, New York (1981).

[Com84] R. A. Comroe and D. J. Costello, Jr., "ARQ Schemes for Data Transmission in Mobil Radio Systems," *IEEE Trans. Veh. Technol.*, **VT-33**, 88–97 (1984).

[Con78] W. J. Conner, "ANTRC-170, A New Digital Troposcatter Communication System," *Proc. IEEE ICC '78*, 40–43 (1978).

[Coo78] G. R. Cooper and R. W. Nettleton, "A Spread Spectrum Technique for Higher Capacity Mobile Radio," *IEEE Trans. Veh. Technol.*, **VT-27**, 264–275 (1978).

[Cox72] D. C. Cox, "Delay Doppler Characteristics of Multipath Propagation at 910 MHz in a Suburban Mobile Radio Environment," *IEEE Trans. Antennas Propag.*, **20**, 625–635 (1972).

[Cox73] D. C. Cox, "910 MHz Urban Radio Propagation: Multipath Characteristics in New York City," *IEEE Trans. Veh. Technol.*, **VT-42**, 104–110 (1993).

[Cox85] D. C. Cox, "Universal Portable Radio Communications," *IEEE Trans. Veh. Technol.*, **VT-34**, 117–121 (1985).

[Cox87a] D. C. Cox, "Universal Digital Portable Radio Communications," *Proc. IEEE*, **75**, 436–477 (1987).

[Cox87b] D. C. Cox, H. W. Arnold, and P. T. Porter, "Universal Digital Portable Communications," *IEEE J. Selected Areas Commun.*, **SAC-5**, 764–773 (1987).

[Cox89] D. C. Cox, "Portable Digital Radio Communications—An Approach to Tetherless Access," *IEEE Commun. Mag.*, **27**, No. 7, 30–40 (1989).

[Cox90] D. C. Cox, "Personal Communications—A Viewpoint," *IEEE Commun. Mag.*, **28**, No. 11, 8–20 (1990).

[Cox91a] D. C. Cox, "A Radio System Proposal for Widespread Low-Power Tetherless Communications," *IEEE Trans. Commun.*, **39**, 324–335 (1991).

[Cox91b] Cox, R. V. et al., "Subband Speech Coding and Matched Convolutional Channel Coding for Mobile radio Channels," *IEEE Trans. Signal Proc.*, **39**, 1717–1731 (1991).

[Cox92a] D. C. Cox, "Wireless Access: An Overview," presented at *PIMRC '92*, Boston, MA (Oct. 1992).

[Cox92b] D. C. Cox, "Wireless Network Access for Personal Communications," *IEEE Commun. Mag.*, **30**, No. 12, 96–115 (1992).

[Coy83] E. J. Coyle and B. Liu, "Finite Population CSMA/CD Networks," *IEEE Trans. Commun.*, **COM-31**, 1247–1251 (1983).

[Coy85] E. J. Coyle and B. Liu, "A Matrix Representation of CSMA/CD Networks," *IEEE Trans. Commun.*, **COM-33**, 53–64 (1985).

[Cro73] W. Crowther, R. Rettberg, D. Walden, S. Ornstein, and F. Heart, "A System for Broadcast Communication: Reservation ALOHA," *Proc. Sixth Hawaii Int. Conf. Syst. Sci*, 371–374 (Jan, 1973).

[Cro76] R. E. Crochiere, S. A. Webber, and J. L. Flanagan, "Digital Coding of Speech in Sub-bands," *Bell Syst. Tech. J.*, **55**, 1069–1085 (1976).

[Cro77] R. Crochiere, "On the Design of Sub-band Coders for Low Bit-Rate Speech Communication," *Bell Syst. Tech. J.*, **56**, 747–770 (1977).

[Cro83] R. E. Crochiere and J. L. Flanagan, "Current Perspectives in Digital Speech," *IEEE Commun. Mag.*, **21**, No. 1, 32–40 (1983).

[CTI88] CTIA Sub-committee for Advanced Technologies, "Users' Performance Requirements," Sept. 1988.

[CTI89] CTIA CDMA Digital Cellular Technology Open Forum, June 6, 1989.

[Cus85] R. H. Cushman, "Third-generation DSP's Put Advanced Functions on Chip," *EDN*, **July**, 59–67 (1985).

[Dau83] D. G. Daut and J. W. Modestino, "Two-Dimensional DPCM Image Transmission Over Fading Channels," *IEEE Trans. Commun.*, **COM-31**, 315–328 (1983).

[Dav89] F. Davarian and J. T. Sumida, "A Multipurpose Digital Modulator," *IEEE Commun. Mag.*, **27**, No. 2, 36–45 (1989).

[Dav91] R. Davis, M. Bensebti, M. Beach, J. P. McGeeham, D. Rickard, C. Shepherd, and S. Wales, "A Comparison of Indoor and Urban Propagation at 1.7, 29, and 60 GHz," *Proc. 41st IEEE Veh. Technol. Conf.*, 289–293 (May 1991).

[Dec88] M. Decina and G. Modena, "CCITT Standards on Digital Speech Processing," *IEEE J. Special Areas Commun.*, **6**, 227–234 (1988).

[Des72] G. A. Deschamps, "Ray Techniques in Electromagnetics," *Proc. IEEE*, **60**, 1022–1035 (1972).

[Dev84] D. M. J. Devasirvatham, "Time Delay Spread Measurements of Wideband Radio Systems Within a Building," *Electron. Lett.*, **20**, 50–951 (1984).

[Dev87] D. M. J. Devasirvatham, "A Comparison of Time Delay Spread and Signal Level Measurements Within Two Dissimilar Office Buildings," *IEEE Trans. Antennas Propag.*, **AP-35**, 319–324 (1987).

[Dev91] D. M. J. Devasirvatham, "Multi-frequency Propagation Measurement and Models in a Large Metropolitan Commercial Building for Personal Communication," *Proc. PIMRC '91*, 98–103 (Sept. 1991).

[Dig82] Digital Equipment Corp. Intel Corp., and Xerox Corp., "The Ethernet: A Local Area Network; Data Link Layer and Physical Layer Specifications," Version 2.0, Nov. 1982.

[Div87] D. Divsilar and M. K. Simon, "Trellis Coded Modulation for 4800–9600 bits/s Transmission Over a Fading Mobile Satellite Channel," *IEEE J. Selected Areas Commun.*, **SAC-5**, 162–175 (1987).

[Div88a] D. Divsilar and M. K. Simon, "The Design of Trellis Coded MPSK for Fading Channels: Performance Criteria," *IEEE Trans. Commun.*, **COM-36**, 1013–1022 (1988).

[Div88b] D. Divsilar and M. K. Simon, "The Design of Trellis Coded MPSK for Fading Channels: Set Partitioning for Optimum Code Design," *IEEE Trans. Commun.*, **COM-36**, 1004–1012 (1988).

[Dix84] R. C. Dixon, *Spread Spectrum Systems*, 2nd ed., Wiley, New York (1984).

[Dos92] B. T. Doshi and P. K. Johri, "Communication Protocols for High Speed Packet Networks," *Computer Networks and ISDN Systems*, **24**, 243–273 (1992).

[Eav77] R. E. Eaves and A. H. Levesque, "Probability of Block Error for Very Slow Rayleigh Fading in Gaussian Noise," *IEEE Trans. Commun.*, **COM-25**, 368–374 (1977).

[Eav87] J. L. Eaves and E. K. Reedy, *Principles of Modern Radar*, Van Nostran Reinhold, New York (1987).

[EDG93] "Cordless PBX," *EDGE On and About AT&T*, **Oct. 4**, 6 (1993).

[EIA54] EIA/TIA, "Dual-Mode Mobile Station/Base Station Compatibility Standard," Interim-Standard 54, Jan. 1992.

[EIA90] EIA/TIA, "Cellular System: Recommended Minimum Performance Standards for 800 MHz Base Stations Supporting Dual-Mode Mobile Stations," TR.45.3, Project No. 2217, incorporating EIA/TIA IS-20 Draft, Electronic Industries Association, Washington, DC, Oct. 1990.

[EIA89] Electronic Industries Association Specification IS-54, "Dual-Mode Subscriber Equipment Compatibility Specification," EIA Project No. 2215, Dec. 1989.

[EIA92] EIA/TIA, "Dual-Mode Mobile Station/Base Station Compatibility Standard," Interim Standard 54, Jan. 1992.

[EIA94a] Electronic Industries Association, "Cellular System Dual-Mode Mobile Station—Base Station Compatibility Standard," EIA/TIA IS-54, Rev. C, June 1994.

[EIA94b] Electronic Industries Association, "Service Description for IS-54 Data Services," June 1994.

[Eln86] S. M. Elnoubi, "Analysis of GMSK with Discriminator Detection in Mobile Radio Channels," *IEEE Trans. Veh. Technol.*, **VT-35**, 71–76 (1986).

[Eph87] A. Ephremides, J. E. Wuselthier, and D. J. Baker, "A Design Concept for Reliable Mobile Radio Networks with Frequency Hopping," *Proc. IEEE*, **75**, 56–73 (1987).

[Fal76a] D. D. Falconer, "Analysis of a Gradient Algorithm for Simultaneous Passband Equalization and Carrier Phase Recovery," *Bell Syst. Tech. J.*, **55**, 409–428 (1976).

[Fal76b] D. D. Falconer and F. R. Magee, "Evaluation of Decision Feedback Equalization and Viterbi Algorithm Detection for Voiceband Data Transmission—Part I," *IEEE Trans. Commun.*, **COM-24**, 1130–1139 (1976).

[Fal76c] D. D. Falconer and F. R. Magee, "Evaluation of Decision Feedback Equalization and Viterbi Algorithm Detection for Voiceband Data Transmission—Part II," *IEEE Trans. Commun.*, **COM-24**, 1238–1245 (1976).

[Fal85] D. D. Falconer et al., "Comparison of DFE and MLSE Receiver Performance on HF Channels," *IEEE Trans. Commun.*, **COM-33**, 484–486 (1985).

[Feh87] K. Feher (Ed.), *Advanced Digital Communications—Systems and Signal Processing Techniques*, Prentice–Hall, Englewood Cliffs, NJ (1987).

[Feh91] K. Feher, "MODEMS for Emerging Digital Cellular-Mobile Radio System," *IEEE Trans. Veh. Technol*, **40**, 355–365 (1991).

[Fer80] P. Ferert, "Application of Spread Spectrum Radio to Wireless Terminal Communications," *Proc. NTC '80*, Houston, TX, 244–248 (Dec. 1980).

[Fis80] M. J. Fischer, "Delay Analysis of TASI with Random Fluctuations in the Number of Voice Calls," *IEEE Trans. Commun.*, **COM-28**, 1883–1889 (1980).

[Fla72] J. L. Flanagan, *Speech Analysis, Synthesis and Perception*, Springer-Verlag, New York (1972).

[Fla79] J. L. Flanagan, et al., "Speech Coding," *IEEE Trans. Commun.*, **COM-27**, 710–737 (1979).

[For72] G. D. Forney, Jr., "Maximum-Likelihood Sequence Estimation of Digital Sequences in the Presence of Intersymbol Interference," *IEEE Trans. Inf. Theory*, **IT-18**, 363–378 (1972).

[For73a] G. D. Forney, Jr., "The Viterbi Algorithm," *Proc. IEEE*, **61**, 268–278 (1973).

[For84] G. D. Forney, Jr., et al., "Efficient Modulation for Band-Limited Channels," *IEEE J. Selected Areas Commun.*, **SAC-2**, 632–647 (1984).

[Fos85] G. J. Foscihini, "Equalizing Without Alternating or Detecting Data," *Bell Syst. Tech. J.*, **64**, 1885–1911 (1985).

[Fra74] L. E. Franks and J. P. Bubrouski, "Statistical Properites of Timing Jitter in a PAM Timing Recovery Scheme," *IEEE Trans. Commun.*, **COM-22**, 913–920 (1974).

[Fra80] L. A. Fratta and D. Sant, "Some Models of Packet Radio Networks with Capture," *Proc. Fifth Int. Conf. Comput. Commun.*, 155–161 (1980).

[Fre91a] T. A. Freeburg, "Enabling Technologies for Wireless In-Building Network Communications—Four Technical Challenges, Four Solutions," *IEEE Commun. Mag.*, **29**, No. 4, 58–64 (1991).

[Fre91b] T. Freeburg, "A New Technology for High Speed Wireless Local Area Networks," *Proc. IEEE Workshop on Wireless LANs*, Worcester, MA, 127–139 (May 1991).

[FS1016] Federal Standard 1016, Telecommunications: Analog to Digital Conversion of Radio Voice by 4,800 bit/second Code Excited Linear Prediction (CELP), National Communications System, Office of Technology and Standards, Washington, DC 20305-2010, 14 February 1991.

[FS1102] Proposed Federal Standard 1102 for Narrowband Digital Land Mobile Radio, Jan. 1993.

[Fuj88] T. Fuji and Y. Kikkawa, "Optical Space Transmission Module," *Natl. Tech. Rep.*, **32**, No. 1, 101–106 (1988).

[Fus90] M. Fusco, "FDTD Algorithm in Curvilinear Coordinates," *IEEE Trans. Antennas Propag.*, **AP-38**, 76–88 (1990).

[G721] CCITT, "32-kb/s Adaptive Differential Pulse Code Modulation (ADPCM)," Recommendation G.721, fascicle III.3, Malaga-Torremolinos, 1984.

[Gan87] R. Ganesh and K. Pahlavan, "Effects of Retransmission and Capture for Local Area ALOHA Systems," *Proc. Conf. Inf. Sci. Syst.*, Baltimore, MD, 272–273 (1987).

[Gan89] R. Ganesh and K. Pahlavan, "On the Arrival of the Paths in Multipath Fading Indoor Radio Channels," *IEE Electr. Lett.*, **25**, 763–765 (1989).

[Gan91a] R. Ganesh, "Time Domain Measurements, Modeling, and Simulation of the Indoor Radio Channel," Ph.D. Thesis, Worcester Polytechnic Institute, May 1991.

[Gan91b] R. Ganesh and K. Pahlavan, "Modeling of the Indoor Radio Channel," *IEE Proc. I: Commun. Speech and Vision*, **138**, 153–161 (1991).

[Gan92] R. Ganesh and K. Pahlavan, "Statistical Characterization of a Partitioned Indoor Radio Channel," *IEE Proc. I: Commun. Speech and Vision*, **139**, 539–545 (1992).

[Gan93] R. Ganesh and K. Pahlavan, "Statistics of Short Time and Spatial Variations Measured in Wideband Indoor Radio Channels," *IEE Proc. H: Microwave, Antennas and Propagation*, **140**, 297–302 (1993).

[Gav82] D. P. Gaver and J. P. Lehoczky, "Channels that Cooperatively Service a Data Stream and Voice Messages," *IEEE Trans. Commun.*, **COM-30**, 1153–1162 (1982).

[Ger83] A. Gersho and V. Cuperman, "Vector Quantization: A Pattern-Matching Technique for Speech Coding," *IEEE Commun. Mag.*, **21**, No. 9, 15–21 (1983).

[Ger90] I. A. Gerson and M. A. Jasiuk, "Vector Sum Excitation Linear Prediction (VSELP) Speech Coding at 8 kbps," *Proc. ICASSP '90*, Albuquerque, NM, 461–464 (Apr. 1990).

[Gev89] J. Gevargiz, P. K. Das, and L. B. Milstein, "Adaptive Narrowband Interference Rejection in a DS Spread Spectrum Intercept Receiver Using Transform Domain Signal Processing Techniques," *IEEE Trans. Commun.*, **37**, 1359–1366 (1989).

[Gfe79a] F. R. Gfeller, H. R. Miller, and P. Vettiger, "Infrared Communications for In-House Applications," *Proc. IEEE COMPCON*, Washington, DC, 132–138 (1979).

[Gfe79b] F. R. Gfeller and U. Bapst, "Wireless In-House Data Communication Via Diffuse Infrared Radiation," *Proc. IEEE*, **67**, 1474–1486 (1979).

[Gfe81] F. R. Gfeller, "Infranet: Infrared Microbroadcasting Network for In-House Data Communication," IBM Research Report, RZ 1068 (#38619), April 27, 1981.

[Gfe82] F. R. Gfeller, "Infrared Microbroadcasting for In-House Data Communications," *IBM Tech. Disclosure Bull.*, **24**, 4043–4046 (1982).

[Gib93] J. D. Gibson, "Speech Signal Processing," Section 14.2 in *The Electrical Engineering Handbook*, CRC Press, Boca Raton, FL, (1993).

[Gil91] K. S. Gilhousen et al., "On the Capacity of a Cellular CDMA System," *IEEE Trans. Veh. Technol.*, **VT-40**, 303–312 (1991).,

[Gio85] A. A. Giordano and F. M. Hsu, *Least Square Estimation with Applications to Signal Processing*, Wiley, New York (1985).

[Git73] R. D. Gitlin, E. Y. Ho, and J. E. Mazo, "Passband Equalization for Differentially Phase-Modulated Data Signals," *Bell Syst. Tech. J.*, **52**, 219–238 (1973).

[Git82] R. D. Gitlin, H. C. Meadors, and S. B. Weinstein, "The Tap-Leakage Algorithm: An Algorithm for the Stable Operation of a Digitally Implemented, Fractionally-Spaced Adaptive Equalizer," *Bell Syst. Tech. J.*, **61** 1817–1839 (1982).

[Gla89] A. S. Glassner, *An Introduction to Ray Tracing*, Academic Press, New York (1989).

[God80] D. Godard, "Self-Recovering Equalization and Carrier Tracking in Two-Dimensional Data Communication Systems," *IEEE Trans. Commun.*, **COM-28**, 1867–1875 (1980).

[God81] D. Godard and D. Pilost, "A 2400 bps Microprocessor-Based Modem," *IBM J. Res. Dev.*, 17–24 (1981).

[Goo87] D. J. Goodman and A. A. M. Saleh, "The Near/Far Effect in Local ALOHA Radio Communications," *IEEE Trans. Veh. Technol.*, **VT-36**, 19–27 (1987).

[Goo89] D. J. Goodman, R. A. Valenzuela, K. T. Gayliard, and B. Ramamurthi, "Packet Reservation Multiple Access for Local Wireless Communications," *IEEE Trans. Commun.*, **COM-37**, 885–889 (1989).

[Goo90] S. H. Goode, H. L. Kazecki, and D. W. Dennis, "A Comparison of Limiter-Discriminator, Delay and Coherent Detection for $\pi/4$-QPSK," *Proc. 40th IEEE Veh. Technol. Conf.*, Orlando, FL, 687–694 (May 1990).

[Goo91a] D. J. Goodman, "Trends in Cellular and Cordless Communications," *IEEE Commun. Mag.*, **29**, No. 6, 31–40 (June 1991).

[Goo91b] D. J. Goodman and S. X. Wei, "Efficiency of Packet Reservation Mulitple Access," *IEEE Trans. Veh. Technol.*, **VT-40**, 170–176 (1991).

[Goo92] D. J. Goodman, "Wireless Personal Communications Networks," *1992 Optical Society of America and IEEE OFC '92*, Plenary Session paper TuA1, San Jose, CA (Feb 1992).

[Gri85] D. W. Griffin and J. S. Lim, "A New Model-Based Speech Analysis/Synthesis System," *Proc. ICASSP '85*, 513–516, Tampa, FL (March 1985).

[Grz75] C. J. Grzenda, D. R. Kern, and P. Monsen, "Megabit Digital Troposcatter Subsystem," *Proc. IEEE NTC '75*, New Orleans, LA, 28-15–28-19 (Dec. 1975).

[GSM89] Group Speciale Mobile (GSM) Recommendation GSM 06.10, "GSM Full Rate Speech Transcoding," Technical Report Version 3.2, ETSI/GSM, July 1989.

[GSM91] GSM Recommendation 05.05, "Radio Transmission and Reception," ETSI/PT 12, Jan. 1991.

[Hab89] I. M. Habab, M. Kavehrad, and C.-E. W. Sundberg, "ALOHA with Capture Over Slow and Fast Fading Radio Channels with Coding and Diversity," *IEEE J. Selceted Areas Commun.*, **6**, 79–88 (1988).

[Hai92] J. L. Haine, P. M. Martin, and R. L. A. Goodings, "A European Standard for Packet-Mode Mobile Data," *Proc. PIMRC '92*, Boston, MA (Oct. 1992).

[Ham86] J. L. Hammond and P. J. P. O'Reilly, *Performance Analysis of Local Computer Networks*, Addison–Wesley, Reading, MA (1986).

[Har74] H. Harris, T. Saliga, and D. Walsh, "An All Digital 9600 bps LSI Modem," *Proc. NTC '74*, 279–284 (Dec. 1974).

[Har89] J. C. Hardwick and J. S. Lim, "A 4800 bps Improved Multi-band Excitation Speech Coder," *Proc. IEEE Workshop on Speech Coding for Telecommun.*, Vancouver (Sept. 1989).

[Har91] J. C. Hardwick and J. S. Lim, "The Application of the IMBE Speech Coder to Mobile Communications," *Proc. ICASSP '91*, Toronto, 249–252 (May 1991).

[Har92] P. H. Harms, J. F. Lee, and R. Mittra, "A Study of the Nonorthogonal FDTD Method Versus the Conventional FDTD Technique for Computing Resonant Frequencies of Cylindrical Cavities," *IEEE Trans. Microwave Theory Tech.*, **MTT-40**, 741–746 (1992).

[Har93] S. Hara et al., "Multicarrier Modulation Techniques for Broadband Indoor Wireless Communications," *Proc. PIMRC '93*, Japan, 132–136 (Sept. 1993).

[Has79] H. Hashemi, "Simulation of the Urban Radio Propagations," *IEEE Trans. Veh. Technol.*, **VT-28**, 213–225 (1979).

[Has93a] H. Hashemi, "The Indoor Radio Propagation Channel," *Proc. IEEE*, **81**, 943–968 (1993).

[Has93b] H. Hashemi, "Impulse Response Modeling of Indoor Radio Propagation Channels," *IEEE J. Selected Areas Commun.*, **11**, 967–978 (1993).

[Hat80] M. Hata, "Empirical Formula for Propagation Loss in Land-Mobile Radio Services," *IEEE Trans. Veh. Technol.*, **VT-29**, 317–325 (1980).

[Hat88] T. Hattori, A. Sasaki, and K. Momma, "Emerging Telephony and Service Enhancement for Cordless Telephone Systems," *IEEE Commun. Mag.*, **26**, No. 1, 53–58 (1988).

[Hau94] T. Haug, "Overview of GSM: Philosophy and Results," *Int. J. Wireless Inf. Networks*, **1**, 7–16 (1994).

[Hay91] V. Hayes, "Standardization Efforts for Wireless LANs," *IEEE Network Mag.*, **5**, No. 6, 19–20 (1991).

[Hei92] G. H. Heilmeier, "Personal Communications: Quo Vadis," *Digest of 1992 IEEE Int. Solid-State Circuits Conf.*, Plenary Session paper WA1.3, 24–26, (Feb. 1992).

[Hel84] G. Held, *Data Compression*, Wiley, New York (1984).

[Hey86] D. P. Heyman, "The Effects of Random Message Sizes on the Performance of the CSMA/CD Protocol," *IEEE Trans. Commun.*, **COM-34**, 547–553 (1986).

[Hir79] K. Hirade, et al., "Error-Rate Performance of Digital FM with Differential Detection in Land Mobile Radio Channels," *IEEE Trans. Veh. Technol.*, **VT-28**, 204–212 (1979).

[Hir84] M. Hirono, T. Miki, and K. Murota, "Multilevel Decision Method for Band-Limited Digital FM with Limiter Discrimintor Detection," *IEEE Trans. Veh. Technol.*, **VT-33**, 114–122 (1984).

[HNS92] Hughes Network Systems, Proposal for E-TDMA, 1992.

[Ho94] C. M. P. Ho, T. S. Rappoport, and M. P. Koushik, "Antenna Effects on Indoor Obstructed Wireless Channel and a Deterministic Image-Based Propagation Model for a In-Building Personal Communication Systems," *Int. J. Wireless Inf. Networks*, **1**, 61–76 (1994).

[Hol83] R. Holland, "Finite Difference Solutions of Maxwell's Equations in Generalized Non-orthogonal Coordinates," *IEEE Trans. Nucl. Sci.*, **NS-30**, 4689–4691 (1983).

[Hol92a] T. Holt, K. Pahlavan, and J. F. Lee, "A Graphical Indoor Radio Channel Simulator Using 2D Ray Tracing," *Proc. PIMRC '92*, Boston, MA, 411–416 (Oct. 1992).

[Hol92b] T. Holt, "A Computer Graphic Package for Indoor Radio Channel Simulation Using a 2D Ray Tracing Algorithm," M.S. Thesis, Worcester Polytechnic Institute, May 1992.

[Hol92c] T. Holt, K. Pahlavan, and J. F. Lee, "A Computer Package for Indoor Channel Simulation Using a 2D Ray Tracing Algorithm," presented at 17th IEEE Annual Conference on Local Computer Networks, Minneapolis, MN, Sept. 1992.

[Hon92] W. Honcharenko, H. L. Bertoni, J. L. Dailing, J. Qian, and H. D. Yee, "Mechanisms Governing UHF Propagation on Single Floors in Modern Office Buildings," *IEEE Trans. Veh. Technol.*, **41**, 496–504 (1992).

[How90a] S. Howard and K. Pahlavan, "Doppler Spread Measurements of the Indoor Radio Channel," *IEE Electr. Lett.*, **26**, 107–109 (1990).

[How90b] S. J. Howard and K. Pahlavan, "Autoregressive Modeleing of the Indoor Radio Channel," *IEE Electr. Lett.*, **26**, 816–817 (1990).

[How90c] S. J. Howard and K. Pahlavan, "Measurement and Analysis of the Indoor Radio Channel in the Frequency Domain," *IEEE Trans. Instr. Meas.*, **39**, 751–755 (1990).

[How91] S. J. Howard, "Frequency Domain Characteristic and Autoregressive Modeling of the Indoor Radio Channel," Ph.D. Thesis, Worcester Polytechnic Institute, Worcester MA, May 1991.

[How92] S. J. Howard and K. Pahlavan, "Autoregressive Modeling of Wideband Indoor Radio Propagation," *IEEE Trans. Commun.*, **COM-40**, 1540–1552 (1992).

[IEE802.10] IEEE Working Group 802.10, "Secure Data Exchange Standard," 1993.

[IEE802.11] IEEE Working Group 802.11, P802.11-93/20b0, Update of Wireless LAN Medium Access Control (MAC) and Physical Layer (PHY) Specifications, March 1994.

[IEE86] IEEE Communications Society, Special Issue on Office Automation, *IEEE Commun. Mag.*, **24**, No. 7, (1986).

[IEE88a] IEEE Vehicular Technology Society, Special Issue on Mobile Radio Propagation, *IEEE Trans. Veh. Technol.*, **37**, 3–72 (1988).

[IEE88b] IEEE, Special Issue on Voice Coding for Communications, *IEEE J. Selected Areas Commun.*, **6**, 225–456 (1988).

[IEE92] IEEE, Special Issue on Speech and Image Coding, *IEEE J. Selected Areas Commun.*, **10**, 793–976 (1992).

[IEEE88] IEEE Special Issue on the Intelligent Network, *IEEE Commun. Mag.*, **26**, No. 12, 1–87 (1988).

[IEEE91a] *Proc. IEEE Workshop on Wireless LANs*, Worcester, MA (May 1991).

[IEEE91b] *Proc. IEEE/IEE Symp. on Personal, Indoor and Mobile Radio Commun. (PIMRC '91)*, London, UK (Sept. 1991).

[IEEE92] *Proc. IEEE/IEE Symp. on Personal, Indoor and Mobile Radio Commun. (PIMRC '92)*, Boston, MA (Oct. 1992).

[Ike91] F. Ikegami, T. Takeuchi, and S. Yoshida, "Theoretical Prediction of Mean Field Strength for Urban Mobile Radio," *IEEE Trans. Antennas Propag.*, **39**, 229–302 (1991).

[ISO87] International Standards Organization, "Protocol for Providing the Connectionless-Mode Network Service," Publication ISO 8473 (1987).

[ISO91] ISO CD 11172-2, "Coding of Moving Pictures and Associated Audio—Part 2," Nov. 1991.

[ISO92] ISO/IEC DIS 10918-1, "Digital Compression and Coding of Continuous-Tone Still Images," Jan. 2, 1992.

[Iye93] V. Iyengar and J. Michaelides, "Performance Evaluations of RLPs (Radio Link Protocol) for TDMA Data Services," TIA Contribution TR45.3.2.5/93.03.30.10, Chicago, March 30, 1993.

[Jak71] W. C. Jakes, Jr., "A Comparison of Specific Space Diversity Techniques for Reduction of Fast Fading in UHF Mobile Radio Systems," *IEEE Trans. Veh. Technol.*, **VT-20**, 81–92 (1971).

[Jak74] W. C. Jakes (Ed.), *Microwave Moible Communications*, Wiley, New York (1974).

[Jam72] R. T. James and P. E. Muench, "AT & T Facilities and Services," *Proc. IEEE*, **60**, 1342–1349 (1972).

[Jay84] N. S. Jayant and P. Noll, *Digital Coding of Waveforms*, Prentice–Hall, Englewood Cliffs, NJ, 1984.

[Jay90] N. S. Jayant, "High-Quality Coding of Telephone Speech and Wideband Audio," *IEEE Commun. Mag.*, **28**, No. 1, 10–20 (1990).

[Jay92] N. S. Jayant, "Signal Compression: Technology Targets and Research Directions," *IEEE J. Selected Areas Commun.*, **10**, 796–818 (1992).

[Jen65] F. A. Jenkins and H. E. White, *Fundamentals of Optics*, 4th ed., McGraw–Hill, New York (1965).

[Jer92] M. C. Jeruchim, P. Balaban, and S. Shanmugan, *Simulation of Communication Systems*, Plenum, New York (1992).

[JTC93] Joint Technical Committee for PCS, Service Requirements Description, 1993.

[JTC94] Joint Technical Committee of Committee T1 R1P1.4 and TIA TR46.3.3/TR45.4.4 on Wireless Access, "Draft Final Report on RF Channel Characterization," Paper No. JTC(AIR)/94.01.17-238R4, Jan. 17, 1994.

[Kah78] R. E. Kahn, S. A. Gronemeyer, J. Burchfiel, and R. C. Kunzelman, "Advances in Packet Radio Technology," *Proc. IEEE*, **66**, 1468–1496 (1978).

[Kam81] M. A. Kamil, "Simulation of Digital Communication Through Urban/Surburban Multipath," Ph.D. Dissertation, EECS Department University of California, Berkeley, CA, June 1981.

[Kan80] H. Kaneko and T. Ishiguro, "Digital Television Transmission Using Bandwidth Compression Techniques," *IEEE Commun. Mag.*, **18**, No. 7, 14–22 (1980).

[Kav85] M. Kavehrad and P. McLane, "Performance of Low-Complexity Channel Coding and Diversity for Spread Spectrum in Indoor Wireless Communication," *Bell Syst. Tech. J.*, **64**, 1927–1965 (1985).

[Kav87] M. Kavehrad and P. J. McLane, "Spread Spectrum for Indoor Digital Radio," *IEEE Commun. Mag.*, **25**, No. 6, 32–40 (1987).

[Kay81] M. Kaya, K. Ishizuka and N. Maeda, "High Speed Data Modem Using Digital Signal Processor," *Proc. IEEE ICC '81*, 14.7.1–14.7.5 (June 1981).

[Kei89] G. E. Keiser, *Local Area Networks*, McGraw-Hill, New York (1989).

[Ker75] D. R. Kern, P. Monsen, and C. J. Grzenda, "Megabit Troposcatter Subsystem," *Proc. IEEE NTC '75*, New Orleans, 28.15–28.19 (1975).

[Kil92] J. A. Kilpatrick, "Update of RAM Mobile Data's Packet Data Radio Service," *Proc. 42nd IEEE Veh. Technol. Conf.*, 898–901 (1992).

[Kin91] K. Kinoshita, M. Kuramoto, and N. Nakajima, "Development of a TDMA Digital Cellular System Based on the Japanese Standard," *Proc. 41st IEEE Veh. Technol. Conf.*, 642–649 (1991).

[Kle74] L. Kleinrock and F. A. Tobagi, "Carrier Sense Multiple Access for Packet Switched Radio Channels," *Proc. IEEE ICC '74*, 21B-1–21B-7 (June 1974).

[Kle75a] L. Kleinrock, *Queueing System, 1: Theory*, Wiley, New York (1975).

[Klc75b] L. Kleinrock and S. S. Lam, "Packet Switching in a Multiaccess Broadcast Channel: Performance Evaluation," *IEEE Trans. Commun.*, **COM-23**, 410–423 (1975).

[Kle75c] L. Kleinrock and F. A. Tobagi, "Packet Switching in Radio Channels, Part I: Carrier Sense Multiple-Access Modes and Their Throughput-Delay Characteristics," *IEEE Trans. Commun.*, **COM-23**, 1400–1416 (1975).

[Kle84] L. Kleinrock, "On a Self-Adjusting Capability of Random Access Networks," *IEEE Trans. Commun.* **COM-32**, 40–47 (1984).

[Kob71a] H. Koboyashi, "Application of Hestens–Stiefel Algorithm to Channel Equalization," *Proc. IEEE ICC '71*, 21.25–21.30 (1971).

[Kob71b] H. Koboyaski, "Simultaneous Adaptive Estimation and Detection Algorithm for Carrier Modulated Data Communication Systems," *IEEE Trans. Commun. Technol.*, **COM-19**, 268–280 (1971).

[Kob77] H. Kobayaski, Y. Onozato, and D. Huynh, "An Approximate Method for Design and Analysis of an ALOHA System," *IEEE Trans. Commun.*, **COM-25**, 148–157 (1977).

[Kon74] A. G. Konheim and B. Meister, "Waiting Lines and Times in a System with Polling," *J. ACM*, **21**, 470–490 (1974).

[Kon93] S. Konidaris, "Overview of the Status of RACE," *Prco. IEEE GLOBECOM- '93*, 311–316, Houston (Nov. 1993).

[Kro92] P. Kroon and K. Swaminathan, "A High-Quality Multirate Real-Time CELP Coder," *IEEE J. Selected Areas Commun.*, **10**, 850–857 (1992).

[Kum74] K. Kummerle, "Multiplexer Performance for Integrated Line- and Packet-Switched Traffic," ICCC, Stockholm, 1974.

[Kup82] F. Kuperus and J. Arnbak, "Packet radio in a Rayleigh channel," *Electr. Lett.*, **18**, No. 12, 506–507 (1982).

[Kwa75] R. K. Kwan, "The TELESAT TDMA Field Trial," *Proc. ICDSC-3*, Kyoto, 135–143 (1975).

[Lam74] S. S. Lam, "Packet Switching in a Multi-Access Broadcast Channel with Application to Satellite Communications in a Computer Network," Ph.D. dissertation, Computer Science Department, University of California, Los Angles (March 1974).

[Lam75] S. S. Lam and L. Kleinrock, "Packet Switching in a Multiaccess Broadcast Channel: Dynamic Control Procedures," *IEEE Trans. Commun.*, **COM-23**, 891–904 (1975).

[Lam80] S. S. Lam, "A Carrier Sense Multiple Access Protocol for Local Networks," *Computer Networks*, **4**, 21–32 (1980).

[LaR84] B. LaReau, "IR Unit Runs Computer Remotely," *Electronic Week*, **Nov. 12**, 71–75 (1984).

[Law91] M. C. Lawton, R. L. Davies and J. P. McGeehan," A Ray Launching Method for the Prediction of Indoor Radio Channel Characteristics," *PIMRC '91*, King's College London (U.K.), 104–108 (Sept. 1991).

[Law92] M. C. Lawton and J. P. McGeehan, "The Application of GTD and Ray Launching Techniques to Channel Modelling for Cordless Radio Systems," *Proc. 42nd IEEE Veh. Technol. Conf.*, Denver, Co, 125–130 (1992).

[Leb89] M. Lebherz et al., "Calculation of Broadcast Coverage Based on a Digital Terrain Model," *Proc. IEEE ICAP '89*, 355–359 (1989).

[Leb92] M. Lebherz, W. Wiesbeck, and W. Krank, "A Versatile Wave Propagation Model for the VHF/UHF Range Considering Three-Dimensional Terrain," *IEEE Trans. Antennas Propag.*, **40**, 1121–1131 (1992).

[Lee82] W. C. Y. Lee, *Mobile Communications Engineering*, McGraw–Hill, New York (1982).

[Lee86] W. C. Y. Lee, *Mobile Communications Design Fundamentals*, H. W. Sams, Indianapolis, IN (1986). [Also see 2nd ed., Wiley, New York (1993).]

[Lee89] W. C. Y. Lee, *Mobile Cellular Telecommunications Systems*, McGraw–Hill, New York (1989).

[Lee93] J. F. Lee, "Numerical Solutions of TM Scattering Using an Obliquely Cartesian Finite Difference Time Domain Algorithm," *IEE Proc. H*, **140**, No. 1, 23–28 (1993).

[LeG91] D. LeGall, "MPEG: A Video Compression Standard for Multimedia Applications," *Commun. ACM*, **34**, 47–58 (1991).

[Les88] A. Lessard and M. Gerla, "Wireless Communications in the Automated Factory Environment," *IEEE Network Mag.*, **2**, No. 3, 64–69 (1988).

[Leu86] V. C. M. Leung and R. W. Donaldson, "Effects of Channel Errors on the Delay Throughput Performance and Capacity of ALOHA Multiple Access Systems," *IEEE Trans. Commun.* **COM-34**, 497–502 (1986).

[Lev76] A. H. Levesque, "Block-Error Distributions and Error Control Code Performance in Slow Rayleigh Fading," *Proc. NTC '76*, Houston, 24.4.1–24.4.4 (1976).

[Lev93] A. H. Levesque and P. J. Kush, "Evaluation and Correction of the Fading Model Used in the Error Mask Generator Computer Program," Contribution No. TR45.3.2.5/93.02.08.06, TIA/TR45.3 Data Services Task Group, February 1993.

[Li87] V. O. K. Li, "Multiple Access Communications Networks," *IEEE Commun. Mag.*, **25**, No. 6, 41–48 (1987).

[Lia85] J. K. Liang, R. J. P. DeFigueiredo, and F. C. Lu, "Designing of Optimum Nyquist, Partial Response, Nth Band, and Nonuniform Tap Spacing FIR Digital Filters Using Linear Programming Technique," *IEEE Trans. Circuits Syst.*, **CAS-32**, 386–392 (1985).

[Lin83] S. Lin and D. J. Costello, Jr., *Error Control Coding: Fundamentals and Applications*, Prentice–Hall, Englewood Cliffs, NJ (1983).

[Lin84] S. Lin, D. J. Costello, Jr., and M. J. Miller, "Automatic-Repeat-Request Error-Control Schemes," *IEEE Commun. Mag.*, **22**, No. 12, 5–17 (1984).

[Lio91] M. Liou, "Overview of the $p * 64$ kbits/s Video Coding Standard," *Commun. ACM*, **34**, 60–63 (1991).

[Liu89] C. L. Liu and K. Feher, "Noncoherent Detection of $\pi/4$-Shift QPSK Systems in a CCI-AWGN Combined Interference Environment," *Proc. 39th IEEE Veh. Technol. Conf.*, San Francisco, 83–94 (May 1989).

[Liu90] C. L. Liu and K. Feher, "A New Generation of Rayleigh Fade Compensated $\pi/4$-QPSK Coherent Modems," *Proc. 40th IEEE Veh. Tech. Conf.*, Orlando, 482–486 (May 1990).

[Log76] H. L. Logan and G. D. Forney, "A MOS/LSI Multiple Configuration 9600 bps Data Modem," *Proc. IEEE ICC '76*, 48.7–48.12 (June 1976).

[Lon68] A. G. Longley and P. L. Rice, "Prediction of Tropospheric Radio Transmission Over Irregular Terrain. A. Computer Method-1968," ESSA Technical Report ERL 79-ITS 67, U.S. Government Printing Office, Washington, DC, July 1968.

[Lor79] R. W. Lorenz, "Theoretical Distribution Functions of Multipath Fading in a Mobile Radio and Determination of Their Parameters by Measurement," Deusche Bundespost Forschungsinstitut, Technischer Bericht, FI 455 TBr 66, March 1979 (in German).

[Luc65] R. W. Lucky, "Automatic Equalization for Digital Communication," *Bell Syst. Tech. J.*, **44**, 547–588 (1965).

[Luc66] R. W. Lucky, "Techniques for Adaptive Equalization of Digital Communication," *Bell Syst. Tech. J.*, **45**, 255–286 (1966).

[Luc68] R. W. Lucky, J. Salz, and E. J. Weldon, Jr., *Principles of Data Communication*, McGraw–Hill, New York (1968).

[Lue84a] R. J. Luebbers, "Finite Conductivity Uniform GTD Versus Knife Edge Diffraction in Prediction of Propagation Path Loss," *IEEE Trans, Antennas Propag.*, **AP-32**, 70–76 (1984).

[Lue84b] R. J. Luebbers, "Propagation Prediction for Hilly Terrain Using GTD Wedge Diffraction," *IEEE Trans. Antennas Propag.*, **AP-32**, 951–955 (1984).

[Lyo75] D. L. Lyon, "Envelope-Driven Timing Recovery in QAM and SQAM Systems," *IEEE Trans. Commun.*, **COM-23**, 1327–1331 (1975).

[Mac77] F. J. MacWilliams and N. J. A. Sloane, *The Theory of Error-Correcting Codes*, North-Holland, Amsterdam (1977).

[Mac79] V. H. MacDonald, "The Cellular Concept," *Bell Syst. Tech. J.*, **58**, 15–41 (1979).

[Mac91] R. Macario (Ed.), *Personal and Mobile Radio Systems*, P. Peregrinus, Ltd., London (1991).

[Mag73] R. F. Magee and J. G. Proakis, "Adaptive Maximum Likelihood Sequence Estimation for Digital Signaling in the Presence of ISI," *IEEE Trans. Inf. Theory*, **IT-19**, 120–124 (1973).

[Mag82] B. S. Maglaris and M. Schwartz, "Optimal Fixed Frame Multiplexing in Integrated Line- and Packet-Swtiched Communication Networks," *IEEE Trans. Inf. Theory*, **IT-28**, 263–273 (1982).

[Mak75] J. Makhoul, "Linear Prediction: A Tutorial Review," *Proc. IEEE*, **63**, 561–580 (1975).

[Mar85] M. J. Marcus, "Recent U.S. Regulatory Decisions on Civil Use of Spread Spectrum," *Proc. IEEE GLOBECOM '85*, New Orleans, LA, 16.6.1–16.6.3 (Dec. 1985).

[Mar87a] M. J. Marcus, "Regulatory Policy Considerations for Radio Local Area Networks," *IEEE Commun. Mag.*, **25**, No. 7, 95–99 (1987).

[Mar87b] S. L. Marple, *Digital Spectral Analysis with Applications*, Prentice–Hall, Englewood Cliffs, NJ (1987).

[Mar91] M. Marcus, "Regulatory Policy Considerations for Radio Local Area Networks," *Proc. IEEE Workshop on Wireless LANs*, Worcester, MA, 42–48 (May 1991).

[Maz75] J. E. Mazo, "Optimum Timing Phase for an Infinite Equalizer," *Bell Syst. Tech. J.*, **54**, 189–201 (1975).

[McK91] J. W. McKown and R. L. Hamilton, Jr., "Ray Tracing as a Design Tool for Radio Networks," *IEEE Network Mag.*, **5**, No. 6, 27–30 (1991).

[McL88] P. J. Mclane et al., "PSK and DPSK Trellis Codes for Fast Fading, Shadowed Mobile Satellite Communication Channels," *IEEE Trans. Commun.*, **COM-36**, 1242–1246 (1988).

[Med83] J. S. Meditch and C. T. Lea, "Stability and Optimization of the CSMA and CSMA/CD Channels," *IEEE Trans. Commun.*, **COM-31**, 763–774 (1983).

[Med93] J. Meditz, "Development of Custom Coded Blocks in Signal Processing Workstation (SPW) Software for Indoor Radio Propagation Modeling," M.S. thesis, ECE Department, Worcester Polytechnic Institute, May 1993.

[Men94] T. H. Meng et al., "Video Compression for Wireless Communications," Department of Electrical Engineering, Stanford University, Jan. 1994.

[Mes87] D. G. Messerschmitt, "Advanced Digital Communications: Systems and Signal Processing Techniques," Chapter in *Echo Cancellation*, Prentice–Hall, Englewood Cliffs, NJ (1987).

[Met76a] J. J. Metzner, "On Improving Utilization in ALOHA Networks," *IEEE Trans. Commun.*, **COM-24**, 447–448 (1976).

[Met76b] R. Metcalfe and D. Boggs, "ETHERNET: Distributed Packet Switching for Local Computer Networks," *Commun. ACM*, **19**, 395–404 (1976).

[Mic85] A. M. Michelson and A. H. Levesque, *Error-Control Techniques for Digital Communication*, Wiley–Interscience, New York (1985).

[Mic91] Microcell Report, January 1991.

[Mil92] L. B. Milstein et al., "On the Feasibility of a CDMA Overlay for Personal Communications Networks," *IEEE J. Selected Areas Commun.*, **10**, 655–668 (1992).

[Moc83] Y. Mochida et al., "VLSI High Speed Data Modem," *Proc. IEEE GLOBECOM '83*, 45.8.1–45.8.6 (1983).

[Mod81] J. W. Modestino, V. Bhaskaran, and J. B. Anderson, "Tree Encoding of Images in the Presence of Channel Errors," *IEEE Trans. Inf. Theory*, **IT-27** 677–697 (1981).

[Moh89] M. L. Moher, "TCMP-A Modulation and Coding Strategy for Rician Fading Channels," *IEEE J. Selected Areas Commun.*, **7**, 1347–1355 (1989).

[Mon77] P. Monsen, "Theoretical and Measured Performance of a DFE Modem on a Fading Multipath Channel," *IEEE Trans. Commun.* **COM-25**, 1144–1153 (1977).

[Mon84] P. Monsen, "MMSE Equalization of Interference on Fading Diversity Channels," *IEEE Trans. Commun.*, **COM-32**, 5–12 (1984).

[Mor92] G. Morrison, M. Fattouch, and H. Zhohloul," Statistical Analysis and Autoregressive Modeling of the Indoor Channel," *Proc. Int. Conf. Univ. Pers. Commun.*, Dallas (Oct. 1992).

[Mot87] A. J. Motley, "Advanced Cordless Telecommunication Service," *IEEE J. Selected Areas Commun.*, **SAC-5**, 774–782 (1987).

[Mot88a] A. J. Motley and J. M. P. Keenan, "Personal Communication Radio Coverage in Buildings at 900 MHz and 1700 MHz," *IEE Electr. Lett.*, **24**, 763–764 (1988).

[Mot88b] A. J. Motley, "Radio Coverage in Buildings," *Proc. Natl. Commun. Forum*, Chicago, 1722–1730 (Oct. 1988).

[Mul73] K. H. Muller, "A New Approach to Optimum Pulse Shaping in Sampled Systems Using Time-Domain Filtering," *Bell Syst. Tech. J.*, **52**, 723–729 (1973).

[Mur81] K. Murota and K. Hirade, "GMSK Modulation for Digital Mobile Radio Telephony," *IEEE Trans. Commun.*, **COM-29**, 1044–1050 (1981).

[Nak60] M. Nakagami, "The *m*-Distribution—A General Formula of Intensity Distribution of Rapid Fading," in *Statistical Methods of Radio Wave Propagation*, Pergamon Press, Elmford, NY (1960).

[Nak90] N. Nakajima et al., "A System Design for TDMA Mobile Radios," *Proc. 40th IEEE Veh. Technol. Conf.*, 295–298 (May 1990).

[Nam84a] C. Namislo, "Analysis of Mobile Radio Slotted ALOHA Networks," *IEEE J. Selected Areas Commun.*, **SAC-2**, 583–588 (1984).

[Nam84b] C. Namislo, "Analysis of Mobile Radio Slotted ALOHA Networks," *IEEE Trans. Veh. Technol.*, **VT-33**, 199–204 (1984).

[Nat88] J. E. Natvig, "Evaluation of Six Medium Bit-Rate Coders for the Pan-European Digital Mobile Radio System," *IEEE J. Selected Areas Commun.*, **6**, 324–331 (1988).

[Nel84] R. Nelson and L. Kleinrock, "The Spatial Capacity of a Slotted ALOHA Multihop Packet Radio Network with Capture," *IEEE Trans. Commun.*, **COM-32**, 684–694 (1984).

[Net80] A. N. Netravali and J. O. Lim, "Picture Coding: A Review," *Proc. IEEE*, **68**, 366–406 (1980).

[Nob62] D. Noble, "The History of Land-Mobile Radio Communications," *Proc. IRE*, *Veh. Commun.* 1406 (May 1962).

[O'Ne80] J. B. O'Neal, Jr., "Waveform Encoding of Voice-Band Data Signals," *Proc. IEEE*, **68**, 232–246 (1980).

[Occ77] B. Occhiogrosso et al., "Performance Analysis of Integrated Switching Communications Systems," *Proc. NTC '75*, Los Angles (Dec. 1977).

[Och89] H. Ochesner, "DECT-Digital European Cordless Telecommunications," *Proc. 39th IEEE Veh. Tech. Conf.*, 718–721 (May 1989).

[Oga94] K. Ogawa, K. Kohiyama, T. Koboyashi, "Toward the Personal Communication Era," *Int. J. Wireless Inf. Networks*, **1**, 17–27 (1994).

[Ogo81] S. Ogose and K. Murota, "Differentially Encoded GMSK with 2-bit Differential Detection," *Trans. IECE Japan*, **J64-B**, 248–254 (1981).

[Oku68] Y. Okumura, et al., "Field Strength And Its Variability in VHF and UHF Land-Mobile Service," *Rev. Electr. Commun. Lab.*, **16**, 825–873 (1968).

[Pad84] P. E. Padgett, "Cordless Telephone: Present and Future," *Proc. Natl. Commun. Forum*, 151–155 (1984).

[Pad94] R. Padovani, "The Capacity of the CDMA Cellular: Reverse Link Field Test Results," *Proc. Int. Zurich Semin. Digital Commun.*, Zurich, Switzerland, 56–66 (March 1994).

[Pah79] K. Pahlavan, "Channel Measurement for Wideband Digital Communication Over Fading Channels," Ph.D. thesis, Worcester Polytechnic Institute, Worcester, MA, June 1979.

[Pah80] K. Pahlavan and J. W. Matthews, "Performance of Channel Measurement Techniques Over Fading Channels," *Proc. IEEE NTC '80*, 58.5.1–58.5.5 (1980).

[Pah85a] K. Pahlavan, "Comparison Between the Performance of QPSK, SQPSK, QPR, and SQPR Systems Over Microwave LOS Channels," *IEEE Trans. Commun.*, **COM-33**, 291–296 (1985). (Also presented at *IEEE GLOBECOM '83*.)

[Pah85b] K. Pahlavan, "Wireless Communications for Office Information Networks," *IEEE Commun. Mag.*, **23**, No. 6, 19–27 (1985).

[Pah85c] K. Pahlavan, "Wireless Data Communication Techniques for Indoor Applications," *Proc. IEEE ICC '85*, Chicago, 3.1.1–3.1.7 (1985).

[Pah87] K. Pahlavan, "Spread Spectrum for Wireless Local Networks," *Proc. IEEE PCCC*, Phoenix, 215–219 (1987).

[Pah88a] K. Pahlavan, "Wireless Intra-Office Networks," *ACM Trans. Office Inf. Syst.*, **6**, 277–302 (1988).

[Pah88b] K. Pahlavan, "Signal Processing in Telecommunications," Chapter 22 in C. H. Chen (Ed.), *Handbook of Signal Processing*, Marcel Dekker, New York (1988).

[Pah88c] K. Pahlavan and J. L. Holsinger, "Voice-Band Data Communication Modems—A Historical Review: 1919–1988," *IEEE Commun. Mag.*, **26**, No. 1, 16–27 (1988).

[Pah89] K. Pahlavan, R. Ganesh, and T. Hotaling, "Multipath Propagation Measurements on Manufacturing Floors at 910 MHz," *IEE Electr. Lett.*, **25**, 225–227 (1989).

[Pah90a] K. Pahlavan and M. Chase, "Spread-Spectrum Multiple-Access Performance of Orthogonal Codes for Indoor Radio Communications," *IEEE Trans. Commun.*, **COM-38**, 574–577 (1990).

[Pah90b] K. Pahlavan and S. J. Howard, "Statistical AR Models for the Frequency Selective Indoor Radio Channels," *IEE Electr. Lett.*, **26**, 1133–1135 (1990).

[Pah90c] K. Pahlavan and J. W. Matthews, "Performance of Adaptive Matched Filter Receivers Over Fading Multipath Channels," *IEEE Trans. Commun.*, **COM-38**, 2106–2113 (1990).

[Pah93] K. Pahlavan, S. J. Howard, and T. A. Sexton, "Decision Feedback Equalization of the Indoor Radio Channel," *IEEE Trans. Commun.*, **COM-41**, 164–170 (1993).

[Pah94] K. Pahlavan, "Wireless Local Area Networks," Chapter 11 in *Personal Communication*, Gardiner and West (Eds.), Artech House, Norwood, MA (1994).

[Par89] J. D. Parsons and J. G. Gardiner, *Mobile Communication Systems*, Blackie, Glasgow, (1989).

[Par92] K. Parsa, "The Mobitex Packet-Switched Radio Data System," *Proc. IEEE PIMRC '92*, Boston, MA, 534–538 (1992).

[Pas79] S. Pasupathy, "Minimum Shift Keying: A Spectrally Efficient Modulation," *IEEE Commun. Mag.*, **17**, No. 4, 14–22 (1979).

[PCW91] *PC Week*, Feb. 4, 1991.

[Pet72] W. W. Peterson and E. J. Weldon, Jr., *Error-Correcting Codes*, 2nd ed., MIT Press, Cambridge, MA (1972).

[Pra88] R. Prasad and J. Arnbak, "Enhanced Throughput in Packet Radio Channels with Shadowing," *IEE Electr. Lett.*, **24**, 986–988 (1988).

[Pre91] W. H. Press et al., *Numerical Recipes in C*, Cambridge University Press (1991).

[Pri58] R. Price and P. E. Green, "A Communication Technique for Multipath Channels," *Proc. IRE*, **46**, 555–570 (1958).

[Pri83] R. Price, "Further Notes and Anecdotes on Spread-Spectrum Origins," *IEEE Trans. Commun.*, **COM-31**, 85–97 (1983). [Also reprinted in C. E. Cook et al. (Eds.), *Spread-Spectrum Communications*, IEEE Press, New York (1983).]

[Pro89] J. G. Proakis, *Digital Communications*, 2nd ed., McGraw–Hill, New York (1989).

[Qui93] R. R. Quick, Jr., and K. Balachandran, "Overview of the Cellular Packet Data (CDPD) System," *Proc. IEEE PIMRC '93*, Yokohama, Japan, 338–343 (1993).

[Qur77] S. H. Qureshi and G. D. Forney, "Performance and Properties of a T/2 Equalizer," *Proc. NTC '77*, 11.1.1–11.1.14 (Dec. 1977).

[Qur84] S. H. Qureshi and H. M. Ahmed (Eds.), *VLSI Signal Processing*, Selected Reprint Series, IEEE Press, New York (1984).

[Qur85] S. H. Qureshi, "Adaptive Equalization," *Proc. IEEE*, **73**, 1349–1387 (1985).

[Rab78] L. R. Rabiner and R. W. Schafer, *Digital Processing of Speech Signals*, Prentice–Hall, Englewood Cliffs, NJ (1978).

[Rah93] M. Rahnema, "Overview of the GSM System and Protocol Architecture," *IEEE Commun. Mag.*, **31**, No. 4, 92–100 (1993).

[Ram87] B. Ramamurthi and M. Kavehrad, "Direct-Sequence Spread Spectrum with DPSK Modulation and Diversity for Indoor Wireless Communications," *IEEE Trans. Commun.*, **35**, 224–236 (1987).

[Ram94] P. A. Ramsdale, "Personal Communications in UK—Implementation of PCN Using DCS-1800," *Int. J. Wireless Inf. Networks*, **1**, 29–36 (1994).

[Rap89] T. S. Rappaport, "Characterization of UHF Multipath Radio Channels in Factory Buildings," *IEEE Trans. Antennas Propag.*, **37**, 1058–1069 (1989).

[Rap90] T. S. Rappaport, S. Y. Seidel, and R. Singh, "900 MHz Multipath Propagation Measurements for US Digital Cellular Radio Telephone," *IEEE Trans. Veh. Technol.*, **39**, 132–139 (1990).

[Rap91a] T. S. Rappaport, "The Wireless Revolution," *IEEE Commun. Mag.*, **29**, No. 11, 52–71 (1991).

[Rap91b] T. S. Rappaport, S. Y. Seidel, and K. Takamizawa, "Statistical Channel Impulse Response Models for Factory and Open Plan Building Radio Communication System Design," *IEEE Trans. Commun.*, **COM-39**, 794–807 (1991).

[Rap92] T. S. Rappaport and D. A. Hawbaker, "A Ray Tracing Technique to Predict Path Loss and Delay Spread Inside Buildings," *Proc. IEEE GLOBECOM '92*, 649–653 (Dec. 1992).

[Ray84] D. Raychauduri, "ALOHA with Multipacket Messages and ARQ-Type Retransmission Protocols—Throughput Analysis," *IEEE Trans. Commun.*, **COM-32**, 148–154 (1984).

[Ric48] S. O. Rice, "Statistical Properties of a Sine Wave Plus Random Noise," *Bell Syst. Tech. J.*, **27**, 109–157 (1948).

[Rob73] L. Roberts, "Dynamic Allocation of Satellite Capacity Through Packet Reservation," *Proc. AFIPS Conf.*, **42** (June 1973).

[Rob75] L. G. Roberts, "ALOHA Packet System with and without Slots and Capture," *ACMSIGCOM Comput. Commun. Rev.*, **5**, 28–42 (1975).

[Ros92] J-P. Rossi and A. J. Levy, "A Ray Model for Deterministic Radio Wave Propagation in Urban Area," *Radio Sci.*, **27**, 971–979 (1992).

[Ros93] J-P. Rossi and A. J. Levy, "Propagation Analysis in Cellular Environment with the Help of Models Using Ray Theory and GTD," *Proc. 43rd IEEE Veh. Technol. Conf.*, 253–256 (1993).

[Rub79] I. Rubin, "Message Delays in FDMA and TDMA Communication Channels," *IEEE Trans. Commun.*, **COM-27**, 769–777 (1979).

[Rus91] A. J. Rustako, Jr., N. Amitay, G. J. Owens, and R. S. Roman, "Radio Propagation at Microwave Frequencies for Line-of-Sight Microcellular Mobile and Personal Communications," *IEEE Trans. Veh. Technol.*, **40**, 203–210 (1991).

[Rus92a] C. M. Rush, "Summary of Conclusions of the 1992 World Administrative Radio Conference," *IEEE Antennas Propag. Mag.*, **34**, 7–14 (1992).

[Rus92b] C. M. Rush, "How WARC '92 Will Affect Mobile Services," *IEEE Commun. Mag.*, **30**, No. 10, 90–96 (1992).

[Sac92] A. Sacuta, "Data Standards for Cellular Telecommunications—A Service-Based Approach," *Proc. 42nd IEEE Veh. Technol. Conf.*, Denver, 263–266, (May 1992).

[Sal73] J. Salz, "Optimum Mean-Square Decision Feedback Equalization," *Bell Syst. Tech. J.*, **52**, 341–1373 (1973).

[Sal82] A. C. Salazar and V. B. Lawrence, "Design and Implementation of Transmitter and Receiver Filters with Periodic Coefficient Nulls for Digital Systems," *Proc. IEEE ICASSP '82*, Paris, 306–310 (May 1982).

[Sal87a] A. M. Saleh, A. J. Rustako A, and R. S. Roman, "Distributed Antennas for Indoor Radio Communications," *Proc. IEEE ICC '87*, Seattle, 3.3.1–3.3.5 (1987).

[Sal87b] A. M. Saleh and R. A. Valenzuela, "A Statistical Model for Indoor Multipath Propagation," *IEEE J. Selected Areas Commun.*, **SAC-5**, 128–137 (1987).

[Sal91] A. Salamasi and K. S. Gilhausen, "On the System Design Aspects of Code Division Multiple Access (CDMA) Applied to Digital Cellular and Personal Communications Networks," *Proc. 41st IEEE Veh. Technol. Conf.*, St. Louis, 57–62 (May 1991).

[Sas93] H. Sasaoka and Y. Omori, "Multi-Carrier 16QAM System in Land Mobile Communications," *Proc. PIMRC '93*, Yokohama, Japan, 84–88 (Sept. 1993).

[Sch66] M. Schwartz, W. R. Bennett, and S. Stein, *Communication Systems and Techniques*, McGraw–Hill, New York (1966).

[Sch82] R. A. Scholtz, "The Origins of Spread-Spectrum Communications," *IEEE Trans. Commun.*, **COM-30**, 822–853 (1982).

[Sch84] D. Schaum et al., "MSE Mobile Subscriber Equipment," *Army Commun.*, **9**, No. 3, 6–22 (1984).

[Sch85] M. R. Schroeder and B. S. Atal, "Code-Excited Linear Prediction (CELP): High Quality Speech at Very Low Bit Rates," *Proc. ICASSP '85*, Tampa, FL, 937–940 (Mar. 1985).

[Sch89] C. Schlegel and D. J. Costello, "Bandwidth-Efficient Coding for Fading Channels," *IEEE J. Selected Areas Commun.*, **7**, 1356–1368 (1989).

[Sei92] S. Y. Seidel and T. S. Rappaport, "914 MHz Path Loss Prediction Models for Wireless Communications in Multifloored Buildings," *IEEE Trans. Antennas Propag.*, **40**, 207–217 (1992).

[Sei93] S. Y. Seidel, K. R. Schaubach, T. T. Tran, and T. S. Rappaport, "Research in Site-Specific Propagation Modeling for PCS System Design," *Proc. 43rd IEEE Veh. Technol. Conf.*, Secaucus, NJ, 261–264 (May 1993).

[Ses91] N. Seshadri and Y.-K. Kim, "A Channel Simulator to Generate Error Masks with Channel State Information," TIA Contribution TR45.3.2.3.2/91.10.22.06, TR45.3 Subcommittee on Digital Cellular Standards, Speech Working Group, Oct. 1991.

[Sex89a] T. A. Sexton and K. Pahlavan, "Channel Modeling and Adaptive Equalization of Indoor Radio Channels," *IEEE J. Selected Areas Commun.*, 7, 114–121 (1989).

[Sex89b] T. A. Sexton, "Channel Modeling and High Speed Transmission Performance for the Indoor Radio Channel," Ph.D. dissertation, Worcester Polytechnic Institute, Worcester, MA, Aug. 1989.

[Sha48] C. E. Shannon, "A Mathematical Theory of Communication," *Bell Syst. Tech. J.*, **27**, 379–423, 623–656 (1948). [Reprinted in book form with postscript by W. Weaver, University of Illinois Press, Urbana, IL (1949).]

[Sha49] C. E. Shannon, "Communication in the Presence of Noise," *Proc. IRE*, **37**, 10–21 (1949).

[Sha59] C. E. Shannon, "Probability of Error for Optimal Codes in a Gaussian Channel," *Bell Syst. Tech. J.*, **38**, 611–656 (1959).

[She75] A. R. Sherwood and I. A. Fantera, "Multipath Measurement Over Troposcatter Paths with Application to Digital Transmission," *Proc. NTC '75*, New Orleans, 28.1–28.5 (Dec. 1975).

[She92] S. Sheng, A. Chandrakasan, and R. W. Broderson, "A Portable Multimedia Terminal," *IEEE Commun. Mag.*, **30**, No. 12, 64–75 (1992).

[Sim84] M. K. Simon and C. C. Wang, "Differential Detection of Gaussian MSK in a Mobile Radio Environment," *IEEE Trans. Veh. Technol.*, **VT-33**, 307–320 (1984).

[Sim85] M. K. Simon et al., *Spread Spectrum Communication*, Computer Science Press (1985).

[Skl88] B. Sklar, *Digital Communications—Principles and Applications*, Prentice–Hall, Englewood Cliffs, NJ (1988).

[Soh87] K. Sohraby, M. L. Molle, and A. N. Venetsanopoulos, "Comments on 'Throughput Analysis for Persistent CSMA Systems'," *IEEE Trans. Commun.*, **COM-35**, 240–243 (1987).

[Ste65] S. Stein, "Theory of a Tapped Delay Line Fading Simulator," *Record, First IEEE Annu. Commun. Conv.*, Boulder, CO, 601–607 (June 1965).

[Ste87] S. Stein, "Fading Channel Issues in System Engineering," in Special Issue on Fading and Multipath Channel Communications, *IEEE J. Selected Areas Commun.*, **SAC-5**, 68–89 (1987).

[Ste90] R. Steele, "Deploying Personal Communication Networks," *IEEE Commun. Mag.*, **28**, 12–15 (1990).

[Ste92] R. Steele (Ed.), *Mobile Radio Communications*, Pentech Press, London (1992).

[Ste93] R. Steele, "Speech Codecs for Personal Communications," *IEEE Commun. Mag.*, **31**, No. 11, 76–83 (1993).

[Ste94] D. G. Steer, "Coexistence and Access Etiquette in the United States Unlicensed PCS Bands," *IEEE PCS Mag.*, **1**, 36–43 (1994).

[Sun87] C-E. W. Sundberg, W. C. Wong, and R. Steele, "Logarithmic PCM Weighted QAM Transmission Over Gaussian and Rayleigh Fading Channels," *IEE Proc.*, **134**, pt. F, No. 6, 557–570 (1987).

[Suz77] H. Suzuki, "A Statistical Model for Urban Radio Propagation: Multipath Characteristics in New York City," *IEEE Trans. Commun.*, **COM-25**, 673–680 (1977).

[T1P92] Technical Sub-Committee T1P1, "Systems Engineering Working Document for Personal Communications (Draft 4)," Jan. 1992.

[Taf75] A. Taflove and M. E. Morris, "Numerical Solution of Steady-State Electromagnetic Scattering Problems Using the Time-Dependent Maxwell's Equations," *IEEE Trans. Microwave Theory Techn.*, **MTT-23**, 623–630 (1975).

[Tak85] H. Takagi and L. Kleinrock, "Throughput Analysis for Persistent CSMA Systems," *IEEE Trans. Commun.*, **COM-33**, 627–638 (1985).

[Tak87] H. Takagi and L. Kleinrock, "Correction to 'Throughput Analysis for Persistent CSMA Systems'," *IEEE Trans. Commun.*, **COM-35**, 243–245 (1987).

[Tan88] A. S. Tannenbaum, *Computer Networks*, 2nd ed., Prentice–Hall, Englewood Cliffs, NJ (1988).

[Tas86] S. Tasaka, "Dynamic Behavior of a CSMA-CD System with a Finite Population of Buffered Users," *IEEE Trans. Commun.*, **COM-34**, 576–586 (1986).

[Tay92] J. T. Taylor, "PCS in the U.S. and Europe," *IEEE Commun. Mag.*, **30**, No. 6, 48–50 (1992).

[TIA92] Telecommunication Industry Association (TIA), Project No. PN-2759, "Cellular System Dual-Mode Mobile Station—Base Station Compatibility Standard" (IS-54, Rev. B), Jan. 9, 1992.

[TIA93a] Telecommunication Industry Association (TIA), Project 25 PN-3124, "Common Air Interface," May 19, 1993.

[TIA93b] TIA, TR45.3.2.5, "Radio Link Protocol for Asynchronous Data Service," 1993.

[TIA94] TIA, Project No. PN-3123, "Async Data and Fax," and Project No. PN-3306, "Radio Link Protocol 1," both issued November 14, 1994.

[Tie93] E. Tiedemann, "Data Services for the IS-95 CDMA Standard," Proc. PIMRC '93, Yokohama, Japan, Sept. 1993.

[Tob75] F. A. Tobagi and L. Kleinrock, "Packet Switching in Radio Channels, Part II: The Hidden-Terminal Problem in Carrier Sense Multiple Access and the Busy-Tone Solution," *IEEE Trans. Commun.*, **COM-23**, 1417–1433 (1975).

[Tob76] F. A. Tobagi and L. Kleinrock, "Packet Switching in Radio Channels, Part III: Polling and (Dynamic) Split Channel Reservation Reservation Multiple Access," *IEEE Trans. Commun.*, **COM-24**, 832–845 (1976).

[Tob80a] F. A. Tobagi, "Multiaccess Protocols in Packet Communication Systems," *IEEE Trans. Commun.*, **COM-28**, 468–488 (1980).

[Tob80b] F. A. Tobagi and V. Hunt, "Performance Analysis of Carrier Sense Multiple Access with Collision Detection," *Comput. Networks*, **4**, 245–259 (1980).

[Tre82] T. E. Tremain, "The Government Standard Linear Predictive Coding Algorithm: LPC-10," *Speech Technol.*, **1**, No. 2, 40–49 (1982).

[Tuc91] B. Tuch, "An ISM Band Spread Spectrum Local Area Network: WaveLAN," *Proc. IEEE Workshop on Wireless LANs*, Worcester, MA, 103–111 (May 1991).

[Tur72] G. L. Turin et al., "A Statistical Model of Urban Multipath Propagation," *IEEE Trans. Veh. Technol.*, **VT 21**, 1–9 (1972).

[Tur80] G. L. Turin, "Introduction to Spread Spectrum Anti-multipath Techniques and Their Application to Urban Digital Radio," *Proc. IEEE*, **68**, 328–353 (1980).

[Tur84] G. L. Turin, "The Effects of Multipath and Fading on the Performance of Direct-Sequence CDMA Systems," *IEEE J. Selected Areas Commun.*, **SAC-2**, 597–603 (1984).

[Twa85] W. Twaddell, "Modem IC's," *Electr. Design News*, **March**, 160–172 (1985).

[Ung74] G. Ungerboeck, "Adaptive Maximum Likelihood Receiver for Carrier Modulated Data Transmission Systems," *IEEE Trans. Commun.*, **COM-22**, 624–636 (1974).

[Ung76] G. Ungerboeck, "Fractional Tap-spacing Equalizer and Consequences for Clock Recovery in Data Modems," *IEEE Trans. Commun.*, **COM-24**, 856–864 (1976).

[Ung82] G. Ungerboeck, "Channel Coding with Multilevel/Phase Signals," *IEEE Trans. Inf. Theory*, **IT-28**, 55–67 (1982).

[Ung87] G. Ungerboeck, "Trellis-Coded Modulation with Redundant Signal Sets," *IEEE Commun. Mag.*, **25**, No. 2, 5–21 (1987).

[Van77] P. J. VanGerwen, N. A. M. Verhoeckx, H. A. VanEssen, and F. A. M. Snijders, "Microprocessor Implementation of High-Speed Data Modems," *IEEE Trans. Commun.*, **25**, 238–250 (1977).

[Var88] P. Vary et al., "Speech Codec for the European Mobile Radio System," *Proc. ICASSP '88*, 227–230 (Apr. 1988).

[Vit66] A. J. Viterbi, *Principles of Coherent Communication*, McGraw–Hill, New York (1966).

[Vit67] A. J. Viterbi, "Error Bounds for Convolutional Codes and an Asymptotically Optimum Decoding Algorithm," *IEEE Trans. Inf. Theory*, **IT-13**, 260–269 (1967).

[Vit79] A. J. Viterbi, "Spread Spectrum Communication—Myths and Realities," *IEEE Commun. Mag.*, **17**, No. 5, 11–18, (1979).

[Vit85] A. J. Viterbi, "When Not to Spread—A Sequel," *IEEE Commun. Mag.*, **23**, No. 4, 12–17 (1985).

[Vit91] A. J. Viterbi, "Wireless Digital Communication: A View Based on Three Lessons," *IEEE Commun. Mag.*, **29**, No. 9, 33–36 (1991).

[Vit93] A. J. Viterbi and A. M. Viterbi, "Erlang Capacity of a Power Controlled CDMA System," *IEEE J. Selected Areas Commun.*, **11**, 882–890 (1993).

[Vit94] A. J. Viterbi, "A Vision of the Second Century of Wireless Communication," *Int. J. Wireless Inf. Networks*, **1**, 3–6 (1994).

[Voi77a] W. D. Voiers, "Diagnostic Evaluation of Speech Intelligibility," in *Speech Intelligibility and Recognition*, M. E. Hawley (Ed.), Dowden, Hutchinson, and Ross, Stroudsburg, PA (1977).

[Voi77b] W. D. Voiers, "Diagnostic Acceptability Measure for Speech Communication Systems," *Proc. ICASSP '77*, 204–207 (1977).

[Wal91] G. K. Wallace, "The JPEG Still Picture Compression Standard," *Commun. ACM*, **34**, 31–43 (1991).

[War83] H. Ware, "The Competitive Potential of Cellular Mobile Telecommunications," *IEEE Commun. Mag.*, **21**, No. 8, 16–23 (1983).

[Wat78] K. Watanabe, K. Inoue, and Y. Sato, "A 4800 bit/s Microprocessor Data Modem," *IEEE Trans. Commun.* **COM-26**, 493–498 (1978).

[Wei71] S. B. Weinstein and P. M. Ebert, "Data Transmission by Frequency-Division Multiplexing Using the Discrete Fourier Transform," *IEEE Trans. Commun. Technol.*, **COM-19**, 628–634 (1971).

[Wei78] C. J. Weinstein, "Fractional Speech Loss and Talker Activity Model for TASI and for Packet-Switched Speech," *IEEE Trans. Commun.* **COM-26**, 1253–1257 (1978).

[Wei84a] L.-F. Wei, "Rotationally Invariant Convolutional Channel Coding with Expanded Signal Space—Part I: 180°," *IEEE J. Selected Areas Commun.*, **SAC-2**, 659–671 (1984).

[Wei84b] L.-F. Wei, "Rotationally Invariant Convolutional Channel Coding with Expanded Signal Space—Part II: Nonlinear Codes," *IEEE J. Selected Areas Commun.*, **SAC-2**, 672–686 (1984).

[Wei93] D. Weissman, A. H. Levesque, and R. A. Dean, "Interoperable Wireless Data," *IEEE Commun. Mag.*, **31**, No. 2, 68–77 (1993).

[Wid66] Widrow, B, "Adaptive Filters, I: Fundamentals," Stanford Electronics Laboratory, Stanford University, Technical Report 6764-6, Dec. 1966.

[Wil84] G. F. Williams and A. Leon-Garcia, "Performance Analysis of Integrated Voice and Data Hybrid-Switched Links," *IEEE Trans. Commun.*, **COM-32**, 695–706 (1984).

[Wil91] S. A. Wilkus, "Standards and Regulatory Aspects of Wireless Local Communications," *Proc. IEEE Workshop on Wireless Local Area Networks*, Worcester, MA, 23–33, (May 1991).

[Wim92] K. A. Wimmer and J. B. Jones, "Global Development of PCS," *IEEE Commun. Mag.*, **30**, No. 6, 22–27 (1992).

[Win85] J. H. Winters and Y. S. Yeh, "On the Performance of Wideband Digital Radio Transmission Within Buildings Using Diversity," *Proc. IEEE GLOBECOM '85* New Orleans, 32.5.1–32.5.6 (1985).

[Wol78] J. K. Wolf, "Efficient Maximum Likelihood Decoding of Linear Block Codes. Using a Trellis," *IEEE Trans. Inf. Theory*, **IT-24**, 76–81 (1978).

[Wol82] J. K. Wolf, A. M. Michelson, and A. H. Levesque, "On the Probability of Undetected Error for Linear Block Codes," *IEEE Trans. Commun.*, **COM-30**, 317–324 (1982).

[Woz65] J. M. Wozencraft and I. M. Jacobs, *Principles of Communication Engineering*, Wiley, New York (1965).

[Yan93a] G. Yang, S. Li, J. F. Lee, and K. Pahlavan, "Computer Simulation of Indoor Radio Propagation," *Proc. PIMRC '93*, Yokohama, Japan (Sept. 1993).

[Yan93b] G. Yang, K. Pahlavan, and J. F. Lee, "A 3D Propagation Model with Polarization Characteristics in Indoor Radio Channels," *Proc. IEEE GLOBECOM '93*, Houston, (Nov. 1993).

[Yan94a] G. Yang and K. Pahlavan, "Performance Analysis of Multicarrier Modems in an Office Environment Using 3D Ray Tracing," *Proc. IEEE GLOBECOM '94*, (Dec. 1994).

[Yan94b] G. Yang, "Performance Evaluation of High Speed Wireless Data Systems Using a 3D Ray Tracing Algorithm," Ph.D. Thesis, Worchester Polytechnic Institute, June 1994.

[Yan94c] G. Yang, K. Pahlavan, and T. J. Holt, "Sector Antenna and DFE Modems for High Speed Indoor Radio Communications," *IEEE Trans. Veh. Technol.* **43**, 925–933 (1994).

[Yee66] K. S. Yee, "Numerical Solution of Initial Boundary Value Problems Involving Maxwell's Equations in Isotropic Media," *IEEE Trans. Antennas Propag.*, **AP-14**, 302–307 (1966).

[Yee93] N. Yee, J.-P. Linnartz, and G. Fettweis, "Multi-carrier CDMA in Indoor Wireless Radio Networks," *Proc. PIMRC '93*, Yokohama, Japan, 109–113 (Sept. 1993).

[Yeg91] P. Yegani and C. D. McGillem, "A Statistical Model for the Factory Radio Channel," *IEEE Trans. Commun.*, **COM-39** 1445–1454 (1991).

[Yeg93] P. Yegani and C. D. McGillem, "FH-MFSK Multiple Access Communications System Performance in the Factory Environment," *IEEE Trans. Veh. Technol.* **42**, 148–155 (1993).

[Yen85] C. S. Yen and R. D. Crawford, "The Use of Direct Optical Beams in Wireless Computer Communications," *Proc. IEEE GLOBECOM '85*, New Orleans, 39.1.1–39.1.5 (1985).

[Yon86a] A. Yongacoglu, D. Makrakis, and K. Feher, "1-Bit Differential Detection of GMSK with Data-Aided Phase Control," *Proc. IEEE ICC '86*, 57.8.1–57.8.5 (1986).

[Yon86b] A. Yongacoglu et al., "A New Receiver for Differential Detection of GMSK," *Proc. IEEE GLOBECOM '86*, 1039–1044 (Dec. 1986).

[Yon88] A. Yongacoglu, D. Makrakis and K. Feher, "Differential Detection of GMSK Using Decision Feedback," *IEEE Trans. Commun.*, **36**, 641–649 (1988).

[Zdu89] K. J. Zdunek, D. Ucci, and J. LoCicero, "Throughput of Nonpersistend Inhibit Sense-Multiple Access with Capture," *Electr. Lett.*, **25**, 30–32 (1989).

[Zel77] R. Zelinski and P. Noll, "Adaptive Transform Coding of Speech Signals," *IEEE Trans. Acoust. Speech Signal Proc.*, **ASSP-25**, 299–309 (1977).

[Zha89] K. Zhang, K. Pahlavan, and R. Ganesh, "Slotted ALOHA Radio Networks with PSK Modulation in Rayleigh Fading Channels," *IEE Electr. Lett.*, **25**, 412–413 (1989).

[Zha90] K. Zhang and K. Pahlavan, "An Integrated Voice/Data System for Mobile Indoor Radio Networks," *IEEE Trans. Veh. Technol.*, **39**, 75–82 (1990).

[Zha91] K. Zhang and K. Pahlavan, "The Effects of Capture on CSMA Local Radio Networks with BPSK Modulation in Rayleigh Fading Channels," *IEEE MILCOM '91 Conf. Rec.*, 1172–1176 (1991).

[Zha92] K. Zhang and K. Pahlavan, "Relation Between Transmission and Throughput of Slotted ALOHA Local Packet Radio Networks," *IEEE Trans. Commun.*, **40**, 577–583 (1992).

INDEX